FAO Fisheries Series No. 5, Volume III

Mammals in the seas

Volume III

General Papers and Large Cetaceans

Selected papers of the
Scientific Consultation on the Conservation and Management
of Marine Mammals and their Environment

FAO ADVISORY COMMITTEE ON
MARINE RESOURCES RESEARCH
WORKING PARTY ON MARINE MAMMALS

with the cooperation of the
UNITED NATIONS ENVIRONMENT PROGRAMME

FOOD AND AGRICULTURE ORGANIZATION OF THE UNITED NATIONS

Rome 1981

The designations employed and the presentation of material in this publication do not imply the expression of any opinion whatsoever on the part of the Secretariats of the Food and Agriculture Organization of the United Nations and the United Nations Environment Programme concerning the legal status of any country, territory, city or area or of its authorities, or concerning the delimitation of its frontiers or boundaries.

Bibliographic entry:

FAO Advisory Committee on Marine Resources Research
Working Party on Marine Mammals (1981).
FAO Fish. Ser., (5) Vol. 3: 504 p.
Mammals in the seas. Volume 3. General papers. Large cetaceans.
Selected papers of the Scientific Consultation on
the conservation and management of marine mammals
and their environment. Bergen, 1976.
Marine mammals. Marine ecology. Fishery management.
Potential Yield. Stock assessment. Cetacea.

P-43
ISBN 92-5-100513-3

Printed in Italy

PREPARATION OF THIS VOLUME

This volume contains the corrected versions of those papers selected for final publication, the originals of which were distributed to participants in and contributors of the Scientific Consultation on the Conservation and Management of Marine Mammals and their Environment, held in Bergen, Norway, in 1976.

The papers contained in the present volume were edited by Ms J. Gordon Clark. The abstracts were edited by Mr J. Goodman, and the bibliographic citations by Ms G.A. Soave.

The volumes in this series are being published as follows:

MAMMALS IN THE SEAS

FAO Fisheries Series No. 5

Volume 3

General papers and large cetaceans

Selected papers of the Scientific Consultation on the Conservation and Management of Marine Mammals and their Environment

Volume 4

Small cetaceans, seals, sirenians, and otters

Selected papers of the Scientific Consultation on the Conservation and Management of Marine Mammals and their Environment

Contributors to the Bergen Scientific Consultation on the Conservation and Management of Marine Mammals produced some 140 formal papers (some of which were reprints from existing publications) and a number of working group documents. Of these some 70 have been selected for inclusion in Volumes III and IV and several others have been published elsewhere. A large proportion received preliminary editorial attention from Mr John K. Goodman, who also was responsible for the preparation of the abstracts.

The papers published here have been written by mathematical scientists, field observers, economists, fisheries biologists, conservationists and anthropologists. Their content spans modelling of population dynamics, anatomical details, conservation aspects and considerations of the unseen influence on scientists and managers alike. They are variable in length and in scope. They provide a much needed overview of marine mammal status, management problems and science, and reflect the lack of knowledge in many areas. Cetaceans large and small share pride of space with the pinnipeds, while the sirenians and marine otter occupy regrettably few pages.

In editing this varied assembly, I have tried only to ensure that the language was good and the structure clear, and to correct any inconsistencies that appeared within papers. A fairly obvious order was chosen: general papers first, covering management objectives and modelling techniques, followed by papers on large cetaceans (Volume III); then papers on small cetaceans, seals, sirenians and otters (Volume IV). Within each group, papers are arranged into species status papers, generally in some form of geographical order, and other more general issues including specific population modelling techniques.

One problem will be familiar to many readers. The terms population, stock, breeding population, herd, school and other such varied usage of the English language tend to assume as many meanings as there are authors, and sometimes more. While I do not believe that I have succeeded in standardizing throughout, I have in general used the terms in the following way:

- *Discrete breeding population:* that which is thought to be or is an independent genetic unit (see paper by Rørvik and Jonsgård);
- *population:* those animals of the species inhabiting a defined geographical area, but not necessarily discrete breeding populations nor containing complete discrete breeding populations;
- *stock:* any desired grouping of animals, usually for management purposes (thus the Area IV stock of fin whales is a stock for which a quota may be set, but it may or may not represent a discrete breeding population);
- *herd:* a group of small cetaceans, possibly containing several discrete breeding populations; and
- *school:* a group of small cetaceans travelling together.

I would like to thank authors for their painstaking checking of references and of final proofs; the extra work involved for those authors for whom English is not a mother tongue is often not appreciated, and I sincerely hope that the text we have agreed between us reflects in English what they would have wished to say in their own languages.

JOANNA GORDON CLARK

GENERAL PAPERS

DOLPHINS DO; DUGONGS DON'T

J.W. Meeker

Abstract

The effects of the emotions of scientists on their work and on management policies need to be understood. Investigation of this problem should include study of the motivations of scientists, of emotional bias and of related aspects of language and mental imagery.

Résumé

Il est nécessaire de comprendre l'effet de l'état émotionnel des chercheurs sur leurs travaux et sur les politiques d'aménagement. Les recherches portant sur ce problème devraient comprendre une étude des motivations des chercheurs, des préjugés émotionnels et des liaisons entre le language et les images mentales.

Extracto

Es necesario llegar a entender los efectos que tienen las emociones de los científicos en su trabajo y en las políticas reguladoras. Las investigaciones sobre este problema han de incluir estudios de las motivaciones de los científicos, de sus propensiones afectivas y de los problemas de lenguaje e imágenes mentales.

J.W. Meeker
Berkeley Center for Human Interaction, 1850 Scenic Avenue, Berkeley, CA 94709, USA

Introduction

The native people of Papua New Guinea, it is said, avoid killing dolphins. They recognize important correspondences between the dolphins and themselves, so they honour the human rights of their marine relatives. Dugongs are as closely related to Papuans as dolphins are, one would suppose, but no such protections are extended to them. Rather, dugongs are killed in good conscience because their resemblance to people is not recognized by Papuan perception or custom. Since dolphins do seem like people, they are spared; since dugongs don't, they die.

Emotional influences

Management policies in scientifically informed societies are presumably more rational than those governed by mythic intuition, but emotional and intuitive states remain influential in virtually every human mental process, no matter how refined. Political leaders, conservationists, and scientists are human beings with emotional susceptibilities as real as those of Papuan natives. Papuans are perhaps more likely to admit that their feelings influence their thoughts and actions.

In his opening remarks, the Secretary, Sidney Holt, commented that sea mammals do not require management, but that humans do. Seas and their creatures have managed for millenia to evolve and to preserve complex systems, many of which have only recently come under threat from the exploiting and polluting activities of an industrialized human era. People have always looked to the sea to satisfy emotional and economic needs, but lately they have done so vigorously enough to threaten the sea's potential for meeting those needs. It is an unstated presupposition of this consultation that only economic and scientific (and perhaps political) considerations are significant to the management of human activities concerning sea mammals. No attention has been paid to the motivating feelings, ideas and states of mind which also contribute to scientific enterprises and management policies. Management of such human and private matters is probably unwise and impossible, but it at least seems reasonable to seek understanding of the roles they play in scientific work.

Effects on scientists

Everyone wants to appear rational, at least in public and scientific affairs. Scientists, more than most others, feel the need to live up to an image of themselves as not only rational, but virtually emotion-free. Yet in any honest appraisal, most scientists would have to admit that emotional elements are present at many levels of scientific activity and that they have their effects. Zoologists often feel affection for the animals they study, as botanists feel the beauty of plants. Natural environments are sources of emotional stimulation for scientists who work within them. In many cases the desire to be in contact with attractive aspects of nature is a root motive for the choice of a scientific career or a research project. At least as much as artists, scientists feelingly appreciate the aesthetics of form that they encounter at every level, from microscopy to global ecology. Science, like art, is an imaginative activity requiring emotional and aesthetic stimulation for its success.

At simpler levels, it has been noticeable that scientists feel about themselves according to the relative "status" of the animals they study: dolphin researchers walk with a jaunty stride and speak with pride of their work, while scientists studying dugongs sound semi-apologetic about the indelicate names or reputations of the animals they study.

It seems wise simply to admit that ideas, feelings and internal states of mind can be important factors in scientific study, and that they must be understood and accounted for wherever possible. This seems especially necessary in an area as culturally and emotionally rich as the study of marine mammals.

Conversations at this conference have demonstrated that basic human feelings are often associated with the study of marine mammals. I have learned that graduate students interested in doing research on whales, for instance, are sometimes selected for their emotional stability because of the very powerful emotions whales tend to elicit among those who observe them regularly.

Many less-private non-scientific factors are likely to influence scientific results, even though they may exert their influences at subjective or subconscious levels. Every scientist is a part of some cultural tradition which conveys powerful messages through its myths, structure of language, and accepted imageries which guide both thought and feeling. All are heirs of philosophical and social value systems which place positive or negative worth upon certain kinds of thought and experience. All have sex and gender to nudge mental life in various possible directions. The methodologies of scientific investigations, of course, are designed to overcome such influences and to promote rational objectivity. Yet anyone attending international conferences cannot help but be struck by the strong disagreements over scientific issues which seem to follow national or cultural lines. Many of the women attending this consultation have also pointed out that male bias appears to have dominated scientific inquiry and to have distorted some scientific facts (such as the understanding of reproductive behaviour in pinnipeds).

Research needs

It is an interesting and important question to what extent the feelings of the workers affect the products of science.

Research is therefore needed into human consciousness and human emotions as they relate to mammals in the seas. In a general sense, a better understanding of the inner lives of people concerning marine mammals would be helpful toward better programmes of public information and education. For instance, sensational authors and film producers should not go unchallenged by the scientific community when they convey false information and arouse public sentiments concerning sea mammals.[1]

The effect of emotional factors on the institutions of science and the conduct of scientific research into marine mammals should also be understood. It would be useful to know more about the inner motivations of young scientists as they select careers relating to marine mammals, and of those senior scientists who guide their studies and who establish research priorities and goals. Once engaged in research work, teaching, or in the formulation and administration of public policy, scientists cannot assume that they are immune to non-scientific influences, although they rarely confront these as part of their work, and even more rarely are able to cope with them effectively.

Conduct of such research

Research of this kind may seem dangerous or improper to many scientists, for it implicitly challenges the claim to impersonal objectivity that many scientists believe in and

---

[1] The supposedly human-like mentality of cetaceans has long been a strong argument in favour of their preservation, and babyfaced fur seals or appealing sea otters arouse human sympathies whenever they appear. Walt Disney's sentimental laws of animal behaviour are demonstrated by a continuous bombardment of film and video-tape showing how winsomely human most animals are, and how deserving they therefore are of human affection. Alternatively, animals are presented as images of horrible jaws ready to munch mankind at every opportunity.

wish to preserve. That claim has, of course, been seriously challenged by the highly credible investigations in recent decades of historians and philosophers of science, psychologists and sociologists, and of many working scientists who have themselves recognized the multiplicity of internal and external forces which contribute to their scientific results. Because of the controversy and potential threats inherent in any such research, it will be important that it be conducted with great tact and care, and with scrupulous concern for the validity of its conclusions. Yet it should be pursued, else scientists who are not aware of non-scientific influences upon their work will be helpless before those influences.

Several examples are given below of basic questions and topics which should be systematically pursued to assess human mental and emotional characteristics as they relate to the study of marine mammals. To elaborate and refine these questions, and to determine the methods and techniques by which they might be explored, a small interdisciplinary team should be assembled soon. Participants should represent the sciences, social sciences, and the humanities, and should have demonstrated their ability to do meaningful work beyond disciplinary boundaries. Their task should be to describe and to propose detailed research into these issues and related ones which may arise from their deliberations.

1. What are the significant aspects of language, imagery and myth which influence human attitudes toward marine mammals and which characterize their expression? The study should summarize the long history of marine mammals in mythology, art and literature, and as represented in historical and current formulations of language and public expression.

2. What are the problems of intellectual and emotional bias which affect the interpretations of information concerning marine mammals (e.g., political or economic bias, sexual bias, social bias, cultural/national bias, etc.)? This study should concentrate upon current attitudes found among scientists and others whose responsibilities include decisions concerning marine mammal research and policy.

3. Identification of logical or illogical presuppositions underlying programmes of research and management. Initial analysis of the effects of unacknowledged premises upon scientific, commercial and conservation policies. The study will require comparative analysis of stated and unstated motives and purposes where it is possible to discern both.

4. Identification and analysis of the roles played by emotions in the conduct of scientific research. What effects, if any, do scientists' emotional experiences of marine mammals have upon the products of their research? What are the consequences of suppressing or ignoring the importance of such experiences?

The major objectives of the inquiries proposed should be to understand and to make explicit the non-scientific influences which act upon relationships between humans and marine mammals. Where there are significant differences between the natural conditions of sea mammals and the mental roles they play in human thought, ways should be sought to minimize these differences or at least to promote awareness of their consequences. Careful studies of this nature could help scientists and managers to deal more effectively with the fact that sea mammals exist in the human imagination as surely as they exist in the seas.

ECONOMIC ASPECTS OF RENEWABLE RESOURCE EXPLOITATION AS APPLIED TO MARINE MAMMALS [1]

C.W. CLARK

Abstract

A simple Gordon-Schaefer bio-economic model of the Antarctic fishery for blue whales is used to illustrate the usual course of exploitation of common-property resources: the early establishment of a harvest capacity above that which the resource can sustain, leading to large initial profits but depleting the resource and endangering the long-term wellbeing of the industry by making rational levels of exploitation difficult to achieve later on. In the model, the object of whaling is assumed to be to maximize the present value of the discounted net economic revenue, thereby accounting for the compromise between short-term economic benefits and the possibility of future benefits; population size and catching effort are permitted to vary over time, compared to static models in which equilibrium position is assumed beforehand. The model gives costs and revenues corresponding to various size levels to which the population is reduced. The total present value — the sum of the net (instantaneous) depletion revenue and the present (discounted) value of the net sustainable revenue — is dominated by the former if the discount rate is large but by the latter when it is small; the population size at which this value is maximized depends, in part, on the ration of the discount rate to the biological growth rate.

For a population of 100 000 whales with a maximum sustainable yield (MSY) of 75 000, an intrinsic growth rate of 0.05 and each whale worth U.S. $ 10 000, the net depletion revenue is greatest when the number of whales is reduced to 20 000. The net sustainable revenue is greatest when there are 85 000 whales and the total present value is greatest at 44 000 and 31 200 whales for annual discount rates of 5 % and 10 % respectively. This demonstrates the importance of the initial depletion stage, the revenue from which, when added to the sustainable revenue, lowers the population size at which profit is maximized from above that producing MSY to well below it. If one wishes to harvest from a population of whales already depleted to 10 000 animals, the total present values are greatest at the same numbers of whales as above but are much less and one must wait at least 25 years for the population to recover. In such a case, it is the length of the recovery period rather than the revenues from depletion which place the maximum total value at a level below that producing MSY.

The model only includes the variable operation costs; it can be modified to include capitalization costs by incorporating parameters representing rates of investment and depreciation, total effort capacity and the level of effort actually employed. For a population of 150 000 blue whales, there is a rapid (instantaneous) expansion of catching capacity — possible because of large profits and economic discounting — which exceeds the sustained yield requirements and reduces the number of whales to 29 100 after 20 years. Eleven years later, when the population has recovered somewhat and depreciation has reduced catch capacity, new investment is justified and a sustainable yield can be taken from an equilibrium population of 43 500 whales.

7

The analysis indicated that economic incentive for conservation may be quite small. Despite this, and with little or no attention to the economic aspects of the exploiting industries, management bodies seem to have assumed that such incentive would automatically ensure that the resource is conserved.

An international Marine Mammal Authority, with rights to marine mammal resources and control over their exploitation, could help to resolve the problems cited here, achieving conservation of the animals, economic efficiency and an equitable distribution of benefits to the world community.

Résumé

Un simple modèle bio-économique de Gordon-Schaefer de l'industrie antarctique de la pêche des baleines bleues est utilisé pour illustrer le processus habituel d'exploitation des ressources appartenant au patrimoine commun: l'établissement initial d'un potentiel d'exploitation supérieur à celui que la ressource peut supporter, entraînant d'abord des profits importants mais appauvrissant la ressource et mettant en danger à long terme la situation de l'industrie en rendant difficile la réalisation ultérieure des niveaux rationnels d'exploitation. Dans le modèle, on prend comme hypothèse que l'objectif de la chasse à la baleine consiste à maximiser la valeur actuelle du revenu économique net actualisé, ce qui réalise un compromis entre des profits économiques à court terme et la possibilité de profits futurs; on permet à la taille de la population et à l'effort de capture de varier avec le temps, par opposition aux modèles statiques, où l'on suppose *a priori* que la position est équilibrée. Le modèle fournit les coûts et les profits correspondant à divers niveaux de taille auxquels la population est réduite. La valeur totale actuelle — la somme du profit net de la réduction (instantané) et de la valeur actuelle (actualisée) du profit net soutenu est dominée par le premier élément si le taux d'actualisation est important, mais par le second quand il est élevé; la taille de population pour laquelle cette valeur est maximisée dépend, en partie, du rapport entre le taux d'actualisation et le taux de croissance biologique.

Pour une population de 100 000 baleines avec un rendement maximal soutenu (MSY) de 75 000, un taux de croissance intrinsèque de 0,05 et une valeur individuelle de U.S. $ 10 000, le profit net de la réduction est le plus grand quand le nombre de baleines est réduit à 20 000, le profit net soutenu est le plus grand quand on a 85 000 baleines et la valeur totale actuelle est la plus grande à 44 000 et 31 200 baleines, pour des taux annuels d'actualisation, respectivement, de 5 et 10 %. Ceci démontre l'importance du stade initial de réduction dont le profit, quand il s'ajoute au profit soutenu, abaisse la taille de la population pour laquelle le profit est maximisé et la fait passer d'un point supérieur à celui du MSY à un point bien inférieur. Si l'on désire exploiter une population de baleines descendue à 10 000 animaux, les valeurs actuelles totales sont les plus grandes pour les mêmes nombres de baleines que ci-dessus mais sont très inférieures et l'on doit attendre au moins 25 ans pour que la population se reconstitue. Dans ce cas, c'est la durée de la période de reprise plutôt que les profits tirés de l'exploitation qui place la valeur totale maximale à un niveau inférieur à celui qui produit le MSY.

Le modèle ne comprend que les coûts d'exploitation variables; il peut être modifié pour comprendre les coûts d'investissement de capital en y incorporant des paramètres représentant des taux d'investissement et d'amortissement, la capacité de l'effort total et le niveau d'effort réellement appliqué. Pour une population de 150 000 baleines bleues, il existe une expansion rapide (instantanée) de la capacité de capture — rendue possible par les profits importants et l'actualisation économique — qui dépasse les besoins de rendement soutenu et réduit le nombre de baleines à 29 100 après 20 ans. Onze ans plus tard, quand la population a quelque peu repris et que l'amortissement a réduit la capacité de capture, un nouvel investissement est justifié et un rendement soutenu peut être obtenu d'une population en équilibre de 43 500 baleines.

L'analyse a indiqué que l'encouragement économique à la conservation peut être minime. Malgré cela et en prêtant peu ou pas d'attention aux aspects économiques des industries exploitantes, les organismes de gestion semblent avoir pensé que cet encouragement garantirait automatiquement la conservation de la ressource.

Une Autorité des mammifères marins internationale, dont les droits s'étendraient aux ressources de mammifères marins et au contrôle de leur exploitation, pourrait faciliter la solution des problèmes mentionnés plus haut, en réalisant la conservation des animaux, l'efficacité économique et une répartition équitable des bénéfices dans la communauté mondiale.

Extracto

Utilizando el modelo bioeconómico de Gordon-Schaefer de la pesquería de ballena azul del Antártico se muestra cómo procede ordinariamente la explotación de recursos de propiedad común: inicialmente se introduce en la pesquería capacidad de explotación superior a la que el recurso puede sostener, lo que se traduce en abundantes beneficios iniciales pero acaba por empobrecer los recursos y poner en peligro la buena marcha de la industria a largo plazo, al hacer difícil conseguir más tarde niveles racionales de explotación. En el modelo se supone que el objeto de la caza de ballenas es elevar al máximo el valor presente de los beneficios económicos netos actualizados, estableciendo así un compromiso entre los beneficios económicos a breve plazo y la posibilidad de beneficios futuros; se permite una variación del volumen de la población y del esfuerzo de pesca con el pasar del tiempo, a diferencia de los modelos estáticos, en los que se supone de antemano una situación de equilibrio. El modelo indica los costos y beneficios correspondientes a distintos volúmenes de población. A la hora de determinar el valor presente total (la suma de los beneficios netos (instantáneos) de una explotación consuntiva (depleción) y el valor presente (actualizado) de los beneficios netos sostenibles), predomina el primero de esos elementos si el índice de actualización es grande, y el último, si es pequeño; el volumen de población con el cual se consigue elevar ese valor al máximo depende, en parte, de la relación entre el índice de actualización y el índice de crecimiento biológico.

En una población de 100 000 ballenas con un rendimiento máximo sostenible (RMS) de 75 000, un índice intrínseco de crecimiento de 0,05 y un valor de 10 000 dólares EE.UU. por cada ballena, los beneficios netos de la explotación consuntiva (depleción) alcanzan la cifra máxima cuando el número de ballenas se reduce a 20 000, los beneficios sostenibles netos la alcanzan cuando las ballenas son 85 000, y el valor presente total alcanza su cifra máxima con 44 000 y 31 200 ballenas, con índices anuales de actualización del 5 y el 10 por ciento, respectivamente. Esto demuestra la importancia de la fase inicial de explotación consuntiva (depleción), cuyos ingresos, una vez añadidos al rendimiento sostenible, hacen que el volumen de población con el que se obtiene un máximo de beneficios se reduzca de una cifra superior a la necesaria para obtener el RMS a otra muy inferior a ella. Si se desea explotar una población de ballenas ya reducida a 10 000 animales, los valores presentes totales alcanzan su cifra máxima con el mismo número de ballenas indicado más arriba, pero resultan mucho menores, y es necesario esperar al menos 25 años para que la población se recupere. En ese caso, lo que hace que el valor total máximo se coloque a un nivel inferior al que produce el RMS es la duración del período de recuperación, más que los beneficios de la explotación consuntiva.

El modelo incluye solamente los costos variables de operación, pero puede modificarse para incluir los costos de capitalización, incorporando parámetros que representen los índices de inversión y depreciación, la capacidad total de esfuerzo y el nivel de esfuerzo aplicado de hecho. Para una población de 150 000 ballenas azules, se produce una rápida expansión (instantánea) de la capacidad de captura (debido a los grandes beneficios y a la actualización económica) superior a la que es necesaria para obtener un rendimiento sostenido, que reduce el número de ballenas a 29 100 al cabo de 20 años. Once años más tarde, cuando la población se ha recuperado algo y la depreciación ha reducido la capacidad de captura, están justificadas nuevas inversiones y puede obtenerse un rendimiento sostenible a partir de una población en equilibrio de 43 500 ballenas.

El análisis indica que los incentivos económicos para la conservación pueden ser muy pequeños. A pesar de esto, y prestando muy poca o ninguna atención a los aspectos económicos de las industrias explotadoras, los organismos reguladores parecen haber supuesto que esos incentivos asegurarían automáticamente la conservación de los recursos. Un organismo internacional para los mamíferos marinos, que ejerciera derechos sobre ellos y controlara su explotación, podría ayudar a resolver los problemas aquí citados, consiguiendo conservar las poblaciones y logrando una mejor eficiencia económica y una distribución equitativa de los beneficios entre la comunidad mundial.

C.W. Clark
Department of Mathematics, University of British Columbia, Vancouver, British Columbia V6T 1W5, Canada

Introduction

The terms of reference of *Ad Hoc* Group IV of the FAO/ACMRR Working Party on Marine Mammals require a systems analysis of "the relation of marine mammals both to relevant aspects of the marine ecosystem and to the human community". Among the important relations with the human community are the following:

(1) direct exploitation for economic purposes;

(2) indirect economic effects on other marine resources;

(3) indirect economic effects associated with alternative uses of habitat (e.g., pollution, shipping, etc.);

(4) aesthetic and "moral" considerations.

This paper will address the first 3 of these relations, that is, the various economic aspects of marine mammal exploitation. The author does not feel qualified to discuss the fourth type of relation, and in any event it appears that many arguments as to the "morality" of exploitation are largely based on economic considerations over the long-term horizon.

Common-property resources

The mobility of marine mammals is such that most populations do not remain within the area of jurisdiction of any single nation.[2] Management of these resources thus requires international cooperation. In the absence of such cooperation, experience shows that these "common-property" resources tend to become severely overexploited from an economic, and frequently also from a biological, standpoint.

---

[1] Revised version of paper prepared for FAO/ACMRR Working Party on Marine Mammals, *Ad Hoc* Group IV. November 1975.

[2] In the event of a 200-mile Economic Zone, some additional stocks may lie within the jurisdiction of a single nation, at least as far as exploitation is concerned.

In many cases the exploiting industry establishes a capacity during the initial stages of development that far exceeds the level required on a sustained-yield basis. The latter phenomenon, which is particularly severe when resource stocks are allowed to become depleted, greatly complicates the subsequent problem of achieving rational levels of exploitation. The history of marine mammal exploitation provides many examples of this process, whales and fur-bearing seals being notable examples.

From an economic point of view it would appear that the rational development of a renewable resource industry would proceed along entirely different lines. The profits derived from the early stages of exploitation (which are often quite substantial) would in part be devoted to obtaining the necessary scientific data so that future exploitation could be based on a sound knowledge of the resource potential. The past approach to marine resource exploitation has usually been just the opposite. The initial profitability of the industry, often resulting from some development in technology or from a rise in demand, leads to a "bonanza" outlook which attracts a large number of eager participants. This process, carried progressively from one resource stock to the next, while it may produce large profits over the short run, can be extremely detrimental both to the resource populations themselves, and to the long-term wellbeing of the exploiting industries. In most cases, therefore, an economic incentive exists for establishing suitable agencies and institutions to regulate the exploitation of the resource. Unfortunately, however, the nature of this economic incentive is often seriously misunderstood.

Sacrificing temporary economic benefits

Conservation has often been defined as the "wise" use of resources. Implicit in this definition is the idea that long-term benefits from a renewable resource stock should not be destroyed for the expediency of short-term gains. The obverse proposition is equally significant: resource conservation requires that some of the potential gain from immediate exploitation must be foregone, or sacrificed, in order to maintain the possibility of future benefits. Until recently this fundamental aspect of resource conservation has been largely neglected, in spite of frequent reference to the "evils" of short-term economic planning. In the case of marine mammals, particularly whales, past exploitation policies influence population levels for many years in the future, and the "sacrifice" principle of conservation assumes special significance. (The principle also applies to other aspects of resource exploitation, including indirect effects. Environmental pollution, for example, may generate short-term benefits while causing long-term disbenefits).

An illustration: blue whales

An extremely simplified bio-economic model of the Antarctic blue whale fishery will be used to illustrate the above concepts.[3] The biological model is that of M. Schaefer (1957), with economic components introduced by H.S. Gordon (1954). The equations are

$$\frac{dx}{dt} = rx \left(1 - \frac{x}{K}\right) - qEx \quad .. \quad (1)$$

$$\Pi(x, E) = pqEx - cE \quad .... \quad (2)$$

where

- x = Number of whales in population at time t (years)
- r = intrinsic growth rate

---

[3] More detailed studies of the economics of whaling appear in Clark (1976, 1976a, 1977).

K = carrying capacity
E = effort
q = catchability coefficient
$\Pi (x, E)$ = net economic revenue (per unit time)
p = price of one whale
c = cost of one unit of effort.

In this illustration the following parameter values will be used:

r = .05
K = 150 000
q = 1.0 (a normalization condition on units of E)
p = US$ 10 000
c = US$ 2×10^8

These values are thought to be reasonable approximations for the Antarctic blue whale fishery, at least insofar as the present model is concerned (Clark, 1977).

The object of management is assumed to be the maximization of the present value of discounted net economic revenue, viz [4]

$$PV = \int_0^\infty e^{-\delta t} \pi (x, E) \, dt \quad \dots \dots \quad (3)$$

where

δ = continuous discount rate (per annum)

This objective (rather than the classical MSY-type objective) automatically incorporates the trade-off between short-term and long-term benefits as discussed above. It also incorporates the main variable costs of whaling (but neglects fixed capital costs: see below).

It is important to realize that the above

formulation of the optimal exploitation problem *does not presume an equilibrium position a priori*. That is to say, both the population level $x = x (t)$ and the effort level $E = E (t)$ are permitted to vary over time, subject to the equation (1). The freedom of a dynamic model of this kind greatly increases its realism as a descriptive model of the (self-regulated) whaling industry.

The figures given in Table 1 indicate how the above model behaves. In order to simplify the calculations, the following additional simplifying assumptions are made. First we consider the case in which the *existing* population level is $x_0 = 100\,000$ blue whales, well in excess of the MSY level of 75 000. (The case of a depleted existing population, of 10 000 whales, is shown in Table 2). Next we assume that management decides to reduce the present population level to a new level x_1. The costs and revenues corresponding to various levels of x_1 are given in Table 1.

The initial "depletion" stage produces a gross revenue R_1 and a cost C_1, given by [5]

$$R_1 = p (x_0 - x_1) \quad \dots \dots \dots \dots \quad (4)$$

$$C_1 = c \int_{x_1}^{x_0} \frac{dx}{x} \quad \dots \dots \dots \dots \quad (5)$$

(Expression (5) can be deduced from equations (1) and (2) as follows: we have harvest $h = qEx$. With $q = 1$, the cost of a unit harvest equals

$$cE = \frac{ch}{qx} = c/x.$$

Thus the total cost of harvesting from x_0 to x_1 is given by the integral in (5)).

The net revenue from depletion is equal to $R_1 - C_1$. It will be observed from Table 1

Table 1. Costs, revenues and present values corresponding to a model of the Antarctic blue whale fishery[1]

Equilibrium Population Level (x_1) (Thousand whales)	Gross Depletion Revenue (R_1)	Depletion Cost (C_1)	Net Depletion Revenue (Π_1)	Total Sustainable Revenue (R_s)	Sustained Cost (C_s)	Net Sustainable Revenue (Π_s)	Present Value Net of Sustainable Revenue (PV (Π_s)) $\delta = 5\%$ $\delta = 10\%$		Total Present Value (TPV) $\delta = 5\%$ $\delta = 10\%$	
				US$ millions						
100	0.	0.0	0.0	16.7	3.3	13.3	266.7	133.3	266.7	133.3
90	100.	21.1	78.9	18.0	4.0	14.0	280.0	140.0	358.9	218.9
80	200.	44.6	155.4	18.7	4.7	14.0	280.0	140.0	435.4	295.4
70	300.	71.3	228.7	18.7	5.3	13.3	266.7	133.3	495.4	362.0
60	400.	102.2	297.8	18.0	6.0	12.0	240.0	120.0	537.8	417.8
50	500.	138.6	361.4	16.7	6.7	10.0	200.0	100.0	561.4	461.4
40	600.	183.3	416.7	14.7	7.3	7.3	146.7	73.3	563.4	490.0
30	700.	240.8	459.2	12.0	8.0	4.0	80.0	40.0	539.2	499.2
20	800.	321.0	478.1	8.7	8.7	0.0	0.0	0.0	478.1	478.1
10	900.	460.5	439.5	4.7	9.3	−4.7	−93.3	−46.7	346.2	392.8

[1] Initial stock level = 100 000 whales.

that this net revenue achieves a maximum value at $x_1 = 20\,000$ whales; this is the "break-even" population level, at which the unit harvest cost c/x_1 equals the unit price p. Depletion below this level is never worthwhile, even in the short run. (Of course the possible influence of fin whales and other species, which may lead to further "incidental" harvests of blue whales, is here neglected; Clark, 1977).

Next, the gross sustainable revenue and the corresponding sustained harvest cost are given by

$$R_s = prx_1 \left(1 - \frac{x_1}{K}\right) \quad \ldots\ldots\ldots \quad (6)$$

$$C_s = cE_s = cr\left(1 - \frac{x_1}{K}\right) \quad \ldots\ldots \quad (7)$$

From Table 1 we observe that, whereas R_s is maximized at $X_1 = 75\,000$ (the MSY point), *net* economic yield (i.e., economic rent),

$\pi_s = R_s C_s$, is maximized at a higher population level, $x_1 = 85\,000$. It might thus appear that the economically optimum yield occurs at a higher population level than the "biological" optimum, or MSY point. Note, however, that this argument applies only to the sustainable yield phase, *and does not include the initial depletion phase*. Only if this phase is neglected can we be certain that the optimal x_1 exceeds the MSY point.

The present value of the sustained future net economic yield is given by

$$PV\ (\pi_s) = \int_0^\infty e^{-\delta t}\ \pi_s\ dt = \frac{\pi_s}{\delta} \quad (8)$$

and the total present value is then the sum of this and the net depletion revenue;

$$TPV = PV\ (\pi_s) + \pi_1 \quad \ldots\ldots\ldots \quad (9)$$

These values are shown in the final columns of Table 1, for annual discount rates of 5 and

13

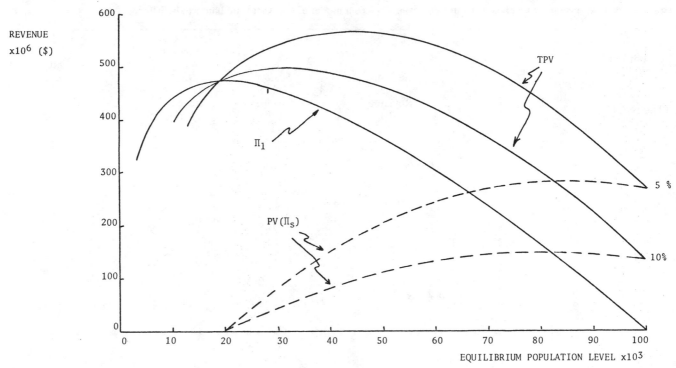

FIG. 1. - Net depletion revenue (π_1), present value of sustained yield (PV (π_s)), and total present value (TPV) for blue-whale model. Initial population level = 100 000.

10 % respectively. The results are also indicated graphically in Fig. 1. The population levels $x_1 = x_1^*$ that maximize total present value are, respectively

$$x_1^* = 44\,000 \text{ for } \delta = 5\,\% \qquad\qquad (10)$$
$$x_1^* = 31\,200 \text{ for } \delta = 10\,\%$$

Note that these levels are far below the MSY level of 75 000 whales. The optimum levels for other rates of discount are given in Table 3.

The relative importance of the net depletion revenue π_1 is clear from these figures. From equations (8) and (9) we have

$$\text{TPV} = \frac{\pi_s}{\delta} + \pi_1 \qquad\qquad (11)$$

Thus when the discount rate δ is small, the benefits from sustained yield, i.e., π_s/δ, dominate the total present value, whereas when δ is large, the depletion benefits π_1 dominate. For

the case of the blue whale resource, the influence of δ is very strong within the *normal* range of discount rates, say 0-20 % per annum. This is explained by the fact that the whale population itself possesses a low rate of growth — only 2.5 % per annum at the MSY level $x = 75\,000$.

In general the optimal equilibrium population level x_1^* is given by

$$x_1^* = \frac{K}{4} [\lambda + \sqrt{\lambda^2 + \mu}] \qquad\qquad (12)$$

where

$$\lambda = \frac{c}{pqK} + 1 - \frac{\delta}{r}$$

and

$$\mu = \frac{8\,c\,\delta}{pqKr}$$

14

Table 2. Sustainable net revenues, recovery times and total present values corresponding to a model of the Antarctic blue whale fisheries

Equilibrium Population Level (x_1) (Thousand Whales)	Net Sustainable Revenue (Π_s)	Recovery Time (T) (years)	Total Present Value (TPV)	
			$\delta = 5\%$	$\delta = 10\%$
	US$ millions		*US$ millions*	
100	13.3	66.6	9.5	0.2
90	14.0	60.9	13.3	0.3
80	14.0	55.5	17.5	0.5
70	13.3	50.1	21.8	0.9
60	12.0	44.7	25.7	1.4
50	10.0	38.9	28.6	2.0
40	7.3	32.5	28.8	2.8
30	4.0	25.1	22.9	3.3
20	0.0	15.3	0.0	0.0
10	−4.7	0.0	−94.0	−47.0

[1] Initial stock level = 10 000 whales.

Notice that x_1^* depends upon the *ratio* δ/r of the economic discount rate and the biological growth rate.

In the above illustration it has been supposed that the initial depletion phase occurs "instantaneously."

In practice, the whaling fleet will possess a maximum effort capacity, and depletion will occupy a finite length of time. Our analysis suggests, however, that optimal depletion should occur as rapidly as possible; this is readily seen to be correct, since delayed revenues must be discounted.

It should be noted from equation (12) that the optimal equilibrium population level x_1^* does not depend upon the initial level x_0.

In the case that $x_0 < x_1^*$, i.e. the population is initially in a depleted state relative to the optimum, the optimal policy (according to our model) requires a moratorium until the population recovers to the optimal level x_1^*. This situation is illustrated by the figures in Table 2, which utilize the same parameter values as before, except for

$$x_0 = 10\,000 \text{ whales}$$

In Table 2, as in Table 1, it is supposed that some "target" population level x_1 is selected, and that management imposes a moratorium until the population has increased to x_1, after which a sustained yield is maintained. The net sustained economic revenue π_s is as before, $\pi_s = R_s - C_s$. Recovery time T is given by

$$T = \frac{1}{r}\ln\frac{x_1(K - x_0)}{x_0(K - x_1)} \quad \ldots\ldots (13)$$

The present value of the sustained yield beginning at time T is therefore equal to

$$TPV = \int_T^\infty e^{-\delta t}\pi_s\,dt = \frac{\pi_s e^{-\delta T}}{\delta} \quad \ldots (14)$$

Values of TPV for $\delta = 5\%$ and 10% are shown in the final columns of Table 2. These present values achieve a maximum at the same population levels x_1^* as given by equation (12), i.e.,

$$x_1^* = 44\,100 \text{ and } x_1^* = 31\,200$$

respectively.

15

Table 3. Optimal equilibrium populations and optimal sustainable yields for a model of the Antarctic blue whale fishery

Annual Discount Rate (δ)	Optimal Population Level (x_1^*)	Optimal Sustained Yield (Y^*)
0 %	85 000 whales	1 842 whales
1 %	74 000 »	1 875 »
3 %	56 100 »	1 755 »
5 %	44 100 »	1 566 »
10 %	31 200 »	1 235 »
15 %	27 200 »	1 106 »
20 %	25 000 »	1 042 »
+ ∞	20 000 »	867 »

In the present case it is not the potential revenues from depletion, but rather the length of the recovery period, that influences x_1^* and forces it below the MSY level.[6]

Capital costs [7]

The above model incorporates only variable costs (operating costs) associated with the search for and capture of blue whales. In order to incorporate capitalization costs, the model can be modified as follows. Let

Q(t) = total effort capacity at time t

We then suppose that the actual level of effort employed is constrained by the condition

$$0 \leq E(t) \leq Q(t) \quad \dots \dots \dots \dots \quad (15)$$

As a first approximation we may suppose that

Q (t) represents the number of catcher vessels in existence at time t. Let

I (t) = rate of investment in catcher fleet
γ = rate of depreciation.

Then Q (t) satisfies

$$\frac{dQ}{dt} = I(t) - \gamma Q \quad \dots \dots \dots \dots \quad (16)$$

Finally, let us assume that whaling vessels are completely specialized and have no resale or scrap value in the event that they are no longer needed in the whaling industry;[8] thus

$$I(t) \geq 0 \quad \dots \dots \dots \dots \dots \dots \quad (17)$$

The net present value expression is now

$$\text{Net PV} = \int_0^\infty e^{-\delta t} [\pi(x, E) - c_I I] \, dt \quad (18)$$

where

c_I = unit cost of effort capacity

For the purpose of illustration we suppose that

$c_I = \$4 \times 10^8$
$\gamma = .15$

Maximization of the net present value expression (18) requires the specification of two optimal policies: an optimal investment schedule I (t) and an optimal effort schedule E (t) (and correspondingly an optimal harvest quota schedule Y (t) = qE (t) × (t)). An analytic solution to this problem has been obtained by Clark, Clarke, and Munro (1977). For the case δ = 10 %, the result is shown in Fig. 2.

In this Figure, t = 0 represents the be-

---

[6] An alternative way of looking at the foregoing problem is discussed in the Appendix.

[7] This section, which was added subsequently to the presentation of the paper at the Bergen Conference, incorporates recent theoretical results of Clark, Clarke, and Munro (1977).

[8] See Clark, Clarke, and Munro (1977) for less restrictive forms of this "irreversibility" hypothesis.

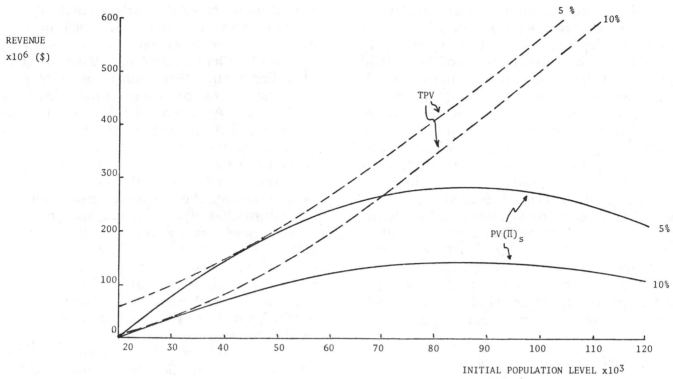

FIG. 2. - Present value of sustainable yield (PV (π_s)), and total present value (TPV) as functions of initial population level (see Appendix).

ginning of the Antarctic pelagic whaling industry, so that

$$x(0) = 150\,000 \text{ whales}$$

A rapid initial expansion of capacity [9] at t = 0 culminates in a total fleet capacity of 72 catcher vessels. The whale stocks are rapidly reduced, reaching a minimum level of 29 100 whales after 20 years' time. As a consequence of depreciation of the fleet capacity, sustainable yield at this stock level can no longer be taken unless fleet capacity is maintained by further investment. However, because of the depleted state of the stocks, such new invest-

ment is not warranted. Ultimately (at t = 31 years) the stocks recover to a level x_2^* at which new investment costs are warranted, and a state of equilibrium is finally achieved. The long-run equilibrium x_2^* can be calculated from equation (12) by replacing variable cost c by the total cost:

$$c_{total} = c + (\gamma + \delta)\,c_I \quad \dots\dots\dots \quad (19)$$

For the assumed parameter values we obtain

$$x_2^* = 43\,500 \text{ whales.}$$

It may seem surprising that the profit-maximizing development path of the whaling industry is as complex as this. But the rationale is fairly clear: because of the large profits available from the initial stages of exploitation, and because of the discounting phenomenon, a large fleet capacity is built up

---

[9] Our model actually permits an infinite rate of expansion at t = 0. If in fact the expansion occurs over a period of several years, it can be seen that the optimal built-up capacity will be somewhat smaller than given here.

17

initially. In fact the optimal fleet capacity exceeds the long-run optimum, even taking account of depreciation. During a temporary period, capacity is thus in excess of the optimal sustained yield requirements, but since whaling vessels are assumed to have no alternative uses, it is economically feasible to continue using them even though this results in further (temporary) depletion of the whale stocks. Ultimately, however, the transient effects of the initial phase disappear, and a long-run optimal equilibrium is achieved at a stock level x_2^* determined by the biological and economic parameters. (Note: the fleet capacity resulting from this calculation is significantly lower than that actually experienced in the Antarctic whaling industry. However, if the analysis is extended to include fin whale stocks as well, more realistic fleet capacities emerge.)

Implications of the model

The foregoing analysis, albeit severely oversimplified, has important implications for the management of whale populations, and also of other marine resources, particularly those possessing low rates of biological growth. The analysis indicates that the economic incentive for conservation of such resources may be quite minimal, as far as the commercial industry is concerned. The economic values of the whaling industry, of course, may not coincide with long-term economic values of society as a whole. For example, it is often argued that the private rate of interest (the "opportunity cost" of private capital) exceeds the socially optimal time-preference rate (Feldstein, 1964). Similarly, society may be willing to absorb greater risks than individual industries.

International institutions such as the International Whaling Commission (IWC) seem to have been established on the assumption that the economic interests of the industry would, if properly channelled, automatically ensure the conservation of the resource. Unfortunately, this supposition has been sup-

ported by much of the early theoretical literature on resource economics, which until recently has been based upon a static analysis. On the other hand, economists have correctly identified many other short-comings of present institutions governing marine resource exploitation. As a general rule, these institutions and their regulations have been established with little, if any, consideration of the economic aspects of the industry. Thus, even in cases where biological conservation has been successful, the economic performance of the industry has often remained marginal.

A proposal arising from the 1975 United Nations Conference on the Law of the Sea in Geneva may point the way to the ultimate resolution of these difficulties. This proposal concerns the establishment of a "Sea-bed Authority", which would possess property rights in the resources of the sea-bed (primarily manganese nodules). The Authority would be empowered to issue licences for and to collect royalties from the exploitation of these resources. Significant revenues which would accrue to the Authority could be devoted to scientific research, and to other beneficial activities. Thus the principal social benefit of such an Authority would lie in the fact that the economic benefit of a common property resource would accrue to the world community at large (through the Authority), rather than either accruing to a select community of exploiting countries, or being dissipated through unregulated competition.

It is possible to contemplate a similar institution – A "Marine Mammal Authority" – with rights to marine mammal resources. If properly constituted, such an Authority could prove highly successful in achieving both biological conservation and economic efficiency, as well as an equitable distribution of benefits. There might be some danger that the Authority would employ high discount rates (which in fact may be justified in the economies of developing countries, for example) to approve the biological overexploitation of some populations. However, if this danger is recognized it should be possible to formulate the charter of the Authority in such a manner that severe overexploitation is not permitted[10]. Other than this proviso, the Authority should be granted the widest possible powers to control the exploitation of marine mammals as it sees fit. The Authority could lead to the solution of the basic problems of conservation, efficiency, and distribution of benefits — problems that have plagued the management of common property resources throughout history.

Appendix

The following alternative method of comparing the static (equilibrium) and dynamic versions of the Gordon-Schaefer model was suggested by K.R. Allen (see Fig. 2). Let parameter x denote the current population of Antarctic blue whales. Then PV (π_s/δ) (solid curve) is the present value of net economic revenue, assuming that the population is maintained at the level x. On the other hand, TPV (dashed curve) is the maximum present value that can be achieved by a harvest policy allowing adjustments to the present population level (see equations (9), (14)). It is clear that

$$\text{TPV} \leq \text{PV} (\pi_s)$$

with strict inequality except in the case that $x = x_1^*$. The difference between the ordinates of the two curves represents the depletion revenue (when $x > x_1^*$), or simply the excess value of a recovery policy (when $x < x_1^*$.

---

[10] Several potential costs that might be associated with serious depletion have been neglected in our model. For example, uncertainty (both biological and economic) should be considered. In practice, such considerations will usually imply additional constraints on the level of depletion (Gulland, 1976).

References

CLARK, C.W., A delayed-recruitment model of popula-
1976   tion dynamics, with an application to baleen whale populations. *J. Math. Biol.*, 3:381-91.

—, Mathematical bioeconomics: the optimal manage-
1976a   ment of reversible resources. New York, Wiley-Interscience.

—, The economics of whaling: a two species model. *In*
1977   New directions in the analysis of ecological systems, Part 1., edited by G.S. Innis. *Simul. Counc. Proc. Ser.*, 5(1):111-9.

CLARK, C.W., F.H. CLARKE and G.R. MUNRO, The op-
1977   timal exploitation of renewable resource stocks: problems of irreversible investment. *Resour. Pap. Univ. B.C. Dep. Econ.*, (8):48 p.

FELDSTEIN, M.S., The social time-preference rate in
1964   cost-benefit analysis. *Econ. J.*, 74:360-79.

GORDON, H.S., The economic theory of a common-pro-
1954   perty resource: the fishery. *J. Polit. Econ.*, 62:124-42.

GULLAND, J.A., Management of marine mammals. Pa-
1976   per presented to the Scientific Consultation on the Conservation and Management of Marine Mammals and their Environment, Bergen, Norway, 31 August-9 September 1976. Rome, FAO, ACMRR/MM/SC/82:7 p.

SCHAEFER, M.B., Some considerations of population
1957   dynamics and economics in relation to the management of marine fisheries. *J. Fish. Res. Board Can.*, 14:669-81.

MAXIMUM SUSTAINABLE YIELD
AND ITS APPLICATION TO WHALING

S.J. HOLT

Abstract

The author presents a discussion on whaling regulations from the International Convention for the Regulation of Whaling, 1946, to the special meeting of the Scientific Committee of the IWC in December 1974. He explains the nature of maximum sustainable yield (MSY), which describes the property not of an animal population, but of a mathematical model which is applied to data from the population. The model may or may not be based on, or derived from, those data. In its simple form, the model assumes that, for a given value of population size, there is a catch which can be taken regularly, for an indefinite period. Although MSY has commonly been referred to as a "biologically determined criterion" for sound resources management, this is misleading, since MSY is a function at least of the sizes of animals which first become liable to capture, as well as of the population sizes determined by the overall rate of exploitation. All applications of the MSY concept have assumed reversibility, so that if intensified whaling overshoots the MSY level, it is only necessary to reduce the intensity sufficiently and for long enough for the stock to recover. This assumption is challenged. The author suggests an interim strategy for the formulation of improved goals in the management of wild stocks, which includes the acceptance of the concept of MSY-stock level, not as the sole objective, but as a lower bound to permissible exploitation, and as part of a multi-faceted definition of "conservation". Appendices include a discussion of the relations between the stock levels and fishing (whaling) effort for MSY and those for maximum net yield, "commercial extinction" and "biological extinction"; the consequences of discounting values of future catches; baleen whales in the Southern Hemisphere treated as a single resource; ineffective catch quotas, and supporting texts. A bibliography is included.

Résumé

L'auteur examine les règlements relatifs à la chasse à la baleine depuis la Convention internationale de 1946 jusqu'à la réunion spéciale, en décembre 1974, du Comité scientifique de la Commission internationale baleinière. Il explique ce qu'il faut entendre par rendement maximal équilibré, notion correspondant non à une population animale, mais à un modèledmathématique qui est appliqué aux données relatives à cette population. Le modèle peut être fondé ou non sur ces données. Sous sa forme simple, le modèle repose sur l'hypothèse selon laquelle, pour un chiffre donné de la population, il existe un taux de capture qui peut êtreréaliséconstammentpendantunepériodeindéfinie.Bienquel'onaitcommunémentqualifié le rendement maximal équilibré de "critère déterminé biologiquement" pour un bon aménagement des ressources, cette définition risque d'induire en erreur, étant donné que ce critère est fonction tout autant de la taille des animaux susceptibles d'être capturés que du volume de la population déterminé par le taux global d'exploitation. Toutes les applications de la notion de rendement maximal équilibré présupposent la réversibilité, de sorte que si l'intensification de la chasse à la baleine excède le niveau du rendement maximal équilibré, il suffit de réduire cette intensité suffisamment et pendant assez longtemps pour que le stock se reconstitue. Il est douteux que cette hypothèse soit fondée. L'auteur propose une stratégie intérimaire pour élaborer de meilleurs objectifs d'aménagement des stocks sauvages; cette stratégie admettrait la notion d'un niveau des stocks correspondant au rendement maximal équilibré, non comme objectif unique, mais comme limite inférieure d'une exploitation admissible et dans le cadre d'une définition complexe de la "conservation". On trouve en annexe du document une discussion concernant les rapports entre le niveau des stocks et l'effort de pêche (ou plutôt de chasse) à la baleine permettant d'atteindre le rendement maximal équilibré et ceux qui permettraient d'atteindre le rendement net maximal ("extinction commerciale" et "extinction biologique"), les conséquences de l'actualisation de la valeur des prises futures, les mysticètes considérés comme ressource unique dans l'hémisphère sud, les contingents de capture inefficaces, auxquels sont joints des textes de référence. Une bibliographie accompagne le document.

Extracto

El autor presenta un análisis de las regulaciones balleneras, desde el Convenio Internacional para la Regulación de la Pesca de la Ballena, de 1946, hasta la reunión especial del Comité Científico de la Comisión Ballenera Internacional, celebrada en diciembre de 1974. Explica la naturaleza del rendimiento sostenible máximo (RSM), que describe las características no de una población animal, sino de un modelo matemático que se aplica a los datos de la población. Dicho modelo puede o no basarse en dichos datos o derivarse de los mismos. En su formulación simple, el modelo parte del supuesto de que, para un determinado valor de tamaño de la población, existe una captura que puede obtenerse regularmente, durante un período indefinido. Aunque comúnmente se ha hecho referencia al RSM como un "criterio determinado biológicamente" para una ordenación válida de los recursos, esta definición se presta a equívocos ya que el RSM es una función, por lo menos, de los tamaños de los animales susceptibles por primera vez de captura, así como de la magnitud de la población determinada por el índice global de explotación. Todas las aplicaciones del concepto de RSM han asumido la reversibilidad, de modo que si la pesca más intensa de ballenas rebasa los límites del nivel RSM, basta únicamente con reducir la intensidad lo suficiente y durante todo el tiempo necesario para que la población se recupere. Esta afirmación es objeto de impugnación. El autor propone una estrategia provisional para la formulación de metas perfeccionadas en la ordenación de poblaciones salvajes, lo cual incluye la aceptación del concepto de nivel de población RSM, no como objetivo único, sino como un límite más bajo para una explotación permisible, y como parte de una definición plurifacética de conservación. En los apéndices figura un análisis de las relaciones existentes entre los niveles de población y el esfuerzo de pesca (pesca de la ballena) para el RSM y las correspondientes a un rendimiento neto máximo, la "extinción comercial" y la "extinción biológica"; las consecuencias de descontar los valores actuales de las capturas futuras; las ballenas de barba del hemisferio austral consideradas como un recurso singular; los cupos ineficaces de captura; y los textos de referencia. Se adjunta una bibliografía.

S.J. Holt
Threshold Foundation, 7 Regency Terrace, Elm Place, London SW7, England

This paper was prepared for the Airlie Workshops in 1975, and revised for the Bergen Scientific Consultation 1976.

"Scientific analyses should include an assessment of their own limitations. It is incumbent upon the practitioners of science to develop procedures which can express the information content of their prediction in a readily understandable manner. The effort may require as much thought and ingenuity as the development of the original prediction".[1]

The International Convention for the Regulation of Whaling, 1946, provides in its Article V (2) (b), that regulations with respect to the conservation and utilization of whale resources "shall be based on scientific findings". The Convention established the International Whaling Commission (IWC), with power to set up committees. One such committee is the Scientific Committee which deals with the above provision. The other standing committee, to which all Member States may nominate members, is the Technical Committee; it is usually composed of the Commissioners themselves; it acts in effect as a committee of the whole Commission and gives most attention to another requirement of Article V, namely that regulations "shall take into consideration the interest of the consumers of whale products and the whaling industry", especially, in practice, the latter.

From early days the Scientific Committee has usually been well aware when whale species and stocks were being overexploited, and has issued warnings, pleas and suggestions as to the levels of permissible catch. It was not until the early sixties, however, that numerical and algebraic models began to be applied systematically to predict the quantitative consequences of proposed regulations. Such applications to the blue and humpback whales in the Antarctic, which were already greatly depleted, and to the fin whale, then in process of

rapid depletion, seemed convincing both to scientists and Commissioners. This temporarily silenced the disagreement among scientists which had previously contributed to delays in taking remedial action. Agreement by consensus on "best estimates" for whale stocks became a major objective of the Scientific Committee. This has, I suggest, over the years, led the scientists concerned to underemphasize the limitations of the methods and data used. It has not been easy to examine these limitations critically because few scientists are occupied full-time (or even most of their time) with this matter, research funds and facilities are chronically inadequate and the rules of the game include unwritten rules of international courtesy.

While understandable, this has regrettable consequences, particularly so as scientific attention moves from the obviously depleted stocks to those which have only been lightly or moderately exploited. It was perfectly obvious, by the time the blue whale had been reduced so much that it was rarely seen, that it could not support continued hunting. If appropriate action had been taken at that time with respect to the fin whale we should not now be arguing about whether it, too, needs complete protection. According to Gambell (1974) "the future for the whales looks bright" because "there is a prospect of management being grounded on rational policies", yet at this time gambles are still being taken with the most important stocks which remain − sperm whales − as well as with the smallest baleen whale, the minke (Holt, 1974). I believe this results from overconfidence in the methods of manipulation data compounded by unrealistic demands for precise scientific advice.

The 1946 Convention does not refer to maximum sustainable yield (MSY) nor to the concept of sustainable yield. The Preamble to the Convention does, however, refer to achieving optimum levels of whale stocks as rapidly as possible, and Article V, para. 2, lays down that amendments to the Schedule should be such as are necessary, *inter alia*, "to provide for ... *optimum utilization* of the whale resour-

[1] Eric Kraus "The unpredictable environment", *New Scientist*, 12 September 1974.

ces", although no definitions are given. The first explicit reference to the sustainable yield criterion for management is in the Resolution adopted by the IWC at its 12th meeting in 1960. This established the Special Committee of Three Scientists to study and report "on the level of sustainable yield that can be supported by the Antarctic whale stocks" and "on conservation measures that would increase this sustainable yield so that it may be brought to an optimum". In its First Interim Report, submitted to the Commission in June 1961, the Special Committee still does not refer to MSY, although it mentions the methods of analysis "to determine optimum yield" (Chapman *et al.*, 1962). In its Final Report (Chapman *et al.*, 1964) the Special Committee used the term "optimum level" to mean the stock level corresponding with MSY. The Committee presented assessments of current sustainable yields in numbers of fin, blue, humpback and sei whales in the Antarctic, but remarked that "consideration must be given to assessing stock levels in units more clearly related to economic needs, taking into account both the yield of economically valuable products and the relative cost of production ... related to the catch per unit effort". An "unofficial" group of whale specialists in a meeting in 1970 continued to use the term "optimal stock size" as the size giving MSY (implicitly, in numbers) (Schevill, 1974).

Meanwhile, in 1966, the 1958 Geneva Convention on the Conservation of the Living Resources of the High Seas had come into force. Its Article 2 contained the following definition of MSY drafted in 1955 by the UN International Technical Conference on the Conservation of the Living Resources of the Sea: "Conservation means the aggregate of measures rendering possible the optimum sustainable yield ... so as to secure a maximum supply of food and other marine products". This definition of conservation has been substantially included in practically all draft articles on fisheries submitted to the 1974-75 UN Conference on the Law of the Sea.

At its Annual Meeting in June 1974, the IWC decided to categorize whale stocks for management purposes. The operative clauses of the decision, which identify 3 categories of stocks for management purposes refer to, but do not define, MSY in the following terms:

"*Initial management stocks* ... may be reduced in a controlled manner to achieve MSY levels or optimum levels;

Sustained management stocks ... should be maintained at or near MSY levels and then at optimum levels;

Protection stocks ... which are below the level of *sustained management* stocks and should be fully protected".

The full text of this "decision" is appended (Appendix 2). The advisory role of the Scientific Committee is defined therein and should be noted; in particular that the measures to be applied to Initial Management Stocks be such as are "necessary to bring such stocks to MSY (and then optimum) level *in an efficient manner and without risk* of reducing them below such level". It should be noted also that the term "optimum yield" is used — but not defined — in the decision, and apparently implies that the corresponding "optimum level" of stock would be somewhat higher than that giving the MSY.

At a special meeting in December 1974, the Scientific Committee attempted to respond to the request from the Commission for "advice on the criteria which should be used in defining (the 3) categories of whale stocks". Since it was made clear that species which were already fully protected ("Protected Species": Blue, Pygmy Blue, Humpback, Right Whales and Gray Whales) were automatically to be designated as Protection Stocks, the Committee discussed the "criteria" having in mind certain "difficult" cases of deciding whether particular rather lightly (or recently) exploited stocks or species would become categorized as under "initial" or "sustained" management, and whether some other stocks which have been more heavily exploited for a longer time would be under "sustained management" or "protection". In the latter case, the difficulty proved to be to interpret the term "near" (MSY) levels; in the former case "near" is also a problem, but the difficulties come mainly from the very feeble knowledge about species (particularly the minke whale) which have only recently become the object of intense pelagic whaling. "Near" could conceivably be defined scientifically in terms of statistical errors of estimates of MSY (though such a definition would not necessarily be acceptable – or even useful – to the Commission itself) but these errors are unmeasured, and are perhaps unmeasurable. Sensing this, the Scientific Committee has sought ways of giving, in effect, blurred definitions of some stock categories, and devises a proposal for a graduated range of whaling effort on stocks below the "best" estimates of the MSY level.

Another general problem involves taxonomy. The IWC and the national (Norwegian) agency which provides statistics of catch and whaling effort to the IWC – the Bureau of International Whaling Statistics (BIWS) – have a long history of lumping different taxa. Important cases have been the lumping of the blue whale and pygmy blue whale subspecies, and of the sei whale and Bryde's whale. However, although the question of the validity of the pygmy blue whale as a distinct sub-species impeded for a year or two the protection of blue whales in the late sixties, much more serious questions are the identification within each species of distinct breeding stocks, as biologically defined, and of stocks as "defined" for purposes of the IWC decision, as those "units which can be most effectively managed individually".

The purpose of the 1946 Convention was "to provide for the proper conservation of whale stocks and thus make possible the orderly development of the whaling industry". To this end the IWC reviews and amends each year a "Schedule" of regulations. It is constrained to base its actions on scientific findings. The possible actions are of several kinds and include closing certain areas ("sanctuaries") to whaling, establishment of minimum size limits, restricting capture of lactating females, opening and closing whaling seasons,

applying different rules to land stations and pelagic expeditions, and special rules to aboriginal peoples catching whales on a subsistence basis for local consumption as human food. The actions the Commission considers most important are, however, the setting of overall catch quotas for each season. These are now defined as the total numbers of each species of large whale that may be taken in various areas; in the case of sperm whales the quotas are also established for each sex. A species may be totally "protected" throughout all or part of its range; this is effectively the same as setting zero quotas, although there are legalistic differences between the 2 actions. It is well-known that until recent years the Commission set quotas for the Antarctic in terms of "Blue Whale Units" (BWU) – a weighted average of numbers of 4 types (6 taxa) of baleen whales.[2]

Since the sixties the quota (whether in BWU or by species) has been divided between those IWC nations which conduct pelagic whaling by agreements concluded concurrent with, but formally outside, IWC meetings – and there is mutual interaction between these debates and Commission debates on quotas. Each nation – and whaling company – will, among other things, have made prior calculations of what catches are needed in the coming season to at least pay the costs of, and yield a modicum of profit to, the expeditions (a factory ship plus a number of catcher boats) that it wishes to send out. As the total number of expeditions has decreased over the years, it becomes more difficult to adjust the whaling effort to an exact quota – one expedition less, or more, can mean a quota allocation not reached, or reached too easily and shared too thinly. Furthermore, although the BWU is not now used, each whaling enterprise continues to make its operational decisions on the basis of the expectation of catches of a number of

---

[2] BWU = 1 blue whale = 2 fin = 6 sei = 2.1/2 humpbacks – based on the oil yield from the different species.

species of baleen whales differing greatly in size, as well as of sperm whales. It is evident therefore that at governmental level the debate on a quota for 1 species — say, fin whale — is influenced considerably by expectations of what might be acceptable for another, such as minke.

Lastly, the limited applicability of IWC decisions should be noted. They apply, of course, only to the vessels operating under the flags of Member States of the Commission; schedule amendments require a three-quarters majority decision and are not binding on a state which has invoked the "90-day objection clause" in the Convention; and there has long been controversy over the application of quota decisions equally to land stations as to pelagic expeditions. A considerable amount of whaling is carried out by non-member states, and certain members participate in joint whaling enterprises with nonmembers, the vessels of which fly the flags of the latter.[3]

The term "MSY" describes a property not of an animal population but of a mathematical model which is applied to data from a population. The model may or may not be based on, or derived from, those data. In this simple form — and that is the form which has been, so far, applied to whales — the model assumes that for a given value of population size there is a catch (the sustainable yield — SY) which can be taken regularly, for an indefinite period. Thus a curve of SY against population size can be computed; SY is zero for an unexploited population, and also for an extinct population; somewhere in between in the greatest SY — the MSY.

In whales both population size and catch have for this purpose been defined as numbers of animals. The population size is variously defined as total number, or the number of mature animals, or the recruited number — that is, the number of animals having a size, distribution and behaviour such that they are liable to capture by the whaling effort, as it is deployed and regulated. Evidently, if "recruitment" occurs several years after birth, the number of "recruits" (which is the most important factor in sustaining the catch at a certain level) will depend on the size of the mature population several years before, rather than during, the year of catch; this time-lag of dependence of SY on population size can readily be — and has been — incorporated into the simple models applied to whales. Consideration of it led Gulland and Boerema (1973) to define another quantity, the *replacement yield* (RY), as "that catch which, if taken (in a given year) will leave the abundance of the exploitable (i.e., recruited) part of the population at the end of the year the same as at the beginning". This led, further, to a definition of (a) *equivalent sustainable yield* (ESY) from a population which has been changing, "as the SY from a population of the same abundance (or with the same abundance of the exploited phase) which has remained at this level of abundance for a long time"; and (b) *maintainable yield* (MY) as "the largest catch that can be maintained from the population, at whatever level of stock size, over an indefinite period". MY is equal to the corresponding SY for populations at levels below that for MSY, and to MSY for populations at or above this level. The following observations about MSY apply fully to these other "properties" of the simple model, few of them are new. Serious criticisms of the concept have been made more often by scientists than by whaling and fisheries administrators despite the fact that MSY is often referred to as a scientifically based management objective (see, for example, Beverton and Holt, 1957).

The MSY and the population which will sustain it may be, and variously are, defined in absolute or in relative terms. Thus MSY can be expressed as a multiple (or fraction) of the current yield, or of the SY from a population of the current size and vice-versa. The MSY stock level may likewise be expressed as a number or as a fraction (or multiple) of the current stock level, or as a fraction of the stock

---

[3] For a detailed historical account of whaling management, see Vamplen, 1972.

level before exploitation started. This latter statistic is of crucial importance in discussion of whaling regulations because series of data for catch-per-unit-whaling effort are often available, and these are assumed to indicate the trend of stock level from soon after the beginning of exploitation. The simplest mathematical model based on the logistic curve of population growth, gives a symmetrical parabolic SY to population curve, so that the MSY stock level is half the initial level. Modified models give skewed curves, usually such that this ratio is greater than 0.5, commonly 0.7-0.8.[4] Others give curves, skew or symmetrical, which are more flat-topped than the parabola (see Appendix 1). The existence of this range of "plausible" simple models would alone make the level for MSY difficult to identify, either in absolute or relative terms, but there are other difficulties.

The 1974 IWC decision recognized that "management of whale stocks should also include such consideration as total weight of whales ...". This recognition followed presentation of evidence (Holt, 1974a) that, notwithstanding the application of size limits, the average size of each species and sex of whales in catches has declined over the period of intensive exploitation. One expects to observe such a trend especially in situations where the exploited animals are rather long-lived, and continue to grow substantially after "recruitment", as is the case for all large whales, but especially for the male sperm whale. The effect has been obscured because whales are not easily and routinely weighed; their lengths are measured and mean lengths change in the same direction as mean weight – but only in proportion to the cube root of the weight, and hence less noticeably. The changes are in any case evident in the statistics for oil yield per whale, but these statistics are affected also by changes in the relative utilization of oil and meat. Although there is some resistance, both among whale biologists and administrators,[5] to the regulation of whaling to maximize weights rather than numbers, the whaling industry has always based its activities and policies on consideration of quantities (and hence total values) of products. The BWU was based on relative oil yields of different species and, although the inter-specific size range is wider than the intra-specific range, the latter is not negligible. In the special case of sperm whales, the males of which are much larger than the females and are distinguishable from them before capture, the calculation of total quotas includes a "weighting factor" to allow for the size difference.[6]

In all models studied the curve of SY (weight) against population size, whether the latter is expressed in number or weight, is skewed to the right relative to the corresponding curve of SY (number); MSY (weight) corresponds with a relatively higher stock size than that which gives MSY (number). The SY (weight) curve is also generally a little more flat-topped than the curve of SY (number). These differences are not very important for the baleen whales as they are at present exploited[7] but they are significant – a shift in population size for MSY in sperm whales of from 7 to 18 %. This is essentially because in sperm whales a greater proportion of the growth in weight occurs after "recruitment"; this proportion is regulated by the setting of minimum size limits. So, if the catching of whales is to continue to be regulated on the basis of MSY calculations, it would be better from *all* points of view to aim at MSY (weight) rather than MSY (number). This would result in higher value from the resources, for less

---

[4] Fish stocks, on the other hand, are said more often to be described by curves skewed the other way, with a ratio < 0.5.

[5] One reader of the first draft of this paper commented "some people will not agree to thinking of whales as meatballs". Another said however, "we should not think of whales and other wild mammals merely as statistics".

[6] Not now (i.e., since 1976).

[7] Except perhaps for the minke whale, for which no minimum size limit has been established by IWC.

whaling effort, and higher stock levels (see Appendix 3).

MSY has commonly been referred to in international discussions as a "biologically determined criterion" for sound resource management. This is misleading. MSY is a function at least of the sizes of animals which first become liable to capture, as well as of the population size as determined by the overall rate of exploitation. Models which have been applied indicate that the yield in weight-per-recruit has a maximum for a finite fishing mortality rate (i.e., exploitation rate

$$E = \frac{F}{F + M} < 1$$

only if the ratio of size (weight) at first capture to final size is less than a critical value

$$= \left(\frac{3}{3 + M/K} \right)^3$$

(Beverton and Holt, 1966). The catch, if *all* surviving animals are caught as soon as they reach that critical size, is theoretically equal to the maximum biomass reached naturally by a year-class during its lifetime; this quantity may have some biological significance, perhaps because the critical size may bear some relationship with — may even be close to — the mean size (and age) at sexual maturity (Holt, 1962). But where — as in whales — the number of recruits is closely related to the numbers of parents such a simple interpretation of yield per recruit models cannot be applied. There could perhaps be calculated a combination of size at first capture and exploitation rate (less than 1) in whales which would predict a unique MSY for each stock; unfortunately, models in which a simple yield-per-recruit equation is combined with a linearly density-dependent reproduction rate, have not yet been used to explore this possibility. In any case it is sure that:

(a) the size limits for whales have so far been

set by the IWC without any reference to such models, and

(b) calculations of curves of SY against whale stock size have always "taken as given" the size at recruitment, equal to the size limit in force, if any.

Another reason why MSY should not be regarded as a "biological" criterion is that since the logistic curve of population growth was first applied in the way outlined above, the argument for managing a fishery so as to hold the stock at the MSY level has been made, implicitly or explicitly, in terms of the *efficiency* of the catching operation. Thus, if the stock is reduced and held, by fishing, below that level the fishing effort is being "wasted" because the *same* catch could be sustained with *less* effort by allowing the stock to recover. So MSY stock level is not so much a level at which to aim as a level below which we should avoid going. Furthermore, once the idea of *sustained maximum efficiency* is introduced there is no reason to prefer the MSY level to any other higher level. Graham (1935, 1939) argued for restoring overfished stocks to the MSY level or above, not to increase net profits but to allow the same fishermen to take the same catch while working much less hard.[8] Such humanistic views have been unfashionable for some decades but the time may be ripe to reconsider and develop them. "Overfishing" is now seen to waste not only the working lives of fishermen, but also the limited supply of non-renewable resources of fuel and materials which are expended in seeking and capturing fish — and whales. A purely economic approach to this problem has not been fruitful in interna-

---

[8] "The benefit of efficient exploitation lies more in economy of effort than in increase in yield, or preservation of future stocks, though both of these purposes may also be served".

tional affairs but, provided we are wary of the consequences of "artificial" price structures and of high discount rates, such an approach gives some insight in to the behaviour of the resource-exploitation system with respect to conservation. Meanwhile, our economic system, having led us to be profligate with non-renewable resources, is now leading us to "mine the biota" too (Woodwell, 1974).

Curves of *net value* of sustained catch (i.e., value minus costs of catching) against stock size have maxima at higher stock levels than the MSY level (Figs. 1 and 2). Thus in steady states there is an overall economic advantage in holding the stock rather higher than the MSY level; this would be advantageous also to each operating unit which there by obtains a higher sustained catch-per-unit of effort (Gulland, 1968; Chapman *et al.*, 1965); (see also Appendix 1). This idea promised to offer an alternative to the MSY concept as an objective of fishery management when it was put forward by Gordon (1953), Beverton and Holt (1956) and others. Unfortunately:

(a) there seems no way to get international agreement on the economic criteria, particularly before there is agreement on allocation of catches among participants; and

(b) this simple economic model takes no account of *discounted* present values of future catches.

It was the recognized difficulty — if not impossibility — of getting agreement on economic criteria that led negotiations before the 1958 Geneva Conference to concentrate on physical measures of desirable gross yield, the argument being that to aim at this would favour the more efficient operators. Clark (1973) and others have shown that for resources having net growth rates of the order found in whales, and with current levels of discount rates, attainment of an economic "optimum" may not require *sustained* harvesting at all (Appendix 5). This must lead us to examine

the roots of our beliefs concerning the "proper" harvesting of natural renewable resources.

A beginning was made, as far as marine fisheries are concerned, by Rothschild (1971) who, however, quotes the views of R. Dorfman and of a Panel of the U.S. Commission on Marine Science, Engineering and Resources, to the effect that a suitable economic goal might be the attainment of "efficiency involving distribution of products to the people that need them, full utilization of resources and high consideration of the needs of the consumer". Rothschild's paper is concerned more with procedures than with goals, but if we can agree on goals his approach will be useful in attaining them. He states that "conservation of resources is essentially an allocative process and the calculus that is conventionally used in fisheries is not a powerful technique for allocation". Unfortunately, he does not — at least in the paper quoted — pursue this idea. In particular, he does not identify the most important aspect of the "allocative process" involved in management for sustainable yields — that is, the allocation of benefits as between the present and future generations of producers and consumers. He does suggest that in necessarily "taking a broader view temporally as well as spatially of fishery management" we should seriously engage in "technological forecasting". No doubt this is desirable, and good information flow within the sub-system "whales and whaling" and from outside it, combined with intelligent analysis aimed at predicting as well as evaluating technological changes, could improve management.[9] The

_____

[9] Scientific advice regarding a proposed regulation of fishing has not always been inadequate in that, although the resource is treated as a dynamic system, the fishery is often treated as if it were static. For example, assessments of the benefits of introducing a higher size limit or larger size of trawl mesh may assume that the fishing effort would remain unchanged. Yet, to the extent that the proposed measure is effective in increasing the stock size, so more effort would be attracted into that fishery; it is not difficult to predict approximately how much. In general, I believe one should assume

29

present need may, however, be for forecasting at another level — what values will our descendents place on whales, what properties will they value, how will they view matters of resource ownership, how will they value the ecosystem of which the whales are a part; what needs will they feel for the unique products, if any, obtained from whales? Just as, in relation to technological forecasting, Rothschild, quoting J.R. Bright, observed that "a radical new technological advance is more visible to society first in written words, then in ... more effective material forms, long before it achieves widespread usage", so we may now perhaps perceive the seeds of change on which social forecasting could be based. The support, in several countries and from many quarters over recent years for the movement for a moratorium on commercial whaling, notwithstanding the opposition of most of the scientists concerned,[10] indicates the need to look, as ecologists, at the whole in order to comprehend the parts, and reminds us that development of resource management criteria is an ideological and general problem, rather than merely a specialized scientific and technical one. Nevertheless, the criteria must be capable of practical application, and it is difficult to visualize any for which world application would not involve scientific analysis.

The above comments on the existence or otherwise of the "M" in MSY and on the appropriate measure of the "Y" are maybe valid, but they are not new (see for example, among many other critical publications, Gulland, 1968). They are also relatively trivial; the con-

cept of MSY can be modified to accommodate them — at least in theory — although the history of the IWC (and of some other international fishery bodies) shows how difficult it is and how long it takes to get acceptance of such innovations. The validity of the "S" in MSY is, however, more difficult to appraise. For any level of yield to be sustainable, it is necessary for the net rate of increase of stock (recruitment or reproduction plus growth minus natural mortality, all expressed as a ratio to population size) to increase with decreasing population size, and vice-versa, at least over some range of population size. The practice in making whale stock assessments has been to assume that natural mortality is not age-specific and is density independent. Thus the density-dependence is sought in the reproduction (or recruitment) rate. In some whale species the reproduction rates are rather better known than for most exploited wild animal stocks, although there are some technical problems and some controversy over interpretations.[11] Reliable information about the changes in reproduction rate with population size or density is, however, scant. Such changes can be inferred from the changes in population structure as exploitation proceeds; these inferences, taken with estimates of maximum feasible reproduction rates (percentage pregnancies, number of calves, etc.) have led to fairly firm statements about whether a certain catch from a much reduced stock will exceed

---

that when one aspect of the industry is regulated, the industry will always, if the regulation has a significant effect, tend to change in such a way as partially to negate the effect. This has happened several times in pelagic whaling — reduced quotas for certain species as they become depleted encourage whaling on regulated species so that the total value of catch-per-unit of effort is maintained at profitable levels.

[10] Supporters of a moratorium have been accused of "being emotional" about whales. Apart from the question of whether emotion is sinful, one observes that scientists are as liable as anyone to make ideological statements when they think they are giving "objective" scientific advice.

[11] The reader is referred to Cushing (1974) for a succinct description of the problem in marine fisheries. He identifies the stock-recruitment relationship as the central puzzle of fisheries research, and states that great collapses of "managed" fisheries "are attributable to the unstated concept that fishing could continue until recruitment was seen to fail. Then failure can be attributed either to natural causes or to fishing; and stocks collapsed while scientists disputed the 2 possibilities". He adds that failure of "management" has been no less than that of science. As in many other situations the devil is in the "unstated concepts" as well as in some of the stated ones, such as "The best regulation of fishing is to leave it alone until there is evidence of adverse effects". See Gulland (1971) for a general discussion of the consequences of acting on such null hypotheses.

the replacement yield, i.e., whether it should lead to a decline or increase in stock in the short term. Tests of such predictions have not been made systematically. In theory, such tests should be easy to make because the catches are recorded, and good effort data provide measures of stock abundance (catch-per-unit effort). In practice, given the likely rate of change in a whale stock – or even the maximum possible rate, when a depleted stock is protected – the abundance estimates are too variable. They are affected by the fact that the whaling effort is expended, in the same area and period, variably on several species; its effectiveness depends on the weather, and its efficiency changes for a number of technological reasons, few of which have been taken into account. The abundance estimates could and should be much improved, but we cannot expect too much from them in attempting to measure the density dependence of reproduction or in testing the assumption of constant natural mortality.

In recent years it has been suggested, mainly on the basis of work by R. Gambell and C. Lockyer on fin whales, that the mechanism of density-dependence is in the change of growth rate in young animals (resulting from density-caused changes in available food) which leads to changes in the age of sexual maturity. This hypothesis – with supporting data – is now being examined critically by the FAO/ACMRR Working Party on Marine Mammals. My concern here, however, is to see how close a determination of the reproductive relationships brings us to understanding of the MSY problem. This is examined in Appendix 4. My conclusion is that even with far better data than are ever available for wild animal populations the level of stock corresponding with MSY could not be determined with any useful accuracy, although a value for MSY itself might be inferred. For example, Doi (1973), having discussed the Antarctic sei whale, "emphasizes that it is possible to estimate the maximum sustainable yield, to diagnose the population status and to determine the management

course judging from a reproduction curve". Unfortunately, this is not so.

According to the simple idea of SY, the net rate of increase should be zero in the initial stock, just before exploitation starts. But this has never been tested. It seems to me quite possible that the abundances of whale stocks were changing in the period before intense exploitation, and that their population structures at the time were, therefore, not characteristic of steady states. Such changes could be cyclic or long-term trends, and related to climatic trends and cycles or other "external" factors, or to man's previous exploitation of other species. Furthermore, the factors causing such changes could usually be expected to continue their effects after exploitation started. The existence of steady states is something to be tested, not assumed; as Bradley (1972) has written: "In a world of dynamic change, it is constancy, not stability, that requires explanation". It is also necessary, for example by simulation, to examine the likely errors of applying steady-state models to real situations in which the stocks might have been changing naturally.

While numerical calculations of whale stock changes have, as mentioned in section 7, included time lags between birth and recruitment, little account has been taken of the fundamental properties of models in which such time lags have been incorporated. Thus, Cook (1965), May (1973) and May et al. (1974) have shown that such models, even variants of the simple logistic, display steady states under certain conditions, but stable-limit cycles, damped oscillations or even wild behaviour under other conditions. Marchessault, Saila and Palm (1976) have recently formulated and examined some of the properties of a simple model incorporating a constant recruitment delay, and applied it to a lobster population. Their model is very close to the one which is implicit in some of the numerical calculations for whale stocks. The model predicts in effect that if the age of recruitment (or maturity) is *under*estimated (in the lobster application the cases compared are actually recruitment at 5

31

years and at 0 years, i.e., the "delay" model is compared with the simple "instantaneous" logistic) the amount of fishing effort required to obtain MSY is overestimated, and so is the MSY itself. Thus, if the age of recruitment is underestimated, scientific advice based on a delay-logistic model would encourage quite serious overexploitation, whether regulation is by effort control or catch quota. The MSY stock level is not affected by the time day, although it must always be borne in mind that the delay model predicts the relationship between sustainable yield and *recruited* or *mature* stock, not the total stock. The relevance of this to a critique of whale assessments is that the age-determination, and hence estimation of the ages at recruitment and at sexual maturity and full reproductive activity has been somewhat controversial. While there have in recent years been important amendments to the age determinations from ear-plug rings, age determination for some toothed and baleen whales are not yet sure, and for the blue whales there are practically no data, since they were greatly reduced in numbers before the ear-plug method was invented.

All applications of the MSY concept have assumed reversibility so that if intensified whaling "overshoots" the MSY level it is only necessary to reduce the intensity sufficiently and for long enough for the stock to recover. It would not grow back along the sustained yield curve, but would approach it as the population structure tended to the steady-state one, at an overall rate given by the assumed linear dependence of net increase rate on total stock abundance or density. I find several difficulties with this assumption. Firstly, it has been said that greatly depleted whale stocks have not always recovered as the model would predict (Talbot, 1974) [12] after they have been pro-

tected. This could be for many reasons: because there have been small permitted catches of the same order of magnitude as the sustainable yield of the depleted stocks; because the models may be erroneous in ignoring a number of properties of the populations, apart from their age structure – one such property is social structure, disrupted in the course of depletion; because the area of distribution may contract as the total population size decreases, thus tending to maintain *higher* densities – and lower net increase rates – than the simple model would predict. [13] Considering how little is yet known about the social behaviour of cetaceans and their physical capabilities, it seems possible that the simple stock models could lead us astray in our quest for the "real" MSY. Perhaps the most likely reason for non-recovery, or recovery at other than the predicted rate, is however the occupancy of an ecological niche by a competing species – perhaps another whale species. This has been suggested to occur between the blue and minke whales and the fin and sei whales.

The "S" in MSY stands not for *sustained* but for *sustainable*, that is, it specifies a future option: if we do certain things now, someone else can, if he wishes, take certain yields in the future. If he does not – the model implies – the "quality" of the resource will deteriorate (the stock will grow toward its asymptotic size, and in doing so lose some of its "productivity"). This leads to the idea that it is "beneficial" for a stock to be reduced, by harvesting, to some fraction of its initial size, and that it is

[12] It is difficult to pin down the facts about this. Gray whales have recovered; humpbacks seem to have increased in some areas; right whales seem at least to be holding their own in the Southern Hemisphere. We do not know what is happening in most cases, because such data as there are on sightings are highly variable.

[13] In this connexion, I do not agree with Talbot's (1974) general criticism that "what is 'known' of the population dynamics of whales is inferred from fish and other animals". It is inevitable that models and methods developed in one field will be applied in another; in fact, however, the earliest application of the "logistic" or "sigmoid" model was to whales (Hjort, Jahn and Ottestad, 1933). One way in which the applications of "surplus stock" models to whaling and to fishing have differed is that whereas in the former yield is expressed as a function of stock size, in the latter it is treated as a function of the fishing mortality rate. This is really only a matter of style but it does set the mind on a different track when considering alternative management actions.

wrong to "waste" a renewable resource by leaving it unharvested.[14] It is not, however, reasonable simply to assume that future generations will place the same values on a resource as we do. This cuts both ways: one IWC delegate responding to an argument that future generations may "need" food, including whale meat, more than we do, said that *they* will decide what they need and how to get it; therefore, why should we give priority to leaving whales in the sea? Nevertheless, the implications of the assumption need to be explored with as much technological, economic and social forecasting as practicable.[15] For whaling, this at least includes looking at the likely availability of similar products from other sources, the relative costs of fuel (pelagic whaling, like distant-water fishing, involves high consumption of non-renewable energy sources) and the evolution of human dietary habits. But should we not go further and try to anticipate qualitatively new "values" for natural "resources" whether as tourist attractions, unique objects of scientific research, companions of man, symbols of our atonement for past sins against nature or other? (McIntyre, 1974). Since we can glimpse such values but not evaluate them, our present actions might be judged in terms of the number of future options they leave open. Intuitively, to refrain from interference now is likely to leave more options than would drastic interference. Our intuition may be wrong; natural processes we do not yet understand may remove some of these options tomorrow; nevertheless, failing other guidance, we live each day as if tomorrow will come.

So the MSY criteria could be judged in terms of its success in restraining interference and thus, hopefully, balancing a variety of future options against some products now. In this respect it scores fairly well, although maximizing weights rather than numbers, and maximizing sustainable net economic yield would be better. Suggested alternative criteria would need to be judged against these virtues of MSY. If we believe that man has a long future before him and will behave no less wisely in future than now, we should give more weight to future options than to present uses — provided that those uses are not necessary for human survival which, in the case of whales, they clearly are not.

One criticism of the simple stock models is that they are deterministic rather than *stochastic*. We are concerned always with the evaluation of risks, and risks cannot be expressed deterministically. This is important in the application of models to fish stocks which fluctuate greatly, mainly because of variability in the survival of young, but not so much to animals which bear 1 or at most 2 young at a time and care for them, and in which the natural mortality is not mainly caused by predation.[16]

The 1974 IWC "decision" concerns stocks, not species of whales. Although it seems to impose an operational definition of stocks for management purposes, that definition is apparently tautological and the discussion by the Scientific Committee continues to be concerned with the identification of biologically distinct populations and with the degree, time and place of any mixing between them. Knowledge of these identities and processes is still very scant, and as far as the measurement of SY and the location of MSY are concerned the practical questions are:

(a)   what may be wrong with regulating the catch as if from a single stock when in reality there are several more or less separate ones?

---

[14] This latter theory has often been invoked in fishery discussions by those who wish to be free to exploit the resources off the coasts of other countries which do not have the technical or economic capacity to exploit them.

[15] We should not overlook the fact that values have changed drastically even in the recent history of whaling. It is not many years since European nations valued baleen whales practically only for their oil, or for their oil and baleen plates.

---

[16] Few stochastic models have been formulated. Except for the simple case of exponential growth (Mann, 1973) they are mathematically complex.

(b)   what may be wrong with regulating catches as if from several separate stocks when there is really only 1 (or there is substantial mixing between several)?

In case (a) we can be sure that the yield will only by chance be the sum of the SYs or MSYs of the separate stocks; usually some stocks will be unwittingly "overexploited", others "underexploited". However, at least 1 compensating mechanism exists in the whaling operations. If, for example, the stocks inhabit different (even if overlapping) areas — as is usually assumed — then the underexploited ones will increase relative to the overexploited ones, presumably become more dense, and thus attract relatively more attention from whalers. Such compensation is, however, limited by the fact that the attention of whalers is not determined solely — or even mainly — by those densities; in particular, they are simultaneously seeking other species of whales. One way of reducing the risks of excessive whaling effort on (unrecognized) stocks is to allocate the total quota by areas; such action has been proposed by the IWC Scientific Committee.

It may be noted that essentially the same problem existed with the BWU in the Antarctic, and was the strongest reason for its eventual abolition. One might ask why the "compensating mechanism" did not operate to protect the blue whale? It probably did, to a certain extent.[17] However, given a certain biomass density of blue whales in the sea and an equal density of fin and other baleen whales, the whalers who are permitted to take a certain catch (in BWU or total weight) prefer to chase and capture fewer of the large species than more of the smaller ones.[18]

A policy of dividing up a total quota by areas in case these contain different stocks, can lead to a different classification of a species, according to the IWC categories. This is true also where there is evidence of separate stocks, but if — as is usually the case — there are no separate measurements of their parameters, so that to all of them are applied the same overall parameter values, then the same species might be classified in one area as "sustained management" and in another area as "protected". But it is likely that in the former area the species had not been reduced by whaling so far below its initial abundance because it has a lower initial abundance in that area and thus attracted less effort; nevertheless, its small residual number may be less able to sustain a continuing catch than in the other area, and its very existence might even be threatened by continued whaling. This and related problems have led E. Mitchell and others to suggest a rider to the 1974 decision, to the effect that no stock should be reduced to a level at which its existence is threatened or from which it is likely not to be able to recover even if it were no longer exploited. A term suggested for such a level is "minimum allowable stock", defined in absolute, not relative, terms. Unfortunately, no one has a clear idea how such a number could be estimated; it would have to be defined probabilistically and in relation to transient rather than steady states, and therefore cannot be regarded as another point on the SY-stock curve. Furthermore, probabilities of extinction or non-recovery surely also depend on population properties other than number. Nevertheless, as an additional criterion — to ensure that management does not restrain the overall impact of whaling while permitting qualitative changes in the ecosystems — it might be worth consideration.[19]

---

[17] Gulland (1968a) pointed out that, paradoxically, the BWU — at the very time it was most under attack as a conservation measure — did operate to "protect" the fin whale relative to the sei whale, because the meat from 3 sei whales was more valuable than the meat from 1 fin whale (BWU = 1 blue = 2 fin = 6 sei = 2.1/2 humpbacks).

[18] Except when, as in the case of sei and fin whales in the BWU, the weighting factor, having been based originally

on relative oil yields, is quite inappropriate when other properties become more highly valued — such as meat quantity and quality. Decline in mean size of intensely exploited species also unbalanced the weighting factors.

[19] See the arguments for protection of blue and humpback whales, in Chapman et al., 1964 — Supplementary Report.

A common approach to empirical determination of the shape of a single species SY curve is to plot estimates of net increase rate against population size, fitting a straight or curved line as the data indicate, or some variant of this procedure. The method depends critically on measures of abundance (or density) derived from catch-per-unit effort data, except when, rarely, direct measures are available, e.g., from sighting or echo-survey. In handling such data the attempt is usually made to define "effort" such that it is strictly proportional to the instantaneous fishing mortality coefficient. In particular, trends in efficiency (usually, it is assumed, increasing) of fishing units must be allowed for. These require analyses of technological factors such as vessel size, power, and search and catching equipment, and this has been done, partially, in connexion with pelagic whaling. Physical factors such as hours of daylight and weather are also important, at least with respect to the variability of the effective effort, but also in certain circumstances with respect to trends. Thus, changes in the latitudinal distribution of Antarctic pelagic whaling and the dates of the season, which are related to changes in hours of daylight and prevailing summer weather, have not yet been adequately taken into account in whale assessments. However, a bigger problem, which has certainly *not* so far been resolved, is how to apply the same total effort data to a number of species caught concurrently, when the abundances and values of individuals of those species show differing trends over the period of study. To the extent that the geographical distributions of the various species differ, the problem is partially resolved by calculating catch-per-unit effort area by area. Much more could be done if whaling effort figures were split into "searching time" and "capture time" but such information is not available. Even if it were available, the decisions of whalers, which are manifest in performance data, have over the past 40 years, been modulated in a complex way, by changing values of different products and by the existence of IWC regulations (such as protection of certain species) as well as by changes in the relative abundances and sizes of the various species of baleen whales and of the sperm whale. Thus, the calculated trends in abundance of each species, even when all practicable adjustments have been made, must be rather uncertain. This remains true even where calculations take account of whale sightings, especially when, as is usually the case, the sightings used are those reported by whaling expeditions.[20] It may well be that the total catch (in weight) of *all* species, per unit adjusted effort, is a better indication of whale biomass than the catches-per-unit effort by species are of species abundances; indeed, Beverton and Holt (1957) suggested that "paradoxically (the 'sigmoid curve method') is perhaps most satisfactory when applied to the behaviour of several species at once, or even of an entire community as done by Graham and Baerends". Rough application to the data of total Antarctic baleen catch in weight-per-unit effort indicates, as might be expected, that these stocks are, as a group, far below the level of MSY (Appendix 5).

The question posed above of managing 1 stock believing it to be several, is a special case of a more general one, which is now assuming importance in the IWC: that is, how are catches to be regulated if different species, simultaneously hunted, are biologically interdependent? It is tempting, if one has become used to thinking in terms of steady-states, to construct models for interdependent species or stocks, to give an SY surface of n + 1 dimensions, which may have one or more peaks on it. This could readily be done, leaving us only (?) with the problem of estimating the set of perhaps n (n − 1) parameters representing the interdependences. Although it would be interesting to see the results of such a theoretical exercise, I think it unlikely that such a multi-species model would resemble

---

[20] A very recent analysis both of sightings data and the division of effort between searching and catching is given by Schweder, 1974.

reality. Most interactive models predict cyclic population behaviour, as in the simple predator-prey model of Volterra.[21] This is true also if the common prey of 2 competing predators is treated dynamically, and a 3-element model may lead – at best – to predictions of cycles and trends to which the MSY concept could not be applied. This will not help us decide whether to take krill out of the Antarctic or baleen whales, or both and in which proportions, or neither (Bondar and Bobey, 1974). Nor will it help us approach sperm whaling rationally in relation to possible intensive exploitation of ocean squids in the future. Nor can such modelling be an adequate response to those who have suggested that perhaps by reducing drastically the number of minke whales in the Antarctic we may facilitate the more rapid recovery, under protection, of one of its competitors, the blue whale. Present knowledge of the structure and behaviour of marine ecosystems is so limited that the analytical methods of population dynamics at present give poor guidance for ecosystem management. Recent work on some ecosystem properties such as "resilience" and "stability" sustains some hope that such properties will be predictable from studies of the major components or the relations between these (Holling, 1973) but we are far from being able to apply such understanding to marine resource management.

Another approach to management of marine populations and ecosystems might be to consider the flows of materials and energy through them. Could we specify a safe limit to the proportion of the total biological production or biomass that may be removed (in terms of current knowledge), with the proviso that the pattern of removal be such as not to change the ecosystem qualitatively? Or can studies of metabolic efficiency help define management objectives? For example, it has been suggested that whales, being warm-blooded, large and long-lived, may be biologically efficient in the sense of maintaining stability at the expense of great energy consumption for maintenance, but for the same reasons highly inefficient as intermediaries for man's use of the biological productivity of Antarctic waters (Steele, 1974). We would need to take into account, however, the energy required to gather that production, which in the case of whales comes from within the system and in our case, at present, from outside it. The provisions and definitions of the U.S. Marine Mammal Protection Act of 1972 (see summary in Appendix 6) have, as yet, no explicit counterpart in international agreements. Some of the provisions do not seem to be incompatible with possible interpretations of some existing conventions, nor with the above suggestions. I suggest the following interim strategy for the formulation of improved goals in the management of wild stocks:

(a) accept the concept of MSY-stock level, not as a sole objective but as a lower bound to permissible exploitation, and as part of a multi-faceted definition of "conservation"; in setting this bound take account of the great uncertainty of estimation of any MSY stock level including the various biases noted in the preceding sections;

(b) modify the yield definitions in "conservative" directions – e.g., by weight rather than numbers, net yield rather than gross yield;

(c) calculate MSY of stocks and species separately and in combination, and select that application which would be most "conservative". (This might lead to the classification of all the Antarctic baleen whales as "a protection stock" in terms

---

[21] Renewed interest in the problem of estimating the parameters in interactive models is indicated by the publication, as this paper was being revised, by Swatz and Bremmermann (1975). The authors discuss the best estimates, and their errors, for parameters of a Lotka-Volterra system, and logistic growth of 4 competing species. See also May *et al.* (1974).

of their present combined biomass as a fraction of their "initial" biomass);

(d) test the assumption of initial steady states and apply an MSY concept appropriately modified in accordance with indications of natural trends and cycles;

(e) examine critically the application of MSY in each case with respect to possible or likely options of future uses and values of the resource; predict, as well as present data and methods permit, relevant technical, economic and social trends;

(f) examine each application with respect to the problems of providing adequate scientific advice according to the guidelines given by ACMRR (FAO, 1974);

(g) in connexion with (f) above, link management for high sustainable yields with the adequate financing of research, not only on the populations being exploited but on the ecosystems of which they are part, as well as on the properties of whales, other than those for which they are at present most valued. This would ensure that present beneficiaries are obliged to put back some of their benefits not only to improve present use patterns but also to give scientific information relevant to future options.

It may be difficult, if not impossible, to get general agreement on some of the modifications of the MSY concept outlined above, particularly the criterion of net yield rather than gross yield. Operators and states have been unwilling to make public, for international use, any economic data such as statistics of costs pertinent to this matter. This problem is tied up with the question of resource "ownership"; if there were a single corporate "owner" (custodian or trustee) of the whale resources some such difficulties could be resolved (Gul-

land, 1968b). Furthermore, nations and peoples who do not at present exploit the resources and those that do not now consume the products, will presumably give a relatively higher value to future options than do the present exploiters. Their interest in the resource should, therefore, be supported, since in a sense they might act as agents, in the present, on behalf of the future generations both of man and whales.[22] This is one reason why the present UN Conference on the Law of the Sea, in which practically all nations are participating, should be a valid forum for the reconsideration of the objectives of conservation and management. So far, in the preparatory work, and in the first session, attention has been focused almost entirely on the question of who gets the benefits, and this is seen as a matter of jurisdiction. Equitable allocation is certainly a necessary though not sufficient condition for realizing long-term benefits from living marine resources.[23] Jurisdiction, research, international advice, and apparently reasonable conservation measures, did not

---

[22] Cetologists also are protentially a powerful lobby for the future, at least when the continuation of their research does not depend indirectly or directly on the present whaling industry for funding or materials, or both. Research now needs to be funded with future values in mind; it is difficult to see how to ensure that without substantial funding from international sources (as well as from those who have profited from past and present exploitation) both for an expanded programme and for continuing some current activities.

[23] "There are several identifiable valid objectives or goals of international fisheries management, including optimization of biological, economic and social yield. None of these goals can be achieved, however, unless there exists a jurisdictional base for management permitting authorities to regulate the disposal of resources". So "the overall goal should be to maximize opportunities of individual states ... to achieve benefits from living marine resources with respect to which they have exclusive or shared management rights" (A.S.I.L., 1974). No hint of a "common heritage" here, but I suppose such "management rights" as are shared by all states would, collectively, form that agreed principle. Unfortunately, the valuable study quoted, drawn up by a group composed, it seems, exclusively of lawyers and economists, reveals little awareness of the subtleties of marine ecology and the uncertainties and difficulties of stock predictions. It recognizes, however, that present research and scientific training efforts are quite inadequate to the demands arising from modern fishing activities.

save the Peruvian anchoveta fishery from collapse. Antarctic whaling is regulated, national allocation of catches are made, and competition is regulated *within* participating nations as well as between them. The resources have been "appropriated" – if only temporarily by present participants – that is, in effect, if not formally by the elimination of other competitors, by possession of the "veto" in the IWC, and by the reduction of stocks to levels so low that it is in the interest of no one else to start whaling. All this has not assured a future for that industry.

Perhaps the problem of whale management – even of marine fishery management generally – is one of the sub-set of "insoluble" problems identified by Crowe (1969). There are problems relating to "the commons", concerning which natural scientists "pass technically insoluble problems over to the political and social realm for solution" and conversely, social scientists, seeing no political solutions, defer to future technical advances. Nevertheless, in attempting to reformulate international conservation objectives, fishery scientists are acting in the hope that the political problems will not, in the lon grun, prove completely insoluble. I doubt, however, if they can construct viable alternative objectives without incorporating some of the findings of other disciplines – some such as general ecology, close to their own; others far removed.

It is perhaps appropriate to close this paper with 2 quotations. Bradley (1973) attempted to "evolve from relevant insights in behavioural theory, economics and political studies a regulatory system similar to the systematic inter-relations that exist in the eco-systems of the natural environment". He observes that:
"In the past, the belief persisted that an increased technological capacity to alter the environment brought increased control. This belief, far from dead, is a manifest delusion. First, the difficulty in devising any physical control system lies not only with generating enough power but also with generating enough of the proper kinds of information.

Since the environmental resource system is a system, any change in a given parameter will have numerous, unpredictable repercussions throughout the system; so even if the effect of the intervention is to bring under control the variable directly affected, the total system is likely to be less predictable than before, while all the learned skills based on the former given are depreciated. Further, interventions and the further interventions to which their unexpected results tend to lead, are likely to be self-multiplying. The rate of change increases at an accelerated speed, without a corresponding acceleration in the rate at which further responses can be made; and this brings ever nearer the threshold beyond which control is lost".

David Bella (Bella, 1974; Bella and Overton, 1972) has expressed more briefly a related idea:
"We are faced with an *environmental predicament*: our ability to modify the environment increases faster than our ability to foresee the effects of our activities", and, further, "the most significant aspect of this predicament may be our inability to foresee our own reactions to the actions that we are capable of taking".

Acknowledgements

Brian Rothschild, John Gulland, Geoffrey Kesteven, Saul Saila, Ray Dasmann, Basil Parrish and Victor Scheffer, among others, kindly commented on earlier drafts of this paper. The views expressed here are, of course, my own.

Afterthought

In this paper I have criticized the theory and practice of fishery science as applied to conservation of living marine resources, especially whales. My intention was to stimulate discussion among colleagues, especially at the "MSY Workshops" at Airlie, Virginia, for which the paper was primarily written. I felt justified in taking a critical position because I have been involved in, and a party to, the formulation of scientific advice which seems, in retrospect, not always as good as it might have been.

However, it must be said that "scientific advisers", especially at the international level, labour under considerable disadvantages. They act as detectives, not only of the properties of natural systems — if that were not difficult enough — but also of the behaviour and intentions of the industry they are trying to assist.

Information about the economic and other considerations which determine the tactical deployment of pelagic whaling effort, for example, would reveal the meanings of abundance indices which are now obscure. Many of the mathematical contortions in which we engage in efforts to understand the dynamics of natural populations and of fisheries are necessary not so much because of nature's secrecy, but because of man's. Scientific advisers are too often expected, collectively, to offer economically and socially meaningful advice, while denied access to the kinds of information on which operational decisions are based.

Twelve years after the desirability of such information was formally recognized by the IWC, its Scientific Committee could only report in 1975 lamely, "we have no information on this aspect at present" referring to economic criteria which might be applied to determine an "optimum" management policy.

Uncertainty lies in this area as much as in our understanding of ecosystems, yet it need not. Now, however, the rules of the game do not favour the future.

In revising this paper for publication, I have taken into account some of the points raised in the Airlie Workshops.

Appendix 1

Relations between the stock levels and fishing (whaling) effort for MSY and those for maximum net yield, "commercial" extinction and "biological" extinction.

1. In the simplest model, the yield curve is a symmetrical parabola:

$$Y = 4P(1 - P) \qquad \text{(Fig. 1)}$$

where P is stock size, as a fraction of initial size

$$(P_{MSY} = 0.5)$$

Y is sustainable yield, as a fraction of MSY.

2. Define cost of whaling to maintain a given P as proportional to F, the fishing mortality coefficient, which is in turn proportional to $(1 - P)$. Write net yield as $y = Y - a(1 - P)$, where a is the ratio of the cost of exerting a unit of effort to the value of a unit of catch. The curve of net yield against stock size has a maximum at $P_1 = 0.5 + a/8$.

3. Define the stock level for zero net yield to which the unregulated industry will "naturally" tend ("commercial extinction") so that the yield at the point, $Y_0 = a(1 - P_0)$

but $Y_0 = 4P_0(1 - P_0)$

so that $a = 4P_0$

Hence the curve of net yield is a squeezed and displaced symmetrical parabola (Fig. 1), with maximum at

$$P_1 = \frac{1 + P_0}{2}$$

and a net yield at that level of $y_{max} = 1 - P_0$.
The ratio

$$\frac{y_{max}}{y_{MSY}} = \frac{(1 - P_0)^2}{1 - 2P_0}.$$

This ratio is 1.067 for $P_0 = 0.2$, for example; that is the maximum net yield would be nearly 7 % greater than the net yield at MSY. The yield at the level of y_{max} is $Y = 1 - P_0^2 = 0.96$ in the example,

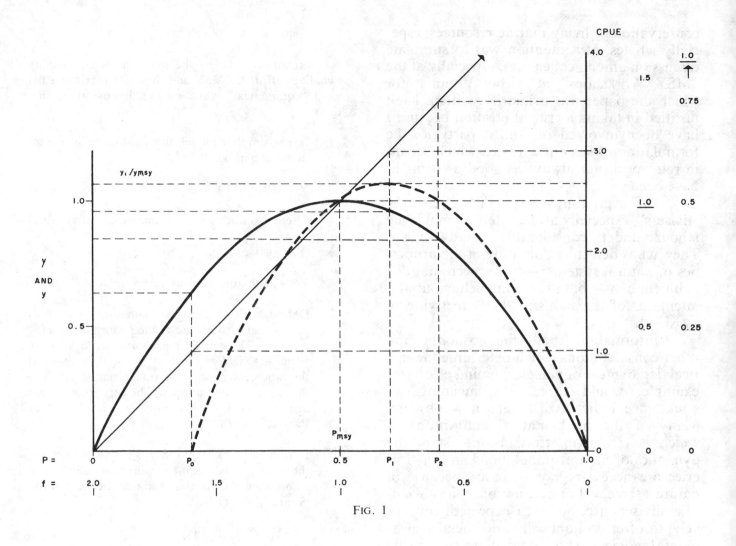

FIG. 1

so, in this case, if one were prepared to accept 4 % less than the MSY, one could gain 7 % in net yield. This would be obtained with only $(1 - P_0)$ of the effort required for MSY, i.e., 20 % less, if $P_0 = 0.2$. (See also 6 below).

4. If the catch-per-unit effort is proportional to P, then we have the following relation

$$\frac{\text{cpue at } y_{max}}{\text{cpue at MSY}} = 1 + P_0$$

and

$$\frac{\text{cpue at } y_{max}}{\text{cpue at } y_0} = 0.5\left(1 + \frac{1}{P_0}\right)$$

In the example of Fig. 1 the gain in c.p.u.e. by

moving from the MSY point to the y_{max} point is 20 %.

5. Similar relations are found for models in which the curve of sustainable yield against stock size is asymmetrical (see 7 below).
In general

$$P_{MSY} \leq P_1 \leq (1 - P_0) P_{MSY} + P_0$$

That is, P_1 is always greater than P_{MSY} by an amount somewhat less than

$$100 \cdot P_0 (1/P_{MSY} - 1) \%$$

Fig. 2 illustrates a situation in which P_{MSY} is displaced 16 % to the right.

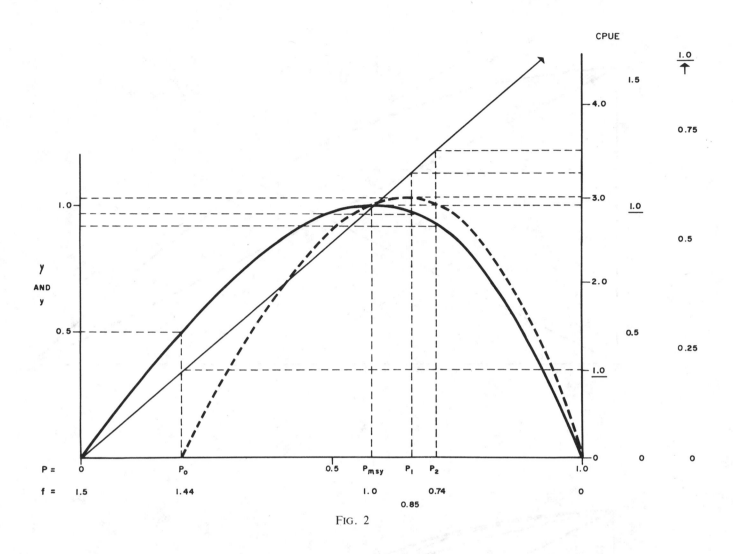

F$_{IG}$. 2

If we treat, for example, all Antarctic baleen whales as a single "stock" for which there is a unique sustainable yield curve, and we know that whaling continues to be profitable at a certain stock level, we can readily estimate an upper limit for the amount by which the stock for maximum net yield would be above that for MSY (see Appendix 5).

6. We may further deduce what stock level, $P_2 > P_1$, will give an SY such that the *net* yield is equal to that which would be obtained at MSY level. For the symmetrical parabola this is $0.5 + P_0$. In our example $0.5 + P_0 = 0.7$, i.e., 40 % higher than at MSY. The SY at this stock level, Y_2, is $(1 - 2P_0)$ $(1 + 2P_0)$, in the example 0.84, i.e., 16 % less than MSY, but the catch-per-unit effort is 40 % higher.

In the skew yield curve of Fig. 2, P_1 is 15 % higher than P_{MSY} and gives a net yield 2.5 % higher than at MSY. The stock that will give the same net yield as at MSY is 23 % higher than P_{MSY}; the gross yield at that level is 0.91 of MSY, i.e., 9 % less.

Lastly, note that for the symmetrical parabola, the SY level at the threshold of "commercial extinction" is $4P_0 (1 - P_0)$ of MSY; in the example 64 % of MSY. For the skew curve it is less (50 % of MSY).

7. Assume the curve of SY against P can be represented by the expression:

$$Y = P (1 - P^{n-1})/b \quad \text{(e.g., Pella and Tomlinson, 1969).}$$

41

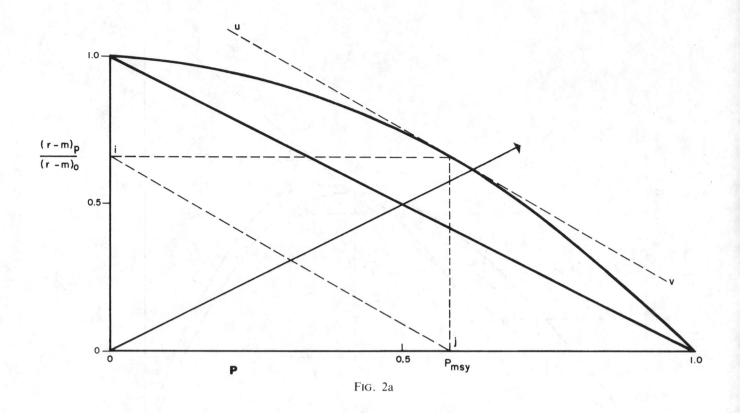

FIG. 2a

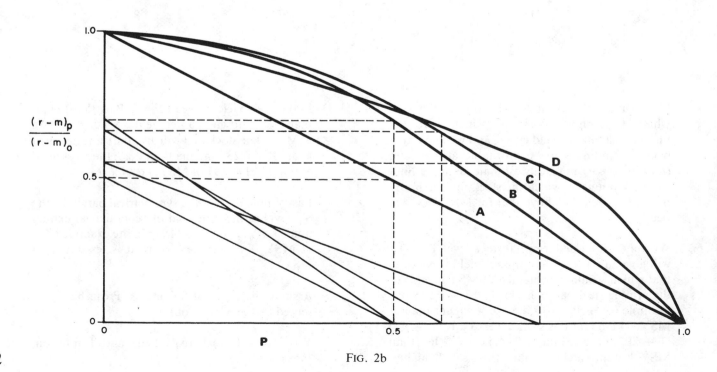

FIG. 2b

The effect of skewness on deductions concerning the relation of y_{max} to MSY can then be examined. In the above equation, as n increases above 2 the SY curve becomes skewed increasingly to the right. This curve has a maximum at

$$P_{MSY} = \left(\frac{1}{n}\right)^{1/n-1}$$

and putting, as before $Y_{MSY} = 1$, we find

$$b = P_{MSY} - P_{MSY}^n = P_{MSY}\left(\frac{n-1}{n}\right) = \frac{n-1}{n^{n/n-1}}$$

Now writing

$$y = Y - P_0 F$$

gives

$$y = (P - P_0)(1 - P^{n-1})/b$$

Define

$$_1y = y/y_{MSY} = \frac{(P - P_0)(1 - P^{n-1})}{(P_{MSY} - P_0)(1 - P_{MSY}^{n-1})}$$

$_1y$ has a maximum at P_1, given by the expression

$$nP_1^{n-1} = P_0(n-1)P_1^{n-2} + 1$$

For the case $n = 3$ we find

$$P_1 = \frac{P_0 + \sqrt{P_0^2 + 3}}{3}$$

from which

$$\frac{P_1}{P_{MSY}} = \frac{P_0 + \sqrt{P_0^2 + 3}}{\sqrt{3}} \cong \frac{1 + P_0}{\sqrt{3}}$$

and

$$_1y_{max} = \frac{3(P_1 - P_0)(1 - P_1^2)}{2(1/\sqrt{3}P_0)}$$

The following table summarizes the properties of the model for $n = 3$ ($A = +0.15$), for various values of P_0

P_0	P_{MSY}	P_1	P_2	Y_1	Y_2	$_1y_{max}$
.1	.577	.612	.645	.994	.979	1.006
.2	.577	.648	.716	.977	.907	1.033
.3	.577	.686	.788	.944	.776	1.105
.4	.577	.726	.863	.892	.572	1.304
.5	.577	.768	.939	.818	.288	2.132

8. Numerical solutions for a range of other values of n and P_0 are shown graphically in Fig. 3. A linear scale is used for an index of skewness:

$$A = 2P_{MSY} - 1$$

This index is zero for $n = 2$, and has a range $0 \to 1$ for $n = 2 \to \infty$ over the calculated range

$$P_1 \cong P_{MSY} + P_0(1 - P_{MSY}).$$

Suppose, for example, MSY is thought to be obtainable from 0.7 of the initial stock; the curve for $n = 6$ would be appropriate. Then, for $P_0 = 0.2$, we find $P_1 = 0.74$, that is 5 % higher than P_{MSY}.

At that stock level the sustainable yield is 0.990, i.e., 1 % less than MSY, and $_1y_{max} = 1.01$.

Lastly, for $_1y = 1$ we find $P_1 = 0.77$ (10 % higher than P_{MSY}) and $Y = 0.964$.

9. We now consider the changes in whaling *effort*, corresponding with movements along the SY and Y curves.

If $\quad Y. = FP$
then $\quad F = (1 - P^{n-1})/b$

and specifically,

$$F_{MSY} = (1 - P_{MSY}^{n-1})/b$$

It is convenient to define effort, f, as the fraction or multiple of the effort which is required to sustain MSY.

So $\quad f = \dfrac{F}{F_{MSY}}$

In the context of this paper, three critical values of f may be calculated:

$f_e \neq 0$ leading to $Y = 0$ i.e., extinction of the stock

$f_0 (\leq f_e)$ leading to $y = 0$ i.e., "commercial extinction"

43

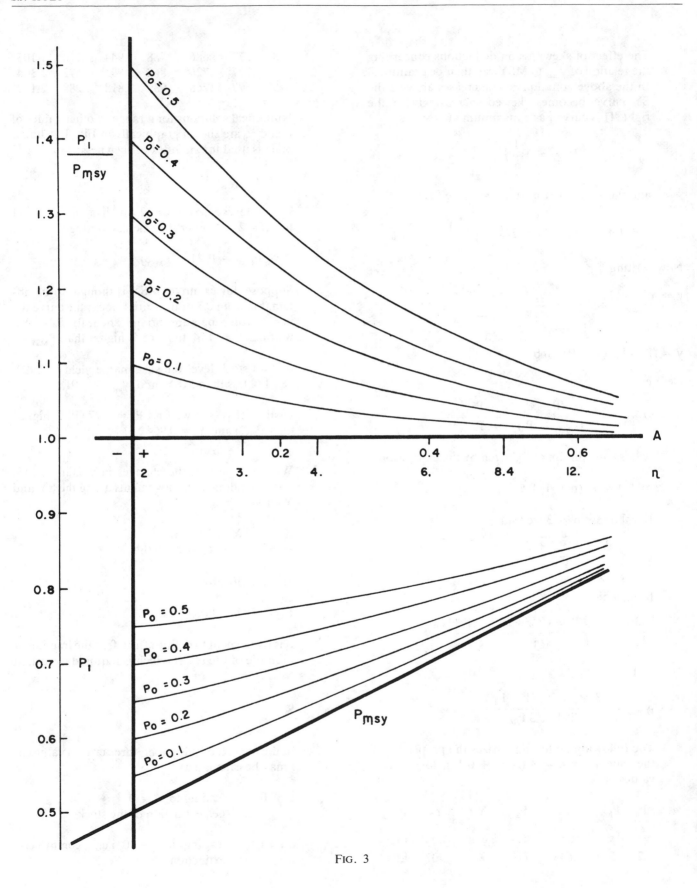

FIG. 3

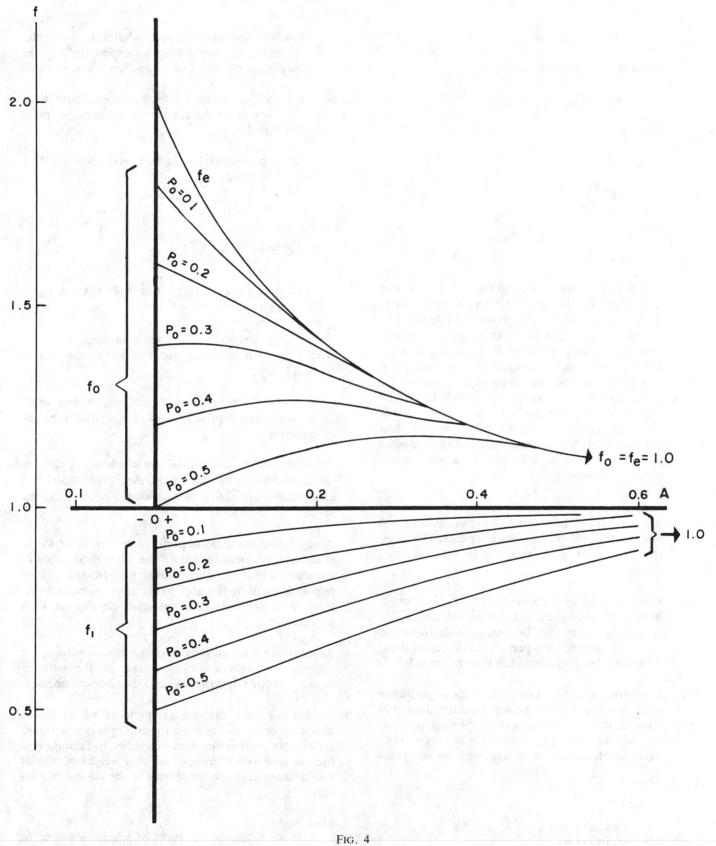

FIG. 4

f_1 leading to y_{max}

From these definitions we have:

$$f_e = 1/(1 - P_{MSY}^{n-1}) = n/n - 1$$

from which

$$P_{MSY} = (f_e - 1/f_e)^{f_e-1}$$

$$f_0 = f_e(1 - P_0^{n-1})$$

and

$$f_1 = f_e(1 - P_1^{n-1}) = n/n - 1(1 - P_1^{n-1})$$

Values of these are plotted in Fig. 4. It will be seen that theoretically stock extinction corresponds with an effort double that giving MSY or less. For the moderately skewed SY curve (n = 3) extinction occurs with an effort 50 % greater than for MSY[1]. This is not to suggest that we may place any reliability on this deterministic model for predicting what is essentially a probability; the examples are given to show that if the curve of yield against stock size is skewed to the right then with relatively little increase in effort beyond what is needed to obtain MSY the stock will decline rather precipitously.

"Commercial extinction" occurs, of course, at lower f values; how much lower is shown for 4 values of P_0. Fig. 4 also shown for these P_0 values, the relative effort f_1 required to maximize net yield.

10. For n < 2 the yield curves are negatively skewed, as appears to be the case for some fish stocks. For n = 1 the expression becomes the derivative of the well-known Gompertz growth curve (Fox, 1970) and has a maximum at P = 1/e = 0.368.

11. I will not attempt to examine here the implications of departures from the simple assumptions that value and costs are proportional, respectively, to catch and to effort. One direction of bias is, however, worth noting, as it may apply to any prac-

---

tically *unique* product from whales, for example, sperm oil. In such a case, we expect the price to vary inversely with the production. Then we find:

(a) y_{max} is given by higher stock levels than would be deduced from the constant price model; and

(b) "commercial extinction" is delayed as effort increases and $f_e \rightarrow f_0$.

Appendix 2

DECISION OF THE 1974 ANNUAL MEETING OF IWC

"The International Whaling Commission
Noting that whale stocks are a common concern to mankind

Concerned that some species of great whales are at present considerably depleted below their optimum population levels

Recalling that the historic decline in whale populations occurred not only because of excessive exploitation, but also because knowledge was inadequate to protect the species

Motivated by the need to preserve and enhance whale stocks as a resource for future use when food needs of the world will be greater because of increased human population and by the need to maintain marine ecosystems in a well-balanced condition capable of high productivity

Taking into consideration the long-range interests of the consumers of whale products and of the whaling industry as cited in Article V. 2 of the Convention, and

Recognizing that the management of whale stocks should be based not only on the concepts of maximum sustainable yield in numbers by species, but should also include such considerations as total weight of whales and interactions between species in the marine ecosystem.

Decides that
1. It shall classify all stocks of whales into one of

---

[1] But a model which gives a finite SY for F → ∞ is unrealistic, e.g., Bell *et al.*, 1973.

three categories according to the advice of the Scientific Committee.

(a) *Initial management stocks* which may be reduced in a controlled manner to achieve MSY levels or optimum levels as these are determined;

(b) *Sustained management stocks* which should be maintained at or near MSY levels and then at optimum levels as these are determined;

(c) *Protection stocks*, which are below the level of "sustained management stocks" as described in (b), which should be fully protected.

2. The Committee should define stocks for this purpose, as the units which can be most effectively managed individually.

(a) Commercial whaling shall be permitted on "Initial Management Stocks" subject to the advice of the Scientific Committee as to measures necessary to bring the stocks to the MSY level and then optimum level in an efficient manner and without risk of reducing them below this level.

(b) Commercial whaling shall be permitted on "Sustained Management Stocks" subject to the advice of the Scientific Committee.

(c) There shall be no commercial whaling on species or stocks classified as "Protection Stocks" including those species listed for full protection in the current schedule.

3. *Also decides* to implement this resolution by

(a) Requesting the Scientific Committee to provide advice on the criteria which should be used in defining categories of whale stocks which should be treated as in Section 1 above, this advice to be provided as soon as possible with a view to its incorporation in the Schedule.

(b) Directing the Scientific Committee to arrange to provide the Commission with annually updated advice on these criteria and on the allocation of stocks to the categories.

(c) Making all necessary amendments to the Schedule not later than the 27th meeting of the Commission".

Appendix 3

(From background paper to Airlie Workshop)

NUMBER OR WEIGHTS?

From the tables of yield functions of Beverton and Holt (1966) may be derived the mean weights of individuals in the catch, as a function of the size at recruitment (relative to final size) of the exploitation rate, E, and of the ratio of the natural mortality coefficient to the growth parameter, K. These mean weights can be expressed relative to, for example, the initial mean weight (i.e., for $E \to O$) or the mean weight for a value of E corresponding with the effort required to give MSY in accordance with the logistic or other model. In general, the former are more useful, since the relative decline in mean size can readily be observed as exploitation intensifies; for our present purpose the latter is, however, preferable.

Taking the Antarctic fin whale as an example, parameter values used by Allen (1974) are

M = 0.04
K = 0.26 from which M/K = 0.15
C (ratio of length at recruitment to final length) = 0.8

The required values of mean weight may be obtained by interpolating in the tables for

M/K = 0.25 and M/K = 0

and are summarized below:

P	Y	$\bar{W}$	$Y \times \bar{W}$	f
0.05	1.0000	1	1.000	1
0.51	.9996	1.0022	1.0018	.98
0.52	.9984	1.0045	1.0029	.96
0.53	.9964	1.0067	1.0031	.94
0.54	.9936	1.0089	1.0025	.92
0.55	.9900	1.0112	1.0011	.90
0.56	.9856	1.0134	0.9988	.88
0.58	.9744	1.0179	0.9918	.84
0.68	.8704	1.0303	0.8968	.64

47

So, although there would not be much gain in sustainable yield of products by maximizing weight rather than numbers caught (less than 1 %), the weight maximum could be obtained with 6 % less effort than MSY (number). Further, the same total weight as that obtained from MSY (number) could be sustained with 12 % less effort. Calculation shows that a sustainable catch of only 1 % less than MSY (weight) could be taken with 16 % less effort than needed to take MSY (number); and 10 % less with 36 % less effort. With further decreasing effort the catch falls rapidly. So, if fin whales were considered in isolation, the net economic yield (discount rate zero) would probably be obtained in the region of P = 0.7, f = 0.6.

The above calculations have been made assuming, as Allen did, that the curve of sustained yield in numbers is symmetrical. If P_{MSY} = 0.6 as Obsumi assumed (n = 3.6), the same calculations can be made and the estimates of relative losses and benefits are of the same order. The SY (weight) curve is not only skewed slightly to the right of the SY (number) curve but it is slightly flatter in the region of its maximum, so that one could afford to keep the stock relatively rather large without risk of forfeiting catch.

This suggests that apart from permitting depleted stocks to recover more than the MSY (number) policy would allow, it would be wise, from both biological and economic points of view, if any "controlled reduction" of "initial management stocks" of, for example, minke whales, were limited to, say 75 % of the initial stock size. But even the MSY (weight) of minke whales may be enough to pay for the costs of catching them, if they alone are caught. Thus it would be better not to reduce these stocks, at least until the stocks of depleted species have had a chance to recover. Otherwise there is bound to be a virtually overwhelming economic incentive to take catches increasingly exceeding the sustainable yield as the stock density declines, with serious danger of "overshooting" even the MSY (number) level.

Any steady-state mean weight will, if exploitation is unselective above the minimum legal size, be reached only after a transient period. In whaling, however, it appears there is, at least initially, some selection in favour of larger individuals. These 2 factors work in opposite directions, but do not necessarily "cancel out". Bearing in mind these uncertainties, the changes in mean weight which the yield tables predict in relation to the rate of exploitation are of the same order of magnitude as those observed in the Antarctic baleen whales. The male sperm whales of the southern hemisphere seem, however, to have declined in mean size more than would be expected, but no more than could be accounted for by selective capture of larger individuals.

Appendix 4

In the process of assessing whale stocks, as with fish, one examines "reproductive relationships". By this is usually meant the determination of a curve relating the total numbers of animals, or of the mature animals, to the numbers of recruits derived from them; this is referred to in the literature of fishery science as the "stock recruitment problem". Since the assessment of whale stocks is mainly a matter of estimating and predicting numbers, a knowledge of the reproductive relationship in each species and stock should permit calculation of limiting values of sustainable yield and even the order of magnitude of MSY, provided we are content to assume that the natural mortality is not also density dependent. What is not generally realized is how precise this knowledge would need to be for predicting the MSY stock level with useful accuracy. Thus very closely similar stock recruit curves can lead to widely differing conclusions as to the level of stock, relative to "initial" stock size, which will give MSY. This is illustrated in Fig. 5.

In Fig. 5a are shown 2 stock recruit curves. The scales and parameters of these are adjusted so that the number of recruits each predicts for the unexploited stock (P = 1) and a stock half that level (P = 0.5) are the same; recruitment from P = 0.5 is assumed to be three quarters of that from P = 1. Curve A represents the expression 1/R = a + b/P which has been commonly discussed in the literature (e.g., the Report of the

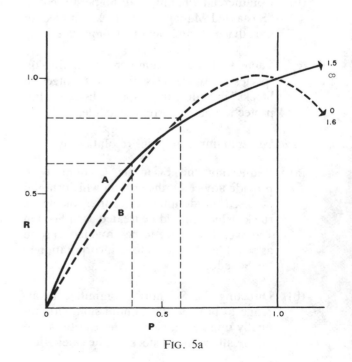

FIG. 5a

IWC Committee of Four Scientists). It is asymptotic to a relative value R = 1.5.

Curve B represents the expression

$$R = P(1.66 - 0.66 P^2)$$

This is the function which may be deduced from the yield model illustrated in Fig 2 (i.e., n = 3) if the natural mortality coefficient is assumed to be constant, independent of density. This function happens to have a maximum at P = 0.91. Thereafter, recruitment is predicted to decline, reaching zero at a *theoretical* stock level 58 % bigger than the "real" range $O \leq P \leq 1$.

The curve for the function corresponding with the symmetrical logistic with constant M [R = 2P – P²] is not shown; it lies between curves A and B, has a maximum at P = 1 and reaches R = O at P = 2. I do not know of data for any marine animals which would allow one to discriminate between curves A and B.

Fig. 5b shows the corresponding yield curves. Curve B is the same as in Fig. 2 except that the vertical scale is adjusted so that SY = 1 for P = 0.5. The vertical scale of Curve A is similarly adjusted. The sustained yield is, from the definition of the stock recruitment curves, the same in all cases – including for the symmetrical logistic – at P = 0.5.

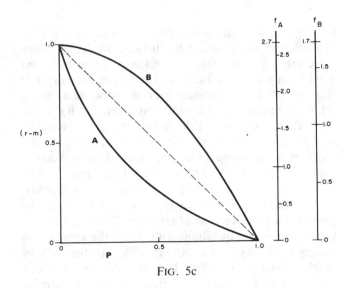

FIG. 5c

It is evident that the MSY predicted from A is 7.2 % higher than that predicted from the logistic; MSY for B is 2.6 % higher. Given similar stock recruitment curves the MSY cannot alter much. MSY is obtained, however, from very different stock levels – 0.37 for case A and, of course, 0.58 for case B. The initial stock sizes and the SY at half initial stock are the same in all cases, yet the MSY stock levels differ over a range ± 25 %. Real data would not permit even that poor degree of discrimination. These differences would be foreseen if it were possible to calculate r = R/P or better (r – M), and plot it, rather than R, against P. The graph for case A is strongly concave upward, having, at P = 0.5 a value of (r – M) = 0.25 of the maximum (r – M) for P = 0 (Fig. 5c). The range of change of r between P = 1 and P = 0 is also different: 3-fold in case A, 1.66 times in case B, and 2 times in the case of logistic. Yet over the critical range which might be observed, say 0.3 < P < 0.7, the r values differ only by, at most, about 5 %.

From this I think one must conclude that a calculated value of stock level for MSY depends practically entirely on the particular function used to "smooth" the data. Furthermore, when data are highly variable one is tempted to decide that the only reasonable line to draw through them is a straight one. This procedure is sufficient to lead to a particular MSY level; to $P_{MSY} = 0.5$ if the data analysis involves, implicitly or explicitly, a regression of estimates of (r – M) on estimates of P. In fact, if an exploited whale stock became stabilized at about half its initial size it would be practically impossible to predict whether the stock should be allowed to increase, or reduced still further in order to obtain MSY. Only an experimental approach

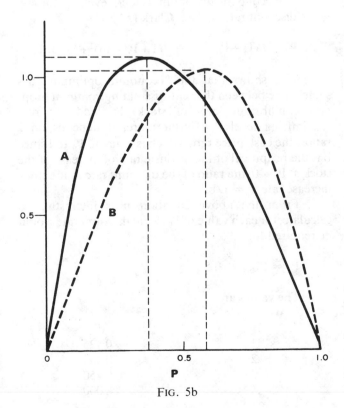

FIG. 5b

49

to the problem would reveal this, but either way the gain in sustained yield would be difficult to detect. Since there are other advantages in taking a given sustained yield from a larger rather than from a smaller stock it is clearly better to graduate recruit stock data, using arbitrary functions, with functions more like B than A.

Data for some fish stocks suggest a concave (r − M) curve, and for some mammals, including the southern hemisphere fin whale, a convex curve. I do not know of data for any other large whales, but it seems reasonable to assume, pending evidence to the contrary, that for these mammals an MSY stock level would be greater than half the initial numbers. There are 2 − at least − important questions concerning the approach to managing these stocks to obtain MSY. The first is "how much greater than half?". The second is how can one determine this in a lightly exploited stock. To examine these questions, focus attention on the bottom right-hand corner of Fig. 5c. Bear in mind that the scale of (r − M) is the same as the scale of the *whaling effort* which is required to stabilize the stock at any given level. This is illustrated on the right-hand side of Fig. 5c showing the effort, f, on scales such that f = 1 for the level of MSY. At the top of these scales are shown the levels of f, relative to that needed to obtain MSY, which, *if maintained*, lead to extinction of the population. Of course, in practice, such levels would not, for economic reasons, be maintained, except possibly if a depleted species or stock were left unprotected while whaling continued on more abundant species or stocks in the same region.

The slope of the (r − M) curve near the initial state determines the possible range of the MSY stock level. In the example, that slope for curve B is − 2. Whatever the exact shape of such a curve (provided its slope changes monotonically) the MSY level is between 0.5 and 0.75.

If the slope at the bottom righthand corner were, say − 3, we could establish that MSY level is in the range 0.67 − 0.83, and so on. In general, the upper limit of MSY stock level, for convex curves, is 2 + 0.5/Slope. The lower limit is either 0.5 or 1 + 1/Slope, whichever is higher.

Two measures are sufficient to locate the stock status on the diagram. That is to say, if one had an index of initial stock abundance, and the ratio between the whaling efforts which stabilize the stock at 2 levels somewhat below the initial level, a first estimate of the range of the MSY level is possible, at least in theory.

At first glance it might seem worth exploring the application of such an approach. In practice, different kinds of information might be combined and one should be able to make analyses which do not require waiting for stabilization, provided that the age compositions of the catches are known. The basic requirement would be good and prompt measures of the changing stock size from sightings and effort. However, the discrepancies noted in estimates by different methods used so far suggest that methods are not yet adequate for this purpose.

Appendix 5

THE CONSEQUENCES OF DISCOUNTING PRESENT VALUES OF FUTURE CATCHES

In the model given in Appendix 1, paras. 2 and 3, the discount rate, d, is zero. Clark (1973) has used the same basic model (a symmetrical, parabolic yield curve, and constant price, and costs per unit effort) to study the consequence of d ≠ 0.

When d → ∞, then $P_1 → P_0$.

The economically optimal stock level, P_d, for any given discount rate is, from Clark (1974)

$$P_d = \tfrac{1}{4}\,[1 + P_0 - db + \sqrt{(1 + P_0 - db)^2 + 8P_0db}]$$

This shows that the "economic optimum" lies somewhere between the rent-dissipating common property equilibrium level of stock, P_0, and the rent maximizing level P_1. Furthermore, it depends on 2 ratios: the cost/price ratio which is equal to P_0 and since b is the reciprocal of the intrinsic rate of increase of the stock at P → 0, the ratio of the discount rate to the stock increase rate, r = 1/b.

From the above equation it is interesting to calculate, for each value of P_0, what discount rate would be required for

$$P_d = P_{MSY} = 0.5$$

The values are:

P_0	db
.1	0.125
.2	0.333
.3	0.750
.4	2.000

To take a concrete example, if for the Antarctic fin whale

$$r = (R_0 - M) = (.12 - .04) = .08,$$

then we find that the economic optimal population, P_d, will be less than P_{MSY} when the discount rate d exceeds

$$\begin{array}{ll} & 1\% \text{ for } P_0 = 0.1 \\ \text{or} & 2.7\% \text{ for } P_0 = 0.2 \\ \text{or} & 6\% \text{ for } P_0 = 0.3 \\ \text{or} & 16\% \text{ for } P_0 = 0.4 \end{array}$$

Clark (1974) suggests that in this case $P_0 \simeq 0.1$ but even if it were considerably higher it is clear that the economic optimum as here defined would be at a smaller stock level than P_{MSY}.

Now this is an oversimplified treatment of the fin whale case. At least 2 other factors come into play. One is the extent to which the costs of whaling are not proportional to the *fin whale* stock level, because other species are also being caught. Then $P_d = 0$ whenever $d > r$, so naturally the fin whale would, if not protected, be hunted practically to extinction. Another factor is the fixed cost of exerting *any* whaling effort — for example the cost merely of moving ships from the Northern Hemisphere to the Antarctic whaling grounds.

This affects the price-cost ratio, and hence tends to increase P_0 to P_0^1. It does not affect the position of the economic optimum stock if $d = 0$ in the proportion

$$\frac{P_1 - P_0}{P_1 - P_0^1}$$

Appendix 6

BALEEN WHALES IN THE SOUTHERN HEMISPHERE TREATED AS A SINGLE RESOURCE

Gambell (1974) has tabulated the calculated biomass of southern hemisphere stocks of baleen whales, summarized below, with the catches taken in the same period. It has been suggested that the blue whale and the minke whale overlap substantially in their diets and are competitors for a limited food supply. (That food is a limiting factor in the Antarctic is presumed from the fact that the early growth rate of fin whales has increased as they have been depleted, so that they now become sexually mature much earlier than before.)

Here I make the very extreme assumption that the minke has attained its recent abundance, starting from an initial very small population, and benefiting from the "surplus" of krill resulting from the decline of the blue whale; furthermore, that other competitors, not necessarily whales, have not so benefited. Thus, were it not for the intensive catching in the seventies, the minke might now be in a period of increase, although according to the SY-model the blue whale should be, relatively, in an even more active period since it has been protected.

The new recruitment over a decade is roughly the consequence of reproduction of the stock at the level existing at the beginning of the decade or a few years earlier. Thus, these figures suggest a combined MSY of a little over 7 million tons per decade from a baleen whale biomass of the order of half the initial level.

The present level is less than 40 % of MSY level ($P < .2$). P_0 is presumably somewhat higher than 0.2, because the profitability of Antarctic whaling is maintained by catches of sperm whales, which in weight in the decade 1961-70 totalled 1.3×10^6 tons. However, the 1970 stock available legally to whalers is not 8.1 but 7.4×10^6 tons, because blue and humpback whales are protected. Furthermore, since 1970 the minke stock has been intensively exploited − it is perhaps already less than one million tons; whaling, as now regulated, appears still to be profitable so that, even taking account

Table 1. Whale stock (tons × 10^{-6})

	1920	1930	1940	1950	1960	1970
Blue	17.1	14.3	3.6	1.6	0.4	0.5
Fin	22.3	21.2	15.2	13.7	6.3	4.4
Sei	2.7	2.7	2.7	2.7	2.7	1.5
Humpback	1.5	1.0	0.6	0.6	0.2	0.2
Minke [1]	0.5	0.6	0.8	1.0	1.2	1.5
Total	44.1	40.0	22.9	19.6	10.8	8.1
	(1) [3]	(.91)	(.52)	(.44)	(.24)	(.18)
Catch	9.0	21.7	8.7	15.9	7.5	
R [2]	4.9	4.6	5.4	7.1	4.8	

[1] Assume exponential threefold increase from small initial stock to Gambell's figure for 1970.

[2] R = difference between catch and decline in stock.

[3] Proportion of 1920s level.

of the sperm whale catch (the significance of which varies greatly among countries whaling in the Southern Hemisphere), P_0 is probably, under present price cost conditions, in the range 0.15 to 0.2.

Appendix 7

INEFFECTIVE CATCH QUOTAS

That the establishment of catch quotas does not necessarily place an effective restraint on whaling is shown by the recent history of such quotas for pelagic whaling in the North Pacific (Table 1).

Table 1. Quotas and catches for North Pacific fin, sei and Bryde's whales, 1967-75

Year	Fin		Sei & Bryde's	
	Quota	Catch	Quota	Catch
1967		2 272		6 116 (63)
68		1 882		5 910 (171)
69	1 600	1 276	[6 116]	5 247 (89)
70	1 453	1 012	5 541	4 643 (39)
71	1 308	802	4 710	3 912 (919)
72	1 046	758	3 768	2 528 (201)
73	650	460	3 000	2 580 (724)
74	350	413	3 000	2 641 (1 362)
75	300		2 000	

First note that the catches of both "species" (sei and Bryde's whales are different species, but because they were not readily distinguishable by gunners, were treated as one in whaling regulations) decline continuously and at the same rate both before and after the beginning of the quota system in 1969. In the last column are given, in parentheses, the catches of Bryde's whale, as identified and reported after capture, and which are included in the total "sei" catches; so the real catch of sei whales declined even in the last 2 years.

It is evident that the quota was not reached, for either species, in any year of regulation, except when the fin quota was exceeded in 1974 by 18 %. (The regulations permit an excess of up to 10 % on 1 species provided there is "an appropriate reduction" in the catch of the other species); furthermore, the catch was on no occasion, other than the exception mentioned, more than 88 % of the quota, and it averaged 77 %.

Now for a catch occasionally not to reach a catch limit is not in itself wrong or unexpected. When a quota is determined the national industries have to make appropriate operational decisions, including how many units — expeditions and catchers — to deploy. But to fail to reach the quotas so consistently and by so much is a strong indication that the quotas were being set far above the real availability of whales.

Whaling quotas are determined in June each year for the following season, that is while the North Pacific whaling season is in progress and thus before current performance can be assessed. There is, therefore, an additional uncertainty regarding the state of the stocks, which does not exist for Southern Hemisphere whaling. This might be offered as explanation of the excessive quotas, were it not for the fact that the "regulated" catch in any 1 year was, more often than not, less than the

Table 2. North Pacific indices of abundance, relative to 1966 (from Wada, 1974)

Year	Fin			Sei (and Bryde)		
	cpue	cpue "adjusted"	sightings	cpue	cpue "adjusted"	sightings
1967	0.69	0.58	0.80	1.62	0.99	1.25
1968	0.56	0.78	1.00	1.67	1.45	1.41
1969	0.27	0.59	0.78	0.99	0.92	0.58
1970	0.25	0.36	0.86	0.86	0.66	0.88
1971	0.27	0.48	0.60	0.69	0.59	0.69
1972	0.24	0.50	0.65	0.65	0.52	0.52
1973	0.21	0.53	0.69	0.66	0.59	0.58

Table 3. North Pacific sei whale abundance, relative to 1968 (from Tillman, 1974)

Year	Sightings	Modified deLury method (0.06 M 0.08)
1963		2.0
1967		1.5
1968	1.00	1.0
1969	0.85	
1970	0.64	
1971	0.50	
1972	0.38	
1973	0.41	
1974		0.9

quota for the *following* year, although the quotas themselves were continuously declining. In fact, next year's quota was set each June at a level lower than the previous season's catch, but higher than the catch eventually taken in the year of decision.

It is difficult to conclude other than that the quotas were having no managing or conserving effect except perhaps to facilitate the division of the catch between participating countries by restraining the competition between them. An outside observer might be forgiven for wondering whether the entire "regulation" was not a facade simply for the systematic "phasing down" of North Pacific whaling.

This, and similar experience elsewhere, suggests a novel approach to management – that is, to penalize countries or enterprises in some way (e.g., by proportionately reducing the quota for the following year) to the extent that they do not succeed in taking the quotas they themselves have negotiated.

It would be interesting to know what was happening to the stocks while this was going on. Unfortunately, the different indices of stock size are highly variable, and not entirely consistent with each other. They are, however, summarized in Tables 2 and 3. It seems that both "species" continue to decline but at what rate is not at all clear; if the quotas had been reached, the declines would have been significantly faster.

Appendix 8

Species and population stocks which are, or may be, in danger of depletion or extinction as a result of man's activities, should not be permitted to diminish beyond the point at which they cease to be a significant functioning element in the ecosystem of which they are a part ... or to diminish below their optimum sustainable population (OSP). Marine mammals, being resources of great international significance, aesthetic and recreational as well as economic, should be protected and managed with the primary objective of maintaining the health and stability of the marine ecosystem. Where consistent with this primary objective, the goal should be to obtain an optimum sustainable population keeping in mind the optimum carrying capacity (OCC). "Depletion" is defined as a decline of the number of individuals within a species or population stock to a significant degree over a period of years, or a decline which, if it continues or is likely to resume, will lead to the species being classed as "endangered" (and therefore protected under the 1969 Act) or is the state in which the stock is below the OCC of its environment. "Conservation" and "management" mean the collection and application of biological information for the purposes of increasing and maintaining numbers at the OCC, including the entire scope of activities that constitute a modern scientific resources programme, including, but not limited to, research, census, law enforcement and habitat improvement, as well as, where appropriate, periodic or total protection, or regulated taking, of species or populations. "Taking" means to harass, hunt, capture or kill, or to attempt to do so. The OCC is the ability of a given habitat to support the OSP in a healthy state without diminishing the ability of the habitat to continue that function. The OSP is the number of animals which will result in the maximum productivity of the population or species. "Stock" means a group of the same species or smaller taxa, in a common spatial arrangement, that interbreed when mature.

References

ACMRR (FAO), The scientific advisory function
1974   in international fishery management and development bodies. Supplement 1 to the Report of the Seventh Session of the Advisory Committee on Marine Resources Research. *FAO Fish. Rep.*, (142) Suppl. 1:14 p.

ALLEN, K.R., Some properties of the Schaefer model for
1974   baleen whale populations. Paper presented to
        the IWC Scientific Committee, December
        1974, Working Paper (2):15 p.

American Society of International Law, Principles or a
1974   global fisheries management regime. *Stud.
        Transnatl. Legal Policy*, (4).

BELLA, D.A., Fundamentals of comprehensive envi-
1974   ronmental planning. *Eng. Issues*, 100(E.11):
        17-35.

BELLA, D.A. and W.S. OVERTON, Environmental plan-
1972   ning and ecological possibilities. *J. Sanit.
        Eng. Div. ASCE*, 98(SA3):579-92.

BEVERTON, R.J.H. and S.J. HOLT, The theory of fishing.
1956   *In* Sea fisheries: their investigation in the
        United Kingdom, edited by M. Graham.
        London, Arnold, pp. 372-441.

–, On the dynamics of exploited fish populations. *Fish.
1957   Invest. Minist. Agric. Fish Food G.B. (2 Sea
        Fish.)*, (19):533 p.

–, Manual of methods for fish stock assessment. Part 2.
1966   Table of yield functions. *FAO Fish. Tech.
        Pap.*, (38) Rev. 1:67 p.

BONDAR, B. and P.J. BOBEY, Should we eat krill? *Ecol-
1974   ogist*, 4(7):265-6.

BRADLEY, M.D., Analysing human and general eco-
1972   systems. *Environ. Aff.*, 2(2):303-13.

–, Decision-making for environmental resources man-
1973   agement. *J. Environ. Manage.*, 1:289-302.

CHAPMAN, D.G. *et al.*, First interim report of the Special
1962   Committee of Three. *Rep. IWC*, (13).

–, Special Committee of Three Scientists. Final report.
1964   *Rep. IWC*, (14):40-106.

–, Report of the Committee of Four Scientists. *Rep.
1965   IWC*, (15):47-50 (see especially pp. 49-50).

CLARK, C.W., The economics of over-exploitation.
1973   *Science, Wash.*, 181:630-4.

–, When should whaling resume? 23 p. (MS).
1973a

–, Fisheries management: maximum sustained yield
1974   vs. optimal sustained yield. 28 p. (MS).

COOK, L.M., Oscillation in the simple logistic growth
1965   model. *Nature, Lond.*, 207:17 July.

CROWE, B., The tragedy of the Commons revisited.
1969   *Science, Wash.*, 166(909):1103-7.

CUSHING, D.H., A link between science and manage-
1974   ment in fisheries. *Fish. Bull. NOAA/NMFS*,
        72(4):859-64.

DOI, T., The theoretical treatment of the reproductive
1973   relationship between recruitment and adult
        stock. *Rapp. P.-V. Réun. CIEM*, 164:341-9.

GAMBELL, R., The unendangered whale. *Nature, Lond.*,
1974   250:454-5.

–, A review of reproduction parameters and their
1974a   density dependent relationship in southern
        hemisphere sei whales. Paper presented to the
        Special Meeting of IWC Scientific Commit-
        tee, December 1974, Working Paper (9):13 p.

GORDON, H.S., An economic approach to the optimum
1953   utilization of fishery resources. *J. Fish. Res.
        Board Can.*, 10(7):442-57.

GRAHAM, M., The modern theory of exploiting a fishery
1935   and application to North Sea trawling. *J.
        Cons. CIEM*, 10:264-74.

–, The Sigmoid curve and the over-fishing problem.
1939   *Rapp. P.-V. Réun. CIEM*, 108(1):57-66.

–, The fish gate. London, Faber, 196 p.
1943

GULLAND, J.A., The concept of the maximum sustain-
1968   able yield and fishery management. *FAO
        Fish. Tech. Pap.*, (70):13 p.

–, The concept of the marginal yield from exploited
1968a   fish stocks. *J. Cons. CIEM*, 32(2):256-61.

–, The management of Antarctic whaling resources. *J.
1968b   Cons. CIEM*, 31(3):330-41.

–, Science and fishery management. *J. Cons. CIEM*,
1971   33(3):471-7.

GULLAND, J.A. and L.K. BOEREMA, Scientific advice on
1973   catch levels. *Fish. Bull. NOAA/NMFS*,
        71(2):325-35.

HJORT, J., G. JAHN and P. OTTESTAD, The optimum catch
1933   essays on population. *Hvalrad. Skr.*, 7:29-127.

HOLLING, C.S., Resilience and stability of ecological
1973   systems. *Ann. Rev. Ecol. System.*, 4:1-23

HOLT, S.J., The application of comparative population
1962   studies to fisheries biology. *In* The exploita-
        tion of natural animal population, edited by
        E.D. Le Cren and M.W. Holdgate. Oxford,
        Blackwell Scientific Publications, pp. 51-71

—, Whales: conserving a resource. *Nature, Lond.*,
1974   251:336-7.

—, The maximum sustainable yield in weight and re-
1974a   lated matters. Paper presented to the Special
        Meeting of IWC Scientific Committee, De-
        cember 1974. Working Paper (12):14 p.

HOLT, S.J. *et al.*, Report of the Working Group on bio-
1974   metrics and management. *In* The whale prob-
        lem: a status report, edited by W.E. Schevill.
        Cambridge, Mass., Harvard University Press,
        pp. 14-20.

MANN, S.H., On the optimal size for exploited natural
1973   animal populations. *Oper. Res.*, 21(3):672-6.

MARCHESSEAULT, G.D., S. SAILA and W.J. PALM, Delay-
1976   ed recruitment models and their application to
        the American lobster fishery. *J. Fish. Res.
        Board Can.*, 33(8):1779-87.

MAY, R.M., Stability and complexity in model ecosys-
1973   tems. Princeton, N.J., Princeton University
        Press.

MAY, R.M. *et al.*, Time delays, density-dependence and
1974   single-species oscillations. *J. Anim. Ecol.*,
        43:747-70.

McINTYRE, J. (Ed.), Mind in the waters. New York,
1974   Scribner's.

OHSUMI, S., Conversion of sustainable yield curve by
1974   numbers into that by weight in the fin and
        sperm whale populations. Paper presented to

the Special Meeting of IWC Scientific Com-
        mittee, December 1974. Working Paper
        (24):10 p.

ROTHSCHILD, B.J., A system review of fishery manage-
1971   ment with some notes on the tuna fisheries.
        FAO Fish. Tech. Pap., (106):33 p.

SCHEVILL, W.E. (Ed.), The whale problem: a status re-
1974   port. Cambridge, Mass., Harvard University
        Press, 419 p.

SCHWEDER, T., Transformation of point processes: ap-
1974   plications to animal sighting and catch prob-
        lems with special emphasis on whales. 183 p.
        (mimeo).

STEELE, J.H., The structure of marine ecosystems. Ox-
1974   ford, Blackwell Scientific Publications, 128 p.

SWARTZ, J. and H. BREMMERMANN, Discussion of
1975   parameter estimation in biological modelling:
        algorithms for estimation and evaluation of
        the estimates. *J. Math. Biol.*, 1:241-57.

TALBOT, L., The great whales and the International
1974   Whaling Commission. *In* Mind in the waters,
        edited by J.M. McIntyre. New York, Scrib-
        ner's, pp. 232-6.

TILLMAN, M.F., Re-examination of population estimates
1974   for the North Pacific sei whale. Doc. 29 to
        IWC Scientific Committee Meeting, La Jolla,
        December 1974 (mimeo).

VAMPLEW, W., The evolution of international whaling
1972   controls. *Marit. Hist.*, 2(2):123-39.

WADA, S., Indices of abundance of large-sized whales in
1974   the N. Pacific in 1973 whaling season. IWC
        Doc. SC/26/9:37 p. (mimeo).

WOODWELL, G.M., Biotic energy flows. *Science, Wash.*,
1974   183:367.

SOME ECONOMIC ASPECTS OF MARINE MAMMAL MANAGEMENT POLICIES: THE FUTURE AND THE DISCOUNT RATE

C. Price

Abstract

The reduction of all future values of a resource to a single present value, according to an annual discount rate, may give these future values too little worth and, as in the case of marine mammals, thereby encourage the intense hunting of a species, even to extinction, to maximize present benefit. Economists justify the discounting of future values by the principles of the time preference rate, by which benefits are chosen for their immediacy and their certainty, and the rate of return on investment. A discount rate can be chosen according to other criteria which may better predict future changes in value; classes of factors that can be considered are the possibility of world destruction, taste change, change in world circumstances, technological change and uncertainty. Analysis of the future food value of whales using these criteria indicates that a large discount rate is probably not justified. Discounting the future aesthetic value of marine mammals, generally experienced vicariously, seems to be supported by some criteria but not by others. The scientific values of marine mammals are not repeatable, like their material and aesthetic ones — there is a limited amount to be found out about them — but there is a continual scientific educational value and so discounting in this case is uncertain. It is possible that a zero discount rate, because it allows greater worth to be placed on future values than on present ones, could represent evaluations of the future based on the above alternative criteria. A zero discount rate could accommodate decisions to manage for optimal population size, in which annual benefit, rather than the present value of the discounted future, is maximized.

Résumé

La réduction de toutes les valeurs futures d'une ressource à une valeur présente unique, selon un taux annuel d'actualisation, peut donner à ces valeurs futures une trop faible importance et, dans le cas des mammifères marins, encourager la chasse intensive d'une espèce, allant même jusqu'à l'extinction, pour maximiser les bénéfices actuels. Les économistes justifient l'actualisation des valeurs futures par les principes du taux de préférence temporelle, établissant que les bénéfices sont choisis pour leur caractère immédiat et certain, et par le taux de rapport des investissements. Un taux d'actualisation peut être choisi selon d'autres critères qui peuvent mieux prédire les futurs changements de valeur; les classes de facteurs qui peuvent être considérés comprennent, de préférence, le changement des situations mondiales, les changements technologiques et l'incertitude. L'analyse de la future valeur alimentaire des baleines, faite d'après ces critères, indique qu'un taux important d'actualisation n'est probablement pas justifié. L'actualisation de la future valeur esthétique des mammifères marins, généralement ressentie indirectement, semble être étayée par certains critères mais pas par d'autres. Les valeurs scientifiques des mammifères marins ne sont pas infinies, comme les valeurs matérielles et esthétiques – ce qu'on peut découvrir à leur sujet est limité – mais il existe une valeur scientifique didactique continue et, dans ce cas, l'actualisation est incertaine. Il est possible qu'un taux d'actualisation nul, du fait qu'il permet d'accorder une plus grande importance aux valeurs futures qu'aux valeurs actuelles, pourrait représenter des évaluations de l'avenir basées sur les critères de remplacement ci-dessus. Un taux d'actualisation égal à zéro pourrait permettre des décisions de gestion visant à obtenir une taille de population optimale où le bénéfice annuel est maximisé, plutôt que la valeur présente du futur actualisé.

Extracto

La reducción de todos los valores futuros de un recurso a un valor presente, aplicando un índice anual de actualización, puede hacer que esos valores futuros tengan muy poca importancia y, en el caso de los mamíferos marinos, fomentar así la explotación intensa de una especie hasta llegar incluso a su extinción, para aumentar al máximo los beneficios presentes. Los economistas justifican la actualización de los valores futuros en virtud de dos principios: la preferencia cronológica, en virtud de la cual los beneficios se eligen por su inmediatez y su certeza, y el rendimiento de las inversiones. El índice de actualización puede elegirse según otros criterios, que tal vez permitan predecir mejor los cambios de valor en el futuro; entre los factores que pueden tenerse en cuenta figuran la posibilidad de destrucción del mundo, las variaciones de los gustos, el cambio en las circunstancias mundiales, los cambios tecnológicos, y la incertidumbre. Un análisis del valor alimentario futuro de la ballena utilizando estos criterios indica que probablemente no está justificado aplicar un índice elevado de actualización. La actualización del valor estético futuro de los mamíferos marinos, que en general se percibe indirectamente, parece estar justificada por algunos criterios, pero no por otros. Los valores científicos de los mamíferos marinos no son repetibles, al igual que los valores materiales y estéticos (la cantidad de datos que puede hallarse sobre ellos es limitada), pero sí puede hablarse de un valor educativo científico continuo y, según eso, la actualización presenta cierta incertidumbre. Es posible que un índice de actualización cero, al permitir dar mayor importancia a los valores futuros que a los presentes, pueda representar una evaluación del futuro basada en los criterios alternativos expuestos. Un índice de actualización cero sería compatible con decisiones de ordenación de la explotación encaminadas a conseguir el volumen óptimo de la población, con lo que se incrementarían al máximo los beneficios anuales, y no el valor presente de un futuro actualizado.

C. Price
University College of North Wales, Department of Forestry and Wood Science, Bangor, Gwynedd LL57 2UW, Wales

Introduction

The economist's account of over-exploitation of resources in the common ownership of mankind is simple. The revenue from catching a whale in the deep oceans, for example, accrues entirely to the successful hunter. The consequent cost of a reduced whale population, in terms of diminished breeding rate, increased search time and possibly aggravated danger of extinction, is borne by all whalers, only a minute fraction falling upon those who make the capture. An agreement to restrict, or even temporarily prohibit whaling may clearly be in the long-term interests of all whale-hunters: in the absence of an agreement, however, it is unquestionably in the interest of the individual hunter to catch what he can. For his purposes, there is no merit in abstention while other individuals continue to deplete the stock.

One may object that a point is surely reached when an economically important species is threatened with extinction, at which potential future revenues from the species outweigh the immediate gains of capture, even for the individual hunter. Again, however, the individual, seeing himself powerless to save the species by unilateral action, will not be moved by self-interest to spare the whale merely for another hunter's capture.

Economists recognize that laissez-faire policies achieve inadequate regulation because no market enables mankind in general to recover the costs imposed on it by the beneficiaries of individual capture. This makes it all the odder that the rather similar costs imposed by present exploiters on future hunters are normally ignored or played down by economists. The short-sightedness of hunters indeed appears to be approved by economists' *discounting* treatment of future benefits and costs (that is, a process, reversing compound interest, which converts future values to a present equivalent, dividing them by a discounting factor once for every year of the time interval).

In plain terms, this amounts to saying that the annual value of whales to the future is less than the annual value of the same physical quantity to the present. Furthermore, because of the continuous nature of the decline over time, the total of *all* future values from a continuously productive resource reduces to a finite present equivalent — which is quite a small multiple of annual value at prevailing rates of discount.[1]

This being so, it is entirely possible that a strategy of maximizing immediate benefits by rapid extermination of a species will be analysed as optimal (Clark, 1973).

To scientists and others, it may be far from clear that future values should be lightly dismissed and this extraordinary postulate requires explanation. Economists have 2 legs to stand on in justifying it and, in general, they contrive to stand on both, or to shuffle from one to another. These are the *time preference rate* of individuals and groups, and the rate of return on investment.

Time preference rate

Those aspects of economics which deal with *what ought to be* appeal to the revealed preferences of consumers as the basis for decisions. It is quite evident from their actions that individuals prefer to have good things now rather than later: and if it is a fundamental premise that the consumer is always right,

---

[1] The infinite series

$$\frac{x}{1 + r} + \frac{x}{(1 + r)}2 + \dots \frac{x^\infty}{(1 + r)}$$

which is the total of an annual value, x, in perpetuity discounted at a rate, r, to the present, has the finite sum x/r. Current rates of discount are around 10 %, bringing total present value to only 10x.

there is nothing that can be argued against a discount on the future. However, the philosophical justification of consumer sovereignty hangs heavily on the assumption that choices are made rationally: that the consumer chooses in such a way as to give himself greatest satisfaction. There are features of the choice between present and future which make it unlikely that rationality so defined will invariably operate. The choice situation is asymmetrical, that is, the present and the future are not viewed on an equal footing. The present is chosen, not so much for its prior position in history as for its *immediacy*. If, in the future, it was possible to choose between *consuming then* and *having consumed now*, it is reasonable to expect that again immediacy would be preferred, despite being later in history. Thus, if one were to analyse all the consumer's *preferences*, looking back as well as forward, it would not necessarily appear that he has the premium on early consumption implied by his *choices*, which, history being unalterable, must always look forward.

In addition to the purely irrational element of immediacy, there is a *certainty* about present consumption which it is quite rational for the individual to prefer. The state of the world 100 years hence is of little interest because he does not expect to participate in it. Nevertheless, his preference for resources to be made available now rather than in 100 years imposes costs on future generations.

The need for a longer view for society as a whole, together with an allowance for irrationality, has caused some economists to suggest that governments' discount rate should be lower than the rate applied by individuals (i.e., governments should set a greater weight on the future). Yet governments too have finite lives – shorter, in fact, than individuals – and expedients offering immediate benefits may be given a weight which is irrational even within that short life. Consequently, even if a time preference rate could be consistently deduced from the behaviour of governments or supranational bodies, it would not be surprising if it overstated the proper discount on the future.

Rate of return on investment

The other normal argument for discounting runs something like this. If I have £.Stg. 1 now, I can invest it and at the end of a year recoup (say) £.Stg. 1.10. Therefore, if I am offered anything less than £.Stg. 1.10 at the end of the year in place of the £.Stg. 1 now, I would be better off with £.Stg. 1 now. There is, one would say, a 10 % discount on next year's money value. In the same way, the £.Stg. 1.10 could be reinvested at compound interest over as many years as it is desired to make comparisons. Fair enough (provided my investment is keeping ahead of inflation) if concern is purely with paper or personal economics. It is harder to see what reinvestment means in real terms. How does one set about investing a dead whale at compound interest? Its material components are mostly consumed rapidly. Such consumer durables and capital as may be created from industrial use of whales do yield annual dividends – but for a finite period and with dubious possibilities for reinvestment. One has to think, perhaps, in terms of better-fed scientists working more effectively to devise whale substitutes in an ever-increasing stream.

It is more plausible to imagine the whale, or its products, being sold to someone who would otherwise have spent the money on some other form of immediate consumption: the proceeds then being used to finance investment, either by whaling companies themselves, or by the governments who tax them. There are, however, physical, economic and political limits to how far this process can go (Price, 1973), not least of which will be the demands of shareholders and voters with high time preference rates that not all surpluses shall be reinvested in favour of future consumption. To the extent that an early whale opens up reinvestment opportunities, this must be allowed for (Henderson, 1968), but it is not clear at all that it should be done by discounting at the market rate of return on investment.

An alternative view of discounting

It seems to me that, if discounting has any rational function, it is to represent trends of value change that are expected to be exponential and indefinite — simply a convenient way of generalizing a prediction. The 5 classes of factors I have identified as having general relevance are:

(i) Possibility of world destruction
(ii) Taste change
(iii) Change in world circumstances
(iv) Technological change
(v) Uncertainty.

How do these apply to the value of one whale, taking first the flow of material benefits?

(i) If there is a cumulative annual probability that all potential consumers of whale products will be destroyed by world-wide catastrophe, the overall expected value of future output from the whale stock is commensurately reduced in succeeding years. A similar element applies to the possibility that some factor not under the control of any management policy will make the oceans impossible for whale survival. Applied at the world level, such an element of discount tends toward self-fulfilling prophecy, since if little weight is put on the future, possibilities which risk the existence of a future will be favourably analysed.

(ii) The possibility of a shift in consumer taste, as from animal to vegetable margarines, again progressively reduces the expected value of foods that are currently in the consumer's basket. A past taste, however, may show resurgence, and basic food values are not mutable by such shifts. It is, therefore, only a portion of current value that should be dis-

counted under this head.

(iii) The circumstances in which food production takes place are evidently not static. The rapidly increasing population and declining base of concentrated stock resources both tend to put a higher value on food that can be harvested without great expenditure of resources. This not only affects the future value of a whale population in a general sense, but increases the premium on a relatively abundant population that can be harvested without large fuel expenditure on search.

(iv) Countering these upward influences on value, technological innovation tends to make food more abundant, and so to diminish the value of any additional supply. As this forms one of the few rational bases of economists' arguments for a discount rate (Pearce, 1971), it is worth examining in a little detail what is the nature of the technological forecast involved: whether the technology is to be financed from whale sales or from other sources.

First, consider the means of harvesting the whale's food intake for human consumption. While the whale is the cheapest way of doing so, its value is whichever is less of (a) the net food value, after harvesting cost, or (b) the *additional* cost of harvesting by the next cheapest method, which is presently avoided by using the whale as intermediary. Technological developments, whether using mechanical methods (trawlnets) or biological systems (fish), may lower the cost differential over time, and so the value of the whale population; a reasonable basis for discounting. The time might come when other methods would become more efficient, and the whale made redundant altogether. A forecast of the probability that this would happen again yields a discount factor.

Second, developments in the total world

61

food supply affect the value of the deep ocean resource and even affect whether it is worth exploiting it. Not all increases in world output are directly substitutable for whale products, but improving technologies of interconversion are reducing the significance of the form, relative to the basic content, of food.

(v) It would be enough to consider only these influences on future value if forecasts could be certain. Strangely, uncertainty has been regarded by some economists as a reason for increasing the discount rate. Their argument is, roughly, that the further into the future one proceeds, the less one knows about the situation and the less weight one can put on any given outcome. This is true: but lesser certainty about future values includes, not just increasing possibility of a given fall in value (as a higher discount rcte implies), but increasing possibility of a given rise in value. Furthermore, the value of food resources has a lower limit of zero (unless, for example, whales could plausibly become a positive and irremediable nuisance), while in the context of fast rising population pressing on uncertainly sustainable food production, there is no definite upper limit to possible food values at all.

An overview of all these factors makes it hard to assign a large discount rate to the food value of the whale, especially one as high as currently used rates. Indeed, in a state of uncertainty about the persistence and exponential nature of predicted trends, discounting seems a rather dangerous evasion of the problems of seeing far into the future, and might better be replaced with explicit ranges of predicted value.

Aesthetic and scientific values

If whales or other creatures are to be conserved, it is not for their direct food and material value alone (Helliwell, 1973). Indirect values, or costs, that might arise through the impact of whales on other food-producing systems would be discounted only for reasons like those applying to the direct values.

Other conservation values can be grouped as aesthetic, and scientific: and to both classes the possibility of elimination (of the whale population by exogenous causes, or of human consumers) is an appropriate element of discount. Aesthetic values — using the term in its broadest sense — are of a generally vicarious nature, experienced most often through film, or an "option value" that derives simply from the knowledge of creatures' existence (Krutilla, 1967). There is, perhaps, at the moment an element of "cause célèbre" about the survival of some species of whales, which one might argue is more likely to decrease than increase in the long run and that a possible change in conservation taste justifies a discount. One would also expect diminution of of any sense of *loss* as a species passed toward oblivion: not all extinctions have retained the emotional following of the dodo. But only part of the value is erodable, for a significant element of interest in marine mammals is intrinsic to their nature. Time itself is unlikely to change this.

"Option demand" is a classic public good — that is, something one cannot readily exclude people from enjoying by charging for it, and which is not used up by the process of enjoyment. One consequence is that conservation values of this kind increase as the population partaking in them does. Whether or not this expected increase in values over time is best dealt with by a *negative discount rate*, it should be incorporated by some means. A seemingly inexorable decrease in the reserves of other species and systems of conservation interest might also focus more value onto such as do survive.

It is perhaps in the element of discounting attributed to technologically induced abundance that the gap between material and aesthetic futures is plainest. The *possibility* cannot be denied that technology may make the whale's utilitarian functions redundant: but attempts to replace interest in a natural system by technological substitutes may well provoke doubts about the limits to technology's proper role. If the surge of interest in wildlife can be read as a symptomatic rejection of mechanistic sophistication, it is evidently pointless, perverse and possibly perilous, to attempt substitution of the antidote by more of the same poison. One may argue, indeed, that technology's success is to extend vicarious participation in the natural reality to a greater population, increasing the value of the total conservation interest (even if the recruits enjoy less *individual* benefit than the enthusiastic pioneers). It would rather be the failure of technology, coupled with an increased population pressure on scarce resources, that might lead to reduced aesthetic interest over time, as concern for basic survival crowds out such dispensable pleasures.

Of uncertainty, one may simply say that it is easier to conceive an upper limit to the aesthetic values of marine mammals than one for their contribution to food supplies. Unknown deviation from expected values suggests therefore a smaller downward influence on discount rate than applies to food values.

In summary, there are some grounds for thinking that discount on the future of marine mammals' aesthetic values should be greater than for their material values and some for thinking it should be less; but none for positively asserting that the case for a discount is better than that for a premium.

The scientific value of marine mammals, unlike the infinitely repeatable aesthetic and material values, is in a sense a stock. There is a large but limited amount to be found out about them, specifically, and about the nature of life. Facts uncovered diminish over time the *undiscovered* research potential of the group: perhaps too, there are conjectures to be made about the declining importance of succeeding discoveries (net of the cost of making them) compared with that of earlier findings. This quasi-stock nature affects the case for dis-

counting variously. On the one hand, over time the value of further knowledge remaining to be extracted falls. On the other, a change in the fashions of research does not extinguish any value, but merely passes it over to a later date. Timing is important, in that the earlier the research results are made available, the longer will be the period of benefit: but it is this additional period of benefit, the discovery having been made, rather than the earliness of the discovery itself that is significant, and there are more rational ways than discounting to allow for it.

Again, one may be dubious about the potential of technology to substitute for research findings in biological systems. A limited substitution of one biological system for another as a research base might become possible. Equally, the probable loss of other species and systems enhances the potential of those remaining and tends to increase the annual value of their contribution to knowledge over time.

Uncertainty in this context is a matter of the possibility that some unexpectedly important discovery may result from research in whales. This possibility is part of the stock value of the population and, like the rest, diminishes slowly, perhaps imperceptibly, as the stock of unexplored knowledge declines.

In summary, scientific values are of a rather different kind from the material and aesthetic ones, representing a stock of potential knowledge whose future value is dependent on the prospects for the human race — which can be treated under factors (i) to (v). Once the knowledge is extracted however, the usefulness of the whale's continued existence is extinguished and it is with this in mind that the value of different survival periods of the species in question should be compared. However, there is always a residual value in a living creature as a medium of scientific education, and, though in the case of the whale the annual value would be fairly small, the argument for discounting, as in the case of aesthetic values, is dubious.

Additionally, though the economist is in

no position to comment authoritatively on this, there seems no absolute reason why satisfaction experienced by the whale population itself should not be an argument for its conservation. And, if this is accepted, it is difficult to see why, other than the possibility of world catastrophe, this satisfaction should be discounted.

Discounting and decisions

The relevance of the discounting process is clearest in decisions as to whether any species should be hunted or brought by other means to destruction. Extinction forecloses all future outputs — and costs — so that the rationality of the extinction option depends critically on what value is assigned to the outputs. Conventional views of discounting pose no mechanical problems, since all future values, no matter of what kind, are unambiguously reduced to a single present value. I have suggested that discounting is by no means as easy as that, and indeed that the philosophy is seriously open to doubt, unless it is very carefully related to expected changes in value. Once zero discount rates are contemplated, the awkward possibility has to be faced of indefinitely large present values of conserved species (Nash, 1973). This does not preclude, as Nash suggests, economic analysis. If indefinite values are attainable by limited present abstention, the case is clear for maintaining a viable population of a species. If the maintenance of the population has recurring costs — such as consumption of an otherwise valued food resource — greater than the expected recurring benefits, there may be a case for extinction, but it is not a case often explicitly argued. It is true that most investments take on an enhanced value at zero discount rate, so that the costs of resources diverted from them into conservation have to be increased: Feldstein's (1964) methods adequately cope with this difficulty. The problems of economic analysis without a discount rate are further explored in Price (1976).

The view taken of the future also bears on optimal stocking of a given species. Because of the time taken to increase a population, a large stock is inherently more flexible, should a higher intensity of food extraction from deep-sea sources be made desirable by shortages from other sources. Naturally, the larger population involves smaller costs of search as well, which again could be important in terms of future fuel-saving. But these advantages may be achieved only after the many years of limited exploitation needed to achieve optimality.

Hence, under conventional discounting, little weight is given to the advantages of an optimal population. A reduction toward extinction might be preferred to maintaining the existing population, even if an optimal population *now* would be preferred to extinction. Zero discount rate suggests that the population should be increased, because of the greater weight put on the value of the far distant optimum.

A final perplexity concerns the economic treatment of gene-pool arguments. Varied genetic constitution gives adaptability to a species so that it may continue its utilitarian role in the face of environmental change. There are also matters of scientific and aesthetic interest for which variety is helpful. It is certain that the *incremental* value of additional variety in the population falls as total population increases. But, however tiny the increment of annual value from an additional individual, its value into the perpetual future becomes indefinitely large. Thus, on these grounds only, an optimum population would be as large as possible. The only limitation on population size can be another recurring item — such as encroachment on the habitats of other species and the breeding and conversion-efficiency losses arising from an overlarge population.

Conclusions

It will be plain from the foregoing that the discounting process should offer no easy

64

solution to the problems of attributing values to the future and of reaching decisions affecting them. Rather, it is to expected physical realities that economists and other inquirers must turn if they wish to establish any basis on which discounting is rational. If, as I suspect, a zero discount rate is plausible as the result of the influences on values described, then species with positive net value should be kept from extinction: and this is by no means as obvious a conclusion to economists as it may be to scientists. Secondly, the question of optimal population can be resolved in a steady state sense, as a maximization of net annual flow of benefit rather than of a net present value of all the discounted costs and benefits of moving to the steady state and maintaining it indefinitely. That does not, of course, mean that changes in optimal population with world circumstances are not to be considered. The necessary forecasting, however, should not be left entirely to economists.

References

CLARK, C.W., The economics of over-exploitation.
1973   *Science, Wash.*, 181:630-4.

FELDSTEIN, M.S., Net social benefit calculation and the
1964   public investment decision. *Oxford Econ. Pap.*, (16):114-31.

HELLIWELL, D.R., Priorities and values in nature con-
1973   servation. *J. Environ. Manage.*, 1:85-127.

HENDERSON, P.D., Investment criteria for public enter-
1968   prises. *In* Public enterprise, edited by R. Turvey. Harmondsworth, Penguin.

KRUTILLA, J.V., Conservation reconsidered. *Am. Econ.*
1967   *Rev.*, 57:777-86.

NASH, C.A., Future generations and the social rate of
1973   discount. *Environ. Plann.*, 5:611-7.

PEARCE, D.W., Cost-benefit analysis. London, Macmil-
1971   lan.

PRICE, C., To the future: with indifference or concern?
1973   *J.Agric. Econ.*, 24:393-8.

—, Project appraisal and planning in over-developed
1976   countries. (Unpubl.).

SOME ASPECTS OF EVALUATING LOW-CONSUMPTIVE USES OF MARINE MAMMAL STOCKS

P. COPES

Abstract

Man can benefit from marine mammals by direct harvest of them for food and other commodities, historically their most common use, but one which may lead to over-exploitation, and by indirect recreational and educational use, "low-consumptive" in that few changes are caused to the marine mammal population. A net economic benefit is achieved only to the extent that the goods and services produced are more valuable than those that would result from a different use of the labour and capital employed. In the recent past, it is possible that little net benefit has been obtained from commercial harvesting of marine mammals. Low-consumptive uses, where costs are often modest, could yield substantial net economic benefits, especially in increasingly affluent and crowded societies; net benefits here are those to users, measured by a "consumer's surplus" (the price of an activity subtracted from the amount users would be willing to pay) and the profit made by those offering the service which could not have been made in another way, as well as certain collective benefits to society more difficult to measure, for example, from education. It can be assumed that many persons viewing marine mammals would not gain equal satisfaction from alternative activities. Low-consumptive uses of marine mammals are not normally jeopardized by properly managed harvesting of them and harvesting is not usually affected by low-consumptive use. Research on the net benefits that can be gained from low-consumptive uses in societies already using marine mammals and in developing countries, may be appropriate.

Résumé

L'homme peut profiter des mammifères marins en les exploitant directement pour en tirer de la nourriture et d'autres produits — ce qui, sur le plan historique, a constitué l'utilisation la plus répandue mais peut conduire à la surexploitation — et par une utilisation indirecte récréative ou didactique, les "autres utilisations", ou utilisations à faible consommation dans lesquelles peu de changements affectent les populations de mammifères marins. Un profit économique net n'est réalisé que dans la mesure où les biens et services produits sont plus précieux que ceux qui proviendraient d'une utilisation différente de la main-d'œuvre et du capital employés. Il est possible, ces derniers temps, qu'on ait obtenu un faible bénéfice net de la chasse commerciale des mammifères marins. Les autres utilisations, où les coûts sont souvent réduits, pourraient offrir des profits économiques nets substantiels, spécialement dans des sociétés de plus en plus aisées et peuplées; les profits nets sont ici ceux qui vont aux utilisateurs, mesurés par un "excédent du consommateur" (le prix d'une activité soustrait du montant que les utilisateurs seraient disposés à verser) et le profit réalisé par ceux qui offrent le service, qui n'aurait pas pu être obtenu d'une autre façon, ainsi que certains avantages collectifs pour la société, plus difficiles à mesurer, par exemple ceux qui proviennent de l'éducation. On peut supposer que de nombreuses personnes contemplant les mammifères marins ne tireraient pas une satisfaction égale d'autres activités. Les "autres utilisations" des mammifères marins ne sont pas normalement mises en péril par une exploitation convenablement gérée et l'exploitation n'est pas habituellement affectée par les "autres utilisations". Il peut être utile d'effectuer des recherches sur les profits nets qui peuvent être retirés des autres utilisations dans les sociétés utilisant déjà les mammifères marins et dans les pays en développement.

Extracto

El hombre puede aprovechar los mamíferos marinos explotándolos directamente para obtener alimentos y otros productos – uso este que ha sido el más común a lo largo de la historia, pero que puede llevar a una explotación excesiva — o sirviéndose de ellos, indirectamente, para fines recreativos y educativos (los llamados "usos no destructivos", que apenas determinan variaciones en la población de mamíferos marinos). Sólo se consiguen beneficios económicos netos cuando los bienes y servicios producidos son más valiosos que los que se obtendrían con una utilización diferente de la mano de obra y el capital empleados. En los últimos años es posible que la explotación comercial de los mamíferos marinos haya dado escasos beneficios netos. Los usos no destructivos, en los que los costos son a menudo modestos, pueden dar beneficios económicos netos importantes, especialmente en las sociedades superpobladas y cada vez más ricas de hoy. Los beneficios netos, en este caso, son: los que obtienen los usuarios, que se miden por el "superávit de consumo" (el precio de una actividad deducido de la cantidad que los usuarios estarían dispuestos a pagar); los que consiguen quienes ofrecen esos servicios, que no podrían haberse obtenido de otra manera; y algunos beneficios colectivos para la sociedad, más difíciles de cuantificar, como, por ejemplo, los beneficios educativos. Se puede suponer que muchas personas que tengan ocasión de observar mamíferos marinos no conseguirán satisfacciones equiparables con otras actividades. Los usos no destructivos de los mamíferos marinos no resultan de ordinario amenazados por la explotación, adecuadamente regulada, de esos mamíferos y, a su vez, la explotación no resulta afectada, de ordinario, por los usos no destructivos. Sería oportuno realizar estudios sobre los beneficios netos que podrían obtenerse con los usos no destructivos en las sociedades en que ya se utilizan los mamíferos marinos y en los países en desarrollo.

P. Copes
Department of Economics and Commerce, Simon Fraser University, Burnaby, British Columbia, V5A 1S6, Canada

Uses of marine mammals

Stocks of marine mammals may be used to generate benefits for humans in 2 ways. Historically, the most common form of utilization has been a direct one in which a harvest is taken from the available stocks, with the catch being processed into food and other commodities for human use. As the output of the economic process in this case physically consists of materials removed from marine mammal stocks, this direct utilization may subject the resource to considerable stress. The tendency toward over-exploitation of the stocks and danger of extinction of some species has been well documented.

In contrast to the above are the indirect forms of utilization. They occur most often in the area of recreational and educational pursuits that rely on observation of marine mammals by means of direct viewing or indirect media presentation. Such uses are "low-consumptive" in a material sense. They place negligible physical demands on the resource, requiring at most the capture and removal of a few individuals from wild stocks for display in oceanaria and other facilities. In some instances (e.g., whale watching) there may be no physical demands at all on the stocks[1]. The resource remains intact and, in effect, serves only as a catalyst for the generation of recreational and educational benefits. The output of the process is not a physical product but an educational or recreational service, which may well be of high economic value despite the negligible material sacrifice extracted from the resource.

---

[1] It is possible, of course, that some whale-watching activities will have a disturbing effect which is deleterious to the whale population.

Measurement of values

The significance of marine mammal stock utilization may be measured in at least 3 ways: through "gross value of product", "value added" and "net economic benefit". The first of these simply measures the total value of the output of industries utilizing marine mammals resources. For instance, the receipts obtained from the sale of whale oil, meal and meat may be counted in respect of direct utilization, and gate receipts for whale shows in the case of indirect utilization. These values may then be compared in magnitude with the product values of other industries. It should be recognized, however, that gross values are not an accurate measure of the intrinsic contribution of the resource itself. For the final product or service in every case incorporates contributions made by cooperating factors: manpower used in catching whales or in maintaining oceanaria; capital equipment used in manufacturing oil and meal or in constructing and operating display buildings; fuel used in running vessels, rendering oil and meal and heating and lighting display areas. Gross values of the output of industries utilizing marine mammals are simply an indication of the economic size of the sectors involving marine mammals.

Another problem in using gross product values is that double counting may take place as between industrial sectors. The product value of sealskin coats sold by furriers will incorporate the cost of purchasing pelts from a company operating sealing vessels, which, in turn, will include the cost of obtaining sealing vessels from the shipbuilding industry. If the output of all of these industries is listed as marine-mammal related, double or triple counting of values will occur. A somewhat more refined comparison of values may be obtained by using a form of net measurement that considers only the "value added" by each industry sector. Thus the furrier would be credited only with the value of coats sold minus the cost of pelts. The sealing vessel's contribu-

69

tion would be recorded as the value of pelts sold, minus the cost of equipment, fuel and other materials used. The "value added" measure nets out, for each sector, the cost of materials and services obtained from other sectors. In the final analysis it measures the contribution made by the sector itself, which is reflected in income generated for the workers, owners and investors of the sector concerned.

The third and most refined measure of the contribution made by marine mammals is obtained by calculating the net economic benefits resulting from their utilization. In this context it must be recognized that any labour and capital that is not devoted to capturing, processing or displaying marine mammals, may be applied to other economic endeavours — producing goods and services of other kinds that are close or remote substitutes for products and services obtained from marine mammal stocks. In the final analysis, the use of marine mammals is making a net economic contribution only to the extent that goods and services are being produced that are of higher value than those that would result from the alternative use of the labour and capital employed. In measuring the net economic benefits of any commodity or service obtained from marine mammals, it is therefore necessary to subtract the value of alternative production.

The measurement of net economic benefits, whether they are obtained from · direct or indirect utilization of marine mammals, is a difficult undertaking. However, considering the nature of the resource utilization process and the products obtained, some intuitive comparisons may be made. In the case of direct physical utilization of marine mammal stocks, the net economic benefits consist of the value to consumers of the physical products obtained minus the costs incurred in producing this output. For many marine mammal products (oil, meal, meat) there appear to be reasonably close substitutes available at competitive prices. When this is so, the value to consumers of marine mammal products is simply measured by what they pay for these products and no more, as they have the alter-

native of switching to ready alternatives at equivalent prices. The costs of obtaining marine mammal products are reflected in manpower time, materials and equipment utilized in the production process. The inputs are a measure of the capacity to produce alternative goods of value to society that are thus sacrificed by devoting them to marine mammal catching and processing. Only to the extent that these inputs can achieve a more valuable output in marine mammal harvesting than in other uses are any net economic benefits generated.

It would seem evident, if one considers the withdrawal in recent times of one fleet after another under pressure of sub-marginal returns, that marine mammal harvesting is only marginally viable in economic terms. This is not surprising, in view of the universal experience that the unmanaged exploitation of international marine resources (that are in the nature of a "common property") tends toward dissipation of any net economic benefits, benefits that could be obtained as a "resource rent" if harvesting competition were restrained and effort limited to that yielding best returns. It may be assumed, that under recent conditions of exploitation the harvesting of marine mammal stocks has been conducted at break-even level, yielding little of the surplus returns that would indicate positive net economic benefits. While an economically rational international harvesting regime, limited to optimum effort, might well allow for significant net economic benefits from marine mammal exploitation, little net benefit appears to have been obtained in the recent past.[2]

There are reasons to speculate that low-consumptive uses of marine mammal resources could yield substantial net economic benefits in relation to the scale of utilization.

---

[2] It may well be, however, that during the last few years, when high-seas whaling operations have been severely limited and confined almost entirely to two countries (the USSR and Japan) that some regeneration of net economic benefits has taken place.

Net economic benefits, here, should be measured by the value of the recreational or educational opportunities enjoyed by users minus the necessary economic costs of producing these opportunities. The costs are often modest, consisting of private travel expenses to reach viewing sites, access fees to viewing facilities, or costs of educational materials. Because of the uniqueness of the experience of marine mammal observation, it may be assumed that many users will not experience equivalent satisfaction from alternative activities available to them. The full value to users of marine mammal observation experiences may be estimated by the "willingness-to-pay" criterion, showing how much they would be prepared to sacrifice from their disposable income if they had to in order to gain access. Estimates of willingness to pay may be secured through surveys, including those based on differential expenditure tests. Net benefits to users are measured by how much users would be willing to pay minus how much they actually have to pay. The difference is a net benefit in the form of "consumer's surplus". There may be additional net benefits for operators of oceanaria and observation vessels and for wild animal collectors, to the extent that their net receipts exceed the incomes they could expect to earn in alternative employment of their labour effort and capital investment.

Some benefits in the educational area may be of a collective, rather than individual nature. Educating the populace regarding the natural world may have diffuse benefits for society in making it more harmonious and generally more efficient in creating and enjoying values derived from the natural environment. While one may despair of trying to measure the net benefits thus created, there may be good reasons to speculate that they are real and positive in nature.

It is useful to consider that the low-consumptive indirect utilization of marine mammal resources usually can be achieved with little or no conflict with the direct utilization of stocks through commercial harvesting. Particularly if commercial harvests are kept within the limits of maximum sustainable yields, plenty of individuals will remain for low-consumptive uses. The demands of the latter usually are trivial in their effect on stock abundance for commercial harvesting.

Some unusual opportunities may exist for low-consumptive use of marine mammals through various forms of domestication and task performance. The exceptional intelligence of cetaceans may be an asset in this respect. Retrieval of valuable equipment at sea by trained dolphins is an example. The proposal to use manatees for weed clearance of waterways is another case of domesticated use that may be considered.

In societies that are becoming both increasingly affluent and crowded, the potential benefits from the availability of additional recreational opportunities are substantial. In this context the development of additional facilities to view, or otherwise experience, marine mammal activities may create considerable net benefits for society. As coastal stocks of marine mammals usually are in the nature of common property resources under the jurisdiction of public authorities, a more extensive utilization of these resources for low-consumptive recreational purposes may be considered a matter of public concern. Accordingly, public agencies may be urged to sponsor research designed to ascertain the magnitude of the net benefits to society that are currently being generated by the low-consumptive uses of marine mammal stocks and the potential for generation of further net benefits by the development of additional opportunities in this area. A particular aspect that might be explored at the international level, is the need for educational and research facilities regarding marine mammals in developing countries and the appropriateness of devoting aid funds to the development of such facilities.

ASPECTS OF DETERMINING THE STOCK LEVEL OF MAXIMUM SUSTAINABLE YIELD

S.J. HOLT

Abstract

This paper examines the need for precise knowledge of the reproduction curve in the estimation from biological data of the stock level producing maximum sustainable yield (MSY). To this end, a number of algebraic functions, which have been suggested or used to graduate data for stock size and recruitment, are examined along with their reproduction curves. It seems unlikely that the MSY level could be deduced from these reproduction curves, even with excellent data, with less than a 20-30 % error.

Résumé

L'auteur montre qu'il est nécessaire de connaître avec précision la courbe de reproduction pour estimer, à partir de données biologiques, le niveau de stock assurant le rendement maximal soutenu (MSY). A cette fin, il examine plusieurs fonctions algébriques qu'on a employées — ou suggéré d'employer — pour graduer les données relatives aux tailles de stock et au recrutement, et présente les courbes de reproduction correspondantes. Ces courbes, même en utilisant des données excellentes, semblent peu susceptibles de permettre de calculer le niveau assurant le MSY avec moins de 20 ou 30 pour cent d'erreur.

Extracto

Se examina la necesidad de disponer de conocimientos exactos sobre la curva de reproducción para estimar, a partir de datos biológicos, el nivel de población que permite obtener el rendimiento máximo sostenible (RMS). A tal fin se examinan varias funciones algebraicas que se han sugerido o utilizado para elaborar los datos sobre población y reclutamiento, así como sus curvas de reproducción. Parece poco probable que a partir de esas curvas de reproducción, incluso disponiendo de datos excelentes, sea posible deducir el nivel de rendimiento máximo sostenible con un error de menos del 20-30 por ciento.

S.J. Holt
Threshold Foundation, 7 Regency Terrace, Elm Place, London SW7, England

General theoretical considerations

In the literature of fishery science it has frequently been asserted or implied that if the curve of recruitment against stock size is known then the state of a stock can be assessed with respect to management for maximum sustainable yield (MSY). This is true in theory for numerical assessments if the coefficient of adult natural mortality is not considered to be density-dependent.

Thus Doi (1973) emphasizes "that it is possible to estimate the maximum sustainable yield, to diagnose the population status and to determine a management course, judging from a reproduction curve". It is not commonly realized, however, just how precise knowledge of the reproduction curve has to be for useful estimation of MSY stock level from data. My purpose here is to demonstrate this graphically. To this end I have examined a number of commonly adopted algebraic functions which have been suggested or used

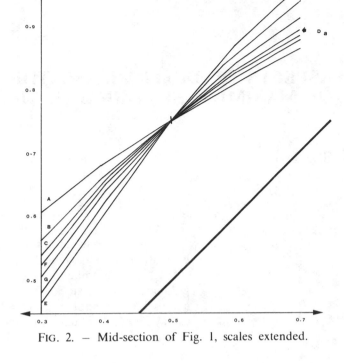

FIG. 2. — Mid-section of Fig. 1, scales extended.

to graduate data for stock and recruitment, and calculated the curves of sustainable yield against stock size that they imply. I have also calculated the stock recruitment functions implied by a number of commonly adopted forms of sustainable yield curve.

The resulting reproduction curves are shown in Fig. 1. For illustration all functions have been reduced to a one parameter form. Stock size, P, is expressed as a fraction of "initial" (unexploited) equilibrium stock. Recruitment, R, is likewise expressed as a fraction of the level of recruitment to the unexploited stock. The value of the single parameter is taken, in each case, such that the recruitment at half the initial stock level is three-quarters of the recruitment at initial level. This is an arbitrary value, but the conclusions are essentially the same if other values are assumed. This value happens, however, to correspond roughly with that implied by a number of whale stock assessments. The natural mortality coefficient is assumed constant.

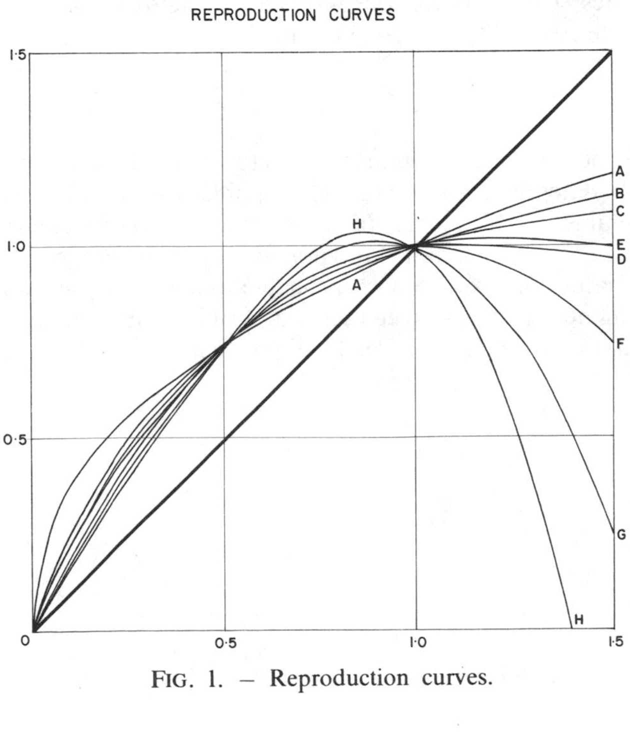

REPRODUCTION CURVES

FIG. 1. — Reproduction curves.

Thus, all the reproduction curves pass through the points:

P = 0	R = 0
P = 0.5	R = 0.75
P = 1.0	R = 1.0

and, of course, predict the same sustainable yield (SY) for P = 0.5.

Except perhaps for the extremes of functions from Cushing, and from Pella and Tomlinson (n = 4) (see legend to Fig. 1) these curves are close to each other in their middle range. It is doubtful if, being fitted to data, discrimination among them would be possible.

The MSY stock level is the value of P for which the tangent to the reproduction curve has a slope of 45°, i.e., is parallel to the diagonal line. In Table 1 the calculated values of P_{msy} are listed, with the corresponding MSY values expressed as a multiple of the MSY for the logistic model, i.e., the SY for P = 0.5.

The possible range of P_{msy} under the particular restraints here applied is zero to 0.5/0.75 = 0.66 (see Chapman, 1973, and Larkin, 1973, for discussion of non-parametric models). It seems unlikely that the MSY level could be deduced from excellent data for a

Table 1. Calculated values of P_{msy} with corresponding MSY multiples for the logistic model

Rep. function	P_{msy}	MSY (relative to logistic)
A Cushing	0.22	1.340
B Beverton & Holt	0.37	1.072
C Chapman	0.41	1.032
D Doi	0.44	1.011
E Ricker	0.45	1.012
F From logistic	0.50	1.000
G From Pella & Tomlinson, n = 3	0.58	1.026
H From Pella & Tomlinson, n = 4	0.63	1.080
	0	2.00
Extremes	0.66	1.33

Table 2. Implications of reproduction function for scales of fishing mortality and effort, and critical values

	$f_{ext}/f_{0.5}$	$f_{msy}/f_{0.5}$	f_{ext}/f_{msy}	$f_{0.5/2}f_{msy}$
A	∞	2.85	∞	0.18
B	4.00	1.45	2.78	0.34
C	2.40	1.26	1.90	0.40
D	2.72	1.50	1.81	0.33
E	2.50	1.12	2.22	0.45
F	2.00	2.00	2.00	0.50
G	1.32	0.88	1.51	0.57
H	1.14	0.86	1.33	0.58

reproduction curve with less than 20 % error and, I suspect, not even within 30 % error in practice. It should be noted that the value of MSY does not (with the exception of the Cushing function) vary more than a few percent; it is always somewhat higher than predicted by the symmetrical logistic, the divergence increasing as P_{msy} moves away from 0.5.

There are, of course, more discriminating methods of analysing stock recruitment data — by, for example, the common method of plotting r − M against P, where r = R/P. Nevertheless, if any mathematical expression, linear or other, is fitted to such data, the level of MSY predicted depends essentially on the properties of that expression. If the data are graduated by eye then the location of the tangent can only be judged very approximately.

Cushing (1973) has examined the variability of a sustainable yield curve derived from the Ricker reproduction function, and using data for northeast Atlantic cod. If he had not assumed this particular form of the curve of R against P but fitted others which give no worse fits to his data (e.g., the logistic) he would have deduced a considerably greater uncertainty in locating the sustainable yield curve, and different values for the sustained fishing mortalities required for MSY or for stock extinction. Cushing plots his sustainable yield against the intensity of fishing; in Table 2 are listed critical values of relative fishing effort for the models here examined.

This table lists the values of fishing effort which would, if sustained, give MSY and those which, theoretically lead to extinction of the stock (f_{ext}) both expressed as ratios of the effort which would hold the stock at half its initial size ($f_{0.5}$). The restraints are the same as in Table 1.

Compare the last column in Table 2 with P_{msy} in Table 1.

Some consequences for whale assessments

Data for whale recruitment present fur-ther difficulties. As recently as 1970, Allen (1973) showed (his Fig. 106) that for the fin whale in the Antarctic there was *no* density dependence of the rate of recruitment; that is, recruits and parent stock were, as near as the data could detect, proportional to each other over a range down to 20 % of the initial stock size. If this is so, P_{msy} is strictly indeterminable, and the SY at any stock size negligible. Allen concludes that the proportionality is an arte-fact due to the use of biased age length keys to estimate age compositions. One wonders if, in a situation where all values of P_{msy} are equally probable over a wide range, it is proper or useful to express the "best estimate" as being in the middle of that range.

References

ALLEN, K.R., Analysis of the stock-recruitment relation
1973   in Antarctic fin whales *(Balaenoptera physalus). Rapp. P.-V. Réun. CIEM*, 164:132-41.

CHAPMAN, D.G., Spawner-recruit models and estima-
1973   tion of the level of maximum sustainable catch. *Rapp. P.-V. Réun. CIEM*, 164:325-32.

CUSHING, D.H., The variability of a yield curve calcu-
1973   lated from a stock and recruitment relation-ship. ICES Document CM 1973/F:29, 5 p. (mimeo).

DOI, T., A theoretical treatment of the reproductive
1973   relationship between recruitment and adult stock. *Rapp. P.-V. Réun. CIEM*, 164:314-49).

HOLT, S.J., The concept of maximum sustainable yield
1975   and its application to whaling. Paper present-ed to the Airlie Workshops on the Concepts of Management and Conservation of Wild Living Resources, ACMRR/MM/IV/4.

LARKIN, P.A., Some observations on models of stock and
1973   recruitment relationships for fishes. *Rapp. P.-V. Réun. CIEM*, 164:316-24.

PAULIK, G.J., Studies of the possible form of the
1973   stock-recruitment curve. *Rapp. P.-V. Réun. CIEM*, 164:302-15.

RICKER, W.E., Critical statistics from two reproduction
1973   curves. *Rapp. P.-V. Réun. CIEM*, 164:333-40.

THE OPTIMIZATION OF MANAGEMENT STRATEGY FOR MARINE MAMMALS

K.R. ALLEN

Abstract

Agreement on objectives in managing marine mammals is difficult. Objectives can be expressed in terms of yield or population level. It is advantageous to use yield because it is the direct object of exploitation; it can be easily measured and it is simply handled in mathematical models. Maximum sustainable yield (MSY) was the first and most widely used objective. Some criticisms of it have mistaken failures in its application for limitations in the concept. Other criticisms have claimed that it does not take account of all factors which affect reproduction and mortality, which it does do implicitly. Only recently has MSY in weight, rather than in numbers of animals, been considered for marine mammals; the difference in population levels producing these is probably much less for marine mammals than for fish. A maximum economic yield can also be formulated; it is produced, for a population in a steady state and in the absence of an economic discount, at a population level above that giving MSY. It is harder to define objectives in terms of optimum population levels. One criterion, that populations not be reduced to a level which will disturb the "health" of the ecosystem, is especially difficult to interpret. Changes already caused by man, natural continuous change caused largely by climate and the possibility that several alternative states could exist in a single system, all make what is meant by a "healthy" ecosystem very unclear. Another suggestion is that populations be kept at a level which can be changed easily so that new objectives of either yield or population level can be met in the future. Analysis of exploitation of competing species or of species one of which eats the other, is especially difficult because of lack of adequate models and data. Such analyses, when they do become possible, must take account of economic as well as biological factors.

Once yield or population level criteria are chosen, the population level which meets these criteria and the pattern of catches and protection necessary to move to this level from the current population level, must be determined. Analysis of the relation of population size to total and net numbers of recruits and to gross and net recruitment rates in two stock-recruitment models with different MSY levels, indicates that the value of MSY and the population level producing it, can probably best be determined from estimates of recruitment rates from age composition data or — though longer time series of data are needed — from estimates of sustainable yield.

In addition to identification of an optimum population and of means to bring the population to this level, other requirements of a satisfactory programme to manage marine mammals can be given. Firstly, there should be safeguards to prevent harm to the population and its ecosystem due to errors in assessment. Next, the population should be brought to its optimum level as rapidly as possible but without major changes in effort that could disrupt the industry. Lastly, in order to improve uncertain estimates of abundance, vital rates and yield, populations should be exploited so that data can be collected over a substantial range of population sizes. The "new management procedure" of the International Whaling Commission, under which the optimum level is, for the time being, defined as that giving MSY, meets these requirements fairly satisfactorily. Its main weakness relates to the last requirement given above.

Résumé

Il est difficile de parvenir à un accord sur les objectifs de l'aménagement des mammifères marins. Les objectifs peuvent s'exprimer sous forme de rendement ou de niveau de la population. Il est avantageux d'utiliser le rendement parce qu'il est l'objet direct de l'exploitation, il peut être facilement mesuré et il est d'un maniement facile dans les modèles mathématiques. Le rendement maximal soutenu (MSY) a été le premier et le plus largement utilisé des objectifs. Certaines des critiques qui l'ont visé ont confondu erreurs d'application et limitations du concept. D'autres critiques ont assuré qu'il ne tient pas compte de tous les facteurs qui affectent la reproduction et la mortalité — alors qu'il le fait implicitement. Ce n'est que récemment qu'on a considéré le MSY en poids, plutôt qu'en nombre d'animaux, pour les mammifères marins. La différence des niveaux de population qui se trouve à la source est probablement bien moindre pour les mammifères marins que pour les poissons. On peut aussi formuler un rendement maximal économique; il est produit, pour une population se trouvant dans un état stable et en l'absence d'une actualisation économique, à un niveau de population supérieur à celui qui donne le MSY. Il est plus difficile de définir les objectifs sous forme de niveau de population optimal. Il est spécialement difficile d'interpréter un critère, à savoir que les populations ne doivent pas être réduites à un niveau mettant en péril la "santé" de l'écosystème. Les changements déjà provoqués par l'homme, le changement naturel continu dû surtout au climat et la possibilité qu'il puisse exister plusieurs états dans un même écosystème contribuent à rendre fort peu clair le concept d'écosystème "en bonne santé". On dit aussi que les populations devraient être maintenues à un niveau qui puisse être changé facilement afin que de nouveaux objectifs de rendement ou de niveau de population soient réalisables à l'avenir. L'analyse de l'exploitation d'espèces concurrentes, ou d'espèces dont l'une se nourrit de l'autre, est spécialement difficile en raison de l'absence de modèles adéquats et de données. Ces analyses, quand elles deviendront possibles, devront tenir compte des facteurs économiques comme des facteurs biologiques.

Quand on a choisi les critères de rendement ou de niveau de la population, il convient de déterminer le niveau de population répondant à ces critères et le mode de capture et de protection nécessaires pour atteindre ce niveau à partir du niveau de population courant. L'analyse de la relation entre l'importance de la population et le nombre total et net de recrues et les taux de recrutement bruts et nets dans deux modèles de recrutement de stock avec des niveaux différents du MSY indique que la valeur du MSY et du niveau de population qui le produit peut probablement être déterminée au mieux à partir des estimations des données par âges ou — bien qu'il soit nécessaire de posséder des séries de données portant sur un temps plus long — à partir d'estimations du rendement soutenu.

Outre l'identification d'une population optimale et des moyens de l'amener à ce niveau, on peut fournir les autres éléments requis pour un programme satisfaisant d'exploitation rationnelle des mammifères marins. En premier lieu, il doit exister des garanties pour éviter que la population et son écosystème aient à souffrir d'erreurs d'évaluation. Ensuite, la population devrait être portée le plus rapidement possible à son niveau optimal mais sans changements majeurs d'effort qui pourraient bouleverser l'industrie. Enfin, pour améliorer certaines estimations incertaines de l'abondance, des taux biologiques et du rendement, les populations devraient être exploitées de telle manière que les données puissent être rassemblées sur une gamme substantielle de tailles de population. La "nouvelle procédure d'aménagement" de la Commission internationale baleinière d'après laquelle le niveau optimal est, pour l'instant, défini comme celui qui fournit le MSY répond à ces demandes de façon assez satisfaisante. Son principal point faible concerne le dernier élément requis mentionné ci-dessus.

Extracto

Es difícil llegar a un acuerdo sobre los objetivos que han de perseguirse con la regulación de la explotación de los mamíferos marinos. Los objetivos pueden expresarse en rendimiento o nivel de la población. Utilizar como objetivo el rendimiento resulta ventajoso, ya que constituye el objeto directo de la explotación, es fácil de medir y su utilización en los modelos matemáticos no resulta complicada. El primer objetivo, y el más ampliamente utilizado, fue el rendimiento máximo sostenible (RMS). Algunas de las críticas que se han hecho a este objetivo han confundido los errores en su aplicación con limitaciones del concepto mismo. Otras críticas dicen que ese concepto no tiene en cuenta todos los factores que influyen en la reproducción y la mortalidad, cosa que sí hace implícitamente. Sólo recientemente se ha aplicado a los mamíferos marinos el RMS en peso, en vez de en números de animales; la diferencia en los niveles de población que permiten obtener esos dos tipos de rendimiento es probablemente mucho menor en el caso de los mamíferos marinos que en el caso de los peces. También se puede formular el concepto de rendimiento económico máximo; en una población estable y sin aplicar principios de actualización económica, dicho rendimiento se obtiene con un nivel de población superior al necesario para obtener el RMS. Más difícil resulta definir objetivos en función del nivel óptimo de población. Uno de los criterios que se utilizan —que la población no se reduzca a un nivel que trastorne el "buen estado" del ecosistema— plantea dificultades especiales de interpretación. Los cambios ya determinados por el hombre, los cambios naturales continuos debidos en buena parte al clima, y la posibilidad de que en un sistema puedan darse varias situaciones alternativas, son todos elementos que hacen que sea muy poco claro lo que se entiende por ecosistema "en buen estado". Otra idea que se sugiere es mantener la población a un nivel que pueda modificarse fácilmente de manera que en el futuro puedan conseguirse nuevos objetivos, bien de rendimiento o de nivel de población. Los análisis de la explotación de especies competidoras, o de especies una de las cuales se alimenta de la otra, resultan especialmente difíciles, por falta de modelos y datos adecuados. En esos análisis, cuando son posibles, hay que tener en cuenta factores tanto económicos como biológicos.

Una vez que se han elegido como criterio el rendimiento o el nivel de población, es preciso determinar el nivel de población que responde a esos criterios y el volumen de captura y las medidas protectivas que son necesarias para conseguir alcanzar ese nivel a partir del nivel presente de población. Un análisis de la relación entre el volumen de la población por un lado, y el número total y neto de reclutas y los índices bruto y neto de reclutamiento, por otro, en dos modelos de reclutamiento con distintos niveles de RMS, indica que probablemente la mejor manera de determinar el valor del RMS y el nivel de población que es necesario para conseguirlo es estimar los índices de reclutamiento a partir de datos sobre la composición por edades o a partir de estimaciones del rendimiento sostenible, aunque en este último caso se necesita una serie cronológica de datos más larga.

Además de indicar el nivel óptimo de población y los métodos que han de aplicarse para que la población alcance ese nivel, son necesarios otros requisitos para que un programa de ordenación de mamíferos marinos resulte satisfactorio. Ante todo, debe haber las salvaguardias necesarias para evitar que la población y el ecosistema en que vive resulten perjudicados por eventuales errores de evaluación. En segundo lugar, hay que elevar la población a su nivel óptimo lo más rápidamente posible, pero sin cambios importantes en el esfuerzo de captura, que podrían trastornar la industria. Por último, para mejorar las estimaciones de la abundancia, los índices vitales y el rendimiento, es preciso explotar la población de manera que sea posible recoger datos con la población a distintos niveles. El "nuevo procedimiento de ordenación" de la Comisión Ballenera Internacional, en el cual el nivel óptimo se define, por el momento, como el que permite alcanzar el RMS, responde bastante satisfactoriamente a estos requisitos, exceptuado el último de ellos.

K.R. Allen
Division of Fisheries and Oceanography, CSIRO, P.O. Box 21, Cronulla, NSW 2230, Australia

Diversity of broad objectives

Extremely diverse pressures affect administrators concerned with management and conservation of natural resources, even within the relatively narrow field of marine mammals. These pressures include political, economic and social factors, and frequently conflict with scientific recommendations. Thus, although we may hope to make substantial progress in analysing the various objectives for which people seek to manage marine mammals and in identifying the consequences implicit in them, there is little reason to believe that general agreement can easily be achieved on a single universally applicable set of objectives which will lead in turn to a set of agreed broad management strategies.

It has recently become almost as fashionable to decry the use of maximum sustainable yield (MSY) as a target in the management of living resources, as it was at one time to use it as the almost universal objective. However, much of the criticism which has been aimed at the use of MSY as a management objective confuses limitations in the concept itself with failures in its application. Many of the failures in management of resources which have been laid at the door of MSY have been either caused by incorrect recommendations made by scientists seeking to apply the concept but hampered by inadequate data, or, probably more often, by the failure of administrators, nationally or internationally, to act on the recommendations of the scientists even when these were correct.

Optimum yield or optimum population level

In seeking a definition of the optimum target for management, the first dichotomy which is faced is between the criteria that relate to the population level and those that relate to the continuing yield. If we could separate these completely, should we say, for example:

the "optimum" population level is that which enables us to take a continuing safe yield which meets the criteria for an optimum yield,

or should we say:

the "optimum" yield is that which can be taken safely and continuously from a population which is maintained at a level which meets the criteria for an optimum population.

In practice, any acceptable solution must take both kinds of criteria into account. Since these approaches are to some extent in conflict, we have therefore to compromise, which makes any generally acceptable solution extremely difficult since it must be based on the balancing of unlike values.

SINGLE SPECIES APPROACH

Yield levels

As a criterion for the management objective, yield has the advantage that in some form it is the direct object of the exploitation of the resource, that it is fairly easily measured in practice and that it can be simply handled in a variety of mathematical models. It is little wonder, therefore, that the first and most widely adopted objective of fisheries management, as it became subject to quantitative study, was the MSY expressed either as the number or overall weight of the animals caught.

MSY IN WEIGHT OR NUMBERS

The choice as to whether number or weight should initially be used has depended to a great extent on the relative significance of growth and reproduction in determining the size of the catch. Animals which can conveniently be caught over only a limited range of sizes and in which the number of recruits is believed to be largely dependent on the size of the parent stock, have tended, as in marine mammals and salmon, to be managed largely on a numerical basis.

Most marine fish, however, may be caught over a wide range of sizes and their recruitment appears to be largely independent of stock size over a major part of the possible range. These have obviously had to be managed from the beginning on a weight basis.

It is only recently that consideration has been given to the possibilities of attempting to manage marine mammals with the aim of maximizing the weight of the catch. This is, of course, only appropriate where the desired product is a fairly fixed proportion of the weight of the whole animal, such as meat or oil. It is probably less appropriate where the product is a unit part of the individual animals, such as a skin, although even here a size-price relationship may be involved.

The difference in population level between that giving the MSY by weight and that giving the MSY by number is probably much less for most marine mammals than it is for fish. For baleen whales, the difference between the 2 MSY levels appears to be only about 5 % or even less and for sperm whales possibly 7-18 %.

FINANCIAL YIELD AND DISCOUNT CONSIDERATIONS

Yield may also be expressed in monetary terms, generally as the difference between the value of the catch and the cost of obtaining it. The relation between financial return and stock size has attracted the attention of many economists and a few main principles are commonly accepted. In stable conditions the profit or economic rent is maximized (maximum economic yield - MEY) at a stock level (MEY level) somewhat above the MSY level. At a lower population level, cost and price balance, and industry "breaks even". This stock level must be below MEY level but does not have any necessary relation to MSY level. For a simple Schaefer-type model, the MEY level is halfway between the break-even point and the unexploited level if cost is assumed to be proportional to effort and fixed costs are ignored. Copes (pers. comm.) has recently shown in a more sophisticated model that, when consumer surplus is taken into account, the economic return is maximized at a level below MEY level.

These conclusions apply to populations in a steady state and assume that the value of future catches is the same as that of present catches. As Clark has shown, if the value of future catches is discounted, the level which will maximize the present value of all future catches depends upon the discount rate adopted but is independent of the starting level of the population. The optimum level in this sense lies between the MEY level, which is the optimum when the discount rate is zero, and the break-even level (BE level) which is optimal for an infinite discount rate.

Table 1 shows the size of the population for each of the various levels discussed in a hypothetical blue whale population similar to that quoted by Clark (1976), i.e., with an initial stock size of 150 000, break-even level of 20 000 and a net recruitment rate at zero population level of 0.02. To enable the calculations to be completed for yield by weight, the break-even level has again been arbitrarily set at 20 000 and growth has been assumed to follow a von Bertalanffy curve with parameters similar to those for fin whales, i.e., $K = 0.26$, $t_0 = 1.5$, $t_r = 5$. The results are rather insensitive to the precise values given to these parameters.

Table 1. Optimum population levels for various yields for a hypothetical blue whale model[1] (Clark, 1975)

	Numerical yield		Yield by weight
Initial population size	150 000		150 000
MEY level	85 000	Range if future	85 000
MSY level	75 000	values are discounted	75 600
Break-even point	20 000		20 000[2]

[1] Clark, 1975.
[2] Set arbitrarily at the same level as for numerical yield.

Population levels

Criteria for defining optimum population levels cannot be identified in nearly such precise terms as can criteria for yields. This is because the benefits which are derived from the living population, either by man or by the ecosystem of which it forms part, are themselves almost impossible to quantify precisely and therefore to relate to population size. There are, however, some general statements about population levels which are probably acceptable to almost everyone.

HEALTHY ECOSYSTEMS

The first, but also the least precise, is that populations should not be reduced to a level which will seriously disturb the "health" of the ecosystem, although there are a number of difficulties associated with such a statement. Apart from the fact that there has probably been no generally acceptable definition of a "healthy ecosystem" as yet, it is also true that a great many marine ecosystems have already been greatly modified by man. Should we therefore arbitrarily decide that, in at least some cases, the present situation is healthy and that we should try to manipulate things so as to maintain it in its present condition? Alternatively, should we say that only the undisturbed ecosystem is "healthy", and that our objective should be to restore ecosystems to their pristine state? The difficulty, of course, with this concept is that, if we take it to its logical extreme, we not only cease to exploit marine mammals but also abandon all our fisheries. A further question arises when some species have benefited from a reduction of competitors which were more attractive as a harvest to man. There is evidence, for example, which suggests that sei whales in the Antarctic may have increased as a consequence of the reduction in numbers of the larger baleen whales. Should we therefore aim at reducing the sei whale population with the object of accelerating the recovery of other species and thus restoring the *status quo*? We cannot ignore the possibility that an ecosystem containing several major species occupying rather similar niches may have several alternative stable states. It is therefore possible that, even if all catching of Antarctic baleen whales stopped now, the system would revert to a condition different from that which prevailed before whaling began: blue or fin whales might not rise again to their original numbers, while sei whales might stabilize at a higher level. Would such a system be any less "healthy" than that existing a century ago and should we take positive steps to try to force it to return to its previous condition? We need to be able to find clear and generally acceptable answers to questions such as these before we can quantify an optimum population level in terms of maintenance of "healthy ecosystems".

The situation is further complicated by the fact that even ecosystems undisturbed by man are probably in a continuous state of change, largely as the result of climatic changes. While the relatively long life and low reproductive rate of most marine mammals probably makes them unsusceptible to short-term changes, there is no reason to suppose that they have been immune from both long-term and medium-term changes, measured perhaps on a time scale of fractions of a century. If this is the case, it becomes both impracticable and pointless to have firm objectives of marine mammal management which are closely tied to current population levels.

MAINTENANCE OF OPTIONS

Another basis for considering optimum population levels which is more easily subjected to quantitative study, is that based on the concept of keeping future options open. This implies that we retain for the indefinite future our ability to manipulate the population so as to move toward new objectives of either yield or population level as they emerge. It implies, in effect, that the population is retained at a level which will ensure that it will not become extinct and will be capable of responding fairly rapidly to desired future changes in its level. The second implication of course involved the first, since any population capable of response to manipulation must be well above any level at which there is a real risk of extinction due to natural causes. The re-

sponse which is of interest in this connexion is the ability to increase in numbers if exploitation is reduced or stopped. Several criteria of this ability are possible. The maximum relative rate of increase in the absence of exploitation would be one possibility. Unfortunately, under most models, this occurs at a population level only just above that at which extinction is a real possibility and it could therefore not be regarded as a useful criterion. A criterion at the other extreme is that the time required to recover to unexploited level in the absence of exploitation should be at a minimum. This condition is of course met only just below unexploited level and is therefore an unrealistic criterion. A third possible criterion is that the rate at which the population would grow in absolute numbers in the absence of exploitation should be at a maximum. This occurs, by definition, at the MSY level, so that a population at this level must be capable of responding rapidly to reduction or cessation of exploitation. Such a population must therefore automatically meet the objective of keeping future options open whether or not it meets any precise criteria which could be set up for this purpose.

It is sometimes suggested that the concept of maintaining populations at MSY level does not take into account such factors as the effects on yield of changes in age structure or social order which will occur in a population at a level different from the unexploited. Where seriously advanced, this argument, when applied to stable conditions, must reveal a misunderstanding of the concept. The essence of this concept is that a population at the numerical MSY level must have, in the steady state, the greatest excess of recruits over natural deaths. When it is correctly applied, therefore, all the factors which influence reproductive and natural mortality rates are taken into account explicitly or implicitly. In practice, at least for marine mammals, it has not been possible to build up recruitment and mortality curves from study of these basic factors. Calculations as to population levels and yields have, therefore, been based either on direct observation of overall mortality and recruitment rates or of the difference between them, or else on theoretical models designed to pass through a few approximately known fixed points. Both approaches include implicitly the effects on reproduction and mortality of such factors as age structure and social order.

MULTI-SPECIES EXPLOITATION

Most of the foregoing discussion has dealt with considerations involved in determining optimum yields or optimum population levels for single species populations. The practical problems become more acute when interacting species are being exploited simultaneously by man. Exploitation may involve competing species or predators and their prey. Among marine mammals the best known example of the first case is provided by the taking of the various species of baleen whales in the Southern Ocean, and perhaps also in the North Pacific; the second case is probably illustrated by the taking of fur seals and Alaskan pollack in the Bering Sea and by baleen whaling and the developing krill fishery in the Southern Ocean. At present we lack adequate models for studying such cases, though we may probably now begin to construct them. Even if conceptually realistic models can be developed, adequate data is probably lacking.

Determination of optimum strategies for multi-species management will have to take account not only of the biological relationships but also of the harvesting costs and the relative values of the various exploited species.

Assessment of population levels

Actual management practice must be based on an assessment of the desired population level and particularly on the relation between this level and the existing level. The best way to approach this problem is to compare separate, independent estimates of the optimum level and the existing level.

An array of methods is now available for estimating existing population levels. In some cases, as for example that of Antarctic fin whales, sufficient data exists to enable the backward extrapolation of the estimates to include the unexploited population level.

The optimum population level may be estimated from the unexploited population size by applying either an appropriate yield curve or an arbitrary relationship between optimum and unexploited population sizes.

METHODS USED BY THE IWC

These methods are currently being used by the Scientific Committee of the IWC, under the "new management procedure" (nmp), which bases the current catch level strictly on the relation between existing population size and MSY level. MSY level is accepted as an interim alternative for the optimum level. In most cases the MSY level has been assessed as a fixed proportion of the original unexploited population level – for most baleen whales MSY level is taken to be 60 % of unexploited level.

The symmetrical parabola stock-recruitment model, in fact, gives MSY at 50 % of the unexploited population level. The arbitrary decision to assume MSY at 60 % of unexploited level was taken partly out of a desire to adopt a more cautious approach and also because it is believed to be likely that the stock-recruitment curve is convex, and thus that the MSY level is higher than 50 %.[1]

For sperm whales, the Committee has used a more sophisticated population model which results in the MSY population level occurring at 55-70 % of the unexploited level de-

---

[1] This is because of the intrinsic limitations in the reproductive rate of whales, which must apply even at the lowest population levels.

pendent upon the values given to the various parameters.

Where estimates of initial and current stock sizes are reliable, this approach is clearly useful. It is, however, extremely dependent upon knowledge of the nature of the stock-recruitment relationship, since it is this which determines the relative MSY level. While mathematical models may be used to explore the consequences of any particular relationship, their use to provide a sound basis for management depends upon the relevant parameters being sufficiently accurately known to enable us to develop models which simulate the real populations. This applies particularly to those parameters which determine the level at which MSY occurs.

EXAMINATION OF
STOCK-RECRUITMENT CURVES

The *Ad hoc* Group I of the Advisory Committee on Marine Resources Research, in its draft report (ACMRR/FAO, 1976) has examined two theoretical relationships which give very different MSY levels (37 % and 58 % of initial stock size). They have also shown that for these models, the curves relating total numbers of recruits to stock size are probably undistinguishable in practice, despite the very great difference in the resulting MSY level.

It is instructive to pursue this comparison further. In addition to total recruitment and sustainable yield – which is the equivalent of net recruitment – the associate gross and net recruitment rates are worth examination. Using the same models as those employed by *Ad hoc* Group I, the 4 curves are shown in sections 1-4 of Table 2, the curves in sections 1 and 2 being the same as those in Figs. 1 and 2 of the Group I report, respectively.

It is apparent that the 2 models differ least in the curves for total recruitment. The sustainable yield curves show considerable differences at middle population levels, while the 2 rate curves show great differences in both

magnitude and slope at low levels, with the difference in slope extending to middle levels.

Such curves for real populations can only be constructed if actual values can be estimated for several population levels. In some cases, the values of the fixed points corresponding to zero and unexploited levels are known on theoretical grounds. The zero values for such points are of course precise, and the natural mortality rate which is the upper limiting value for gross recruitment rate is, in most cases, one of the most accurately known parameters. The values of these fixed points are summarized in Table 2. The Table also includes indices of the sensitivity with which the 2 curves may be distinguished. These are based on the comparison of the mean slopes of the curves over the range between .45 and .55 of unexploited level. The 2 indices are the absolute and relative differences between the slopes for the 2 models.

The curves for total number of recruits are clearly the least sensitive. The differences in their shapes are minor and very precise measurements would be needed to distinguish them. These curves also represent directly the quantity of prime interest and, in addition, have the advantage of fixed points at each end. Therefore, a single measure of the slope in 1 region, i.e., of values at 2 population levels, gives some information as to the position of the maximum, since it tells us whether the maximum is above or below the level at which observations are available. Data over a sufficient range or with sufficient accuracy to provide an estimate of curvature would enable the position of the maximum to be located with some accuracy.

The most important characteristic of recruitment rate curves is their curvature. If they are convex upward, the MSY level is above 50 % of unexploited level and vice versa, while the greater the curvature, the greater the departure from the 50 % level. The upper fixed point is zero for the net recruitment curve and is equal to M for the gross recruitment curve. Thus, a useful approximation to the curve can be obtained from knowledge of the slope and

Table 2. Relation of important variables to population size for two stock-recruitment models with constant natural mortality rate

M	Fixed Points		Sensitivity				Basis of Estimation
	Lower	Upper	Slope A	Slope B	A − B	$\dfrac{A-B}{\frac{1}{2}(A+B)}$	
1. Total Number of Recruits	0	Annual total recruits in unexploited population	.475	.764	.259	.466	(a) Product of gross recruitment rate and absolute population size (b) Change in population size adjusted for natural mortality rate and catch
2. Net Number of Recruits = Sustainable Yield	0	0	-.431	.320	.751	13.532	(a) Change in population size (absolute or relative) adjusted for catch
3. Gross Recruitment Rate	Gross recruitment rate in minimal population	Gross recruitment in unexploited population (= M)	-.427	-.197	.231	.738	(a) Directly from age structure
4. Net Recruitment Rate	Net recruitment rate in minimal population	0	-1.166	-.567	.599	.692	(a) From gross recruitment rate adjusted for natural mortality rate (b) From relative population change adjusted for catch

level at 1 region of the curve. In some cases, it may be possible to calculate, from knowledge of reproductive and juvenile mortality parameters, the maximum possible recruitment rate at low population levels. A useful indication of the shape of the curve could then be obtained with only a single estimate of recruitment rate at an intermediate population level. In this case only, therefore, the construction of curves which will provide useful information on the relative population size at MSY level is not dependent upon the availability of reliable data for at least 2 distinct population levels.

This analysis has indicated that the best approaches to the problem of determining relative population size at MSY level are either directly through estimates of sustainable yield or indirectly through estimates of either gross or net recruitment rates. Table 2 also sets out briefly the most direct methods available for estimating the various quantities considered. Recruitment rates can be measured directly from age composition and do not require the use of long time series to derive each point estimate. Direct measures of sustainable yield can be attempted but are dependent upon reliable estimates of at least relative changes in population size, and therefore upon substantial time series of data. Study of recruitment rates by such means as the analysis of age distributions seems therefore to provide the most promising approach to the identification of MSY and related population levels, while the next best alternative is by direct observation of rates of population change at different levels and rates of exploitation.

The curves which have been discussed are based upon relations which exist for populations which are stable at each level. In practice, the population characteristics will be observed in a changing situation and are subject to lag effects so that the recruitment rates and yields observed at any particular level cannot be inserted directly into such curves. These lag effects are of 2 kinds. The first stems from the fact that recruits into the exploited stock in any one season are derived from the breeding stock a certain number of years earlier. Adjustment for this effect may be made fairly easily. The second source of lag is much more difficult to handle. It is caused by the differences in age composition and social structure which will exist between a stable population of a given size and a changing population of the same size. An expanding population, for example, will have a higher proportion of young animals than a stable one. These differences may in turn cause differences in overall mortality or reproduction rates.

The path to the optimum

If the current population level, the optimum level and the optimum yield are all known, the only question which remains is by what pattern of catches to move to the optimum level. It has long been realized that, if the MSY level is taken to be the optimum level, the procedure to maximize the total of future catches is to catch nothing until MSY level is reached if the population is below that level, and to catch the excess as rapidly as possible if the initial population is above MSY level. More recently, Clark has shown that the same principles apply even when the value of future catches is discounted and allowance is made for costs of catching, provided that it is assumed that there are no economies or diseconomies of scale in harvesting and that demand is infinitely elastic.

In practice, the three critical quantities are known very approximately at the best. In addition, the major disruptions (e.g., short periods of complete closure or abrupt changes in effort) which are involved in the yield maximizing solutions in the all-knowing situation, will commonly create economic or social problems for the industry.

Since some of the difficulty arises from the uncertainty in the population and yield values involved, another important consideration in determining the optimum strategy is the acquisition of the data which will enable the estimates to be improved. We have already seen that determination of the MSY or other optimum level is largely dependent on the availability of data over a substantial range of population sizes. This introduces a further constraint into the development of the ideal strategy. It should allow, and indeed promote, a change of population size over a fairly substantial range and, in order to allow sufficient data to be accumulated, either this change should take place slowly or the population should be held stable for a period of years at each of two or more distinct levels. The situation is further complicated by the fact that at

87

present so much of the data essential to these problems can only be obtained on a population which is being exploited on a fairly substantial scale. On the one hand, calculations based on catch per unit effort remain by far the most effective means of estimating population sizes. On the other hand, data on age composition and reproductive condition which are required for estimating recruitment and mortality rates are also only obtainable from dead animals.

Development of a practical management strategy therefore requires:

(a) that acceptable criteria for the determination of optimum yields or stock levels are identified;

(b) that safeguards are incorporated to ensure that major harm is not done to the population or ecosystem because of errors in assessment;

(c) that the yield of the population is brought to the optimum level, however this is defined, as rapidly as possible while avoiding major discontinuities and particularly reversals in trends of effort;

(d) that it should cause the population to change in size significantly;

(e) that the population should also be subjected to substantial exploitation et each level.

These requirements are to some extent conflicting and in this fact lies one of the fundamental difficulties in developing effective management policies for marine mammals. A particular difficulty arises in the case of populations which have been exploited for some time under conditions which have apparently kept them stable at a level which may be at or near the optimum. It may be impossible to determine whether this level is in fact optimal without changing the regime to bring the population to a significantly different level. The risk is, of course, that the new level may be inferior to that originally prevailing.

THE NEW MANAGEMENT PROCEDURE OF THE IWC

The so-called "new management procedure" of the IWC is a major step forward in that it does involve adoption of a consistent strategy whose effects can be forecast for any given population model. It also goes some distance toward meeting the requirements outlined above. Safeguards against incidental excessive exploitation which might reduce the population below the optimum level are introduced in 2 ways. First, the allowable catch is not allowed to exceed 90 % of estimated MSY at any population level. Second, stocks more than 10 % below the estimated MSY level are fully protected, and permitted catches rise linearly from zero at this level to 90 % MSY at MSY level. If accurately applied, these procedures will allow populations to be brought to the vicinity of the MSY level whatever the starting point. If MSY is accurately known, the level at which the population will ultimately stabilize will be above MSY level at the point where the sustainable yield equals 90 % MSY.

For recruitment curves giving MSY at 50 % of unexploited level, the stable population level giving 90 % MSY is at about 66 % of unexploited level, while if MSY occurs at 60 % of unexploited level, the stable level giving 90 % MSY will be at about 74 % of unexploited level.

Effect of errors in estimates of MSY

If there are errors in the estimation of MSY, or of the population size at which MSY occurs, or of the current population size, stability will still be attained but at a population level dependent upon the nature and magnitude of the errors. The stable condition can be quite easily identified by a graphical approach.

The following abbreviations are used:

MSY: maximum sustainable yield;
SY: sustainable yield;
MSYL: MSY level = theoretical level giving MSY;
EMSYL: estimated MSYL;
AEMSYL: EMSYL adjusted for error in estimate of population size;
OY: optimum yield;
OYL: optimum yield level = population size giving OY;
EOY: estimated optimum yield;
AEOYL: EOY level adjusted for error in estimate of population size;
MPC: maximum permitted catch.

Fig. 1A shows the situation in a population when all variables are accurately known. The dashed curve represents the true relationship between sustainable yield (SY) and population size; the open circle, the correctly known MSY available at the true MSY level; the solid circle is the permissible catch (90 % MSY) at this population level; and the solid line represents the catch permitted at other population levels. Stability occurs at point X where SY and permitted catch are equal.

Fig. 1B shows a case in which both MSY level and MSY are incorrectly estimated but the true population size is correctly known. Stability for this condition will occur again at point X.

Fig. 1C includes also an error in the estimation of the actual population size, causing the population to be over-estimated and thought to be at level P_1 (the estimated MSY level, whether correct or incorrect is immaterial) when it is actually at level P_2. A population at this level (P_2) is therefore the smallest from which the maximum permitted catch (90 % MSY) can be taken. The relation between permitted catch and actual population size is therefore again as shown by the solid line with stability at point X.

The stable situation is therefore com-

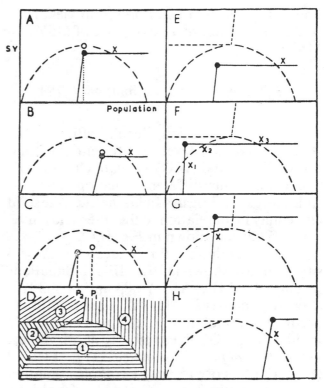

FIG. 1. — Effect of errors in estimates of MSY.

pletely determined by the position of the solid circle in relation to the true yield curve for the population. The circle indicates a yield equal to the permitted fraction — maximum permitted catch (MPC) — currently 90 %, of the estimated MSY at the estimated MSY level (EMSYL) adjusted for any error in the estimate of actual population size (AEMSYL). Any error in actual population estimate must apply to the current population, but it can be extrapolated, perhaps proportionately, to derive AEMSYL from MSYL.

This graphical approach to identification of the stability condition will apply whatever the shape of the stock-yield curve and whatever criterion is used to determine the value of the yield. It is not restricted to the symmetrical curve used for simplicity in the figure, or to the optimum yield being the MSY in accordance with current IWC practice.

In the following discussion the terms

optimum yield (OY) and optimum yield level (OYL) are therefore used in place of MSY and MSYL.

Effect of errors of estimation of OSY

We can now consider the effect on the stable population level of different combinations of EOY and AEOYL relative to true OY and OYL. Four different situations are possible corresponding to 4 fields in which the solid circle may lie in relation to the true yield curve. These fields are shown in Fig. 1D.

The resulting stability levels for these 4 situations are shown in Figs. 1E-1H. Situations placing the MPC-AEOYL point in fields 1 (Fig. 1E) and 4 (Fig. 1H) will cause stability to occur at a population level about *true* OYL. Field 3 (Fig. 1G) gives stability at a level just below the *adjusted estimated* OYL. Field 2 (Fig. 1F) gives rise to a more complex situation with apparently 3 possible levels of stability. Level X_2, however, represents an unstable equilibrium, since a population only slightly below or above this level will move respectively down to level X_1 or up to level X_3.

EFFECTIVENESS OF NMP

If it is accepted that one of the objectives of the management procedure is to maintain the population at a level no lower than the true MSYL, then either of the situations in fields 1 and 4 meets the requirement. That is, a situation which satisfies either, but not necessarily both of 2 conditions is acceptable. These conditions are:

(a) Maximum permitted catch is not greater than the true sustainable yield at the adjusted estimated OYL, i.e., $MPC \lesssim SY$ (AEOYL);

(b) Adjusted estimated OYL is not less than true MSYL, i.e., $AEOYL \gtrsim OYL$.

The fact that it is necessary to comply only with one or other and not with both of these conditions to ensure that the population is maintained at or above true OYL is of considerable importance in assessing any uncertainty allowances which should be applied in deciding upon what yields and population levels the management procedure should be based.

In addition to these 2 sufficient conditions, the situation in field 2 can also bring about stabilization above the true OYL provided that a second requirement is also met. Thus we have a third sufficient condition, i.e.,

(c) Maximum permitted catch is less than sustainable yield at current population level, i.e., $MPC < SY$.

Finally, it may be noted from Figs. 1E and 1H that if the adjusted estimated OYL is greatly in excess of the true OYL, stability is only achieved at high population levels and consequently low sustainable yields. This applies whether the estimated OY used is above or below the true value. Selection of an appropriate estimated OY population level is therefore probably more critical than the selection of an estimated OY in optimizing management strategy.

Thus, the new management procedure appears to meet the first requirement satisfactorily in that a stable population at or above the optimum level will be ultimately achieved provided that any one of conditions (a), (b) and (c) is met. The stable level will never be more than 10 % below estimated optimum yield and will commonly be above it.

The procedure also meets the third requirement fairly satisfactorily in that, provided no abrupt changes in estimates are made, it will bring the population to the desired level as rapidly as natural population growth rates allow while avoiding very severe discontinuities in effort. For populations initially near the unexploited level, an increase in effort by a factor unlikely to be more than 2.0 would take place over several years as the population was reduced to MSY level while maintaining the catch constant at the 0.9 MSY level. For populations below the target level, a more significant increase in effort would be required to achieve the permitted catches as the population rose from 10 % below MSY level to MSY level, but even this would normally be spread over several years.

The fourth requirement, that the procedure should cause the population to change in size significantly, will be satisfactorily met for populations which initially were well away from MSY level. In the case of populations believed to be close to this level, however, little change would occur and thus it would be difficult to assess the accuracy of the model on which the procedure is based.

The final requirement, that in order to provide adequate data for the assessment of the relation between yield and population size, an industry should be permitted to operate over a range of population sizes, will only be met for populations initially well above MSY level, since populations well below this level will be protected until they are within 10 % of it.

Conclusions

It appears, therefore, that the present procedure of the IWC is fairly satisfactory as a management procedure to reach the most obvious objectives, provided that the population parameters are reasonably accurately known. Its chief weakness, particularly as applied to populations already below MSY level, is that it will make it difficult to obtain the data which are needed to enable the estimates, and therefore the procedures themselves, to be amended and improved. It is, of course, based on the assumptions that in the present state of knowledge, numerical maximum sustainable yield is the most practicable definition of optimum yield, that it is only practicable to manage each species individually, and that changes in population size or structure due to natural or man-induced environmental effects are negligible compared to those resulting from exploitation.

A number of studies have been made, from the days of the Committee of Four Scientists onward, on the effects on a given population of various alternative management procedures. Such studies are generally fairly simple and the techniques on which they are based are well-established. Little attention has, however, been given to the self-correcting characteristics of management procedures, i.e., the extent to which they will provide data which will provide a basis for improving them. The foregoing analysis suggests that there is a real need for studies aimed at identifying management procedures which would produce sufficient data to enable the model to be further improved while at the same time providing for adequate protection of the stocks and for the realization of safe and desirable yields from them.

References

CLARK, C.W., Economic aspects of renewable resource
1976   exploitation as applied to marine mammals. Paper presented to the Scientific Consultation on the Conservation and Management of Marine Mammals and their Environment, Bergen, Norway, 31 August-9 September, 1976. Rome, FAO, ACMRR/MM/SC/65:13p.

A NOTE ON THE STRATEGY OF THE MANAGEMENT OF MARINE MAMMALS

J.A. GULLAND

Abstract

This note discusses the medium-term strategies that might be pursued in managing marine mammals. Any such strategy must take into account *inter alia* the uncertainties in the estimates of population parameters (abundance, mortality rates, etc.), possible long-term changes in human priorities (e.g., the possible direct use of krill), and the desirability for year-to-year changes in regulations (e.g., catch quotas) to be reasonably small so as to allow those affected, especially the industry, to be able to adjust their operation.

Management policies should also be formulated so that the events of the fishery provide as much additional information on the stock, so as to provide better estimates of the optimum population level and sustainable catch. It is suggested that present policies are not likely to satisfy these criteria, and alternative strategies are proposed.

Résumé

Cette note traite des stratégies à moyen terme qui pourraient être appliquées à la gestion des mammifères marins. Toute stratégie de ce genre doit tenir compte, entre autres, de l'imprécision relative aux paramètres de population (abondance, taux de mortalité, etc.), des changements possibles à long terme dans les priorités humaines (par exemple, l'éventuelle utilisation directe du krill), et du fait qu'il est souhaitable que d'une année à l'autre les réglementations (notamment sur les contingents de prises) ne subissent que des révisions modérées, de manière à permettre aux secteurs d'aval — spécialement l'industrie — d'ajuster leurs activités en conséquence.

Les politiques de gestion doivent également être formulées en prescrivant que les opérations de pêche fourniront, sur le stock, des informations supplémentaires qui permettront d'estimer plus précisément le niveau optimal de population et le rendement eumétrique.

Il semble que les politiques actuelles ne soient sans doute pas adaptées pour satisfaire ces critères, et de nouvelles stratégies sont proposées.

Extracto

En esta nota se examinan las estrategias a plazo medio que podrían aplicarse para la ordenación de los mamíferos marinos. Toda estrategia debe tener en cuenta, entre otras cosas, el carácter incierto de las estimaciones de los parámetros de población (abundancia, mortalidad, etc.), los posibles cambios a largo plazo de las prioridades que se fija el género humano (por ejemplo, el posible uso directo del krill) y la conveniencia de mantener dentro de límites razonables las modificaciones anuales de las normas establecidas (por ejemplo, los cupos de capturas), para que los sectores afectados, en especial el sector pesquero, puedan ajustar sus operaciones.

Las políticas de ordenación deben formularse de manera que se obtenga de la pesquería toda la información adicional posible sobre población, para poder preparar mejores estimaciones del nivel óptimo de población y de la captura sostenible. Se indica que probablemente las actuales políticas no se ajustan a esos criterios, y se proponen otras posibles estrategias.

J.A. Gulland
Chief, Marine Resources Service, Fishery Resources and Environment Division, Department of Fisheries, FAO, Via delle Terme di Caracalla, 00100 Rome, Italy

Policy, strategy and tactics of management

The management of marine mammals has been so extensively discussed during the past few years that it might appear superfluous to add anything more. However, the discussions have been mainly concerned either with the general principles and policy of management (should we aim at MSY, MEY, etc.) or at the other extreme of the range of problems, i.e., with arguments on what the quota of, say, Antarctic fin whales should be in the forthcoming year. Less attention has been paid to the middle area of general strategy, i.e., the pattern of policies over a period required to achieve a general objective. This note examines some of the strategic aspects of these middle-term problems of management.

UNCERTAINTIES

The impression that might be gained from some of the discussions of principles is that once an objective (say MSY or optimum utilization of the ecosystem) has been determined, it is then possible (and in fact desirable, if not essential) to devise rather rigid rules which will ensure that the objective is met. This could be a dangerous course, especially if the procedure for amending the rules is complicated or lengthy in practice. Our knowledge of population dynamics is certainly not good enough to allow the development of rules that will ensure the long-term achievement of the objectives. The International Whaling Commission itself provides a good example. Its initial objectives, as set out in the documents establishing the Commission, were excellent and, a quarter of a century before the main flood of environmental concern, foreshadowed many of the ideas. Further, the values (in terms of Blue Whale Units) of the quotas originally fixed were quite close to the combined values of the sustainable yields of the stocks at the time that the IWC was established. Unfortunately, the quotas were slightly too high, and the machinery for amending them was somewhat cumbersome. There was, for instance, no suggestion in the early discussions that the figures for the allowable catch given in the Schedule should be revised each year more or less automatically in the light of scientific findings. The result was that, for a critical period while the stocks of fin whales, and the sustainable yields from them, declined in the fifties, the Commission failed to adjust the quotas, so that by the time action was taken in 1965 the stocks were very low.[1]

---

[1] It is of some interest that one reason for this paralysis on the part of the Commission was because its Convention did not allow it to allocate national quotas. This was done to ensure that the Commission was not distracted by economic or political considerations from its task of conservation. The effect was, in fact, less effective conservation!

SAFETY FACTORS

We have to accept that, in practice, there will always be uncertainty about many aspects of marine mammals stocks including the precise value of their abundance, and of the important population parameters. One approach to this uncertainty has been to apply safety factors to all estimates, or at least to the proposed quotas. Since the more dangerous errors are those that overestimate the abundance and the sustainable yield, it has been suggested that action should be taken on figures somewhat below the best estimate of these quantities. For example, assuming that the objective should be MSY, and that this occurs at 60 % of the unexploited population, a "safe" action would be to maintain the population at 70 % of the initial stock to allow for errors in the estimate of the stock, or of the level at which MSY occurs. This approach, of setting a target figure (of a catch quota, or level of stock abundance) deliberately to one side of the best available estimate of that figure does represent an explicit attempt to deal with the fact that few, if any, of the quantities or relations being discussed are known exactly. Put bluntly in terms of population abundance we seldom know at all precisely what the present population is, what it ought to be if our objectives (MSY or whatever) are to be achieved, or how any specific proposed action (e.g., a catch quota for the coming season) will alter the population. It may further be justified by the possibility that certain other objectives, e.g., improved economic returns or greater ecosystem stability would be achieved at a larger population size.

Tactically this is a reasonable procedure, and in many or most present-day situations could lead to better management actions in the short run. In the long run, if reactions are swift and information on that state of the populations is readily available, the most likely result is that the population will be maintained at a level greater than that corresponding to the best estimate or the chosen objective (MSY, MEY or other).

While this errs on the side of caution, it is silly not to attempt to reach the desired objective directly. We can hope that uncertainty will reduce with time and we should be able to design a management strategy that takes advantage of increased knowledge, and design one that helps to increase it. This note examines some of these possible strategies.

STRATEGIES FOR MANAGEMENT

An important advantage of an empirical or flexible approach over a more rigid formulation is that better account can be taken of uncertainties regarding, for example, the objectives to be pursued, or in the current knowledge of the dynamics of the population or system being managed. (Many of the analyses used so far to determine the values of, say, the annual quota required to achieve a given objective, do not take much account of the interaction between the stock of animals being managed and other animals living in the same body of water). For some time the achievement of the maximum sustainable yield was fairly widely accepted as the proper objective. Now other objectives are suggested. Clearly a rigid management regime which would produce the MSY will not be satisfactory if the objective is to maximize the economic return. Yet more drastic changes in regulations would occur if objectives such as maximizing "non-consumptive uses" — whale watching, etc. — were adopted. Even if the objective is unchanged, it may well be that with additional information, a given policy based on the best evidence available in earlier years will be seen to be far from the best one to achieve the chosen objective.

This note rejects the approach of some theoretical discussion of resource management — that no exploitation should occur until the management policy is established in full detail and that this policy should be based on "complete" scientific study, i.e., that there should be a series of separate steps — scientific research; establishment of management policy; and

95

controlled harvesting of the resource. The argument presented here is rather that research, management and exploitation should be considered as a whole, with continuing interaction between each activity. It follows the lines discussed by Walters and Hilborn (1976) and Silvert (1978) in which research and management are interacting processes.

Changes and uncertainties

CONDITIONS FOR EXPERIMENTAL MANAGEMENT

Traditional analyses of populations for proposing management actions have usually been concerned with equilibrium conditions, and single, steady-state values of quantities such as stock abundance, and sustainable yield. For the present purpose we are more concerned with changes, rates and differences — for example, the size of annual changes in a population that might be caused by exploitation or by natural factors, or that could be detected by different methods of analysis. We are also concerned with the speed with which regulations can be changed — e.g., quotas reduced — taking account of the time needed for scientists to carry out the necessary observations, and for the negotiations and administrative actions to be carried out.

If the relative rates at which the population can change, and at which modifications can be introduced to a management regime[2] are favourable, then management can be entirely empirical. That is, if the population changes slowly, but the effects of exploitation can be quickly detected, and action can be taken quickly to counteract these effects, then little sophisticated analysis is needed. For a

given stock level an initial quota (or other regulation) can be set more or less at random and adjusted up or down according to whether the stock increases or decreases from the initial level occurring when the management policy is introduced. A few years of such trial and error will show what catch can be taken that will leave the stock unchanged, i.e., it will determine a sustainable yield from that stock. Then allowing the stock to be reduced a little will show whether a smaller stock will give a "better" sustainable yield, and further trial periods will permit the determination of the "best" sustainable yield. Clearly, analysis using even a simple model (e.g., the parabolic stock/sustainable yield curve) can greatly reduce the period of trial and error required before the "best" position is located. Provided the reaction time of management is quick enough the stock abundance need not change much except in the deliberate search for the "best" level.

In practice, conditions may not be ideal. Exploitation may very quickly have a severe impact on the stock, or if the impact is less than severe, it may be some time before it becomes effective on the exploitable stock (e.g., if recruitment is affected, there will be at least a delay equal to the period between birth and recruitment); the effect may take time to be detected, and once detected there may be delays before the necessary adjustment to catch quotas, etc., can be made. Any of these effects can result, if the initial regulations are wrong, in serious damage being done to the stock.

RELATION TO MARINE MAMMALS

Fortunately, many of the factors relating to marine mammals and their exploitation and management are favourable to a controlled empirical approach. Variations in population abundance (other than caused directly by man) seem to be very small, and the critical rates (net recruitment, natural mortality and fishing mortality within the sustainable range)

---

[2] "Regime" is used here as denoting the set of regulations (catch quotas, closed areas) in force at a particular moment or during a particular year.

are all low — a few percent per year in the case of whales, and probably no more than perhaps 10-15 % per year for smaller marine mammals compared with natural mortality rates in fish stocks which can reach 50 % per annum in small tropical species. Those concerned with management, particularly of whales, have become well aware of the need for prompt action. Thus, once the need for action has been identified, e.g., as a result of a scientific monitoring programme which is capable of detecting when the population has departed from some optimum level, it is likely that action will be taken before the departure from optimum has greatly increased.

A less favourable factor in marine mammals is the critical one of detecting changes in abundance. When this can be done with a high degree of accuracy, then it will be relatively easy and involve no risk of damage to the stock, to feel the way to the optimum harvesting pattern by trial and error. When detection is slow — e.g., the best estimates of absolute or relative abundance in any one year are subject to large sampling error, so that even a large change in population abundance is only detectable in average values taken over a long period — then the error following an unsuccessful trial could involve a serious decline in the stock.

Nevertheless, it seems worthwhile to examine the possibilities of adopting a less rigid and more empirical approach to the exploitation of marine mammals, and especially to consider the form it could take and the practical steps required, including arrangements for specific scientific programmes and for timely implementation of management action.

Constraints and policy considerations

In establishing a flexible strategy, which explicitly acknowledges the likely need to modify at regular intervals the detailed objectives being pursued and the tactical measures (e.g., the values of annual catch quotas) used to pursue these objectives, 3 broad groups of interests need to be taken into account:

(a) short-term interests of the industry and other current users of the resource;

(b) long-term interests of all future users, actual or potential;

(c) the need to obtain better information on the resource.

These groups are likely to conflict to a greater or lesser extent. The conflicts between short-term and long-term interests are well-known, but it may also be noted that strategies to obtain maximum information are likely to involve large and rapid changes in the harvesting pattern, which will conflict with other interests, especially when the quotas (or similar) are drastically reduced. There must therefore be trade-offs between the 3 groups, as well as between different aspects within groups. Well-known simple strategies can be derived as special cases of these trade-offs. Thus MSY will be the preferred strategy if interests are measured only by catch, long-term and short-term interests are combined by giving equal weight to the catch in each year (for an indeterminate distance into the future), and no account is paid to improving information on the resource. In fact, MSY policies are often put forward with the implicit assumption that the resource, or at least the value of the MSY, is known perfectly.

INDUSTRIAL ADAPTABILITY

The returns to short-term or long-term interests are determined by several factors, of which the magnitude of the catch is only one. In a given season commercial interests need to pay attention to such things as costs of inputs, prices of products, etc., as well as to gross production. From year to year the changes in production are as important as the absolute level. Over the last decade whaling interests have found themselves adapting to levels of allowable catch which a few years previously would have been considered catastrophic and unacceptable. This does not imply that their earlier protestations were unjustified. Rather it shows that people and organizations can adapt to most changes if the period for adaptation is adequate. A sudden reduction of half or more in total income in a single year (the implicit result of some of the proposals made to the IWC) is not easily accepted by anyone. Sudden *increases* in catch quotas are not so obviously objectionable, but any sudden feast of this type can lead to industrial or economic indigestion. An important objective within the short-term considerations should therefore be to minimize the year-to-year changes in the regulations, e.g., to ensure that the catch quota is not changed by more than 10 % in any year.

LONG-TERM OBJECTIVES

Long-term considerations have been discussed at length elsewhere. Here it is sufficient to note that there are a number of quite different objectives – including maximizing catch, net economic return, or employment[3], ensuring the stability of the ecosystem, "optimizing the contribution of marine mammals to that ecosystem" (a concept that is not easy to translate into concrete policy) – some of which are less quantifiable than others. It may be assumed that any balance between objectives that might be agreed in 1975 is not the same as that which might be agreed upon in 1985 or 2075. If we cannot be sure what the precise objectives will be in the future, it is useless to try and set precise rules in order to

reach them. However, it should be noted that most objectives will be most easily achieved if the population abundance is high.

RESOLUTION

Discount Rates

The main conflict in fisheries between short-term and long-term interests occurs because most fisheries are open access, so that the individual fisherman has little opportunity to affect his long-term catches. With few exceptions any animal which he does not capture when he can will be caught by someone else in the future, i.e., the trade-off is between his own short-term interests and the long-term interest of someone else. However, even the existence of a single owner of a resource would not resolve all the problems of balancing short-term and long-term interest. Partly this is because the timing of benefits (e.g., the value obtained from harvesting a fin whale) has an influence on the real worth of these benefits. The sooner benefits are available, the more useful (in general) they are likely to be. The economic aspects of timing can be dealt with by using an appropriate discount rate and the value of this rate can influence the management regime adopted to pursue certain objectives. For example, C.W. Clark (1973, 1976) showed that if the felling policy for a long-lived forest is determined by discounting future harvests at some reasonable rate, then the average long-term harvest (assuming instant replanting of trees after felling) is less than if a zero discount rate is used. This is true if all the net income is used outside the forest, though if the discount rate is realistic the economy as a whole should feel better off. However, if the money that is available earlier from cropping trees when still comparatively small, is invested in developing forest technology, applying fertilizer, pest control, etc., it is probable that the average long-term harvest will be increased. The same principles apply to harvesting

---

[3] Maximizing employment is a bad long-term objective, since with a limited resource, it puts a ceiling on the benefits received by each person employed. In the short to medium term, until alternative occupations are developed, it is often an important consideration.

marine mammals. Though it is not so obvious where investment from current harvests can be used productively for improving future whale harvests, taking the wider view there are many places, especially in developing countries, where it would be more desirable to use income generated now rather than having to wait for some undefined period in the future. Even if marine mammals and their ecosystem are viewed in isolation, it is clear that money spent on research now should lead to better management in the future, and the need to finance this research (assuming, as would be reasonable in a logical world, that a proper method of financing this research is from the proceeds from current harvests) provides a reason to give some greater weight to present in favour of future catches. That is, in determining long-term strategy a non-zero discount rate should be used.

Value Changes

Another factor that makes balancing long-term and short-term interests difficult is that the perceived value of a particular whale species can change. For example, a fin whale caught in 1965 was worth more than one caught in 1950 because more use was made of the meat and other products, whereas earlier the oil was the main product. One factor that could greatly change the perceived value of a whale in the future is the use of other elements of the ecosystem. If even the more conservative estimates of the potential harvest of krill from the Antarctic are correct, and the technological and economic problems of large-scale harvesting and use of krill are solved, the harvesting of krill (? 100 million tons per year) will be much more important to mankind than that of Antarctic whales (a maximum of ? 2 million tons). This is likely still to be true even if krill has to be fed through battery chickens (or battery salmon or sole) to be acceptable. The conversion efficiency of some of these animals with present feeds is high (20 % or better). Probably the efficiency

of use of krill would be lower because of the high chitin content, but it would almost certainly still be considerably higher than that achieved through whales. The efficiency of whales as convertors of krill to human food may be somewhat different from the 2 % implied by the figures above but is certainly low. Lockyer (1976) has examined the energy budget in detail and found that the growth efficiency varied between 33 % (for suckling calves) to 1 % (for adults). If we have to give up eating beef so that the grain used for cattle fattening can be used for feeding people, the same may be true in the sea, at one further stage in the food-chain. We can hope that careful krill harvesting and whales can co-exist, but the human valuation of a whale would then include its effect as a predator or competitor, as well as its direct economic or aesthetic value. Again we must accept the possibility of quite radical changes in policy objectives, and hence in at least tactical decisions.

INFORMATION NEEDS

The need to improve the information on the resource has received little attention as a factor in determining management policy. Since the present knowledge of the resources of marine mammals is far from good, it seems highly desirable, other things being approximately equal, to choose management policies that offer the best chance of improving that knowledge. The information is likely to be increased by policies that allow large changes in population. Such changes should generate the density-dependent changes in the parameters of the population (mortality and pregnancy rates, age at maturity, net reproductive rate, etc.) for which estimates are needed to be included in assessment models. In practice most policies currently adopted minimize changes, and therefore do not generate new information. Except for the stocks that were accidentally depleted (and too often even for these the vital data were not collected at the time), we do not know enough about

99

how population parameters vary with abundance to determine what the optimum population size and harvest would be to accord with any given strategy. For example, the belief that MSY occurs at half the unexploited population is only the consequence of one simple model, with little, if any, observational basis, and other beliefs that MSY occurs at 60 % or 80 %, have little, if any, stronger foundation. If we wish to be sure of achieving close to MSY in the future, we should now be allowing the population to vary across the likely range of the optimum level in order to define it adequately. This implies that in some recent years the population will be appreciably different from the optimum and we are again involved in a trade-off between current and future values — to what extent can we accept a less-than-perfect situation now in order to get a closer-to-perfect situation in the future. It is improbable that the "best" trade-off is definable, but some degree of trade-off seems desirable. It does not seem to be the best policy to try to get immediately as close as possible to the estimated "perfect" situation at the expense of better future estimates. A promising form of trade-off is to follow the suggestion that, taking advantage of what appears to be a fair degree of independence of sperm whale stocks in the Antarctic, each of those stocks should be exploited, over the next few years, deliberately at different rates. This question has been recently discussed by Silvert (1978), who points out that in the absence of management actions designed to increase the knowledge of the resource, the exploitation pattern may be frozen permanently in a sub-optimal condition.

A dynamic strategy

Given this background, an outline of a dynamic management strategy can be proposed. It could consist first of a proposed regime (annual catch quotas, etc.) based largely on the best current estimates of population abundance, sustainable yield and present views on what immediate and long-term objectives should be; second, of a well-defined system for making adjustments to this regime each year in the light of experience of the past year.

The type of approach may be made clearer by a hypothetical example. Suppose an unexploited stock is believed to contain 100 000 whales, and the generally accepted objective is MSY, which is estimated to be 3 000 animals, occurring at a stock abundance of 60 000. Further, it can be accepted that there could be errors of factors of up to x 2 in the estimation of the original stock size and of the ratio of MSY to the stock abundance. Further, it may be known that, for operational reasons, changes in the quotas of successive years of more than 500 animals are difficult for the industry to accept.

Neglecting the possibility of error in estimating abundance but accepting some uncertainty in MSY, a reasonable programme might be to reduce the stock fairly rapidly to perhaps 40 000 or 50 000 animals, and then cut the quota to allow a recovery to around the MSY level, which might thereby be quite well estimated, as might be the level necessary to achieve any other objective (e.g., net economic return) that might receive greater attention in the future. The detailed programme might be catches of 5 000 whales for 10 years, reduced in steps of 500 to, say, 3 500 until the planned stock reduction had been achieved (perhaps in a further 10 years, depending on the pattern of sustainable yield as a function of stock), followed by a further phased reduction of quota to 2 000 until the stock is at or close to the MSY level. At this stage or earlier, a further programme could be elaborated to pursue more closely the objective of MSY, or some other objective, or to define more precisely the population parameters. This initial programme, if not adjusted, would run into trouble if the true stock were only 50 000 animals and the MSY only 2.5 % of the population. How much trouble would depend on how quickly the error in estimation is detected and acted on. It could well be that detection would not occur for 5 years and correction involve another year; by then, to prevent the stock falling to the possible danger level would require an undesirably drastic cut in quota. That is, the programme does not guard adequately against the possibility of error. On the other hand, scaling the whole programme down to match the most pessimistic estimates would be unnecessary. An intermediate programme, e.g., quotas of initially 3 000 whales should not involve a danger, even if the worst hypotheses were true, but be closer to the programme that would be appropriate if the best estimates were true. The best value of the catch in such a programme could be defined as the largest that would allow action to be taken to prevent the stock falling to the "danger level" even for the worst possible values of likely population abundance and other parameters, for the slowest likely speed of detecting changes and acting on them, and without requiring that annual quotas change at more than the acceptable speed. Exceptionally (with quick reaction) this may be larger than the catch determined as desirable using the best estimates. In that case no modification of the programme would be needed to guard against disaster.

In the following year the knowledge about the resource should be improved, and the views on immediate or long-term objectives might have changed. This will require consideration of changes in the management programme. An essential element of a dynamic strategy is that these changes — particularly those related to additional knowledge of the resource — should be automatic. That is, the programme set out in previous years (specifically the levels of the catch quotas) would only remain unaltered if there were positive evidence that the estimates of population size, etc., on which they were based were still reasonable. This implies, as suggested by Larkin (1972) that the scientific advice on management (e.g., catch quotas) contains an explicit and detailed prediction of what would happen. In the example here, one part of the prediction would be that the stock would decline by 5 %

per year for the first few years. Further, this prediction should include a note on the precision with which changes in abundance can be detected, from catches per unit effort or otherwise. If this were $\pm$ 10 %, the estimates of population in successive years would be expected to fall within the following ranges (as percentages of the original abundance) 85-105; 80-100; 75-95, etc. If the estimate falls outside this range, then the catch quota is adjusted upward or downward by a unit step of 500 whales. In practice, rather more complicated decision rules would be needed, for example, to take account of trends over one year, or to allow for different sizes of quota adjustments (within the acceptable limit of 500).

The difference in approach suggested here may be seen with respect to some of the current discussions concerning the management of whale stocks in different states ("protected", "initial management" or "sustainable yield"). For protected stocks no immediate change would be suggested for the badly depleted stocks — the faster they are restored the better — but as they approach the "sustained yield" range, a dynamic management policy would tend to favour an earlier re-introduction of moderate harvesting, so as to obtain better information from the lower sizes of population abundance, as well as to avoid the practical economic difficulties involved in a sudden switch in catches from zero to around MSY level. Conversely, for initial management stocks a more cautious approach, i.e., smaller catch quotas, may be desirable for similar reasons — to get more information and to avoid sudden changes in quota, especially when the initial information on the resource is poor.

The biggest difference concerns stocks around the sustained yield level. For these there are often arguments as to whether the catch quota — presumed to be set equal to the sustainable yield level — should be equal to, say, 7 000 or 7 500 animals. Clearly, if the sustained yield is truly 7 000 animals (with a population in the 70-140 000 range), little dam-

101

age would be done to the stock by catching 7 500 for a few years. Equally, if the sustainable yield is truly 7 500, the industry would not — other things being equal — suffer great economic hardship if restricted to 7 000. In fact, other things may not be equal. It is only possible to send an integral number of expeditions to the Antarctic and 7 000 whales may be too much to be harvested conveniently by 2 expeditions, but not provide an economic living for 3. In such a situation, the economic constraints on quota changes may be more complex than "not more than a 500 whale change in any year"; perhaps, for example, they should be equivalent to adding or subtracting 1 expedition and should be decided 1 year in advance.

In these circumstances, both exploiters and conservers should not be concerned if the quota differs from the sustainable yield (or even if it is below or above it), *provided* the procedures exist for automatic adjustment before things depart too far from the "optimum"

situation. For example, if 7 500 whales are definitely more economic as an annual quota than 7 000, the former might be set as the current annual quota, even if the best estimates of sustainable yield were slightly less than 7 500, provided that there was an automatic procedure to reduce the quota to, say, 5 000 (the best economic quota with 1 less expedition) before damage could be done to the stock — say, as soon as the catch per unit effort (adjusted as necessary to allow for changes in efficiency) fell to less than, say, 90 % of the current value. Possibly it would never do so if the present estimate of sustainable yield were too low.

Acknowledgements

I would like to thank S.J. Holt and C.W. Clark for helpful comments on the first draft of this paper.

References

CLARK, C.W., The economics of overexploitation.
1973   *Science, Wash.*, 181:630-4.

—, Economic aspects of renewable resource exploita-
1976   tion as applied to marine mammals. Paper presented to the Scientific Consultation on the Conservation and Management of Marine Mammals and their Environment, Bergen, Norway, 31 August-9 September 1976. Rome, FAO, ACMRR/MM/SC/65:13 p.

LARKIN, P., A confidential memorandum on fisheries
1972   science. *In* World fisheries policy, edited by B.J. Rothschild. Seattle, University of Washington Press, pp. 189-97.

LOCKYER, C., Growth and energy budgets of large baleen whales from the S. Hemisphere. Paper
1976   presented to the Scientific Consultation on the Conservation and Management of Marine Mammals and their Environment, Bergen, Norway, 31 August-9 September 1976. Rome, FAO, ACMRR/MM/SC/41:179 p.

SILVERT, W., The price of knowledge: fishery manage-
1978   ment as a research tool. *J. Fish. Res. Board Can.*, 35(2):208-12.

WALTERS, C.J. and R. HILBORN, Adaptive control of
1976   fishing systems. *J. Fish. Res. Board Can.*, 33(1):145-59.

OBJECTIVES FOR THE MANAGEMENT
AND CONSERVATION OF MARINE MAMMALS

J. Gordon Clark

Abstract

Sixteen possible objectives are identified for the management and conservation of marine mammals, and are grouped as product oriented (4), ethical (3), educational, recreational and scientific (3), ecosystem conservation (1), democratic (1), marine mammalian (3), and evolutionary (1). Several of these objectives appear to be mutually incompatible. Existing management bodies appear to be inadequate to deal with the wide range of objectives as identified, and the current understanding of mammal populations and ecosystems is insufficient for any proper management, with the result that management has failed in its specified objectives, such as maintaining whales at their MSY level. More weight should be given to the non-product oriented objectives of management, and management should be directed toward maintaining a number of specified population characteristics. It is expected that non-product oriented use of marine mammals will increase.

Résumé

Seize objectifs possibles pour la gestion et la conservation des mammifères marins sont recensés et groupés en fonction des critères suivants: orientation vers les produits (4), éthique (3), éducation, récréation et science (3), conservation de l'écosystème (1), démocratie (1), mammifères marins (3) et évolution (1). Plusieurs de ces objectifs paraissent mutuellement incompatibles. Les organismes chargés actuellement de la gestion ne semblent pas en mesure de traiter d'une gamme d'objectifs aussi vaste; par ailleurs, les connaissances actuelles sur les populations de mammifères et les écosystèmes sont insuffisantes pour permettre une gestion appropriée, de sorte que celle-ci n'a pu atteindre les buts précis qu'elle visait, par exemple, le rendement maximal équilibré pour les baleines. Il conviendrait d'accorder plus de poids aux objectifs ne s'identifiant pas à des produits, et la gestion devrait s'orienter vers le maintien numérique de diverses populations aux caractéristiques bien précises. On escompte que l'utilisation des mammifères marins à des fins non productives s'accroîtra.

Extracto

Se señalan 16 objetivos posibles de las actividades de ordenación y conservación de mamíferos marinos y se agrupan en varias categorías: orientados hacia la producción (4), éticos (3), educacionales, recreativos, y científicos (3), ecológicos (1), democráticos (1), relacionados directamente con los mamíferos marinos (3) y relacionados con la revolución (1). Varios de estos objetivos son incompatibles entre sí. Los órganos de ordenación existentes no están en condiciones de ocuparse de la amplia gama de objetivos individuados, y los conocimientos actuales sobre las poblaciones de mamíferos y los ecosistemas no bastan para tomar medidas adecuadas de ordenación, debido a lo cual ésta no ha logrado sus objetivos específicos, como, por ejemplo, mantener las ballenas a su nivel de rendimiento máximo sostenible. Debe concederse más importancia a los objetivos de ordenación no orientados a la producción, y los trabajos de ordenación han de encaminarse al mantenimiento de determinadas características de la población. Se prevé un aumento del aprovechamiento de los mamíferos marinos para fines distintos de la producción.

J. Gordon Clark
Old Post Office, Huntingfield, Halesworth, Suffolk, England

The aims of management must be clarified and the conflicts between various objectives must be identified, together with the areas of research required, before scientists can develop the management strategies needed to satisfy these objectives.

This paper selects the main objectives that are generally agreed to apply to marine mammals, with some emphasis on whales. It discusses some management methods and makes several practical suggestions designed to correct the imbalances that exist between the achievement of certain objectives and others.

Objectives leading to the overexploitation of species or the unwise use of the physical and other resources they provide receive little sympathy. In common with the United Nations Environment Programme, and many nations, organizations and individuals, the author supports the need for a 10-year halt to commercial whaling. However, in writing this paper for the Working Group on Objectives of the Consultation, an attempt has been made to ensure that the data selected and the arguments used are unbiased. As Meeker (1976) points out, everyone has their own prejudices, and those who believe passionately in economic growth, the barter system, free enterprise, state control, the purity of science, or the existence of UFOs, will have equally great prejudices; the important thing is to recognize it. No scientist or administrator working in international resource management can nowadays claim to be non-political.

The few papers on objectives that have been written for this Consultation appear to have considered chiefly product-oriented objectives. To restrict our thinking on management to a view of marine mammals as mere producers of meat, fur or oil, not only offends a very considerable part of human (and animal) life and perception, but also belittles the importance of areas into which science is now proceeding with increasing confidence – areas such as the communicating powers of whales, their ethology and intelligence.

An approach to the natural world which views it simply as a supermarket is likely to lead to poor conservation and management, if only because aspects of nature other than its obvious capacity to produce will be ignored, and with them some of its important properties. The attitude of the exploiter is important. Societies that have been dependent on an animal – for example, the "beaver economies" – have seldom, if ever, overexploited them. They have made maximum use of animal products and have treated the animal in question with a respect amounting almost to worship. We cannot claim, of course, that such an attitude was the only factor that prevented overexploitation – lack of technology may have been another – but it certainly contributed to a balance between predator and prey.

Objectives for marine mammals may be classified as human (anthropocentric), marine mammalian (those that marine mammals might be considered to have for themselves) and evolutionary. Some objectives may instinctively be dismissed as "unrealistic" by those whose training and employment have conditioned them to consider only the greater or lesser constraints of industrial desires. But our idea of "reality" is always changing and last year's impossibility often becomes this year's discussion topic. If scientists never consider the means of attaining new and perhaps unusual objectives, then those objectives "realistically" become harder to attain.

Anthropocentric objectives

These are arranged in order of those objectives concerned with product yield, ethics, education, recreation and science, conservation of ecosystems, and the equitable distribution of benefits from marine mammals.

PRODUCT-ORIENTED OBJECTIVES

Maximization of short-term profits from marine mammal products

Clark (1976) has shown how industry may profit more, given certain combinations of investment, discount rates and renewal rates of the resource, by eliminating a series of populations or species than by harvesting them on a sustainable basis. This is illustrated by present-day pelagic whaling where capital is invested in (ageing) ships taking whales from depleted or decreasing whale populations. The revenue from the *sustainable* yield of a depleted stock appears to be low or non-existent. One delegation to the International Whaling Commission (IWC) has pleaded on grounds of economic necessity for quotas not to be reduced, implying that the science was irrelevant, and illustrating a lack of interest in sustainable yields. Certain shore-based whaling

operations also pursue this objective, when investment in whaling equipment and ships has increased and whaling grounds have moved further out from shore as local stocks have been depleted. Sealing operations in the past have also fallen into this category and, in addition, pirate whaling and sealing operations clearly operate on this principle.

In every case any profit is strictly short-term because the populations are not able to sustain the level of kill demanded (1).

Maintenance of a maximum sustainable yield of marine mammal products

This would ensure that return on investment is steady within the variations of product prices and that return on future investment is assured, as is employment and the conservation of the mammals populations. In theory, the concept of "maximum sustainable yield" (MSY) would allow this objective to be fulfilled, while leaving options open for future product users to continue taking such a yield without having to stop some or all exploitation to allow populations to recover from a depleted state. In practice, this does not happen, as will be discussed later.

The declared objective of the new management procedure of the International Whaling Commission is to maintain whale populations at their MSY level. It should be noted here that this objective requires a very fine adjustment between "maximum" and "sustainable" which we are currently far from achieving in practice. The objective could still be that of some shore-based sealing and whaling operations, where they are the only exploiters of the populations involved (2).

Obtainment of certain essential products from marine mammals at a level commensurate with the need

It is clear that if mammal products are essential, so is conservation of the available

105

populations from which they are taken. Certain groups of people may be said to hold this objective: in particular, discrete cultures dependent to a greater or lesser degree on the marine mammals involved. The Eskimos of Greenland and Alaska have depended in the past on whales, seals, walrus and other mammals for their food, construction materials for their boats and many other products integral to their lifestyles. It does not appear that the populations of these species were affected in the past by the level of kill. Now, however, whale numbers have been reduced by the developed cultures' commercial whaling; local technology has coincidentally changed; local lifestyles are changing, and with them the degree of dependence.

One other group may appear to hold this objective (although in reality it does not): developed societies requiring a particular mammal product for certain "key" aspects of their technology. While it appears that no whale product is at present essential to such a group, we may cite the cases of the Polaris missile makers and the UK leather tanners, who claim to require sperm-whale oil as a lubricant and a softener, respectively. Synthetic substitutes have, in fact, been developed for both purposes, indicating that few, if any, technological requirements are truly "dependent" on such products. It is more accurate to say that the mammal products were available at the particular time the technologists were seeking a particular item.[1]

Need in this context refers therefore to the need of a cultural group as a whole to be able to take certain essential products from the mammal populations. As cultures change the problems of defining such need are considerable (3).

---

[1] A discrepancy appears to arise if the user-group claiming to "need" the product is not responsible for obtaining it. If, as seems to be the case, whalers operate in pursuit of the first objective (maximum short-term profit), then the users requiring a truly essential product will be ill-served. Indeed, as sperm whales have been depleted, the price of sperm-whale oil has risen in the UK (by as much as 72 % in a single 12-month period).

To reduce the take by marine mammals of species from which a product yield is desired

Fishermen often seek to reduce certain marine mammals because they believe that their own fish catch is affected. In some cases (dolphins killed by Japanese fishermen) there is little interest in the product yield from the marine mammals; in others (seals taken in the UK) the products are used. There is generally insufficient evidence to show that the marine mammals, rather than man, are adversely affecting fish populations. Sergeant (1976) has suggested, however, that this objective will become more widely held and that, for example, fin whales will be seen as pests by capelin fishermen.

As Sergeant and Stirling (1976) point out, pressures are moving away from marine mammals to the lower levels in the food web — krill in the Antarctic offers a much greater theoretical yield than baleen whales. Competition from marine mammals is only seen as relevant when the early catch levels become less easy to maintain. Nonetheless, even at this stage the fishery may harm the mammals through their food supply, especially where they have already been reduced to critically low levels or are under heavy exploitation pressure. The failure of the management scheme for the North Pacific fur seal for a period of years may have been caused by a change in killing policy coinciding with a developing Alaska pollack fishery.

Because the potential yield of krill is very much larger than that of Antarctic baleen whales, profits are also likely to be higher. In consequence, the conservation of Antarctic baleen whales is likely to be seen as irrelevant to those taking krill (4).

ETHICAL OBJECTIVES

It is sometimes suggested and often implied, that to hold certain ethical objectives in relation to animals that provide consumer goods is a luxury in which only a well-fed, sentimental, Western-type culture can afford to indulge. This is not true, but even if it were, such objectives must still be valued. It is certainly true that some industrial objectives are those of a well-fed, profit-oriented, industrialized business class: the cases of pelagic whaling and harp and hooded sealing are examples. We should not forget in this discussion that for religious reasons a very large number of people in developing countries do not kill or eat any animals.

No marine mammals should be killed

Either because of a respect for animal life in general or because marine mammals are thought to be particularly interesting in view of their evolutionary kinship to mankind and their marine development, this objective is held by a large number of people. In addition, the appeal of a furry seal or a playful humpback whale is generally greater (perhaps because of their warm blood) than that of a spider or a crocodile (5).

No cruelty should be inflicted on marine mammals

The particularly unpleasant methods used to kill whales — explosive harpoons, and in the case of the minke whale, cold grenades — disgust many people and there is a very strong anti-whaling movement for that reason alone. A further form of cruelty also occurs when social groupings of intelligent and communicative animals are disrupted by exploitation, or when single dolphins and killer whales are kept in captivity.

The effects of cruelty are not limited solely to the animals in question

The behaviour exhibited by a person toward other people, animals and things will af-fect that person's own spiritual and psychological development. Thus American soldiers repatriated from the Vietnam war tended to exhibit violent symptoms, having been affected by the violence they had inflicted on others. To be cruel or violent, whether directly or through an agent, is to encourage cruelty and violence to spread (6).

Animals with intelligence comparable to our own should not be killed

Disregarding the argument that the more intelligent an animal, the more it is likely to suffer, this objective focuses on the special nature of intelligent animals. If, as the anatomical evidence leads one to suppose (as does the capacity for play and continuous communication between individuals), toothed whales are highly intelligent — perhaps as intelligent as ourselves — then the only moral course is not to kill them. This applies particularly to the sperm whale, the orca and some of the dolphins.

An increasing number of scientists, notably those concerned with cetacean behaviour and brain anatomy, hold the view that these animals are indeed comparable to ourselves in intelligence: to kill them would be tantamount to murder in human society. It certainly appears that the direction of evolution has been toward the development of increasing consciousness in mammals as a tool for surviving environmental crises. Mankind has much to learn from the animals which have evolved intelligence and consciousness in an aquatic environment and there is a real possibility of communication with cetaceans.

This whole objective is affected by the "special" feeling that people develop for animals in which they recognize similar traits to their own, and it is thus not surprising that the ability in question is called intelligence, and is measured in ways that we measure our own. It is this same sympathy that makes it as abhorrent for some people to discover that sperm whales are being turned into leather softeners

107

and lipsticks as it was for much of the world to discover once that people were turned into soap (7).

EDUCATIONAL, RECREATIONAL AND SCIENTIFIC OBJECTIVES

People should be allowed undisturbed observation of unexploited marine mammal populations

The enjoyment of watching whales, without the handicap of knowing that they are being exploited, and the study of the natural history of undisturbed whale populations, is increasingly sought by scientists and lay people alike. In addition, much valuable scientific information can be gained from the study of undisturbed whale populations (8).

Marine mammal populations should be available for entertainment

Increasing numbers of films are now being made about whales, their natural history and conservation, and probably earn a larger revenue than the current trade in whale products by whaling industries. In addition, considerable local employment is provided through organized whale viewing by tourists (9).

Educational and scientific benefits from the study of marine mammal populations should be maximized

The study of marine mammals is vital to understanding the ecosystems of which they form a part; and the study of intelligence in cetaceans is of prime importance in understanding the aquatic evolution of intelligence and consciousness. The study of live marine mammals in their habitat is a seriously neglected area of science, and this requires immediate remedy for the proper fulfilment of this particular objective. Depending on the branch of science involved, however, achieving this object may call for very different management strategies – varying from killing large numbers for statistical purposes to trying to communicate with undisturbed whale populations at sea to gain a better understanding of their social behaviour (10).

THE ECOSYSTEM CONSERVATION OBJECTIVE

Marine mammals should be so managed that the ecosystems they inhabit are undamaged

In this objective, marine mammals are seen as an important part of a more important whole. To damage either marine mammals or their ecosystem is to damage both, and the objective reflects the recent extension of awareness of the nature of biological systems. Our understanding of ecosystems is still poor and the implementation of this objective presents a very demanding task given the present state of the natural sciences. It is suggested that to damage an ecosystem is to affect it in such a way that its ability to resist disruptive perturbations is reduced because the homoeostatic mechanisms operating within it are weakened or destroyed.

In general, a system consisting of many pathways for low-level energy flow will tend to be more stable than one consisting of a few high-energy pathways. The consequence of this is that the more diverse the species existing within an ecosystem, the more likely it is to be stable. The objective, therefore, requires that marine mammals should be allowed to return to areas where they once occurred, assisted if necessary by restocking, and that the greatest number of geographically separate populations and the greatest number of species should both be maintained.

A *reductio ad absurdum* is sometimes

raised along these lines: "We cannot let the ecosystem return to its pristine, unexploited state without simultaneously stopping every aspect of exploitation. Should we then stop fishing?" This argument ignores the following points.

If imbalances have been created in ecosystems it is most relevant to pay special attention to recovering the top predators, and particularly those with long generation periods (such as whales). Fish populations – unless very drastically overfished – are more capable of quick recovery, but to re-establish the balance of the Antarctic ecosystem, for instance, it is clear that special attention must be given to whales. In addition, there are a great many areas where fishing has had to be stopped because of exploitation – the recent dramas surrounding North Sea fisheries are strong evidence of this. A halt to fishing in certain areas and for limited time scales is no longer a protectionist's fantasy but an economic necessity.

We are not at present able to predict the consequences of deliberate manipulation of overexploited ecosystems, and currently far more is risked by adopting such a course than by simply attempting to ensure recovery of the most ecologically important species (11).

DEMOCRATIC OBJECTIVE

The benefits from marine mammals should be available to all, and decisions affecting them should be taken by all

Many marine mammals occur in ocean areas which are difficult to reach, and others are exploited on the high seas where little effective control over exploitation can be exerted. Thus, not only are marine mammals' rights to the seas denied but the rights of most of the world to cetaceans are also denied. This conflict is seen, for instance, in the allocation of quotas for whales between Brazil on the one hand, taking whales from the same breeding population as Japan and the USSR on the other hand. Overexploitation by pelagic fleets can reduce the profitability of a developing nation's land station. Canadian and Norwegian kill of harp seals affects the Greenland take.

Argentina is developing a tourist industry around the right whales it has protected within its 200 mile EEZ, but a single pirate whaling ship operating outside that zone could completely eliminate this new industry. Squid fishermen from small Pacific nations may benefit one day from knowing where sperm whales find their food, but if sperm whales are exterminated from that area, there will be no such benefit (12).

Marine mammalian objectives

People can only guess at the objectives held *by* marine mammals *for* marine mammals. We must assume, however, that such objectives include equal rights to: life, unhindered social development, and an undamaged habitat wherever they may have been or might be (13), humane treatment (14), and an equal or greater say in the management of the ecosystem of which they form an integral part (15).

Evolutionary objective

TO MAINTAIN MARINE MAMMAL
GENE POOLS AND DIVERSE
ENVIRONMENTS

Successful evolution needs the right conditions: a large gene pool and varied environmental conditions. The human race may destroy itself – at least as the dominant species – and many land environments and species with it, and the sea may then prove to be

109

then main reservoir for evolution. Marine mammals are already a unique evolutionary phenomenon, by reason of their adaptation of mammalian physiology to marine life. In addition, some of the whales represent a peak of intelligence in marine life (by conventional measurement), and one of the few comparable to our own. Unlike humans, they do not appear to have caused habitat destruction or the extinction of other species. It cannot be stated that an evolutionary objective would aim at ensuring that marine mammals (or at least the intelligent ones) be conserved in as varied a form as possible, but the possibility should be recognized.

The satisfaction of this objective would require (among other things) that genetic entities such as breeding populations should not be eliminated and that marine mammals should exist in habitats as widely diverse as possible. How else would the river dolphins have evolved (16)?

Compatibility of objectives

This section points out the areas where objectives defined earlier may conflict in terms of their impact on the system or species. However, it should be noted that when a method adopted to achieve one agreed aim is faulty, it may cause conflicts with other aims — conflicts that theoretically do not exist. For instance, a careful take from discrete breeding populations would not necessarily interfere with the process of evolution, and that take could theoretically provide a maximum sustainable yield, but if the definition of breeding populations is wrong and the calculation of take for MSY is wrong, then the original evolutionary objective cannot be achieved.

Of the product-oriented objectives, (1), (2) and (4) would appear to conflict with all the other objectives. Maximizing profit and the requirement of maximum sustainable yield from a population may not always conflict with other objectives, but they imply that all other considerations are secondary, including the understanding of marine mammals societies; communication with dolphins; aesthetic, tourist and scientific observations of marine mammals in their mature, unexploited states; the need for maintenance of the ecosystem and for a maximum potential for evolution; and the desire that no marine mammals be killed, or that the killing method should at least be humane.

The desire for an "essential" product from marine mammals (3) conflicts only with objectives involving total protection, or humane killing, or (where populations have already been depleted) with the prevention of extinction. Conflict here arises with the definition of need. Except for subsistence whaling there does not appear to be any whale product that is essential to human beings for food, shelter or clothing — the proportion of protein represented by whale meat in the Japanese diet in 1973 was 0.8 % for instance. Many of the world's food problems are attributable to poor distribution, not to an overall shortage of production, and the worst starvation occurs in countries with little power to import food. Even subsistence cultures are changing so that their populations grow (and with them a possible need for marine mammals meat), but they simultaneously become less dependent on such traditional food sources.

The objective of killing no animals conflicts with all product-oriented and some scientific objectives; that of not killing intelligent animals would make an exception of sperm whales and some toothed whales.

If any exploitation of marine mammals were to occur only as part of an ecosystem exploitation that did not create irreversible changes in that ecosystem; if killing were humane, and exploitation did not affect the widest possible distribution of marine mammal populations (particularly in those areas where people wish to go to see them); if the more intelligent of the species were protected; if all those with an essential need for marine mammal products were able to obtain them; and if all people had a say in the exploitation and management of marine mammals; then the only objectives that would not be satisfied would be those of total protection and those of quick profit or maximum sustainable yield. In addition, the options for future generations would be left wide open.

This may seem unrealistic, but as opined earlier reality is no more than way of looking at life and every individual's reality is different. The consensus reality changes every year, frequently depending on the movement of large sums of money.

It is necessary at this point to put the demand for food in perspective because every conservationist is at some time or other charged with starving the world's partly because we have been spending food capital, not interest. It is also because of the huge discrepancies in goods and income between rich and poor. As with all resource systems, there are natural limits and at present it looks likely that we are approaching those of the sea. Most conventional food fish in the North Sea and North Atlantic have been overexploited, and the Northwest Atlantic capelin fishery may already be beyond its limits. The Bering Sea system is appearently in the same condition and no doubt there are other examples. Man is now contemplating taking krill simply because it is available but there is no evidence that the take will benefit the world's hungry people.

Desirable population levels: appropriate management strategies

Clearly there are some objectives that conflict irreconcilably at present and the best aim must be to resolve these in the fairest way possible. Most existing institutions — such as the International Whaling Commission — have developed as a means of restraining exploitation to levels that satisfy the exploiters, and as such are inadequate to deal with the much wider range of objectives identified

here. One step toward improving this situation would be to ask the United Nations to agree on the objectives that must be satisfied.

Two further factors are needed to guarantee good management: one, that the political structure within which management decisions are made is adequate to balance the conflicting objectives; and two, that the science on which those decisions are based has a sufficiently secure data base and assumptions to avoid errors that could cause irreversible change or other undesirable long-term consequences.

It has been suggested in the past by Gulland that whales should be made the property of the United Nations; Clark (1976) suggests that a Marine Mammals Authority (with similar powers to the Sea Bed Authority originally envisaged for mineral exploitation) should be set up, and he goes on to show that such an authority is more likely to manage successfully the exploitation of cetaceans for products than any individual industry. The case for whale management by a UN body (see Friends of the Earth International, 1976) has been repeated more recently. A further suggestion is that the value of marine mammals of a given species, age, sex or tribal status should be quantified, and the exploiter required to pay accordingly.

Certain general remarks about management apply whatever the political system may be. In the past, we have tried rather too hard and rather too arrogantly to "manage" in fine detail, at a time when our understanding and ability to maintain a consistent policy has been limited. Faulty management can result from one scientist, one manager being replaced within a short period; by overnight changes of industrial requirements and markets; by a country being invaded; a management body replaced; a highly toxic chemical spilled; or by a new fishery starting up and rapidly affecting the whole local system. Detailed, "know-all" management requires consistent and careful follow-up which no one is in a position to give.

The second and perhaps more telling drawback is that current understanding of pop-

111

ulations and of ecosystems is, contrary to popular opinion, in its infancy. The mathematical techniques are sophisticated, but the data they process are not, and even the most thorough mathematical analysis can present a picture that is wildly different from the actual events in the systems being "managed".

Paper after paper admits our current ignorance of whales, and not a few to ignorance of seals. The Scientific Committee of the IWC has reiterated the problems of handling data so poor that they cannot be subjected to proper statistical analysis. Gulland has said: "Put bluntly in terms of population abundance, we seldom know at all precisely what the present population is, what it ought to be if our objectives (MSY or whatever) are to be achieved, or how any specific proposed action (e.g., a catch quota for the coming season) will alter the population". Again: "The belief that maximum sustainable yield occurs at half the unexploited population is only the consequence of one simple model with little, if any, observational foundation" (Gulland, 1976).

Scarcely anything is known of the size of hood seal populations currently being taken for their highly-valued fur and considerable doubts have been expressed as to current management practices for harp seals.

In the case of whales, some of the critical unknowns are:

(i) critical survival levels;

(ii) discrete breeding populations;

(iii) the effects or removing the bulk of the adult population (as is now done) on the survival rate and learning period of young;

(iv) the effects on the social structure of a lowered age at maturity and an increased reproduction rate;

(v) the survival value of sonic communication;

(vi) natural fluctuations in whale populations, and original or existing population levels;

(vii) rates of recovery (if any) for most protected whale stocks, and population levels at which they may stabilize;

(viii) what pain whales feel and how intelligent they may be; and

(ix) whether a maximum sustainable yield may be taken from a discrete breeding population, and if so at what population level.

These unknowns, and more, lead to errors in calculation far greater than the differences in population levels by which the New Management Procedure of the IWC classifies whales (Holt, 1976).

In general, we may add that it is not known what degree of real competition may exist between marine mammals and fishermen. Our knowledge of what fish mammals eat and where they eat it is poor, and it is by no means certain that a decrease in fish eaten by marine mammals would mean an increase available to fishermen.

This introduction has attempted to give the general background for a discussion of desirable population levels and management strategies. While scientists still have to devise the latter, there is now considerable experience of strategies that have not worked, and some suggestions can be made as to where to proceed in future.

Desirable population levels

Consideration of conflicting objectives seems to lead to the conclusion that marine mammal populations should:

(a) be widely distributed;

(b)   contain as many discrete breeding groups as they are capable of;

(c)   be diverse in age groups;

(d)   be sufficient in number to maintain their role in the ecosystem and to withstand environmental changes (such as exploitation of their food supply); and

(e)   some populations of each species should be completely protected.

In addition, no inhumane killing should occur, nor should any of the intelligent species be killed.

Allen (1976) has suggested that stocks should be maintained at their maximum sustainable yield level, because at that level they are most capable of change in numbers. However, this assumes no environmental changes such as krill fishing, and fails to satisfy most of (if not all) the requirements (a) to (e). In addition, the practical application of MSY has failed because of a faulty data base (ignorance of discrete populations or population levels) and of questionable assumptions (for instance, that MSY occurs at 50 or 60 percent of initial population) resulting in MSY estimates for the same whale stocks that vary by thousands of whales from year to year.

In an attempt to devise management strategies that would take account of the need to manage the ecosystem rather than a species in isolation, the Airlie House workshops on the exploitation of wild living resources defined the desirable state of a resource system as one which fulfilled the following three conditions:

1.   physical yields or other derived values are capable of being maximized on a continuing basis;

2.   diversity of present and future options in using the resource are ensured; and

3.   risk of irreversible change and long-term adverse effects resulting from use is minimized.

They added: "Simplistic goals of conservation, including the widely adopted one of assuring maximum sustainable yield, prove to be inadequate". The concern of these workshops was mostly with physical yield, but even within that limited range of objectives MSY was found wanting.

The U.S.A. has proposed to the North Pacific Fur Seal Convention that Optimum Sustainable Population (OSP) should replace the MSY concept, and that this new concept should include considerations of the optimum carrying capacity of the habitat and the health of the ecosystem (Baker, 1976).

Management strategies

Partly as a method of determining the size of the cake that those exploiting unowned resources have to divide between them, the maximum sustainable yield concept has been in wide use in fisheries and marine mammal management. Apart from its inability to satisfy many of the objectives defined in this paper, it has also failed in practice. There are a number of reasons.

It is necessary to make assumptions about the level at which MSY will occur; it is usually applied in ignorance of many important parameters including the true initial and actual population levels; it is not applied to individual breeding populations; time-lags in the system will occur before changes in important variables (such as food supply and social impact effects on recruitment) are reflected in the few variables that are measured; and the critical survival levels of the stocks to which MSY is applied are not known. It is surprising to find proponents of the continued application of MSY continuing to draw neat graphs of yield against population size with little, if any, reference to such significant deficiencies.

It is the objective of the "New Management Procedure" of the International Whaling

113

Commission to bring whale stocks to their MSY level and maintain them there. While quotas for stocks thought to be at that level are set at 90 % of MSY, the variations in estimates of MSY themselves are frequently larger than the quotas set. Little wonder then that, political considerations apart, the IWC has failed to manage whales in a sustainable fashion.

A management step satisfying rather more of the defined objectives would be to follow up the Airlie House recommendations and develop strategies for the exploitations of ecosystems as entities. Certainly this would eliminate the need to reduce marine mammal populations where they appear to compete with man for fish and krill, and it would by definition prevent the occurrence of irreversible changes. To satisfy objective (12) the strategy would have to allow for area-based exploitation and non-consumptive use within the ecosystem. Both Mexican and Californian gray-whale watchers will be adversely affected if pelagic fleets deplete the populations they observe.

Protection has always been a part of management. At least some of the ethical objectives would be satisfied by the creation of marine parks specifically as sanctuaries for entire populations of each marine mammal species. Similar to the early concept of national parks, such areas would allow total protection of the species, its habitat and food supply, while permitting some tourist and scientific observation. This action would be additional to the ecosystem management proposed first, and in the case of cetaceans, it would require setting aside very large corridors of ocean, and identification of the populations' feeding, travelling and breeding areas.

Relatively little management is required if the objective is either total protection or total exploitation – except in the case of total protection when critical habitats still need preserving.

In the case of cetaceans there is a strong practical argument for the cessation of all commercial whaling. Not only are the whales important to the ecosystem and slow to recover from overexploitation, but the presently exploited species are declining in most areas under IWC management, with the possible exception of the minke and sei whales, which may be threatened by krill exploitation in the near future. If scientific expeditions designed to obtain the maximum data possible, with no commercial pressures and with strict control over killing, were to continue under a moratorium, the information thus gathered would undoubtedly be more valuable than the inadequate data at present obtained from the whaling industries themselves. Research would be geared to the full range of objectives, and would thus include population dynamics, social interactions and communication, learning, migration, ecological relationships, intelligence, identification of populations and feeding habits. Gaskin (1976) has articulated and documented the doubts as to whether most protected whale populations are recovering at all, yet such knowledge is essential to our understanding of the ecosystem.

While at this point the murmur of "realists" beating their war drums may be heard, it is reasonable to suppose that if men can be put on the moon, live whales can be studied with success. Two papers (Mohl, Larsen and Amundin, 1976; Whitehead and Payne, 1976) have presented methods of determining size in sperm and right whales without killing them. Payne has also found distinctive patterning on the head and the fluke of the right and humpback whales, respectively, thus facilitating the study of individual behaviour and population migration.

Future trends

It is unlikely that all industrial objectives will be product-oriented in future. Even now, live dolphins are being trained as guards against hostile frogmen in docks, and for recovery of test equipment from the sea bed. Industry may need to communicate with live whales; fishermen may need sperm whales to find the deep-sea squid.

The pattern of exploitation is likely to change. At present, a few countries are responsible for the major part of the take of cetaceans, and probably of all marine mammals. Most of these are developed countries. The cost of operating on dwindling stocks has already affected the economics of pelagic whaling, and so the catches of the more numerous countries whaling from land stations have become more significant. It may be that in future marine mammals will be killed more locally. Where the products are used for local consumption there is some chance that the objective will be to obtain a sustainable yield. But if the present trend continues and they mainly enter international trade (in seal skins, whale meat, whale oil) there will be a trend toward the maximum immediate profit objective.

In the ethical and political fields we are seeing a change in the expression of long-held values, rather than a change in the values themselves. The desire to appreciate marine mammals for their own worth, and a decreasing desire for their products result both in import prohibition on some marine mammals products by a number of countries, and in the creation of cetacean sanctuaries and whale-watching areas for tourists.

The developing countries are now more emphatic in their intention of obtaining a fair share of the world's remaining resources and of restricting the activities of those whose only claim to consume is possession of the technology to do so. The increase in evidence for cetacean intelligence and the likelihood of direct communication with some whales is already affecting some people's values, and this influence is likely to grow fast if we ever learn to communicate with dolphins.

Even within the limited range of objectives concerned only with products, it is the demand for food which will surely dominate in future, given a human population that cannot help but expand considerably in the next 30 years. However, the highest level it may reach is uncertain. We must, therefore, consider whether the depletion of any marine mammal stock, whether for fur coats or for oil to soften fashion leathers, can be justified.

Acknowledgements

I would like to thank Richard Sandbrook of IIED, Angela King and Mick Hamer of Friends of the Earth for their encouragement, contributions and advice; Dr. S.J. Holt for his ideas and comments; Carol and Julian Fox for the final editing and typing work, and all other whose papers have helped me to understand the issues involved a little better. I would also like to thank Doris Dearing for producing from an unintelligible draft a final copy of this paper in a very short space of time.

References

ALLEN, K.R., The optimization of management and
1976   strategy for marine mammals. Paper presented to the Scientific Consultation on the Conservation and Management of Marine Mammals and their Environment, Bergen, Norway, 31 August-9 September 1976. Rome, FAO, ACMRR/MM/SC/57:12 p.

BAKER, R.C., Background and summary of the North
1976   Pacific Fur Seal Conference, Washington, D.C., 1-12 December 1975. Paper presented to the Scientific Consultation on the Conservation and Management of Marine Mammals and their Environment, Bergen, Norway, 31 August-9 September 1976. Rome, FAO, ACMRR/MM/SC/75:2 p.

CLARK, C.W., Economic aspects of renewable resource
1976   exploitation as applied to marine mammals. Paper presented to the Scientific Consultation on the Conservation and Management of Marine Mammals and their Environment,

115

Bergen, Norway, 31 August-9 September 1976. Rome, FAO, ACMRR/MM/SC/65:13 p.

Friends of the Earth International, A proposal for
1976   whales. London, Friends of the Earth.

GASKIN, D.E., The evolution, zoogeography and ecolo-
1976   gy of Cetacea. *Oceanogr. Mar. Biol.*,
         14:247-346.

GULLAND, J.A., A note on the strategy of the manage-
1976   ment of marine mammals. Paper presented to
         the Scientific Consultation on the Conserva-
         tion and Management of Marine Mammals
         and their Environment, Bergen, Norway, 31
         August-9 September 1976. Rome, FAO,
         ACMRR/MM/SC/82:7 p.

HOLT, S.J., Whale management policy. Paper presented
1976   to the Scientific Consultation on the Conser-
         vation and Management of Marine Mammals
         and their Environment, Bergen, Norway, 31
         August-9 September 1976. Rome, FAO,
         ACMRR/MM/SC/99:10 p.

MEEKER, J.W., Dolphins do, dugongs don't. Paper pre-
1976   sented to the Scientific Consultation on the
         Conservation and Management of Marine
         Mammals and their Environment, Bergen,
         Norway, 31 August-9 September 1976. Rome,
         FAO, ACMRR/MM/SC/138:2 p.

MØHL, B., E. LARSEN and M. AMUNDIN, Sperm
1976   whale size determination: outlines of an
         acoustic approach. Paper presented to the
         Scientific Consultation on the Conservation
         and Management of Marine Mammals and
         their Environment, Bergen, Norway, 31 Au-
         gust-9 September 1976. Rome, FAO,
         ACMRR/MM/SC/84:4 p.

SERGEANT, D.E., History and present status of popula-
1976   tions of harp and hooded seals. *Biol. Conserv.*,
         10(2):95-118.

SERGEANT, D.E. and T. STIRLING, Comments on objec-
1976   tives of marine mammal management. Paper
         presented to the Scientific Consultation on the
         Conservation and Management of Marine
         Mammals and their Environment, Bergen,
         Norway, 31 August-9 September, 1976. Rome,
         FAO, ACMRR/MM/SC/Cmt. 1:2 p.

WHITEHEAD, H. and R. PAYNE, New techniques
1976   of assessing populations of right whales
         without killing them. Paper presented to the
         Scientific Consultation on the Conservation
         and Management of Marine Mammals and
         their Environment, Bergen, Norway, 31 Au-
         gust-9 September, 1976. Rome, FAO,
         ACMRR/MM/SC/79:46 p.

SOME ASPECTS OF CETACEAN NEUROANATOMY

P. FORTOM-GOUIN

Abstract

The high degree of anatomical development of the brain of cetaceans, comparable to the development of man's brain, and in certain ways perhaps more advanced than it, suggests that cerebellum of whales and dolphins is generally better formed than in terrestrial mammals, the paleocortex is much smaller and in the cerebral cortex of some species, there are more convolutions and a larger number of neurons than in man, and a large area apparently devoted to association functions. The brain weight of several species is greater than that of man. The influence of the environment may account for the apparently different ways in which whales and dolphins express their intelligence. If the intelligence of these animals is extraordinary, it may be ethically wrong to kill them.

Résumé

Le haut degré de développement anatomique du cerveau des cétacés, comparable au développement du cerveau humain et, sur certains points, peut-être plus avancé que lui, laisse penser que certaines espèces de cétacés recèlent un potentiel d'intelligence comparable à celui de l'homme. Le cervelet des baleines et des dauphins est généralement mieux formé que celui des mammifères terrestres, le paléocortex est beaucoup plus petit et, dans le cortex cérébral de certaines espèces, on trouve des circonvolutions plus nombreuses et un plus grand nombre de neurones que chez l'homme, ainsi qu'une large zone apparemment consacrée aux fonctions d'association. Le poids du cerveau de plusieurs espèces est supérieur à celui de l'homme. L'influence de l'environnement peut expliquer les façons apparemment différentes dont les baleines et les dauphins expriment leur intelligence. Si l'intelligence de ces animaux est extraordinaire, il pourrait être blâmable, sur le plan de l'éthique, de les tuer.

Extracto

El complicado desarrollo anatómico del cerebro de los cetáceos, comparable al del cerebro humano y, desde ciertos puntos de vista, tal vez más avanzado que él, sugiere que quizá algunas especies de cetáceos tengan un potencial de inteligencia comparable al del hombre. El cerebelo de las ballenas y los delfines está en general mejor formado que el de los mamíferos terrestres, la paleocorteza es mucho menor y la corteza cerebral de algunas especies tiene más circunvoluciones y mayor número de neuronas que el hombre, con una amplia zona dedicada aparentemente a las funciones asociativas. El peso cerebral de varias especies es mayor que el del hombre. Las formas aparentemente diferentes en que las ballenas y delfines manifiestan su inteligencia quizá se deban a la influencia del medio ambiente. Si la inteligencia de estos animales es extraordinaria, es posible que matarlos sea éticamente inadmisible.

P. Fortom-Gouin
Institute for Delphinid Research, Box N3531, Nassau, Bahamas

Introduction

The high level of cerebral development by cetaceans is one of their most interesting characteristics, as the brain is the principal anatomic component of "intelligence".

The brains of the 80-odd species of cetaceans present extreme variations of size and complexity. At one end, the sperm whale brain is the largest of any species on earth. At the other end, *Platanista gangetica* brain weighs a mere 150 g.

Brain weight

The average brain weight of the adults of some species is given in the following table:

		g
Cetaceans	Odontoceti:	
	Common dolphin	750
	Bottlenose dolphin	1 500/1 800
	Beluga	2 200
	Risso's dolphin	2 250
	Pilot whale	2 500
	Killer whale	4 500
	Sperm whale	8 000/9 000
	Mysticeti:	
	Gray whale	4 200
	Sei whale	4 700
	Fin whale	7 000
Primates	*Homo sapiens*	1 300

The ratio of brain weight to body weight, also called the encephalization ratio, is lowest in the largest species (blue whale, sperm whale) and highest in the smaller species (harbour porpoise, bottlenose dolphin). This ratio in the bottlenose dolphin is very close to the value in man.

This ratio does not seem to be a very significant interspecies index of "intelligence". Man does not have the highest ratio among land mammals.

Brain anatomy

The cerebellum, which is involved in the coordination of the motor outputs and the processing of some sensory inputs, is highly evolved in most cetaceans, and generally better formed than in terrestrial mammals. It represents a larger percentage of the total brain weight in Mysticeti than in Odontoceti as the following table shows:

	Cerebellum weight (g)	Weight ratios cerebellum/ total brain
Beluga	340	15
Pilot whale	443	18.3
Sei whale	867	19.4
Fin whale	1 575	25.7
Homo sapiens	148	10.1

The cerebrum, made up of the cerebral hemispheres, is folded longitudinally to fit the evolutionary modifications of the skull, so that it is generally higher and wider than it is long.

It is well developed in all cetaceans and presents a large number of gyri and sulci. In smaller odontoceti it is organized into 4 lobes (primates have a 3-lobe arrangement). A *Tursiops* cerebrum is more intricately convoluted than a human brain, with a correspondingly thinner cortex and larger cortex surface. The degree of fissuration and the extent of the neocortex surface, which increase from lower primates to higher primates to man, is generally considered a criterion of evolutionary progress.

The paleocortex, on the contrary, is very small in comparison with land mammals, including primates. It has been called the reptile brain because it is the most primitive structure of the brain, to which evolution has simply added. It has been suggested that this marked reduction in the limbic structures might be responsible for the decidedly hedonistic nature of the dolphins (Pilleri, 1967).

However, some structure show an enormous development: in *Delphinus delphis*, the superior olive and the ventral nucleus of the lateral lemniscus are respectively 150 and 200 times larger than in man. A similar condition is found only in bats (Zvorykin, 1963), which points to their function in ultrasonic perception.

The weight ratio of neocortex to paleocortex, which is sometimes used as an index of evolutionary progress in land mammals, is higher in *Tursiops* than in man.

Cerebrum cortex anatomy

In the cortex of the cerebrum of humans the higher mental functions, such as memory, language, concept formation and recognition are located.

It is therefore interesting to study in more detail the cerebrum cortex anatomy in cetaceans.

NEURON DENSITY AND NEURON NUMBER

It is sometimes said that although cetaceans have larger brains than man, they have fewer neurons. A study by Kraus and Pilleri provides some interesting information which is summarized in the following table (Kraus and Pilleri, 1969).

	Brain weight (g)	Cortex volume (mm^3)	Average neuron density	Total neuron number
Ganges susu	150		55 000	
Pilot whale	2 200	453 000	65 500	30 × 10^9
Fin whale	5 850	1 155 000	13 000	15 × 10^9
Homo sapiens	1 300	181 000	76 500	14 × 10^9

There is a certain margin of error in those numbers (the total neuron number in the human cerebrum cortex has been estimated by various researchers from 10×10^9 to 16×10^9), but it gives a rough comparison.

The fin whale, a mysticete, with a cortex more than 6 times larger than a human, has a slightly larger total cortical neuron number than humans because of a correspondingly lower neuron density.

The cortex of a pilot whale, an odontocete, with a neuron density slightly lower than a human, has twice as many neurons as a human cortex.

The neuron is the equivalent of a multipolar switch in a computer. The total number of neurons is therefore a good index of the complexity of the cerebrum "biocomputer". The above study points to the difference in cortex neuron densities between mysticete and odontocete.

119

NEURON ANATOMY

Could the neurons be more numerous but less efficient? Morgane and Jacobs (1972) have carried out several Golgi analyses of homologous areas of cortex in several species of delphinid and in monkeys and humans and conclude that "the neuron connectivities synaptic geometry and dendritic fields are comparable between dolphins and primates by all applicable criteria". Zvorykin (1963) reached the same conclusion when comparing the temporal cortex of the common dolphin and of man.

CORTICAL LAMINATION

All 6 layers typical of the mammalian brain are found in the cetacean cortex (Jansen and Jansen, 1969; Rose, 1926).

In cetaceans, as in the higher primates, the second layer is the narrowest and the sixth the broadest. The last layer is noticeably broad in all cetaceans (Pilleri and Gihr, 1976). This and other variations in layer thickness await analysis.

REGIONAL DIFFERENTIATION

In a mammal cerebrum cortex, there are histologically different sensory, motor and association areas.

It is the extension of the association areas which, more than anything else, sets man apart from the other primates.

In cetaceans, cytoarchitectonic assessments show high area differentiation, with most of the formations being of the association type. The motor formations, with their characteristically large pyramidal cells are somewhat limited and located in the anterior extreme of the brain (Lende and Akdikmen, 1968); the sensory formations are well developed.

Conclusion and discussion

The neuroanatomy of the cetaceans is in every way comparable to that of higher primates, with specific variations which await analysis. What is more, in several species of larger odontoceti, the brain and the cerebrum neocortex is several times larger than in humans, in terms of weight, total neuron count and size of cortical association areas.

I have said that neuroanatomy is the principal anatomic component of "intelligence". The other anatomic components are the "peripherals" of the biocomputer:

(i) the sensory organs for information input, with the auditory organs having, in cetaceans, the pre-eminence that visual organs have in man;

(ii) the motory organs characterized in cetaceans by the lack of hands with the motory activity channelled mainly into sound production and body movement.

The brain with its peripherals represents the intelligence potential of the species. Interaction with the environment (physical, social, cultural) can programme the biocomputer in very different directions. The influence of the environmental component is enormous: it can turn the same individual into, say, an electronic engineer, or a Bindubu Australian aborigine (who makes no shelter, no clothes and just about no utensils or tools) or even a wolf-child with no verbal language and rudimentary conceptual thinking.

Let us also remember that while the genus *Homo* has been around 2 million-plus years (*Homo abilis olduvai*) and *Homo sapiens* 70 000 years, we never developed any technology more complex than putting a stone on the end of a stick, except in the last 10 000 years.

The study of cetacean neuroanatomy makes a very strong case for accepting the possibility that the intelligence potential of several species is quite comparable to ours. It is probably in the overwhelming influence of the environment (physical, social) and in the different type of motory outputs that we must look for the explanation of the very different (but not inferior) forms that their intelligence potential takes.

For this reason (among several others such as their extraordinary unaggressiveness and even friendliness toward humans) I believe that there is a very grave ethical problem of whether humans, just because they have the necessary technological power, should kill cetaceans at all. Should humans kill sperm whales, with the largest and possibly the most complex brain on earth, to make gear oil, leather softeners, or pet food out of them? The question must be asked.

References

JANSEN, J. Jr. and J. JANSEN, The nervous system of
1969   cetacea. *In* The biology of marine mammals, edited by H.T. Andersen. New York, Academic Press, pp. 175-252.

KRAUS, C. and G. PILLERI, Quantitative Untersuchung-
1969   en über die Groshirnrinde die Cetaceen. *In* Investigations on Cetacea, edited by G. Pilleri. Berne, Brain Anatomy Institute, vol. 1:127-50.

LENDE, R.A. and S. AKDIKMEN, Motor field in the cere-
1968   bral cortex of the bottlenose dolphin. *J. Neurosurg.*, 29:495-9.

MORGANE, P.J., and M.S. JACOBS, Comparative anat-
1972   omy of the cetacean nervous system. *In* Functional anatomy of marine mammals, edited by R.J. Harrison. London, Academic Press, vol. 1:117-244.

PILLERI, G., Considerations sur le cerveau et la com-
1967   portement de *Delphinus delphi. Rev. Suisse Zool.*, 74:665-77.

PILLERI, G. and M. GIHR, The central nervous system of
1976   the Mysticete and Odontocete whales. *In* Investigations on Cetacea, edited by G. Pilleri. Berne, Brain Anatomy Institute, vol. 2:89-126.

ROSE, M., Der Grundplan der Cortextonik beim
1926   Delphin. *J. Psychol. Neurol.*, 32:161-9.

ZVORYKIN, V.P., Morphological substrate of ultrasonic
1963   and locational capacities in the dolphin. *Arkh. Anat. Hist. Embriol.*, 45:3-17.

CETACEAN BEHAVIOUR, LEARNING AND COMMUNICATION

M. OVERLAND

Abstract

Scientific studies of some species of small cetaceans have shown that they are highly intelligent animals capable of learning in advanced ways and among whom social interaction, especially communication, is important. Communication should be taken to include behaviour other than vocalization, is important in so much as it affects behaviour and the relations between animals, and can be expected to be very complex. Although there have not been many behavioural studies of cetaceans in the wild, some work indicates that communication between man and these animals is possible. Two basic kinds of sounds are known to be made by cetaceans — a high frequency one composed of pulsed clicks and used for echo location, and a low frequency one used in communication and including sounds that are peculiar to certain species and individual animals. Interest in the central nervous system of cetaceans has been stimulated especially by the question of the role of the large and highly convoluted cerebral cortex. Although very little is known, it has been suggested that it is concerned with communication, sight and perhaps with "higher" central nervous system functions associated with "intelligence" in man — this latter suggestion seems to be supported by the presence of a large amount of intrinsic nucleic formations in the dolphin thalamus and by the apparently small portion of the cortex associated with primary motor or sensory functions. Olfactory nerves are absent and cutaneous sensibilities are thought to be reduced.

Résumé

Les études scientifiques portant sur certaines espèces de petits cétacés ont montré qu'il s'agit d'animaux hautement intelligents, capables d'apprendre des éléments complexes et chez lesquels l'interaction sociale, spécialement la communication, est importante. Il convient de faire entrer dans les communications le comportement autre que la phonation; ce processus est important dans la mesure où il affecte le comportement et les relations entre animaux et l'on peut compter qu'il est très complexe. Bien qu'il n'y ait eu qu'un nombre limité d'études du comportement des cétacés en liberté, certains travaux indiquent que la communication entre l'homme et ces animaux est possible. On sait que les cétacés émettent deux types essentiels de son, un son à haute fréquence composé de cliquètements pulsés, utilisé pour le repérage par écho et un son à basse fréquence, utilisé pour la communication et comprenant des sons particuliers à certaines espèces et à certains animaux. L'intérêt pour le système nerveux central des cétacés a été stimulé spécialement par le problème du rôle du vaste cortex aux nombreuses circonvolutions. Encore que les connaissances soient très fragmentaires, on a suggéré qu'il est lié à la communication, la vue et peut-être à d'autres fonctions "supérieures" du système nerveux central associées à l'"intelligence" chez l'homme. Cette hypothèse semble être étayée par la présence d'une grande quantité de formations nucléiques intrinsèques dans le thalamus du dauphin et par la portion apparemment réduite du cortex associée aux fonctions

123

primaires motrices ou sensitives. Les nerfs olfactifs sont absents et l'on pense que la sensibilité cutanée est réduite.

Extracto

Los estudios científicos realizados sobre algunas especies de pequeños cetáceos han mostrado que son animales muy inteligentes, capaces de aprendizaje y entre los que la interacción social, especialmente la comunicación, tiene su importancia. La comunicación, que incluye otras manifestaciones además de la vocalización, reviste importancia en la medida en que afecta al comportamiento de los animales y a las relaciones entre ellos, y parece ser que es muy compleja. Aunque no se han realizado muchos estudios sobre el comportamiento de los cetáceos en libertad, algunos trabajos realizados indican que la comunicación entre el hombre y esos animales es posible. Se sabe que los cetáceos emiten dos tipos básicos de sonido: uno de gran frecuencia, formado por chasquidos intermitentes, que utilizan para la localización de objetos por su eco, y otro de baja frecuencia, que se utiliza para la comunicación e incluye sonidos que son peculiares de determinadas especies e individuos. El interés por el sistema nervioso central de los cetáceos se ha visto estimulado especialmente por el problema de la función que desempeña la corteza cerebral, grande y con muchas circunvoluciones. Aunque se sabe aún muy poco, se ha sugerido que en ella radican la comunicación, la vista y quizás algunas funciones "superiores" del sistema nervioso central, asociadas con la "inteligencia" del hombre, sugerencia esta última que parece confirmada por la presencia de gran cantidad de formaciones nucleicas intrínsecas en el tálamo del delfín y por la porción aparentemente pequeña de la corteza cerebral que está asociada con las funciones primarias motoras o sensoriales. No hay nervios olfatorios y se cree que la sensibilidad cutánea es reducida.

Behaviour and learning

The social interactions of cetaceans have not yet received the attention accorded to, for example, the primates, though there is certainly an increasing interest in the interactions between individuals in captive cetacean colonies (Evans and Bastian, 1969). An example of an extensive study of cetacean behaviour by Bateson (1966, 1974) is important because of its approach to communication between individuals. His interest is in the paradigm in which the presence or absence of communication data are meaningful to members of a group in the sense that their behaviour is affected. He stresses that the behaviour itself is communication and that the discourse is about the fundamentals of the relationship itself. He argues that since cetaceans are mammals, their communication will be concerned with *patterns* of relationships, and because they are social animals and have large brains, their communication will be *highly complex*. His first requirement for researching cetacean communication is the identification and classification of the different types of behaviour and their components. For him, the statistical aspects of studying vocalizations are secondary to the behavioural metaphors that cetaceans use. A summary of the behavioural transactions that he observed in a group of captive dolphins is given below:

(i) Slow side-by-side swimming with synchronous movements and respirations; sometimes with flipper contact;

(ii) fast swimming with progressively synchronized jumping;

(iii) coitus;

(iv) threat movements characterized by their suddenness: lunging, head turning with mouth open, fluke or beak striking. Threats are commonly accompanied by vocalization. A threat hierarchy exists and the order is irreversible;

(v) beak-genital propulsion, sometimes with pair reversals;

(vi) body, fluke or flipper rubbing; repetitive or continuous;

(vii) pat-a-cake; alternating dorsal and palmar surfaces of flippers;

(viii) chase; follows threat, precedes coitus and is possibly engaged in as competitive playing;

(ix) rest formation; a slowly circling group pattern formation; 2 of the 7 members were not a part of it, one of whom "wanted" to be; the other remained a loner, "hanging" in the water.

For those attempting ethological studies in the wild, Bateson (1974) stresses that particular attention should be paid to the sleep/rest pattern of the animals, since it relates closely to their overall social organization.

Another point that both Bateson (1974) and Lilly (1977) make is that one should not place oneself in a patronizing position toward the cetaceans as one would do with a dog or a cat. When in the water and interacting with cetacean species, the chance of learning something is greater if respect is paid to the intelligence potentials of these animals.

The behaviour of captive dolphins and whales is also described by Pryor (1973) and Caldwell and Caldwell (1966). Pryor emphasizes that cetaceans are highly social and intelligent animals who use a variety of postures, gestures, movements and sounds in their daily behaviour. In addition, cetaceans demonstrate higher orders of learning: learning by observation, single-trial learning, and the acquisition of learning rules. The Caldwells reviewed the literature on the remarkable care-giving behaviour that cetaceans exhibit toward each other, particularly to juvenile, injured or distressed individuals. There have been many reports on the general organization and behaviour of cetaceans in open waters, their distribution and migration habits and of personal encounters with them from antiquity to the present (e.g., Slijper, 1962; Caldwell, Caldwell and Rice, 1966; Norris, 1974; Norris and Prescott, 1961; McIntyre, 1974; and others too numerous to catalogue).

Scientific studies of cetaceans in their environment are far fewer than are the general accounts mentioned above. One ongoing series of field studies with *Orcinus orca* points in the required direction: Spong, 1974; Spong, Bradford and White, 1970; Spong, Michaels and Spong, 1971, have been studying pods of killer whales in the Johnstone Straits area of British Columbia annually. They have observed this species of whale in their natural habitat and have monitored their behaviour and begun experimental studies on communication with them, using live music and acoustic synthesizers. Similarly, Bott has documented with film an interchange between the killer whale and a music synthesizer. These studies, albeit introductory in nature, demonstrate that meaningful exchange can take place between man and cetacean, in the wild, in an atmosphere of mutual acceptance and cooperation.

Sound communication

Given that some recent work has been started on interspecies communication, we may ask whether there is a sufficient basis of fact concerning the vocalization abilities of cetaceans. There is. Marshall (1974) has summarized the types of sounds that cetaceans have evolved. There are basically two modes of voice, one used for navigation and one for

125

social intercourse. The navigation sound emittance is in the form of "clicks" which are pulsed and may reach frequencies of over 100 kHz. These echo-locating signals are used for finding objects (e.g., terrain and food), scanning them for detail, and are most likely the vocal mode used to calculate distance and speed. The sounds used for social communication are in lower frequencies – the sonic range that man can hear. These are described as "squeaks", "squawks", "barks" or "grunts" and are thought to be used in self-expression. Pure-tone "whistles" are also emitted in the form of "trills", "arpeggios", "glissandos" and so forth. These latter vocalizations are species specific and individual. That the navigational clicks and the whistle signatures can be generated individually or simultaneously points to the sophistication of their sound control system. Of further interest is that it has been shown in studies on the spectral analysis of the killer whale that sex differences can be determined from the sonograms of their emitted signals (Singleton and Poulter, 1967; Poulter, 1968).

In summary, the physical and to some extent the behavioural aspects of cetacean vocalization are well documented (e.g., Caldwell, Caldwell and Hall, 1973; Evans, 1973; Lilly, 1968). Clearly, the perceptual world of cetaceans is dominated by sound.

The nervous system

The perceived status of the central nervous system of the cetacean has increased in recent years, in parallel with the interest in their behaviour and other modes of communication. Although the gross morphological features of the cetacean brain were known at the turn of the century, the first modern comprehensive review including histological details did not appear until 1960. Breathnach (1960) catalogued the general size relationships of the gross brain, the cranial nerves, the spinal cord, the cerebellum, the inferior olive, the diencephalon and the cerebral hemispheres. He points specifically to (i) the absence of the olfactory nerves and related structures; (ii) the large auditory nerve and its end stations; (iii) the apparent reduced cutaneous sensory pathways, and (iv) the most striking characteristic, the high degree of forebrain development, marked particularly by the expansive convoluted cerebral cortex. There has been some dispute in the past as to whether the cerebral cortical mantle of cetaceans is as fully differentiated as in primates. However, the view now is that the neocortex is indeed well differentiated and shows regional differences that are similar to those in the cortex of man.

It is the voluminous cerebral cortex that has stimulated man's inquisitiveness as to the function of the cetacean brain. In his study of the dolphin thalamus, Kruger (1959) discusses the view that thalamic development determines the organization of the cerebral cortex (Rose and Woolsey, 1949). Part of the argument states that an outstanding feature of phylogenetic development is the expansion of cortex that receives projections from "intrinsic" thalamic nuclei. The dolphin thalamus studied by Kruger illustrates this. That is, it has a relatively large amount of intrinsic nucleic formations. In man the intrinsic nuclei along with their respective interconnected cortices are thought to be involved in such "higher" central nervous system functions as language thought and intelligence.

What then does the cetacean cerebral cortex do? Very little is known but the remarkable folding of the cortex with its attendant increased volume of nerve cells presents a morphological base to which serious attention may be given as to its function. To begin with, because of the cetaceans' particularly well developed auditory nerve, their echo-location abilities and their highly individualized vocalizations, attention has been focused on their auditory system. Bullock and Ridgway (1972) recorded from the midbrain, from an auditory centre in the brain stem, and from several cortical sites in unanaesthetized, unrestrained dolphins. Their data have led them to suggest that high frequency ultrasonic tones may be analysed at the midbrain level while the lower frequency sounds, the whistles, are analysed at the cortical level. This is of considerable importance since the whistle frequency range of sound is thought to convey information relevant to social transactions.

Other studies are beginning to shed some light on the primary motor and sensory representations at the cortex (Lende and Akdikman, 1968; Lende and Walker, 1972). It would appear to date that only a relatively small portion of the cortex is marked out primarily for motor or sensory functions. This leaves a vast expanse of cortical region, comparable to the association cortex seen in man, which deals with the so-called higher central nervous system functions.

Whereas most experimental attention has been directed to various aspects of the auditory system, it has recently been found that the visual system of some cetaceans is not as poor as the earlier literature indicated. Spong and White (1971) and White et al. (1971) have concluded that the orca has a visual acuity under water that is roughly comparable to that of a cat in air. Others have also indicated that the bottlenose dolphin can see relatively well in air (Pepper and Simmons, 1973; Dawson, Birndorf and Perez, 1972). The retina has been examined histologically by Perez, Dawson and Landau (1972) who conclude that the dolphin retinal elements are comparable to those of most mammals, are

suitable for visual acuity and may be able to detect colour. No studies have as yet been carried out on the cortical representation of vision but it would be very surprising if it did not occur. Kellog and Rice (1966) have shown that dolphins can solve problems, using visual discrimination. It would seem unlikely that dolphins can do so without visual representation at the cortex.

In summary, we can conclude that the cetacean cerebral cortex receives and integrates auditory, and possibly visual, but not olfactory input, since there are no olfactory nerves. Cutaneous and proprioceptive sensibilities are thought to be present though somewhat reduced (Jansen and Jansen, 1969). In view of the interpretation that sharks' furore behaviour in the presence of blood is related to the centrality of olfaction in their brain development, one can ask whether the dolphins' peaceful nature is related to the absence of olfactory nerves. On the basis of cytoarchitectural studies of the cetacean brain, Jacobs (1974) and Morgane (1974) have been led to propose that these animals have a brain substrate capable of a high degree of cortical activity, which could possibly express itself as "intelligence". On evolutionary grounds, Bunnell (1974) proposes that the expanded neocortical areas could be used to orient the cetaceans toward perceptions and inter- and intra-personal relationships, in contrast to the human brain which expresses itself in inter- and intra-personal relationships and in external action on the environment.

References

BATESON, G., Problems in Cetacean and other mamma-
1966   lian communication. In Whales, dolphins and porpoises, edited by K.S. Norris. Berkeley, University of California Press, pp. 569-79.

−, Observations of a Cetacean community. In Mind in
1974   the waters, edited by J. McIntyre. New York, Charles Scribners & Sons, pp. 146-65.

BREATHNACH, A.S., The Cetacean central nervous system.
1960   Biol. Rev., 35:187-230.

BULLOCK, T.H. and S.H. RIDGWAY, Evoked potentials in
1972   the central auditory system of alert porpoises to their own and artificial sounds. J. Neurobiol., 3:79-99.

BUNNEL, S., The evolution of Cetacean intelligence. *In*
1974   Mind in the waters, edited by J. McIntyre. New York, Charles Scribners & Sons, pp. 52-60.

CALDWELL, D.K., M.C. CALDWELL and D.W. RICE, Be-
1966   havior of the sperm whale. *In* Whales, dolphins and porpoises, edited by K.S. Norris. Berkeley, University of California Press, pp. 677-730.

CALDWELL, M.C. and D.K. CALDWELL, Epimeletic
1966   (caregiving) behaviour in Cetacea. *In* Whales, dolphins and porpoises, edited by K.S. Norris. Berkeley, University of California Press, pp. 755-89.

CALDWELL, M.C., D.K. CALDWELL and N.R. HALL, Ability
1973   of an Atlantic bottlenosed dolphin (*Tursiops truncatus*) to discriminate between and potentially identify individual whistles of another species, the common dolphin (*Delphinus delphis*). *Cetology*, (14):1-7.

DAWSON, S.W., L.A. BIRNDORF and J.M. PEREZ, Gross
1972   anatomy and optics of the dolphin eye (*Tursiops truncatus*). *Cetology*, (10):1-12.

EVANS, W.E., Echolocation by marine Delphinids and
1973   one species of fresh water dolphin. *J. Acoust. Soc. Am.*, 54:191-9.

EVANS, W.E. and J. BASTIAN, Marine mammals commu-
1969   nications: social and ecological factors. *In* The biology of marine mammals, edited by H.T. Andersen. New York, Academic Press, pp. 425-75.

JACOBS, M., The whale brain: input and behaviour. *In*
1974   Mind in the waters, edited by J. McIntyre. New York, Charles Scribners & Sons, pp. 78-83.

JANSEN, J. and J.K.S. JANSEN, The nervous system of Ce-
1969   tacea. *In* The biology of marine mammals, edited by H.T. Andersen. New York, Academic Press, pp. 175-252.

KELLOG, W.N. and C.E. RICE, Visual discrimination and
1966   problem solving in a bottlenose dolphin. *In* Whales, dolphins and porpoises, edited by K.S. Norris. Berkeley, University of California Press, pp. 731-54.

KRUGER, L., The thalamus of the dolphin (*Tursiops trun-
1969   catus*) and comparison with other mammals. *J. Comp. Neurol.*, 111:133-94.

LENDE, R.A. and S. AKDIKMAN, Motor field cortex of the
1968   bottlenose dolphin. *J. Neurosurg.*, 29:495-9.

LENDE, R.A. and W.I. WALKER, An unusual sensory area
1972   in the cerebral neocortex of the bottlenose dolphin, *Tursiops truncatus. Brain Res.*, 45:555-60.

LILLY, J.C., Sound production in *Tursiops truncatus*
1968   (bottlenose dolphin). *Ann. N.Y. Acad. Sci.*, 155:321-41.

—, The mind of the dolphin: a non-human intelligence.
1977   New York, Doubleday.

MARSHALL, P., The ways of whales. *In* Mind in the waters,
1974   edited by J. McIntyre. New York, Charles Scribners & Sons, pp. 110-40.

McINTYRE, J. (ed.), Mind in the waters. New York,
1974   Charles Scribners and Sons.

MORGANE, P., The whale brain: the anatomical basis of
1974   intelligence. *In* Mind in the waters, edited by J. McIntyre. New York, Charles Scribners & Sons, pp. 84-93.

NORRIS, K.S., The porpoise watcher. New York, W.W.
1974   Norton & Co., Inc.

NORRIS, K.S. and J.H. PRESCOTT, Observations on Pacific
1961   Cetaceans of California and Mexican waters. Berkeley, University of California Press.

PEPPER, R.L. and J.V. SIMMONS, Jr., In-air visual acuity of
1973   the bottlenose dolphin. *Exp. Neurol.*, 41:271-6.

PEREZ, J.M., W.W. DAWSON and D. LANDAU, Retinal
1972   anatomy of the bottlenosed dolphin (*Tursiops truncatus*). *Cetology*, (11):1-11.

POULTER, T.C., Analysis of the communication and
1968   echolocation signals of the killer whale, *Orcinus orca* Linnaeus. *In* SRI Report on echo ranging signals. Menlo Park, Stanford Research Institute.

PRYOR, K.W., Behavior and learning in porpoises and
1973   whales. *Naturwissenschaften*, 60:412-20.

ROSE, J.E. and C.N. WOOLSEY, Organization of the
1949   mammalian thalamus and its relationship to the cerebral cortex. *Electroencephalog. Clin. Neurophysiol.*, 1:391-404.

SINGLETON, R.C. and T.C. POULTER, Spectral analysis of
1967   the call of the male killer whale. *IEEE Trans. Audio Electroacoust.*, (AV-15):104-13.

128

SLIJPER, E.J., Whales. London, Hutchinson & Co., Ltd.
1962

SPONG, P., The whale show. *In* Mind in the waters, edited
1974   by J. McIntyre. New York, Charles Scribners &
        Sons, pp. 170-85.

SPONG, P. and D. WHITE, Visual acuity and discrimination
1971   learning in the dolphin *(Lagenorhynchus obli-
        quidens). Exp. Neurol.*, 31:431-6.

SPONG, P., J. BRADFORD and D. WHITE, Field studies of the
1970   behavior of the killer whale *(Orcinus orca).
        Proc. Annu. Conf. Biol. Sonar Diving Mamm.*,
        7:181-5.

SPONG, P., H. MICHAELS and L. SPONG, Field studies of the
1971   behavior of the killer whale *(Orcinus orca).
        Proc. Annu. Conf. Biol. Sonar Diving Mamm.*,
        9:181-5.

WHITE, D. *et al.*, Visual acuity of the killer whale *(Orcinus
1971   orca). Exp. Neurol.*, 32:230-6.

Bibliography

ANDERSON, H.T. (ed.), The biology of marine mammals.
1969   New York, Academic Press.

CALDWELL, M.C. and D.K. CALDWELL, Vocalization of
1966   naive captive dolphins in small groups.
        Science, Wash., 159:1121-3.

HALL, J.D. and C.S. JOHNSON, Auditory thresholds of a
1972   killer whale, *Orcinus orca* Linnaeus. *J. Acoustic
        Soc. Am.*, 51:515-7.

LILLY, J.C., Sonic-ultrasonic emissions of the bottlenose
1966   dolphin. *In* Whales, dolphins and porpoises,
        edited by K.S. Norris. Berkeley, University of
        California Press, pp. 503-9.

ROSE, M., Der Grundplan der Cortextektronik beim
1926   Delphin. *Z. Psychol.*, Leipzig, 32:161-9.

BIOCHEMICAL GENETIC STUDIES, THEIR VALUE AND LIMITATIONS IN STOCK IDENTIFICATION AND DISCRIMINATION OF PELAGIC MAMMAL SPECIES

G.D. SHARP

Abstract

Biochemical genetic studies, using chi-square statistical analyses, are probably more useful to identify differences between species than those between sub-specific groups. The latter require very large, homogeneous samples, often difficult or impossible to obtain, to produce valid analyses. Sexual behaviour can affect homogeneity of characters and distribution patterns can sometimes be difficult to identify and so confuse sampling. It is possible to derive beforehand, from statistical theory, the optimum sample size, determined in part by the degree of rigour desired. It will be rare that precise sampling of marine mammal populations is possible.

Résumé

Les études génétiques biochimiques utilisant des analyses statistiques du chi carré sont probablement plus utiles pour identifier les différences entre les espèces qu'entre les groupes de sous-espèces. Dans ce dernier cas, il faut des échantillons très vastes et homogènes, souvent difficiles ou impossibles à obtenir, pour produire des analyses valables. Le comportement sexuel peut affecter l'homogénéité des caractères et les modes de distribution peuvent parfois être difficiles à identifier, ce qui perturbe l'échantillonnage. Il est possible d'obtenir au préalable, à partir de la théorie statistique, la taille optimale de l'échantillon, déterminée en partie par le degré de rigueur requis. Il est rare qu'un échantillonnage précis des populations de mammifères marins soit possible.

Extracto

Probablemente los estudios genéticos-bioquímicos, utilizando análisis estadísticos, son más útiles para individuar las diferencias entre especies que las diferencias entre grupos subespecíficos. En este último caso, para obtener análisis válidos son necesarias muestras homogéneas muy grandes, que a menudo es difícil o imposible obtener. El comportamiento sexual puede afectar a la homogeneidad de los caracteres, y la estructura de distribución puede resultar a veces difícil de individuar y, por tanto, crear confusión en el muestreo. Es posible obtener de antemano, a partir de la teoría estadística, el volumen óptimo de la muestra, determinado en parte por el grado de exactitud deseado. Sólo en raras ocasiones será posible un muestreo preciso de las poblaciones de mamíferos marinos.

G.D. Sharp
Fishery Resources Officer, Fishery Resources and Environment Division, Fisheries Department, FAO, Via delle Terme di Caracalla, 00100 Rome, Italy

Ardent use of biochemical techniques over the last 20 years, particularly gel electrophoresis, has provided extensive information about the species relationships and subspecific structure of many taxonomic groups. Biochemical genetic data are such that when distinct differences occur, little needs to be done other than write up the "obvious" distinction(s) between contrasted materials and define "groups". The typical picture is one where the farther apart taxonomically two species are, the less likely they are to share any electrophoretic characters, and vice versa. The lower limit to this process is one where species comparisons show no electrophoretic distinctions other than frequency differences between alleles of enzymatic systems. At this level, taxonomy is in constant flux, in that these relationships have been traditionally considered to be sub-specific in nature, resulting in such labels as sibling species, sub-population or race, and deme. Labelling depends upon the particular species group and information available to investigators concerning the ethological and geographical structure of the populations and samples which were used to evaluate the relationships.

Of particular historical interest is the fact that virtually all sub-specific studies in which biochemical polymorphisms have been used to characterize genetically distinctive groups have been carried out on relatively immobile terrestrial species. Few studies of migratory birds, fishes or marine mammals have successfully described sub-specific structure using only biochemical techniques. This is primarily because of the requirements for rigorous statistical evaluation of frequency data. The problem is directly related to statistical methods appropriate to evaluation of frequency or classification data, and to the relative homogeneity of the study species (Sharp, 1972).

Early work using serology for discrimination of marine mammal populations was generally ineffective for any but broad-scale ocean-to-ocean or hemispheric comparisons, again primarily because of the non-parametric nature of data where each individual sample generates only 1 count. Evaluation of biochemical genetic data in diploid organisms results in an accountability for each of 2 genes per locus, which doubles the information per individual as compared to serological data. The few reported studies of mobile marine populations which appear to exhibit distinguishing features (phenotype frequency and gene frequency distributional differences) have been based on *small sample, replicate sampling techniques*. In these studies, a 3-stage (individual/lots/replicates) sampling model was used as opposed to a 2-stage model (individuals/aggregations). A limitation imposed by the 3-stage model is that one must assume something about the relationship of the replicates in order to interpret the data. This method may not be appropriate unless stringent criteria are met. The primary criterion is an interesting dichotomy in that one must assume homogeneity of samples with respect to genetic characteristics. This is really what one is supposed to be evaluating in the rigorous experimental design and seems an illogical *a priori* assumption. Some workers attempt to get around this problem semantically, by definitions of stock or race or sub-population which involve the sampling area, rather than species and hypothetical sub-specific distributions. Indeed, a large part of the taxonomic disarray at present is caused by similar logical confusion where geographical sampling has become relatively continuous over time and where the original descriptions of "species" tended to be rather patchy in a geographical context. The resulting increase in morphological variation with respect to area as compared to the rather sharp breaks between areal "types" in the historical data has caused a trend toward reduction of the number of accepted species, often blurring obvious sub-specific information.

In two separate studies of small cetacean populations in the eastern Pacific Ocean, with which I am familiar (Perrin, 1975; Evans, 1975), racial characterizations have been made using multivariate statistical analyses of skull morphometrics, non-parametric examination of colour patterns and geographic stratification. These two data sets were sufficient for rigorous statistical characterization of the sub-specific groups for the species and components examined. In none of the cases was the sample size large enough to be useful for rigorous biochemical characterization of these groups (unless great differences exist). In an exploratory assay of red blood cell and serum proteins, it was impossible to distinguish between all but one of the racial groups of *Stenella* species for which materials were available, or in some instances the two eastern Pacific species. In some characters (haemoglobins and lactate dehydrogenases), no distinction between several *Stenella* races, *Delphinus*, *Steno*, *Orcinus* or *Tursiops* from the eastern Pacific could be found using various starch gel electrophoresis techniques (Figs. 1 and 2). The non-*Stenella* species listed could, however, be distinguished from electrophoretic data using other enzymatic material. Glutamate oxaloacetate transaminase (GOT), tetrazolium oxidase, and a serum globin observed by

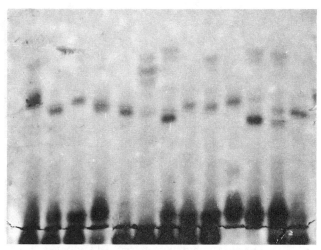

FIG. 2. — The electrophoretic patterns of haemoglobins and unidentified globins from blood preparations of 13 odontocete cetaceans. The haemoglobins are the broad dark band near the bottom of the figure, the globins the more concise bands near the top. The samples from left to right are: *Orcinus orca*, *Steno bredanensis*, *Delphinus delphis* (Costa Rica), *Globicephalus* sp., *Lagenorhynchus obliquidens*, *Lissodelphis borealis*, *Tursiops gilli* (Hawaii), *Stenella attenuata*, *Stenella longirostris* (eastern spinner), *Feresa attenuata*, *Tursiops truncatus*, *Tursiops* X *Steno* (hybrid) and *Steno bredanensis* (mother). Note particularly the obvious additive nature of the hybrid's pattern with respect to the parent species patterns. The physical oxygen binding dynamics of the haemoglobins of several of these species and their electrophoretic similarities have been described elsewhere (Sharp, 1975).

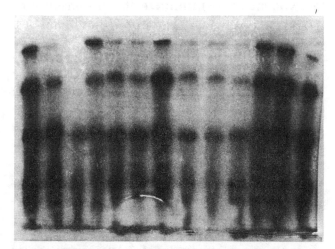

FIG. 1. — The electrophoretic patterns of the serum lactate dehydrogenase of 9 species of odontocete cetaceans indicate a remarkable conservatism as evidenced by their similarity. The only distinguishable pattern is that of *Feresa attenuata*, a species known to differ from most of the others by having fewer chromosomes (Duffield, pers. comm.). The samples in order from left to right are: *Steno bredanensis*, *Globicephalus* sp., *Feresa attenuata*, *Orcinus orca*, *Tursiops truncatus*, *Lagenorhynchus obliquidens*, *Tursiops gilli* (Hawaii), *Tursiops truncatus*, *Tursiops gilli* (Baja California), *Lagenorhynchus obliquidens*, *Steno bredanensis* (mother), *Steno* X *Tursiops* (hybrid) and *Tursiops truncatus*.

non-specific protein staining were adequate for discrimination of these species. A general paucity of biochemical polymorphisms as compared to those reported in terrestrial species is observed in the blood proteins of the delphinid cetaceans. An interesting example of hybridization can be verified from Figs. 2 and 3. A *Steno* mother and a hybrid calf with a proposed *Tursiops truncatus* father are compared and the calf is obviously the biochemical hybrid of the two parent species.

Sub-specific comparisons using biochemical genetic data are extremely dependent upon obtaining *large* samples from known, homogeneous components, and on the presence of polymorphisms. The social behaviour of sub-specific components of some species may seriously affect results of sampling from these species. For example, the occurrence of polygyny or dominance in sexual behaviour can affect phenotype distributions dramati-

FIG. 3. – The electrophoretic patterns of 13 odontocete ceta-cean's tetrazolium oxidase indicate both similarities and gross dissimilarities among the species represented. The order of samples, from left to right, is the same as in Fig. 2. Note again the hybrid's characteristic additive pattern with respect to the parental types.

cally by increasing occurrence of one allele in small deme-like components and can generate phenotype distributions which appear to be in disequilibrium with respect to random mating Hardy-Weinberg equilibrium expectations. The problem of sample homogeneity is one which needs to be examined carefully in all aspects of population differentiation and requires careful logical sampling design.

Examples of "superherds" that are heterogeneous as to race and species are commonly observed in the small cetaceans in the eastern Pacific. These race and species aggregations occur in what appear to be environmentally controlled distributions which, like the weather which governs them, are unpredictable. Overlay this effect with seasonal migration behaviour and serious problems arise in defining geographically rigid distribution patterns. All of this leads to serious problems of resampling and characterizing sub-specific components with transient distributional characteristics.

These problems are typical of the large mobile pelagic species, fish and mammals alike, and as such require a rigorousness in sampling which is often difficult or impossible to attain. For example, a relative merit of ap-

plying biochemical genetic techniques to studies of population structure in marine mammals is that when comparing species it is often possible to compare the individual species from materials which do not require difficult preparations such as flensing, bone cleaning or morphometrics. The problems of using biochemical techniques for sub-specific evaluations are the sheer numerical requirements for the discrimination of genetically distinct components. In such analyses, we are testing the null hypothesis that the frequency of a particular allele in one sample (P_1) is different from that in another (P_2) and we state the following:

$$H_0 : P_1 = P_2$$

where the alternative hypothesis is:

$$H_A : P_1 \neq P_2$$

Rigorous evaluation of these hypotheses necessitates that one decides with some certainty whether or not the hypothesis (H_0) has truly been tested. In statistical terms, for a rigorous test you need to minimize the probability of having made a type II error (β) subject to a fixed low probability of making a type I error (α). Biochemical data (serology or electrophoresis) are classification or non-parametric data and the statistic of choice is chi-square (χ^2) or its approximation (G test).

From statistical theory it is possible to derive optimum sample size requirements for discrimination between components based on:

— the minimum difference (Δ) one is willing to accept as being significant between 2 populations ($P_1 - P_2 = \Delta$);

— the significance level (α) for rejection of H_0;

— the power of the ($1 - \beta$);

— the expected range of values for P.

Most studies concern themselves only with α and some nebulous Δ which turns out to be directly related to the size of samples on hand, rather than having been decided *a priori*. Sample size optimization should be examined before sampling is initiated so as to evolve rapidly a rigorous sampling methodology, as sampling is expensive, time- and energy-consuming and opportunities are often transient.

Where the decision for equal sample size is made, $n_1 = n_2$ and $n_1 + n_2 = N$, so discrimination between 2 samples with unique relative frequency values, P_1 and P_2, for the major allele is dependent upon the range of the relative frequency (P) and the magnitude of the difference between the 2 P values, Δ. The optimum sample size for discriminating differences of magnitude Δ is determined by the rigour desired, and is accomplished by setting the α and β error at predetermined levels. The (α) significance level is generally set by tradition ($\alpha = .05$) and the other is set by the need to know if $P_1 = P_2$ (e.g., budget).

From Chapman and Nam (1968) and Meng and Chapman (1966) sample size can be estimated from the relation

$$\lambda = \frac{\Delta^2 n_1 n_2}{P (1 - P) N} \quad \dots \dots \dots \dots (1)$$

where for the case of equal sample sizes this simplifies to

$$\lambda = \frac{\Delta^2 N}{4 (P) (1 - P)} \quad \dots \dots \dots \dots (2)$$

where λ is the non-centrality parameter for the non-central chi-square distribution and can be found in collections of statistical tables (Ambramowitz and Stegun, 1964) and N is the number of genes to be sampled. Several minimum sample size criteria must be met to satisfy the statistical requirements for unbiased analyses utilizing chi-square methods or the approximations to these methods. A limiting factor is that for valid chi-square tests the classes of expected frequencies should have 5 or more events in them and this limitation can be calculated by the relation

$$\frac{5}{1 - P} \text{ for } P > .50$$

From equation (2) and a table of critical values of the non-central chi-square distribution (λ), values of sample sizes can be calculated which show the effects of varying Δ, P and the $1 - \beta$ (Power), for fixed $\alpha = .05$, on the determination of the optimum sampling for these various characteristics and limitations. In the biochemical genetic studies N_i = individuals per sample $= \dfrac{N}{2}$ then, the relations for determining the sample size is

$$N_i = 2 \frac{(P) (1 - P) (\lambda)}{\Delta^2} \quad \dots \dots \dots (3)$$

The P value used in this equation is best considered to be that P appropriate to the largest sample examined to date or, conservatively, .55 where no prior information exists.

The following tabled values are the number of individuals per sample which would permit the differentiation of 2 populations with the differences of the magnitude Δ for the frequencies listed (P) at $\alpha = .05$ and the appropriate power level. The investigation of more than one genetic system for discrimination of samples imposes even more stringent sample size requirements due to the increasing probability of making decision errors with each additional analysis. These problems are further explored in Sharp, Mobrand and Francis (MS).

Δ P values =	.95	.90	.80	.70	.55	Power
.05	146	276	492	645	760	$1 - \beta = .5$
.10	*50*[1]	69	123	162		$\lambda = 3.84$
.20	*50*	25	31	40	48	
.50	*50*	25	*13*	9	6	

Δ P values =	.95	.90	.80	.70	.55	Power
.05	299	564	1006	–	–	$1 - \beta = .8$
.10	76	141	252	332	389	$\lambda = 7.85$
.20	*50*	27	64	82	99	
.50	*50*	25	*13*	14	16	

Δ						Power
.05	400	756	–	–	–	$1 - \beta = 9$
.10	102	189	337	444	520	$\lambda = 10.51$
.20	*20*	50	85	110	132	
.50	*50*	25	14	20	22	

[1] The values in italics represent lower limits determined from the minimum frequency requirements previously discussed.

Only rarely will marine mammal researchers be afforded the opportunities to sample "rigorously" for population discrimination from a serological or biochemical genetic point of view. Species level discrimination can be accomplished from relatively limited sample sizes, but 20 or more individuals should be examined for contrasting species and for evaluation of polymorphisms within and among species.

Better methods for population discrimination in marine mammals utilizing smaller samples are exemplified by the studies of eastern Pacific *Stenella* and *Delphinus* species (Perrin, 1975; Evans, 1975). My impression is that biochemical and serological techniques and their sampling requirements lend themselves better to species identification problems than to sub-specific studies in studies of pelagic marine mammals.

References

ABRAMOWITZ, M. and J.A. STEGUN, Handbook of math-
1964   ematical functions. *Appl. Math. Ser.*, (55):
1044 p.

CHAPMAN, D.G. and J. NAM, Asymptotic power of
1968   chi-square tests for linear trends in proportion.
Biometrics, 24(3):315-27.

EVANS,   W.E., Distribution and differentiation of
1975   stocks of *Delphinus delphis* (L.) in the northeastern Pacific. Paper presented to the
Scientific Consultation on the Conservation
and Management of Marine Mammals and
their Environment, Bergen, Norway, 31 Au-
gust-9 September 1976. Rome, FAO,
ACMRR/MM/SC/18:72 p.

MENG, R. and D.G. CHAPMAN, The power of chi-square
1966   tests for contingency tables. *J. Am. Statist.
Assoc.*, 61:965-75.

PERRIN, W.F., Variation of spotted and spinner porpoise
1975   (genus *Stenella*) in the eastern Pacific and
Hawaii. *Bull. Scripps Inst. Oceanogr.*, (21):206
p.

SHARP, G.D., Studies of the biochemical genetics of
1972   yellowfin tuna of the eastern Pacific Ocean.
Doctoral dissertation. Scripps Institution of
Oceanography, 75 p.

—, A study of the electrophoretic and O_2 dissociation
1975   properties of the hemoglobins of some
delphinid cetaceans. *Comp. Biochem. Physiol.
(A Comp. Physiol.)*, 51:673-81.

SHARP, G.D., L.E. MOBRAND and R.C. FRANCIS, A tabulation and discussion of the effect of power
and other parameters on sample size for X^2
test of similarity of genetic frequency data
(MS in preparation).

A CENSUS OF CAPTIVE MARINE MAMMALS IN NORTH AMERICA

L.H. CORNELL and E.D. ASPER

Abstract

A census of marine mammals in 193 institutions in Canada and the United States made from 1 July to 20 August 1976 showed 1 135 animals in captivity, 301 in 69 zoos and 834 in 28 oceanaria and aquaria. Of these, 755 were pinnipeds (of which about half were California sea lions, *Zalophus californianus*), 359 smaller cetaceans (mostly bottlenose dolphins, *Tursiops truncatus*), 12 manatees, *Trichechus* spp., and 9 sea otters, *Enhydra lutris*. Nine hundred and eighty-eight were collected specifically for display and related research, most of them before the passage of the Marine Mammal Protection Act of 1972 (MMPA) in the United States, and 147 were animals stranded and later salvaged. The average longevities, by species, of animals acquired before the MMPA and of all animals now in captivity, are given.

More than 130 million people viewed captive marine mammals in 1975, and each year about one million students take part in educational programmes that use captive animals. Four thousand people were employed in 1975 by the 28 oceanaria and aquaria; during the same year, Sea World in San Diego, California, and the oceanaria and aquaria in Florida provided about US$ 25 million and US$ 225 million, respectively, in revenue to their communities (after application of the US economic dollar multiplier and including taxation). Research on captive marine mammals has provided much information relevant to animals in the wild. Captive breeding programmes are still in an early stage: data on 14 species shows 58 animals alive out of 228 births. Canada, Mexico and the United States have strict controls on collection of marine mammals. Of the 191 permits applied for in the United States since 1972 for the collection of marine mammals, 107 have been for the capture of 757 animals for public educational display; 71 of these applications have been for the capture of 757 animals for public educational display; 71 of these applications have been granted, comprising 357 animals or 0.7 percent of all animals for which permits have been issued.

Résumé

Un recensement effectué du 1er juillet au 20 août 1976, portant sur les mammifères marins de 193 institutions du Canada et des Etats-Unis, a révélé 1 135 animaux en captivité, 301 dans 69 zoos et 834 dans 28 océanoriums et aquariums. Sur ce nombre, 755 étaient des pinnipèdes (dont la moitié environ étaient des lions de mer de la Californie, *Zalophus californianus*), 359 des petits cétacés (surtout des dauphins à gros bec, *Tursiops truncatus*), 12 des lamantins, *Trichechus* spp. et 9 loutres de mer, *Enhydra lutris*. Neuf cent quatre-vingt dix-huit étaient rassemblés spécifiquement aux fins d'exposition et de recherche, la plupart avant l'adoption de la loi sur la protection des mammifères marins de 1972 (MMPA) aux Etats-Unis et 147 étaient des animaux jetés à la côte et sauvés par la suite. On donne la longévité moyenne, par espèce, des animaux acquis avant la MMPA et de tous les animaux à présent en captivité.

Plus de 130 millions de personnes ont vu des mammifères marins en captivité en 1975 et chaque année, environ un million d'étudiants participent aux programmes didactiques utilisant des animaux en captivité. Quatre mille personnes étaient employées en 1975 dans les 28 océanoriums et aquariums; cette même année, "Sea World" à San Diego, Californie, et les océanoriums et aquariums de Floride ont fourni des revenus s'élevant respectivement à 25 millions et 225 millions de dollars E.U. à leur communauté (après application du multiplicateur économique E.U., impôts compris). La recherche sur les mammifères marins en captivité a fourni de nombreuses informations applicables aux animaux en liberté. Les programmes de reproduction en captivité n'en sont qu'à leurs débuts: les données relatives à 14 espèces font état de 58 animaux vivants sur 228 naissances. Le Canada, le Mexique et les Etats-Unis ont établi un strict contrôle sur les collections de mammifères marins. Sur 191 demandes de permis déposées aux Etats-Unis depuis 1972 pour la capture de mammifères marins, 107 portaient sur la capture de 757 animaux destinés à être exposés au public dans des buts didactiques; 71 ont été accordés, portant sur 357 animaux, soit 0,7 pour cent du nombre total d'animaux pour lesquels des permis ont été accordés.

Extracto

Un censo de los mamíferos marinos existentes en 193 instituciones de Canadá y los Estados Unidos hecho desde el 1° de julio al 20 de agosto de 1976 reveló la presencia de 1 135 animales en cautividad: 301 en 69 parques zoológicos y 834 en 28 oceanarios y acuarios. De ellos, 755 eran pinípedos (la mitad, aproximadamente, leones marinos de California, *Zalophus californianus*), 359 cetáceos menores (la mayoría tursones, *Tursiops truncatus*), 12 manatíes (*Trichechus* spp.) y 9 nutrias marinas (*Enhydra lutris*). De esos animales, 988 se habían recogido específicamente para exposición e investigaciones, la mayoría de ellos antes de la aprobación de la Ley de Protección de los Mamíferos Marinos de 1972 en los Estados Unidos, y 147 eran animales que habían quedado encallados y habían sido salvados. Se da la longevidad media, por especies, de los animales adquiridos antes de la entrada en vigor de la ley de 1972 y de todos los animales actualmente en cautividad.

En 1975, más de 130 millones de personas tuvieron ocasión de visitar mamíferos marinos en cautividad, y cada año un millón de estudiantes participan en programas educativos en los que se utilizan animales cautivos. En 1975, las personas empleadas por los 28 oceanarios y acuarios eran 4 000; durante el mismo año, el Sea World de San Diego (California) y los oceanarios y acuarios de Florida aportaron a sus comunidades, como ingresos, 25 millones y 225 millones de dólares EE.UU., respectivamente (tras haber aplicado el multiplicador económico correspondiente e incluidos impuestos). Las investigaciones sobre mamíferos marinos en cautividad han facilitado muchos datos aplicables a los animales que viven en libertad. Los programas de reproducción en cautividad están aún en fase inicial: los datos disponibles sobre 14 especies indican que de 228 animales nacidos sobreviven 58. Canadá, México y los Estados Unidos controlan estrictamente la captura de mamíferos marinos vivos. De los 191 permisos solicitados en los Estados Unidos desde 1972 para capturar mamíferos marinos vivos, 107 se referían a la captura de 757 animales para exposición al público; se han aprobado 71 de esas solicitudes, para un total de 357 animales, o sea el 0,7 por ciento de los animales para cuya captura se ha concedido permiso.

L.H. Cornell
Department of Economics and Commerce, Simon Fraser University, Burnaby, British Columbia, V5A 1S6, Canada

E.D. Asper
Sea World of Florida, 7007 Sea World Drive, Orlando, FL 32809, USA

Introduction

Marine mammals have intrigued mankind for thousands of years. They have been revered, sometimes feared, but consistently admired.

The application of modern techniques to water purification and food handling, along with general husbandry techniques learned from other animals, led naturally to the establishment of colonies of wild marine mammals in man-made controlled environments. In their early form the presentations were static and the animal was viewed from above in pools or holding facilities. Later, glass windows allowed underwater viewing. Some of the earlier pinniped displays were in the form of circus acts, since it was obvious that many of these animals could be trained and they adapted easily to a semi-aquatic fresh water environment.

As time passed, displays became more sophisticated, and in the late 1930's, an institution in Florida successfully displayed a colony of bottlenose dolphins in a huge pool. Today, in North America (United States and Canada) there are approximately 200 zoos and aquaria/oceanaria.

In the past, such collections have been blamed for the suspected depletion of marine mammal populations in some areas.

The assembly of statistics on a world-wide basis is an awesome task and would be impossible without a thorough investigation. However, in the United States, statistical data for North America has become far easier to obtain, following the implementation of the Marine Mammal Protection Act of 1972.

Since this Act went into effect, record keeping by the various institutions engaged in public educational display has been more than creditable. The Act itself does not require record keeping except in those instances where the Act would be implemented for the capture and husbandry of marine mammals, i.e., animals collected by permit after December, 1972.

Since December 1972 it has become necessary to apply for a permit if marine mammals are to be collected in the United States. The application should include pertinent records of the institution's recent history and expertise with marine mammals; some of the other data required are the collection site, age, sex and size of the animal to be collected, and the longevity of the existing collection. Upon application, this information becomes retroactive to include even those animals maintained prior to passage of the Marine Mammals Protection Act of 1972.

Because of these requirements, collecting data for this paper was, in most cases, simply a matter of inquiry to the various institutions involved in the educational display of marine mammals.

Method of survey

A total of 193 institutions were contacted from July 1 to August 20, 1976, and all responded. 96 did not have marine mammals on exhibit at the time. Of the 97 institutions possessing or displaying marine mammals, 69 were classified as zoos with or without aquaria, and 28 as oceanaria or aquaria whose principal concern is to display marine mammals.

The major source book for the zoos, aquaria and oceanaria that were contacted within the United States and Canada was the 1975 Directory of the American Association of Zoological Parks and Aquariums (AAZPA). No approach was made to institutions that were known never to have had marine mammals.

In a few instances the statistics were compiled through the appropriate Federal and State agencies and not through the institution itself. In these cases the information was up to date as required by law.

Numbers in captivity

As indicated in Table 1, there are 1 135 marine

Table 1. Public educational display census [1] - USA and Canada

Marine mammals	Marine mammals in zoos	Marine mammals aquaria/oceanaria	Total marine mammals
Family Delphinidae			
Delphinus delphis	0	4	4
Globicephala sp.	0	7	7
Lagenorhynchus obliquidens	0	14	14
Orcinus orca	0	17	17
Pseudorca crassidens	0	2	2
Stenella sp.	0	12	12
Steno bredanensis	0	2	2
Tursiops truncatus (Atlantic)	7	266	273
Tursiops truncatus (Pacific)	0	13	13
Family Monodontidae			
Delphinapterus leucas	0	10	10
Family Phocoenidae			
Phocoena phocoena	0	1	1
Family Platanistidae			
Inia geoffrensis	2	2	4
		Total cetaceans	359
Family Trichechidae			
Trichechus inunguis	1	1	2
Trichechus manatus	1	9	10
		Total sirenia	12
Family Otaridae			
Arctocephalus sp.	12	0	12
Callorhinus ursinus	2	6	8
Eumetopias jubatus	0	12	12
Otaria sp.	9	1	10
Zalophus californianus	193	230	423
Family Odobenidae			
Odobenus sp.	2	8	10
Family Phocidae			
Cystophora cristata	0	1	1
Halichoerus grypus	20	10	30
Mirounga angustirostris	1	31	32
Pagophilus groenlandicus	0	2	2
Phoca sp.	51	164	215
		Total pinnipeds	755
Family Mustelidae			
Enhydra lutris	0	9	9
		Total sea otters	9

[1] August 1976.

mammals presently housed in North American zoos and aquaria/oceanaria. Of these, 301 are in zoos and 834 are in aquaria/oceanaria.

Educational benefits

In 1975, these relatively few marine mammals provided educational displays for over 130 million visitors, over one half of the population of the United States of America. This represents 114 537 visitors per animal per year.

In many of the major aquaria/oceanaria in North America, in addition to the regular educational displays, specially designed classes and educational programmes for students of all ages are currently under way, reaching, on average, one million students per year. The accredited teachers follow curricula, and employ visual aids and learning experiences with resident animals. The experience thus gained should help our future citizens to evolve appropriate management programmes for future generations of both man and marine mammals. At all creditable zoos there are educational programmes, docent and/or volunteer lectures and related programmes which include marine mammals.

Economic and employment benefits

Four thousand persons were employed in 1975 by the 28 aquaria/oceanaria in North America listed in the AAZPA 1975 Directory. In 1976 this number has increased by roughly 15 %.

Detailed figures were available from Sea World in San Diego, California. Nearly 800 people are employed there at the height of the annual season. The annual payroll for this group of employees is almost $4 million. By applying the US economic dollar multiplier this gives a benefit to the community of US$20 million. Rent, property tax and sales tax add up to a further US$1 016 600.00, or over $5 million after the multiplier has been applied, which reaches the community from the City of San Diego's share. While these figures will be different for other aquaria/oceanaria, they do indicate the value to the community of every dollar of such income. In a community such as San Diego, income generated (when the multiplier is applied on a yearly basis) is sufficient to maintain San Diego State University's operating budget for one year. Likewise, smaller aquaria have the ability to provide income for community education in proportion to their size.

The example of Florida is biased by the fact that it is a coastal state with more aquaria/oceanaria than many other states. In 1975, a recession year, 25 million out-of-state visitors came to Florida, generating a revenue of US$1.87 billion. The application of the multiplier suggests that nearly US$9 billion reached Florida communities from these visitors. Seventeen percent of the 25 million visited institutions listed as aquaria/oceanaria by the AAZPA; 1.5 million in-state residents made up the rest of their attendance. Direct revenue from all visitors in 1975 to these institutions was US$38 million, or US$192.5 million in revenue to the Florida communities after application of the multiplier. Taxation on the income thus generated yielded US$6.5 million to the state (US$32.5 million after application of the multiplier).

Relevant income to Florida zoos could not be calculated, due to the difficulty of discriminating revenue generated by marine mammals from that generated by other animals.

Impact on wild populations

Much attention is generally focused upon the impact on wild populations of col-

Table 2. **Percentage of captive marine mammals acquired prior to the Marine Mammal Protection Act, and average longevity by species currently being maintained or displayed** [1]

Species	Current percentage of captive pre-Act marine mammals	Current average longevity of captive pre-Act marine mammals (years)
Tursiops truncatus (Atlantic)	60 %	6.1
Tursiops truncatus (Pacific)	85 %	6.4
Lagenorhynchus obliquidens	91 %	5.3
Orcinus orca	93 %	7.2
Stenella sp.	37 %	7.0
Delphinapterus leucas	50 %	7.2
Inia geoffrensis	100 %	8.5
Globicephala sp.	66 %	7.0
Pseudorca crassidens	100 %	8.0
Steno bredanensis	50 %	12.0
Zalophus californianus	60 %	6.1
Phoca sp.	58 %	7.0
Halichoerus grypus	93 %	6.9
Mirounga angustirostris	32 %	4.4
Eumetopias jubatus	91 %	4.9
Odobenus sp.	90 %	9.9
Otaria sp.	13 %	6.0
Callorhinus ursinus	83 %	8.4
Pagophilus groenlandicus	100 %	3.0
Trichechus inunguis	100 %	8.5
Trichechus manatus	60 %	18.0
Enhydra lutris	56 %	5.8

[1] August 1976.

lecting marine mammals.

Of the 1 135 marine mammals currently held, 988 were collected specifically for public educational display and closely related research. The majority of these were collected prior to the passage of the Marine Mammal Protection Act of 1972. Another 147 were "stranded" or "beached" animals which were salvaged and which recovered sufficiently from their illnesses to allow their maintenance for research and/or display. The survival of stranded animals represents a substantial veterinary medical achievement. Of the 147, the average longevity to date is 3.3 years. Table 2 shows the percentage by species of marine mammals currently displayed or maintained and taken prior to December, 1972, and the average longevity of each species to date.

For some marine mammals currently being maintained, no collection or longevity data were available. Some had been in captivity so long that collection dates had been lost or misplaced. Other records were incomplete because personnel changes over the years had depleted or interrupted the keeping of accurate records, and in a few cases records had never been kept at all.

Longevity figures in this paper were only calculated for those marine mammals whose records were accurately maintained both pre-Act and post-Act. They are therefore conservative.

Table 3 shows, by species, the number and percentage of marine mammals whose records accurately reflect their longevity.

Combined longevity figures show a lower average than pre-Act (before December, 1972) longevity figures. Less than four years have passed since the implementation of the Act, therefore the maximum longevity of animals collected after the Act would be 3 years 8 months. Most post-Act animals have not been

Table 3. Current number of marine mammals being maintained in captivity, percentage of accurate longevity data available, and average longevity of each species [1]

Species	Current number in captivity	Percent accurate longevity data available	Average longevity in captivity (years)
Tursiops truncatus (Atlantic)	273	60 %	5.4
Tursiops truncatus (Pacific)	13	100 %	5.7
Lagenorhynchus obliquidens	14	79 %	4.8
Orcinus orca	17	82 %	6.8
Stenella sp.	12	92 %	3.5
Delphinapterus leucas	10	80 %	3.9
Delphinus delphis	4	100 %	1.0
Inia geoffrensis	4	100 %	8.5
Globicephala sp.	7	86 %	4.3
Pseudorca crassidens	2	100 %	8.0
Steno bredanensis	2	100 %	6.0
Phocoena phocoena	1	100 %	<1.0
Zalophus californianus	423	84 %	4.2
Phoca vitulina	215	83 %	4.6
Halichoerus grypus	30	100 %	6.6
Mirounga angustirostris	32	75 %	2.5
Eumetopias jubatus	12	92 %	4.6
Odobenus sp.	10	100 %	9.0
Arcticephalus sp.	12	75 %	2.0
Otaria sp.	10	80 %	2.2
Callorhinus ursinus	8	75 %	7.2
Pagophilus groenlandicus	2	100 %	3.0
Cystophora cristata	1	100 %	<1.0
Trichechus inunguis	2	100 %	8.5
Trichechus manatus	10	50 %	11.0
Enhydra lutris	9	100 %	3.2

[1] August 1976.

in captivity for so long, and many have only been recently acquired.

Research and breeding

Research and captive breeding are two important activities that stem from maintaining species in zoos and aquaria/oceanaria. Both benefit the wild populations in many ways.

Research on captive animals of all kinds has solved many problems of disease and others with direct relevance to wild populations. Captive animals have lent themselves to research programmes on wild food supplies, nutritional studies, disease control, population dynamics, environmental support and the establishment of proper wildlife reserves, to name but a few fields of study. Equally important is the establishment of captive breeding colonies for zoological specimens, particularly for some endangered species which in zoos are reproducing with greater efficiency than in the wild. It is not also inconceivable that release to the wild of captive-bred stocks could aid the recovery of the species involved.

The breeding programmes for marine mammals now in progress are in the preliminary stages. Many births have been unplanned. It is not inconceivable that future captive stocks of some marine mammals could be

143

Table 4. Births of marine mammals in captivity and number still alive (by species) from accurate current data

Species	Number of births	Number still alive
Tursiops truncatus [1]	107	22
Lagenorhynchus obliquidens	2	0
Delphinus delphis	2	0
Stenella sp.	3	1
Delphinapterus leucas	2	0
Steno bredanensis X *Tursiops truncatus*	1	0
Odobenus r. divergens	2	0
Zalophus californianus	70	15
Phoca vitulina	22	11
Halichoerus grypus	10	5
Cystophora cristata	1	0
Otaria sp.	1	0
Enhydra lutris	2	1
Trichechus sp.	3	3
Total	228	58

[1] The data on *Tursiops* was both collected directly and summarized from the *breeding dolphins present status, suggestions for the future*, edited by S.H. Ridgway and K. Benirschke (Workshop held December, 1975). Marine Mammal Commission Report MMC-76/07, November 1977.

supplied wholly from captive breeding populations.

Table 4 shows, by species, the numbers of births which have taken place in zoos and aquaria/oceanaria. These figures represent breeding statistics from 53 % of the zoos and 67 % of the aquaria/oceanaria surveyed. Most losses were due to stillbirths or occurred before one month of age, reflecting the early state of the art.

Table 4 does not distinguish between animals conceived in captivity and those conceived in the wild. The figures are conservative, since a large amount of data provided from memory also was discarded.

13 species are represented here as well as a *Tursiops truncatus* — *Steno bredanensis* hybrid.

Future impact

No information is available on the past impact on wild stocks of marine mammals collected for public educational display. However, as far as North America is concerned, both the future scale of collection and its potential impact can be estimated, thanks to the rigid controls enacted by Canada, the United States and Mexico on collecting programmes of any kind.

Since 1972, there have been 191 permit applications to the United States Government for the collection of marine mammals for scientific research and public educational display. Of these, 84 requests were for scientific research, representing 57 990 marine mammals. The remaining 106 requests were for public educational display, representing a potential removal from the wild of 737 marine mammals. One additional application combining research and public educational display asked for 120 animals for research and 20 for public educational display.

To date, a total of 137 applications have been granted. 66 were for scientific research, representing a total of 51 049 animals or 93 % of all animals for which permits have been granted. Of these, 4 046 animals are to be sacrificed as indicated in the permit; 45 398 animals are to be taken and released; 1 322 were beached or found dead; and 283 are to be taken and kept alive for research purposes.

71 applications have been granted for public educational display representing 357 animals (or 0.7 %). It is obvious from these figures and from trends of the past few years that the future impact of the public educational display industry on wild population stocks will be negligible.

Acknowledgements

The authors wish to express their gratitude for the tremendous cooperation received from the National Marine Fisheries Service, National Oceanic and Atmospheric Administration, Department of Commerce, Washington, D.C. (Mr Robert Brumstead); State of Florida, Florida Marine Patrol (Major L.W. Shelfer); the zoological membership of the American Association of Zoological Parks and Aquariums, and the cooperation of each director and individual who, without fail, contributed their information and expressed their appreciation for this census. And special thanks to Dr Deborah Duffield and her aides for conducting so thoroughly the survey needed to complete the census.

We wish to acknowledge the cooperation of the following zoological parks, aquaria, oceanaria, Federal and State offices contacted during this census:

Birmingham Zoo
Birmingham, Alabama

Phoenix Zoo
Phoenix, Arizona

Randolph Park Zoo
Tucson, Arizona

Marineland of the Pacific
Palos Verdes Estates, California

Marine World
Redwood City, California

Steinhart Aquarium
San Francisco, California

Sea World, Inc.
San Diego, California

Zoological Society of San Diego
San Diego, California

Santa Barbara Zoological Garden
Santa Barbara, California

San José Zoological Gardens
San José, California

Los Angeles Zoo
Los Angeles, California

Baby Zoo
San José, California

Children's Fairyland U.S.A.
Oakland, California

Oakland Baby Zoo
Oakland, California

San Francisco Zoological Gardens
San Francisco, California

Morro Bay Aquarium
Morro Bay, California

Knowland Park Zoo
Oakland, California

Sacramento Zoo
Sacramento, California

Busch Gardens
Van Nuys, California

Sea Wonders Aquarium and Serpentarium
Crescent City, California

Shipwreck Aquarium
Eureka, California

T. Wayland Vaughan Aquarium
La Jolla, California

Applegate Zoo
Merced, California

Magic Mountain Animal Farm
Valencia, California

Living Desert Reserve
Palm Desert, California

Junior Museum
Walnut Creek, California

Denver Zoological Gardens
Denver, Colorado

Cheyenne Mountain Zoo
Colorado Springs, Colorado

City Park Zoo
Pueblo, Colorado

Stamford Museum & Nature Center
Stamford, Connecticut

Grande Game Farm
Wessuffield, Connecticut

Wild Animal Farm
Willington, Connecticut

Mystic Marinelife Aquarium
Mystic, Connecticut

Brandywine Zoo
Wilmington, Delaware

National Zoological Park
Washington, D.C.

Sea World of Florida
Orlando, Florida

The Gulfarium
Fort Walton Beach, Florida

Ocean World, Inc.
Fort Lauderdale, Florida

Miami Seaquarium
Miami, Florida

Gulf World
Panama City, Florida

Coral Lagoon Resort
Marathon, Florida

Ocean Reef Club
Key Largo, Florida

Sugar Loaf Lodge
Sugar Loaf Key, Florida

South Florida Museum
Bradenton, Florida

Waltzing Waters
Pine Island, Florida

Atlanta Zoological Park
Atlanta, Georgia

Tift Zoo
Albany, Georgia

Sea Life Park
Waimanolo, Hawaii

Honolulu Zoo
Honolulu, Hawaii

Kahala Hilton
Honolulu, Hawaii

Ross Park Zoo
Pocatello, Idaho

Boise City Zoo
Boise, Idaho

John G. Shedd Aquarium
Chicago, Illinois

Chicago Zoological Park
Brookfield, Illinois

Lincoln Park Zoological Gardens
Chicago, Illinois

Plainsmen Zoo
Elgin, Illinois

Robinson Zoo
Springfield, Illinois

Miller Park Zoo
Bloomington, Illinois

Animal Kingdom Zoo
Chicago, Illinois

Glen Oak Zoo
Peoria, Illinois

Niabi Zoological Preserve
Moline, Illinois

Children's Zoological Gardens
Fort Wayne, Indiana

Glen Miller Park Zoo
Richmond, Indiana

Mesker Park Zoo
Evansville, Indiana

Indianapolis Zoological Park
Indianapolis, Indiana

Clifty Acres Zoo Farm
Hanover, Indiana

Washington Park Zoological Gardens
Michigan City, Indiana

Potawatomi Park Zoo
South Bend, Indiana

Des Moines Zoo
Des Moines, Iowa

City Park Zoo
Iowa City, Iowa

Weed Park Zoo
Muscatine, Iowa

Riverside Park & Ralph Mitchell Zoo
Independence, Kansas

Topeka Zoological Park
Topeka, Kansas

Brit Spaugh Park and Zoo
Great Bend, Kansas

Municipal Zoo
Wichita, Kansas

Sedgwick County Zoological Society
Wichita, Kansas

Louisville Zoological Garden
Louisville, Kentucky

Alexandria Zoological Park
Alexandria, Louisiana

Greater Baton Rouge Zoo
Baton Rouge, Louisiana

Louisiana Purchase Gardens & Zoo
Monroe, Louisiana

Audubon Park Zoo
New Orleans, Louisiana

Odenheimer Aquarium
New Orleans, Louisiana

Baltimore Zoo
Baltimore, Maryland

Salisbury Zoological Garden
Salisbury, Maryland

New England Aquarium
Boston, Massachusetts

Sealand of Cape Cod, Inc.
Brewster, Massachusetts

Walter D. Stone Memorial Zoo
Stoneham, Massachusetts

Capron Park Zoo
Attleboro, Massachusetts

Southwick Wild Animal Farm, Inc.
Blackstone, Massachusetts

Franklin Park Children's Zoo
Boston, Massachusetts

Franklin Park Zoo
Boston, Massachusetts

Children's Zoo
Springfield, Massachusetts

Kirkham Aquarium & Zoo
Springfield, Massachusetts

Worcester Science Center
Worcester, Massachusetts

International Animal Exchange, Inc.
Ferndale, Michigan

Animal Kingdom Wildlife Refuge
Byron Center, Michigan

Deer Forest
Coloma, Michigan

John Ball Zoological Gardens
Grand Rapids, Michigan

Potter Park Zoo
Lansing, Michigan

Saginaw Children's Zoo
Saginaw, Michigan

King Animaland Park
Richmond, Michigan

Como Zoo
St. Paul, Minnesota

Minnesota State Zoological Garden
St. Paul, Minnesota

Duluth Zoo
Duluth, Minnesota

Oxbow Park Zoo
Rochester, Minnesota

Jackson Zoological Park
Jackson, Mississippi

Dickerson Park Zoo
Springfield, Missouri

Kansas City Zoological Gardens
Kansas City, Missouri

St. Louis Zoological Park
St. Louis, Missouri

147

Red Lodge Zoo
Red Lodge, Montana

Lincoln Municipal Zoo
Lincoln, Nebraska

Henry Doorly Zoo
Omaha, Nebraska

Lincoln Children's Zoo
Lincoln, Nebraska

Riverside Park Zoo
Scottsbluff, Nebraska

Las Vegas Valley Zoo
Las Vegas, Nevada

Turtle Back Zoo
West Orange, New Jersey

Warner Bros. Jungle Habitat
West Milford, New Jersey

Terry Lou Zoo
Scotch Plains, New Jersey

Space Farms Zoological Park
Sussex, New Jersey

Cohanzick Zoo
Bridgeton, New Jersey

Rio Grande Zoological Park
Albuquerque, New Mexico

New York Aquarium
Coney Island, New York

Living Museum
Buffalo Museum of Science
Buffalo, New York

Central Park Zoo
New York, New York

New York Zoological Park
New York, New York

Seneca Park Zoo
Rochester, New York

Buffalo Zoological Gardens
Buffalo, New York

Staten Island Zoological Society
New York, New York

Burnet Park Zoo
Syracuse, New York

Utica Zoo
Utica, New York

Municipal Zoo
Asheville, North Carolina

North Carolina Zoological Park
Asheboro, North Carolina

Dakota Zoo
Bismarck, North Dakota

Sea World of Ohio
Aurora, Ohio

Cincinnati Zoological Society
Cincinnati, Ohio

Cleveland Zoological Park
Cleveland, Ohio

Columbus Zoological Gardens
Columbus, Ohio

Toledo Zoological Gardens
Toledo, Ohio

Akron Children's Zoo
Akron, Ohio

Oklahoma City Zoo
Oklahoma City, Oklahoma

Depoe Bay Aquarium
Depoe, Oregon

Seaside Aquarium
Seaside, Oregon

Portland Zoological Gardens
Portland, Oregon

Moore Park Zoo
Klamath Falls, Oregon

Pittsburgh Zoological Gardens
Pittsburgh, Pennsylvania

Underground Zoo
Children's Zoo and Aquarium
Pittsburgh, Pennsylvania

Erie Zoo
Erie, Pennsylvania

Elmwood Park Zoo
Norristown, Pennsylvania

Nay Aug Park Zoo
Scranton, Pennsylvania

Columbia Zoological Park
Columbia, South Carolina

Great Plains Zoo
Sioux Falls, South Dakota

Watertown Zoo
Watertown, South Dakota

Wylie Park Zoo
Aberdeen, South Dakota

Opryland, U.S.A.
Nashville, Tennessee

Overton Park Zoo and Aquarium
Memphis, Tennessee

Knoxville Zoological Park
Knoxville, Tennessee

Sea-Arama Marine World
Galveston, Texas

Seven Seas
Arlington, Texas

Abilene Zoological Gardens
Abilene, Texas

Dallas Aquarium
Dallas, Texas

Dallas Zoo
Dallas, Texas

Forth Worth Zoological Park and Aquarium
Fort Worth, Texas

Houston Zoological Gardens
Houston, Texas

Central Texas Zoo
Waco, Texas

Gladys Porter Zoo
Brownsville, Texas

Midland Zoo
Midland, Texas

San Antonio Zoo and Aquarium
San Antonio, Texas

El Paso Zoological Park
El Paso, Texas

Storyland Zoo
Amarillo, Texas

Hogle Zoological Garden
Salt Lake City, Utah

Lafayette City Park Zoo
Norfolk, Virginia

Seattle Aquarium
Seattle, Washington

Northwest Marineland, Inc.
Seattle, Washington

Woodland Park Zoological Gardens
Seattle, Washington

Inland Empire Zoological Society
Spokane, Washington

Good Zoo
Wheeling, West Virginia

Henry Vilas Zoological Park
Madison, Wisconsin

Milwaukee County Zoological Park
Milwaukee, Wisconsin

Racine Zoological Park
Racine, Wisconsin

The Ranch Zoo
Menomonee Falls, Wisconsin

Vancouver Public Aquarium
Vancouver, British Columbia, Canada

Quebec Zoological Garden
Orsainville, Quebec, Canada

The Aquarium
Provincial-Ste-Foy, Quebec, Canada

Montreal Aquarium
Montreal, Quebec, Canada

Calgary Brewery Aquarium
Calgary, Alberta, Canada

Calgary Zoo
Calgary, Alberta, Canada

Storyland Vallery Zoo
Edmonton, Alberta, Canada

Stanley Park Zoo British Columbia
Vancouver, British Columbia, Canada

P.E.I. Wildlife Park
North Rustico, Prince Edward Island, Canada

Granby Zoological Society
Granby, Quebec, Canada

Jardin de Merveilles
Montreal, Quebec, Canada

Assiniboine Park Zoo
Winnipeg, Manitoba, Canada

Bowmanville Zoo
Bowmanville, Ontario, Canada

Sealand of the Pacific
Victoria, British Columbia, Canada

National Marine Fisheries Service
Washington, D.C.

National Marine Fisheries Service
California Regional Office

California Department of Fish and Game
Division of Law Enforcement

Department of Natural Resources
Tallahassee, Florida

THE TRANSITION FROM FISHERIES TO MARINE FARMING AS THE PRIMARY LONG-TERM OBJECTIVE OF MARINE MAMMAL MANAGEMENT[1]

A.V. Yablokov

Abstract

Because of ecological and economic reasons, human nutritional needs and scientific interest, the present reductions in the abundance of nearly all species of marine mammals are extremely undesirable. Improperly regulated exploitation has caused this situation and continues to endanger many species. Experience with other exploited and often endangered animals reveals that the solution to this problem is the deliberate transition from present hunting practices to marine farming, involving some degree of domestication. In some cases, such farming would be based on the production of raw materials for food and light industry, and in other cases — perhaps more important in the future — on the use of marine mammals for recreational purposes. Farming will ensure a sustained yield, perhaps of high economic value, that could be increased further if certain husbandry techniques are found to be applicable. Domestic farming of cetaceans could be conducted using herding and pasturing concepts in the open ocean and perhaps also in sheltered bodies of water. The most productive and profitable marine farming would seem to be using right whales. However, farming of pinnipeds is likely to be easier to begin, with northern fur seals and harp seals offering the best possibilities.

The transition from hunting to farming requires development of special multidisciplinary research programmes, including investigation of population biology, behaviour and behaviour control and ecology. The development of programmes for the improvement and expansion of marine mammal habitats and the intensification of research related to domestication of different species, are also required. Such programmes are likely to become the primary focus of future studies and their effectiveness will be improved by international coordination.

Résumé

Pour des raisons écologiques et économiques, de nutrition humaine et d'intérêt scientifique, la réduction actuelle de l'abondance de presque toutes les espèces de mammifères marins est totalement indésirable. Une exploitation mal gérée a provoqué cette situation et continue de menacer de nombreuses espèces. L'expérience acquise avec d'autres animaux exploités et souvent mis en danger montre que la solution de ce problème réside dans la transition des actuelles méthodes de chasse à la mariculture, avec une certaine domestication. Dans certains cas, cette mariculture se fonderait sur la production de matières premières destinées à l'industrie alimentaire et légère et dans d'autres cas (ce qui pourrait présenter une importance plus considérable à l'avenir) sur l'utilisation des mammifères marins à des fins récréatives. Leur élevage pourra assurer un rendement eumétrique, ayant peut-être une très haute valeur économique, qui pourrait être encore augmenté si certaines techniques d'élevage se révèlent réalisables. L'élevage des cétacés pourrait être réalisé en appliquant des concepts de rassemblement des troupeaux et de pâturage en pleine mer et, éventuellement, dans des étendues d'eau fermées. L'élevage le plus productif et le plus rentable semblerait être celui des baleines franches. Cependant, l'élevage des pinnipèdes se révèlera sans doute le plus aisé, les otaries à fourrure des Pribilov et les phoques du Groenland offrant les meilleures possibilités.

La transition de la chasse à l'élevage demande la mise au point de programmes spéciaux de recherche multidisciplinaire, y compris l'étude de la biologie des populations, de leur comportement, du contrôle du comportement et de l'écologie. L'élaboration de programmes pour l'amélioration et l'expansion des habitats des mammifères marins et l'intensification de la recherche portant sur la domestication de différentes espèces sont aussi nécessaires. Ces programmes deviendront vraisemblablement le point focal des études futures et leur efficacité sera améliorée par la coordination internationale.

Extracto

Por razones de estabilidad ecológica y económica y habida cuenta de las necesidades nutricionales del hombre y del interés biológico de los mamíferos marinos, la presente reducción de las poblaciones de casi todas las especies de esos animales es sumamente negativa. Esta situación ha sido determinada por una explotación mal regulada, que sigue amenazando a muchas especies. La experiencia del pasado con otros animales explotados y a menudo amenazados revela que la solución de este problema es la transición deliberada del sistema actual de caza a la cría, con cierto grado de domesticación de los animales. En algunos casos, ese nuevo sistema de explotación se basará en la producción de materias primas para la alimentación y para la industria ligera y, en otros casos — y quizás esto sea un aspecto más importante en el futuro — en el uso de los mamíferos marinos con fines recreativos. El cultivo de estos animales asegurará un rendimiento sostenido, que puede tener elevado valor económico y podría aumentarse aún más si fuera posible aplicar determinadas técnicas de cría. El cultivo de cetáceos podría realizarse aplicando los conceptos de rebaño y pastoreo en el mar abierto y quizás en masas cerradas de agua; el sistema más productivo y rentable de cultivo de mamíferos marinos parece consistir en la cría de ballenas francas. Sin embargo, es probable que también la cría de pinípedos sea fácil: los que mejores posibilidades ofrecen son los lobos marinos de dos pelos y la foca de Groenlandia.

La transición de la caza a la cría hace necesario preparar programas especiales multidisciplinales de investigación, que incluyan estudios sobre la biología de la población, el comportamiento y el control del comportamiento, y la ecología. Es necesario también preparar programas para mejorar y ampliar el hábitat de los mamíferos marinos e intensificar las investigaciones relativas a la domesticación de diversas especies. Probablemente esos programas estarán en el centro de los estudios futuros, y su eficacia resultaría mejorada con una coordinación internacional.

A.V. Yablokov
Institute of Development Biology, USSR Academy of Sciences, 26 Vavilov Street, Moscow W-334, USSR

For most species of marine mammals, the period of exploitation when commercial catch was limited solely by the technological level of equipment has long passed. This situation seems especially clear with respect to whales: the introduction of the harpoon gun with its explosive harpoon, the stern slipway for hauling whales onto the deck, electronic searching equipment, the compressor for pumping air into the carcasses, and the radio buoy, together with high-powered vessels, have virtually solved the problems of searching, catching and processing the whales at sea. Icebreakers, high-powered sealing vessels and quick-firers, together with aerial reconnaissance, have played the same role with respect to seals. Modern technology has made it possible to hunt every species of marine mammals in every part of the ocean.

Experience shows that commercial considerations have always been more important than those of preservation of resources. There are many explanations for this fact, including the following:

(i) catching regulations have been rather unsatisfactory. It was often the practice to adopt regulations prohibiting or limiting the hunting of certain species only after their numbers had been severely reduced;

(ii) each country has striven for maximum yields;

(iii) inadequate enforcement has allowed violations of regulations;

(iv) man's exploitation of the ocean has had an increasingly negative effect on the breeding potential of various marine mammals (e.g., both stocks of gray whales).

---

[1] Adapted from Russian language version in *Nauka*, Zoological Journal, USSR, 1976.

In addition, it should be noted that one of the main reasons for the gradual depletion of many marine mammals (and also of other ocean resources) is the lack of an owner of the resource; however, this situation is now quickly changing. It is presently generally accepted that the natural resources of the high seas are not exclusively the property only of those countries equipped to utilize them, but belong to all nations. Thus, it seems imperative that an exploitation regime be constructed that will not undermine the natural reproductive capacities of the ocean. One more point to be stressed in this connexion is the increasing extension of national control over coastal waters: all the natural resources of the continental shelf waters are now almost considered national property. The application of this principle to marine mammals (which has occurred only in a few instances) permits, in many cases, national ownership of populations of marine mammals and will result in changes in the geographic distribution of national industries. Whether this will lead to good or bad results depends primarily on the economic level, and orientation, of the countries concerned.

Exploitation of the ocean has resulted in severe reductions in the numbers of practically all species of marine mammals and some species have been found to be on the verge of extinction. The question arises whether we generally need marine mammals in the ocean. What if there had been neither whales nor seals? To answer these questions, one must first consider the general concept of the inter-relationships between man and the environment. From this consideration, it follows that a reduction in species diversity in the natural world must be considered especially undesirable and unprofitable. Many arguments supporting this thesis, ranging from genetic to philosophical, can be found in the voluminous literature on ecology and protection of nature. As regards marine mammals, there are several special arguments justifying their extreme importance to mankind and emphasizing the particular undesirability of the disappearance of even a few species. These arguments include the following:

1. The requirement for high quality products of natural origin for food and other industries will always exist, even in the distant future. To obtain protein products, mankind – progressively growing in number – will necessarily have to utilize the resources of the ocean. Understanding that man as a biological organism must have the full set of required proteins to maintain his normal existence, one cannot but realize that the possibility of living on *Chlorella* or any other monoculture is a rather naive idea (for a review of this subject, see Pokrovsky, 1975).

2. From the general concept of biogeocoenology and Vernadsky's theory of the biosphere (Vernadsky, 1926; Sukachev, 1967; Timofeev-Ressovsky, 1974), there follows an important and sometimes underestimated principle, namely, that only complex natural communities possess the necessary stability for continued existence. Theoretical substantiation of this principle is provided by the works of the mathematicians Lyapunov (1963) and Poletaev (1966). As regards ecological substantiation, it is clear enough without the use of mathematics that within a biogeocoenosis (ecosystem) containing numerous complex links, a temporary decrease in the matter or energy flow in one link can be compensated for by an increase in the others. If the links are few in number, a break in even one of them may result in a catastrophe for the biogeocoenosis. The complexity of a biogeocoenosis depends upon the number of links in the food chain. Cetaceans and pinnipeds always occupy the upper trophic level and it is their presence that determines the complexity of the marine biogeocoenosis. Their removal from the biogeocoenosis will necessarily give rise to pronounced fluctuations in the pro-

ductivity of one or another trophic level of the ocean and may finally lead to a catastrophe in a commercial fishery, which is still the main form of utilization of marine resources. Thus, it is the presence of marine mammals that provides the completeness and stability of the biogeocoenoses of the ocean (for a review of this subject, see Tomilin, 1971; Sokolov, Tomilin and Yablokov, 1974).

3. Owing to the high level of development of their central nervous systems, marine mammals hold a unique position among mammals. A high level of sociability – a feature rather rare in other groups of animals – was shown to be characteristic of many marine mammals (Krushinskaya, 1974). Therefore, further in-depth investigations into these areas are essential to the study of the phenomenon of mind in the evolution of life.

4. Closely related to the preceding point is the recreational role of marine mammals. This role is increasing now and will increase still further in the near future. In addition to oceanaria, sea parks are to be established in the future; these will no doubt open up new possibilities for man to have contact with marine animals. As society continues to develop and its wealth increases recreational problems (i.e., organization of leisure time to enrich our lives) become more vital. Solutions to these problems are in some ways as important as those of nutrition.

It would be a mistake to think that those small cetaceans and pinnipeds that are kept in captivity are the only marine mammals of recreational importance. Fascinating observations of wild groups of gray whales (Yablokov, 1975), sea otters, and northern fur seals in North America, observations of seal rookeries (sealing grounds) in the Antarctic, together with the interesting studies of sperm whales and fin whales in the wild (Cousteau and

154

Diole, 1975) demonstrate clearly that the recreational opportunities presented by marine mammals are impressive.

From the arguments listed above, it is evident that there are a number of reasons why there will always be a need for the presence of marine mammals in the ocean, today as well as in the future. Hence, the problem of allowing marine mammal populations to recover and recolonize their original range increases in importance. Incidentally, this consideration is not confined to the marine mammals, but also applies to terrestrial mammals, in particular the larger ones.

Brief examination of the history of man's exploitation of the biosphere reveals the solution to this problem. Several forms of exploitation of natural resources are known to have existed in the past, the first, more primitive ones having been successively superseded by those more ingenious and beneficial to the development of mankind. The initial forms of more of less modest exploitation were hunting and gathering. The period of fishing followed next and it then proved necessary to establish certain regulations concerning the harvest of animals from nature. These were mainly measures meant to limit catches during the reproduction periods, such as protection of spawning grounds, bans on hunting during reproductive periods, bans on hunting nursing females and young, and so on; all were rather insufficient. As a result of virtually unlimited exploitation, the populations of large animal species were practically reduced to zero: the mammoth, woolly rhinoceros, aurochs, bisons and other hoofed animals, the giant sloth, great auk, and many others. From all of these species, only those domesticated (if only partially) and cultivated were able to survive. The cow, camel, yak, sheep, horse, llama, vicuna, Indian elephant, reindeer, and other animals, together with a wide range of species bred as sources of fur — several dozen species — are only some of those which have been domesticated in the past: antelopes, cheetahs, weasels and others. In the future, all of the biosphere will be maintained at a proper level by man. I

cannot consider this problem completely here (Yablokov, 1973, 1975), but as regards marine mammals, the only way to conserve them in the ocean is through the organization of different forms of marine farming, that is, by real management.

Farming or management is the organization of commercial exploitation of natural resources such that before taking something from nature one must provide for enhanced reproduction of the resource, the amount of produce taken being in accordance with the added amount reproduced. Compared to fisheries practices, marine farming is a higher step in the organization of natural resource exploitation. This transition — from fisheries to farming — is the only sure way to preserve marine mammals, especially commercial species in the ocean. For most species of marine mammals, the actual mechanics of this transition are mostly unknown at present; therefore, it seems imperative that their development should become one of the aims of all future long-term scientific research programmes on marine mammals. This means a re-orientation of all commercial studies: from studying "where and how to take more and how to use the products in the best way" to studying "what must be done today to ensure the harvest tomorrow".

For whales, such practices may in the beginning involve pasturing of individual stocks as is done with reindeer, herds of horses in the steppes of Kazakhstan, or flocks of sheep. Special herdsmen will look after the young, help the animals in periods of bad weather and poor food supplies, and see that the fodder areas are exploited more or less evenly. Hormonal stimulation of females may be applied to increase the reproductive abilities of the stock (the percentage of barren females is known to be rather high sometimes); among other things, some kind of selective breeding programme might prove possible. Each herdsman will know which and how many of his animals can be periodically harvested from the herd and still ensure reproduction on an enlarged scale.

Maintenance of whales in stocks in the open sea might not prove to be the only or even the main form of farming. There have been projects for rearing whales in enclosed lagoons of coral islands and other natural enclosed water areas (Tomilin, 1964). Successful inter-generic breeding in cetaceans has been demonstrated with captive dolphins. Based on rough estimates, a total harvest of 25 whales can be obtained from a pair of adult fin whales over 20 years (including the third generation); a group of 50 pairs of adult fin whales will provide an annual harvest of 25 whales provided that half of the adult females give birth each year and all of the young survive.

Similar estimates illustrating the possible high economic value of such practices can be given; these economic possibilities are supported further by the fact that the reproductive rate of whales and seals is as high in some cases as that of cattle. Additionally, if it proves possible to achieve a predominance of females in herds of marine mammals (as is the case in herds of domesticated large hoofed animals), the effectiveness of such marine farming will increase several-fold. It may be found that some species are more easily domesticated than others; the economic values of different species may also vary. Farming of right whales will be the most productive and profitable.

The formation of several new herds of Greenland and southern right whales from the preserved animals presently counted in the mere hundreds (assuming that these remaining animals could be saved), would be a successful and rather impressive beginning to whale farming. Since these species are comparatively sedentary animals, shepherding would not present any difficulties and it would be quite easy to organize the necessary preparatory research.

Domestic farming of pinnipeds is likely to be much easier to establish, even at the present time. The existence of lake forms of seals (Ladoga, Baikal and Caspian Sea) points to their broad adaptive possibilities. Conversely, it is known that mortality in prenatal to yearling animals can be enormously high, reaching about 30-40 % in some species. Lowering this mortality rate would increase the population in a short period of time. For some species, it would seem possible to develop methods of semi-domestication similar to that of marals (*Cervus elaphus sibiricus* Sev.) in the Altai territory and in the Soviet Far East; no doubt, some varied forms of farming may be found.

Among the Pinnipedia, northern fur seals and harp seals, and sea otters among the Carnivora, offer the best possibilities for farming. However, much of the scientific information necessary to begin such programmes is lacking. Severe drops in some fur seal populations inhabiting the USSR, not yet fully explained, should serve as a proper warning against too confident an attitude regarding the status of these animals.

The aim of the present paper is not to give any conclusive recommendations on the organization of marine mammal farming — that is a task for a number of research groups — but to illustrate the importance of this interesting and promising area which is likely to become the primary focus of future research programmes on marine mammals. It is possible, however, to name some important problems to be encountered in the transition from fisheries to farming.

Historical and zoogeographical analysis shows that several locations are especially promising for future farming programmes. Thus, it would be worthwhile to resume the experiments on semi-domestic rearing of *Pusa caspica*[2] on the Caspian Sea islands which were successfully initiated by B.I. Badamshin, and to consider the establishment of a floating station for long-term, continuous observations of their gathering places. Prospects for the establishment of a large research centre on the White and Barents Seas should be examined. The rich marine mammal fauna found in this

---

[2] These did not include captive breeding.

area not so long ago leads one to consider various possibilities: experimental introduction of walruses, creation of experimental hauling-out grounds for grey seals, continuous contact with the White Sea stock of white whales, and, of course, the development of harp seal management (Nazarenko, 1975). I think that the acclimatization of the sea otter in this area should still be on the agenda. The Azov-Black Sea basin is of special importance for the work on 3 species of dolphins (*Delphinus delphis, Phocoena phocoena* and *Tursiops truncatus*): if the dam in Kerch Strait is built, what is to be done about the Azov common porpoise? The seas of the Soviet Far East basin are of definite interest. There are some data suggesting the existence of small groups of right whales in the Okhotsk Sea: if confirmed these small groups may be large families which could form the basis of a herd tended by a herdsman (Berzin and Kuzmin, 1975), as mentioned above. Until now, the unlikely supposition concerning the possible existence of a small number of Steller's sea cows in inaccessible coastal regions of eastern Kamchatka has not been investigated.

Another interesting area of marine mammal research is the development of amelioration programmes, i.e., the broad improvement of areas having potential for marine mammal farming. For example, the possibility of improving the autumn resting places of seals on the northern sand banks of the Caspian Sea should be investigated; perhaps the creation of artificial islands from old ships' hulls is worth considering. Some successful practical projects have already been taken in this field, including expansion of fur seal resting and breeding grounds. A survey of past programmes and experience throughout the world would be worthwhile.

Special attention should be given to a detailed study of the role of individual species and populations of marine mammals in the marine biogeocoenoses. Present available data are insufficient to discern the necessary quantitative and qualitative roles of marine mammals in the flows of matter and energy through the biotic turnover of ocean, coastal and island waters (Tomilin, 1970).

Domestication of marine mammals is also worthy of special study. The ancient Greeks tried to tame marine mammals and had some success. Semi-tamed killer whales were used by Australian hunters for driving humpback whales several decades ago. Some specimens of seals have been tamed while still in the wild. Recent work with captive dolphins in oceanaria has generated great interest. Trained dolphins released into the open sea come back on signal; such animals have been used as helpers in ocean studies. Marine mammal domestication programmes involve: (i) the determination of species appearing most adaptable to domestication; (ii) development of biotechnical selective breeding methods, and (iii) creation of special farms and various experimental herds, in addition to oceanaria.

At present, with many researchers studying dolphins, the coordination of all activities related to the domestication of the bottlenosed dolphin needs attention. As early as 1964, Tomilin spoke about domestication of this species (Tomilin, 1971). This area of research is closely connected with special studies on the behaviour of marine mammals. Although little work has been carried out in this field (Voronin *et al.*, 1974), such behavioural studies would discern the behavioural controls that are necessary and feasible for the introduction of marine mammals into marine farming.

One of the most important areas of future research is the study of population structure, including the determination of intrapopulation units (i.e., family units) and the relationships among them. Because direct genetic study of whales, dolphins and seals is mostly impossible, investigations of population morphology would seem very useful. Research in this area should be founded upon the concept of phenotypes — discrete, simple patterns exclusive of variations. Useful application of this principles was made when Kleinenberg *et al.* (1969) found that the Far East stock of

157

white whales differs from the North Atlantic stocks by the frequency of division of the 4th or 5th finger. This set of investigations also made it possible to discern family organization of small groups of *Delphinapterus leucas* (each group includes one old female and several of her female offspring and grandchildren). Recently, a similar research plan has been initiated by Berzin on the population morphology of sperm whales. He has found that within small groups of sperm whales, phenotypic colour patterns are often very similar. It is possible to determine, for example, the sisters and father of foetuses. In one harem we now have strong evidence for a single father of several foetuses, although several males were included in this group. This points to the possibility of using different phenotypic principles to study other cetaceans and pinnipeds.

The above list of research areas should not be considered complete. During the transition from hunting to marine farming, many problems will arise requiring the special attention of various experts, such as ecologists, morphologists, geneticists, fishermen and technologists. The extensive bionic studies of marine mammals being conducted at the present time in many countries will certainly contribute much to the solution of these problems; present investigations will especially add to the development of maintenance and breeding methods and toward solutions of related questions of behaviour and physiology.

In conclusion, I would like to present some comparisons from history. In the 18th and 19th centuries, systematics and faunistics were the main fields of marine mammal research. In the first half of the 20th century, ecology and fisheries science received the most attention. At present, broad bionic studies prevail, including investigations on captive maintenance, and related areas of behaviour, functional morphology and physiology.

In the near future — the last quarter of this century — it seems that research directly or indirectly concerned with the problems of passing from fishing to marine mammal management, and especially farming will be dominant (e.g., studies of behaviour, breeding and conservation).

I am sure that consistent development of the transition from commercial hunting to farming will make a great contribution to general animal science and to our understanding of the relation of man to the rest of the biosphere.

References

BERZIN, A.A. and A.A. KUZMIN, Grey and right whales of
1975   the Sea of Okhotsk. *In* Marine mammals. Kiev, Naukova Dumka, vol.1:30-2 (in Russian).

COUSTEAU, J.Y. and F. DIOLE, Sovereign of seas. *Nauka*
1975   *Zhizn'*, (9):110-7.

KLEINENBERG, S.E. *et al.*, Beluga (*Delphinapterus leucas*).
1969   Investigation of the species. Transl. from Russian. Jerusalem, Israel Program for Scientific Translations, vol. 1:376 p.

KRUSHINSKAYA, N.L., Behaviour of marine mammals.
1974   *Itogi Nauki Tekh. (Zool. Bespozv.)*, 6:40-86.

LYAPUNOV, A.A., On controlling systems of living nature
1963   and common understanding of world phenomena. *Probl. Kibern.*, (10):179-93.

NAZARENKO, Yu.I., On possible method of determining
1975   barrenness of Greenland seal. *In* Marine mammals. Kiev, Naukova Dumka, vol. 2:26-8.

POKROVSKY, A.A., Nutrition problem and biosphere. *In*
1975   Methodological aspects of biosphere studies. Moscow, Nauka, pp. 345-65.

POLETAEV, I.A., On mathematical models of elementary
1966   processes in biogeocoenoses. *Probl. Kibern.*, (16):171-80.

SOKOLOV, V.E., A.G. TOMILIN and A.V. YABLOKOV, On
1974   necessity of restoring the abundance of commercial cetaceans in the World Ocean. *Itogi*

Nauki Tekh. (Zool. Bespozv.), 6:9-20.

SUKACHEV, V.N., Biogeocoenology and its modern tasks.
1969   *Zh. Obshch. Biol.*, 28(5):501-9.

TIMOFEEV-RESSOVSKY, N.V., Elementary phenomena of
1974   evolutionary process. *In* Philosophy and theory
        of evolution. Moscow, Nauka, pp. 114-20.

TOMILIN, A.G., About domestication of some mammalian
1964   species. *Tr. VSIZO*, 17(2):97-101.

−, About keeping dolphins in captivity and their behav-
1971   iour. *Byull. Mosk. O-Va. Ispyt. Prir. (Biol.)*,
        76(3):146-57.

VERNADSKY, V.I., The biosphere. Praha-Paris.
1926

VORONIN, L.G. *et al.*, Some data on domestication of the

1974   Black Sea dolphins. *In* Morphlogy, physiology
        and acoustics of marine mammals. Moscow,
        Nauka, pp. 108-22.

YABLOKOV, A.V., From fisheries towards economic prac-
1973   tice. *Priroda*, (1):86-7.

−, Methodological aspects of inter-action between man
1975   and animal world. *In* Methodological aspects
        of biosphere studies. Moscow, Nauka, pp.
        365-71.

−, Grey whale: success and hopes. *Nauka Zhizn'*,
1975a   (9):118-20.

YAKOVENKO, M.L., Determination of the efficiency of the
1975   White Sea seal reproduction from the data of
        aerial photography for perspective. *In* Marine
        mammals. Kiev, Naukova Dumka, vol.
        2:189-91.

LARGE CETACEANS

MAN'S EXPLOITATION OF THE WESTERN ARCTIC BOWHEAD

J. BOCKSTOCE

Abstract

Regular hunting by Eskimos of the western Arctic bowhead whale (*Balaena mysticetus*) began on the Asian side of the Bering Strait around 500 B.C.; by the year 1200 or earlier, following an eastward migration of Eskimos into North America that began by about the year 800, Eskimos from the Bering Strait to the eastern Beaufort Sea were hunting bowheads, taking as many as sixty whales each year — the species was important as food and as a basis for religious and artistic traditions. This situation was changed by the discovery by American whalers in 1848 of the Arctic Ocean whaling grounds — in 1850, more than 200 ships, mainly from New England, were hunting bowheads here, taking most of them on their southward migration along northwest Alaska. By the late 1860's, though still dominating American whaling, catches of bowheads were declining because of overhunting, and walrus (*Odobenus rosmarus divergens*) began to be taken as well. The introduction of steam auxiliary vessels was at first successful but only further reduced the number of whales. After a decline in whale and walrus oil prices from 1875 to 1885, and a rise in the value of baleen, nearly 20 shore-based whaling stations — employing Eskimos and introducing them to the use of the darting and shoulder guns and a cash economy — were begun from 1885 to 1900 to catch the bowheads on their early spring migration northward. This fishery, and the last phase of pelagic whaling, from 1890 to 1910 in the eastern Beaufort Sea, both at first took many whales but by 1900 the size of the bowhead population was seriously reduced. More than 20 000, and perhaps more than 30 000, whales were taken by the pelagic fishery between 1848 and 1910, when the industry was near its end. The Eskimos were forced to return to subsistence hunting: between 1910 and 1935, 35-40 crews, mainly at Point Barrow and Point Hope, took 10-15 whales each year, a slow increase in the annual catch, indicating that the population of whales may have increased somewhat during this time. Since the mid-sixties, with large earnings from new construction projects in Alaska, the number of Eskimo whaling crews and of whales taken has increased greatly — during the spring whaling season of 1977, 26 whales were killed and landed, 2 killed and lost and 77 struck and lost. To determine whether the present population of whales can tolerate this level of harvest and what might be the effects on the whales of offshore oil development, it is important to know the present size of the population and its size before 1948. Problems with counts from the air and the ice edge make estimation of the present size difficult; however, compilation by computer of data from logbooks, accounting for about 20 percent of the voyages to the western Arctic, and of gross seasonal returns from shipping newspapers, can give the information needed on the population before 1848, as well as provide a model for historical studies of other whale populations.

Résumé

La chasse régulière par les Esquimaux de la baleine à fanons de l'Arctique ouest (*Balaena mysticetus*) a commencé sur la côte asiatique du détroit de Bering vers 500 av. J.-C.;

en l'an 1200 de notre ère, ou même plus tôt, suivant une migration vers l'est des Esquimaux en Amérique du Nord qui commença vers 800, les Esquimaux chassaient cette baleine du détroit de Bering jusqu'à l'est de la mer de Beaufort, en capturant jusqu'à 60 par an — l'espèce était importante sur le plan de l'alimentation et sur celui des traditions religieuses et artistiques. Cette situation changea quand, en 1848, les baleiniers américains découvrirent les lieux de chasse de *B. mysticetus*. En 1850, plus de 200 bateaux, provenant surtout de la Nouvelle-Angleterre y chassaient la baleine du Groenland, en capturant la plupart lors de leur migration vers le sud, le long du nord-ouest de l'Alaska. Vers la fin des années 60, encore qu'occupant une place dominante dans la chasse américaine de la baleine, les captures diminuaient par suite de surexploitation et l'on commença à capturer aussi le morse (*Odobenus rosmarus divergens*). L'apparition de bateaux auxiliaires à vapeur permit des succès au début mais ne fit que réduire encore le nombre de baleines. Après le déclin du prix de l'huile de baleine et de phoque de 1875 à 1885, et une hausse de la valeur des baleines, près de 20 stations de chasse à la baleine, installées à terre — employant des Esquimaux et leur apprenant à utiliser le dard et le canon à baleine ainsi qu'à pratiquer une économie basée sur la monnaie — furent créées entre 1885 et 1900 pour capturer les baleines du Groenland lors de leurs migrations vers le nord au début du printemps. Cette pêche et le dernier stade de la chasse baleinière pélagique, de 1890 à 1910 dans la mer de Beaufort, permirent au début la capture de nombreuses baleines mais, en 1900, la taille de la population de *B. mysticetus* était sérieusement réduite. Plus de 20 000 baleines — et peut-être plus de 30 000 — furent capturées par les pêcheries pélagiques entre 1848 et 1910 et l'industrie toucha à sa fin. Les Esquimaux ont été contraints de revenir à leur chasse de subsistance: de 1910 à 1965, 35 à 40 équipages, surtout à Point Barrow et Point Hope, ont capturé de 10 à 15 baleines par an, un lent accroissement des captures indiquant que la population de baleines peut avoir augmenté quelque peu pendant ce laps de temps. Depuis le milieu des années 60, avec les importants revenus provenant des nouveaux projets de construction en Alaska, le nombre d'équipages esquimaux pratiquant la chasse et le nombre de baleines capturées ont grandement augmenté — au cours de la campagne de chasse du printemps 1977, 26 baleines ont été tuées et mises à terre, 2 ont été tuées et perdues et 77 ont été touchées et perdues. Pour savoir si la population actuelle de baleines peut tolérer ce niveau d'exploitation et quels pourraient être les effets du développement de l'exploitation off-shore du pétrole sur les baleines, il est très important de connaître la taille actuelle de la population et ce qu'elle était avant 1848. Les difficultés que présente le recensement aérien et celui qui est fait depuis la limite des glaces posent un problème pour l'estimation de la taille présente; cependant, la compilation sur ordinateur des données provenant des livres de bord, qui représentent environ 20 pour cent des expéditions dans l'Arctique occidental et du montant brut des profits des campagnes fourni par les journaux de la marine marchande, peut donner les informations requises sur la population d'avant 1848 et servir de modèle pour l'étude historique d'autres populations de cétacés.

Extracto

Los esquimales comenzaron a cazar regularmente la ballena de cabeza arqueada del oeste del Artico (*Balaena mysticetus*) en el extremo asiático del Estrecho de Bering alrededor del año 500 a. de J.C.; por el año 1200 o antes, tras la emigración de los esquimales hacia el este, para penetrar en América del Norte, emigración que comenzó hacia el año 800, los esquimales que se encontraban desde el Estrecho de Bering hasta la parte oriental del Mar de Beaufort se dedicaban a la caza de ballenas de cabeza arqueada, llegando a capturar hasta 60 animales al año. Esta especie era para ellos importante como alimento y como base de sus tradiciones religiosas y artísticas. La situación se modificó al descubrir los balleneros americanos en 1848 los cazaderos de ballenas del Océano Artico: en 1850, más de 200 barcos, procedentes principalmente de Nueva Inglaterra, cazaban en esa zona *Balaena mysticetus*, sobre todo durante los movimientos migratorios de esas ballenas hacia el sur, a lo largo del noroeste de Alaska. A finales de los años sesenta del siglo pasado, las capturas de ballena de cabeza arqueada, aunque seguían predominando en el total de las capturas estadounidenses, empezaron a disminuir debido a la explotación excesiva y empezaron a capturarse también morsas (*Odobenus rosmarus divergens*). La introducción de barcos auxiliares de vapor tuvo éxito al principio, pero no hizo más que reducir el número de ballenas. Tras la disminución de los precios del aceite de ballena y morsa entre 1875 y 1885 y el aumento del valor de las barbas de ballena, se crearon entre 1885 y 1900 unas 20 estaciones balleneras en tierra (en las que trabajaban esquimales, a los que se enseñó a utilizar cañones y escopetas lanza-arpones y se habituó a una economía monetaria) para capturar ballenas de cabeza arqueada al comienzo de sus migraciones de primavera hacia el norte. Tanto en esa pesquería, como en la última fase de la caza pelágica de ballenas en la parte oriental del Mar de Beaufort (desde 1890 a 1910), se capturaron inicialmente muchas ballenas, pero para 1900 el volumen de la población se había reducido gravemente. Entre 1848 y 1910, cuando la industria se acercaba ya a su fin, se capturaron en alta mar más de 20 000 y quizás más de 30 000 ballenas. Los esquimales se vieron obligados a volver a la caza de subsistencia. Entre 1910 y 1965, entre 35 y 40 grupos de pescadores, principalmente en Point Barrow y Point Hope, capturaron de 10 a 15 ballenas al año, y quizás el lento aumento de las capturas anuales indique que la población ha aumentado ligeramente durante ese período. Desde mediados de los años sesenta, con los grandes beneficios derivados de los grandes proyectos de construcción en Alaska, el número de grupos de esquimales dedicados a la caza de ballenas y el número de ballenas capturadas aumentó notablemente: durante la campaña ballenera de la primavera de 1977 se mataron y recogieron 26 ballenas, se mataron y perdieron otras dos, y se hirieron y perdieron 77. Para determinar si la población actual de ballenas puede tolerar ese nivel de explotación, y qué efectos pueden tener para las ballenas las prospecciones petrolíferas submarinas, lo más importante es conocer el volumen actual de la población y su volumen antes de 1948. Los problemas que plantea la realización de recuentos desde el aire y desde el borde de los campos de hielo hace difícil estimar el volumen actual; sin embargo, compilando en computadora los datos recogidos en diarios de a bordo que cubren un 20 por ciento, aproximadamente, de los viajes hechos al oeste del Artico y los datos sobre los beneficios brutos de las campañas balleneras que se encuentran en algunos boletines de compañías de armadores, es posible obtener información sobre el estado de la población antes de 1848 y preparar un modelo para la realización de estudios históricos sobre otras poblaciones de ballenas.

J. Bockstoce
Whaling Museum, Old Dartmouth Historical Society, 18 Johnny Cake Hill, New Bedford, MA 02740, USA

Today the survival of the western Arctic bowhead whale (*Balaena mysticetus*)[1] stands in question; the fragility of this population is the legacy of over-exploitation by the commercial whalers of the nineteenth century, but its future may be decided by the Eskimo whalers of northern Alaska.

History of bowhead whaling in the western Arctic

Mankind has been harvesting from the western Arctic bowhead population for at least the last 2 500 years. Around 500 BC hunters on the Asian side of Bering Strait developed the ability to hunt whales regularly and reliably through the introduction of sophisticated harpoon and drag-float equipment from farther south and west on the coast of Asia (Bockstoce, 1973). But it took another thousand years or so for intensive whaling to spread from there into northwestern Alaska, probably because of slow growth in Eskimo populations and inconducive climatological factors. Nevertheless by about AD 800 Eskimos ancestral to today's population in northern Alaska, Canada, and Greenland carried the practice with them as they spread eastwards across the North American Arctic (Bockstoce, 1976), and by AD 1200 or earlier Eskimos from the Bering Strait region to the eastern Beaufort Sea were hunting the western Arctic bowhead with much the same methods and technology that would survive until today.

Using toggling headed harpoons that trailed long lines and towed several inflated sealskin drag-floats, each year these Eskimo whalers may have taken as many as sixty whales. Because of the enormous amounts of food and raw materials derived from these animals — up to fifty tons in a large whale — the bowhead was central to Eskimo culture, both calorifically and spiritually, and the hunt was surrounded by a complex and sophisticated body of practices. In some particularly fortunate areas, such as Point Hope and Point Barrow, the whale harvest was so bountiful that nearly half the winter's food supply could be gathered from 8 weeks of whaling and some walrus hunting; that is, half the winter's food was taken in less than one quarter of the winter. The resulting surplus for these societies contributed to the development of rich artistic, religious and mythological traditions (Bockstoce, 1976; Rainey, 1974).

These whaling societies existed in harmony with their ecosystem, but the harmony was shattered forever when in 1848 Captain Thomas Roys, a Yankee whaleman from Sag Harbor, Long Island, sailed through the Bering Strait and discovered the rich whaling grounds in the Arctic Ocean. He found the bowheads there to be fat, docile, and valuable, yielding an average of 100 barrels of oil (each barrel was 31½ gallons) and 2 000 pounds of baleen per animal. Roys quickly filled his ship and departed for Hawaii to trumpet his discovery. His report quickly set off a stampede of vessels to the western Arctic; in fact, the results of the 1849 season were so good that in 1850 more than 200 ships were cruising in those icy seas, and they quickly established a routine of voyaging that would characterize the earlier phase of this whaling industry.

Sailing from the whaling ports of New England[2] in the autumn, they rounded Cape Horn in the southern summer, and after a short cruise on the west coast of South America, or on Baja California, or in the central

---

[1] The bowhead population of North America's western Arctic is commonly and erroneously referred to as "Bering Sea" population. This term is misleading, for these whales spend a greater amount of time north of Bering Strait, in the Chukchi and Beaufort Sea — where they have been reported throughout the waters of both — than they do in the Bering Sea, where they are only found in its northern half.

[2] In the 1950's a small number of vessels also sailed from Long Island and Australian ports as well as from Bremen, Le Havre, Honolulu, and a few other places.

Pacific, they reached Hawaii in March to take on provisions for their northern cruise. Entering the Bering Sea, they would reach its retreating ice floes about the first of May and then work their way northward through the ice along the Asian shore, occasionally taking a few whales. By then though, the bulk of the whales would already have passed northward into the impenetrable ice of Bering Strait and the Chukchi Sea on their way around Point Barrow to the summer feeding grounds near the Mackenzie River delta in the eastern Beaufort Sea. By the middle of June, when the ships reached Bering Strait through the melting ice, most of the whales were a thousand miles beyond them, safe amid the fastnesses of the polar ice pack. With little hope of finding the whales again, until the ships could reach Point Barrow and intersect the bowheads on their return migration, the whalemen would put into Port Clarence, a spacious and secure harbour near Bering Strait, to rest, repair, and trade with the natives.

By the late 1860's however — although the importance of the western Arctic dominated American whaling activities — the ships had made sufficient inroads on the whale population that catches were declining, and the whalemen often turned to taking walrus (*Odobenus rosmarus divergens*) for their oil in the "middle season", before the ships could reach Point Barrow and the bowheads again. Between about 1868, when walrus hunting began in earnest, and about 1880, when oil prices slumped dramatically and ended the slaughter, it is likely that more than 100 000 walrus were taken.

1880, in fact, became a pivotal year for western Arctic whaling, dictating the character of its later phase. The decade from 1875 to 1885 was marked by a steady decline in whale and walrus oil prices — pressed by cheaper petroleum and fish oils — and a steeper rise in the value of baleen which was in increasing demand from the fashion industry for use in skirt hoops, corset stays and other goods; in those ten years the price of a pound of baleen rose from just over US$1.00 to about US$3.50.

But the bowheads had already been significantly reduced in number, and the industry's response was to introduce steam auxiliary vessels which could manoeuver with greater ease amid the dense ice floes and thus follow the whales into their retreats. Initially the steamers met with great successes, but this innovation only hastened the reduction of the bowhead population — and the demise of the industry (Bockstoce, 1977).

Despite the immediate profitability of the steamers, the whaling merchants were alive to the fact that the greater proportion of the whale population was still able to elude the ships because of its early spring migration along the coast of northwestern Alaska. Consequently between 1885 and 1900 nearly twenty shore-based whaling stations — some very short-lived — were established on the whales' migration path to hunt them in the Eskimo fashion. Like the steam vessels, the stations were quickly successful, but equally important — and unlike the steamers — they were cheap to operate. Their effects were immediate and far-reaching to both the whales and the Eskimos. First, the stations served to reduce further the number of bowheads by increasing the number of crews hunting them three- or four-fold above the level of aboriginal Eskimo hunting; second, because most of the crews were hired from among local Eskimos, the stations ended the thousand-year aboriginal whaling pattern, drawing the Eskimos inexorably into a cash economy and acquainting them with the use of the darting gun and shoulder gun.[3] For a short time these

---

[3] The darting gun and shoulder gun were developed in the nineteenth century by American whalemen to cut down on the loss of wounded bowheads that often escaped into the ice pack towing lines and gear; because they were much more effective than their forerunners, hand lances, they were quickly adopted by the Eskimos. The darting gun, mounted at the end of the harpoon shaft, with the harpoon's toggle iron beside it, was designed to fire a small bomb — set to explode a few seconds later — into the whale simultaneously as it was struck with the iron; the shoulder gun, a thirty-five pound brass smooth-bore, could fire a similar bomb through the air accurately from as far as about ten meters.

shore-based operations greatly increased the number of bowheads being taken, but by 1900 the bowhead population had been cut to the quick and these high catches were short-lived. It is likely that about 1 000 whales were killed by shore whalers from 1880 to 1910.

The last phase of pelagic whaling activity for the western Arctic bowhead took place between 1890 and 1910 in the waters of the eastern Beaufort Sea near the Mackenzie River delta. In 1890 an American whaleman returned to Point Barrow from having spent the winter in the delta and reported seeing large numbers of whales feeding there in the summer; immediately a fleet of steam whalers forced their way east and confirmed his story. The whalers expected to make large catches there but realized that because they were so far from their home port, San Francisco, they would have to winter in the Arctic to avoid the time-consuming trip during the short whaling season. At the start they took a great number of whales, but the whale population had already been so severely reduced that, even though the whalemen pushed on to the extremities of the Beaufort Sea, the great returns quickly ended (Bockstoce, 1976, 1977a).

Curiously enough, it was the whales' very scarcity that ultimately saved them from extinction. It is probable that more than 20 000 − possibly more than 30 000 − whales were taken by the pelagic fishery between 1848 and 1910. As the stock of bowheads was reduced, the greater rarity of baleen drove the price higher and higher, until spring steel, a cheap substitute, was introduced, driving the price of baleen from US$5.00 per pound to less than US$0.50 within 3 years. By 1910 most of the shore stations had been closed, and all that remained of the American whaling fleet in the western Arctic were a few ships going north primarily for the fur trade (Bockstoce, 1977).

For the Eskimos the collapse of the whaling industry meant the end to cash employment or subsidies for their whale hunt and forced a return to subsistence hunting, the third phase in the history of Eskimo whaling. As always, the Eskimos depended on the whale for its great yield of food and raw materials; but, lacking any substantial cash support, the number of whaling crews shrank back to about the level it had been in the aboriginal phase. From 1910 to 1965 only about 10 to 15 whales were taken annually in Alaska; the number of whaling crews also remained relatively steady: roughly 20 at Point Barrow, 10 or 12 at Point Hope, 4 or 5 at St. Lawrence Island, 1 or 2 at Wainwright, and occasionally 1 or 2 at Wales or Kivalina. Since the whale now had no great cash value and whaling equipment, particularly shoulder guns and darting guns, was scarce and expensive, it became difficult to start new whaling crews. To start a new crew, and thus attain the prestigious position of whaling captain, a young man usually had 3 options open to him: to inherit his whaling equipment, to marry into it, or − less commonly − to gain great wealth, enough to purchase it.

Judging from the numbers of whales taken since 1910, it seems that the reduction of hunting pressure eventually allowed the bowhead population to increase somewhat. Without a significant increase in the number of crews the Alaskan Eskimo catch inched up from an average of about 10 per year to about 15 per year between 1910 and 1965.

Modern eskimo whaling

The relatively steady level of the hunting effort characteristic of the subsistence phase ended about 1965 when the number of whaling crews began to increase; this change ushered in the fourth and present phase of Eskimo whaling. Since the mid-1960's the increasing tempo of construction projects in Alaska has made a seller's market for labourers. Unskilled and semi-skilled labourers can now amass several thousand dollars (after taxes) in only a few months' work. Today, though a well-equipped whaling captain may have US$8 000 or US$9 000 invested in his outfit and will require at least US$2 000 more for food and other expendable goods, such sums are well within the reach of any energetic and ambitious worker. The number of whaling crews has increased dramatically; in 1976 it approached twice the number that were active in the early 1960's. Most important, with the rise in crews has come a dramatic rise in the number of whales killed. In 1974 20 whales were killed and butchered, 3 were killed but lost, and at least 28 were struck but lost. In 1976, 48 were killed and butchered, 8 killed but lost, and at least 35 were struck but lost. With the autumn whaling season still to come, in 1977 26 whales had been killed and butchered, 2 killed and lost and 77 struck and lost.

Recommended actions

The recent level of the Eskimo bowhead kill has caused much popular outcry throughout the world that the take is excessive, questioning whether the bowhead population — still severely reduced below pre-1848 — can tolerate such intense pressure. As this paper is being written (August 1977) the issue of whether or not the Eskimo whale hunt will be allowed to continue, whether the USA will accept the decision of the June 1977 Meeting of the International Whaling Commission that no bowheads be killed in the 1977/78 season, is still unresolved.[4]

4 *Editor's note.* In December, at a Special Meeting of the International Whaling Commission, the IWC agreed a quota of 12 bowheads landed or 18 struck, whichever occurred first, for the 1978 spring hunt in Alaska. A resolution was also passed urging control on hunting techniques and equipment, killing or striking of calves and females with calves; calling for a review of the bowheads' status in June 1978 in the light of information resulting from the USA's proposed research programme, with a view to establishing regulations based on the advice of the Scientific Committee (of the IWC), which should include comments on risks associated with different levels of removals from the stock; urging that reduction in take of bowheads should not significantly affect the take of

But beyond the questions of politics and beyond the questions of how best to regulate the Eskimo hunt[5] lies what is surely the central question: the present size of the western Arctic bowhead population and its pre-1848 size. Because of our profound lack of knowledge of this whale and this population, all attempts to estimate its size come near to being guesswork; furthermore, aerial surveys and ice edge counting have been seriously hampered by problems of poor visibility and by the more fundamental question of what percentage of this population the assessors have counted. In the short term these censuses are useful for the most part only in comparison with other annual censuses; it will probably require at least a decade before the body of these data can have real estimative value. Regrettably, the issue will not wait for a decade. Not only the demands of the Eskimo hunt but also the impetus for off-shore oil development in the Bering, Chukchi and Beaufort Seas require quick estimates of the population size as well as delineation of critical habitat areas.

I believe that the only way to gain information on the pre-1848 population level — and it would be both quick and relatively inexpensive — is through a compilation of data from historical sources, for a rich and untapped source of information on the western Arctic bowhead exists in whaling logbooks and other records of the industry. There are about 400 extant logbooks spanning the history of this whaling industry, and they give information on about 600 April-to-October voyages to the western Arctic, covering about 100 000 days of observations. In most cases the following information was recorded for each day: vessel position (in latitude and longitude), visibility, current, and sea, ice and weather

beluga whales; and urging that all necessary measures be taken to preserve the habitat of bowhead and beluga whales.

The clause on protection of habitat was prompted by the possible deleterious effects of future oil, gas and mineral exploitation in the Arctic.

5 For suggestions on possible regulations see Bockstoce, 1977b.

conditions. When whales or walrus were observed, the time, their species, their numbers and their direction of travel were usually noted, and if a kill was made, the size of the animal (expressed in pounds of baleen and ivory and/or barrels of oil) was recorded. Such information can be relatively easily transferred to a computer punch-card and then stored in a retrieval system.

The daily information from logbooks, which accounts for about 20 % of all whaling voyages to the western Arctic, can then be amalgamated with information on the gross seasonal returns from virtually *all* whaling voyages, drawn from the various shipping newspapers published in New Bedford, San Francisco, Hawaii and elsewhere. Thus an estimate can be constructed by mathematical methods to delineate, among other things, the pre-1848 population size, its annual changes, migration patterns, and feeding and breeding

areas (Bockstoce and Botkin, 1976).[6]

It is unfortunate that the western Arctic bowhead is one of the few, if not the only whale population which will lend itself to such a study. Similar studies of other great whale populations will be hampered either because they are wide-ranging (as with sperm whales) or because data is poorly recorded (such as the early bowhead fisheries in the waters adjacent to the North Atlantic). Thus the historical study of the Western Arctic bowhead has this further significance — its results will generate information that may be applied, by comparison, to other great whale populations, for which such detailed information is not available.

[6] *Editor's note.* In March 1978 it was still uncertain to what extent the Eskimos would observe the IWC regulations. The National Marine Fisheries Service indicated its willingness to fund the proposed study.

References

BOCKSTOCE, J., A prehistoric population change in the Bering Strait region. *Polar Rec.*, 16(105).
1973

—, Contacts between American whalemen and the Copper Eskimos. *Arctic*, 28(4).
1975

—, On the development of whaling in the Western Thule Culture. *Folk, Copenh.*, 18.
1976

—, Steam whaling in the western Arctic. New Bedford, Mass., Old Dartmouth Historical Society.
1977

—, A chronological list of commercial wintering voyages to the Bering Strait region and western
1977a Arctic, 1848-1910. *Musk-ox, Winnipeg*, July issue.

—, An issue of survival: bowhead vs. tradition. *Audubon*, September issue.
1977b

BOCKSTOCE, J. and D. BOTKIN, Historical research on the bowhead whale of the western Arctic. Grant proposal submitted to the Marine Mammal Commission, Washington, D.C.
1976

RAINEY, F., The whale hunters of Tigara. *Anthropol. Pap. Am. Mus. Nat. Hist.*, 41(2).
1947

THE HUMPBACK WHALE:
PRESENT KNOWLEDGE AND FUTURE TRENDS IN RESEARCH
WITH SPECIAL REFERENCE TO THE WESTERN NORTH ATLANTIC

H.E. WINN **and** G.P. SCOTT

Abstract

Eleven "prime world populations" of humpback whales (*Megaptera novaeangliae*) can be recognized. It would be worthwhile to estimate their original size and compare these to estimates of their present sizes and to identify stocks within these populations. Stocks can be identified through studies of density distribution, migratory routes, possible song dialects and differences in coloration (especially on the dorsal sides of flippers), morphometry and external tagging; biochemical analyses can be used over a longer period of time. There are 2 subsistence fisheries for humpback whales in the western North Atlantic. About 4 animals, mainly mothers and calves, are taken each year in Bequia, in the Lesser Antilles, from a population of perhaps less than 100 animals. This catch may prevent it from increasing in size. A stochastic Schaefer model, in which growth in abundance is exponential and reproductive and natural mortality rates (derived from other populations) are constant, can be used to describe this fishery. Up to 10 humpbacks are taken off the west coast of Greenland each year. The ecological characteristics of the tropical breeding grounds of humpback whales and the relation of them to original and present population sizes should be described. The energetics of the species, especially in relation to abundance, social behaviour and migration, should be investigated. Historical studies may be important to several areas of research on this species. Interference by man with humpback whales and pressure to hunt them will probably increase in the future, especially as the number of whales grows larger. Plans for sensible exploitation and for the establishment of sanctuaries should be made in time.

Résumé

On peut distinguer onze « populations originelles » de mégaptères (*Megaptera novaeangliae*). Il pourrait être utile d'estimer leur taille originale et de la comparer aux estimations de leur taille présente et d'identifier les stocks au sein de ces populations. Les stocks peuvent être identifiés par l'étude de la distribution de la densité, le chemin des migrations, les dialectes éventuels de sons et les différences de coloration (spécialement sur la face dorsale des nageoires), la morphométrie et le baguage extérieur; les analyses biochimiques peuvent être utilisées sur une plus longue période de temps. Il existe deux pêcheries de subsistance des mégaptères dans l'Atlantique nord-ouest. Environ 4 animaux, surtout des mères et des baleineaux, sont capturés chaque année à Bequia, dans les Petites Antilles, sur une population peut-être inférieure à 100 animaux. Ces captures peuvent l'empêcher de croître en taille. Un modèle stochastique de Schaefer, dans lequel la croissance de l'abondance est exponentielle et les taux de reproduction et de mortalité naturelle (tirés d'autres populations) sont constants, peut être utilisé pour décrire cette pêcherie. On capture jusqu'à 10 mégaptères par an au large

171

de la côte ouest du Groenland. Les caractéristiques écologiques des terrains de reproduction tropicaux des mégaptères et leur relation avec les tailles de population originales et présentes devraient être décrites. Les caractéristiques énergétiques de l'espèce, spécialement en rapport avec l'abondance, le comportement social et la migration, devraient faire l'objet de recherches. Les études historiques pourraient être importantes pour plusieurs secteurs de recherche relatifs à cette espèce. Les interférences de l'homme avec les mégaptères et les pressions exercées pour les chasser augmenteront probablement dans l'avenir, spécialement à mesure que leur nombre s'accroît. Il conviendrait d'établir en temps opportun des plans relatifs à l'exploitation rationnelle et à la création de sanctuaires.

Extracto

Se distinguen 11 « poblaciones mundiales primarias » de ballena jorobada (*Megaptera novaeangliae*). Valdría la pena estudiar el volumen original de cada una de esas poblaciones y compararlo con las estimaciones de su volumen actual, e identificar los grupos existentes dentro de esas poblaciones. Los grupos pueden individuarse estudiando la distribución de densidades, las rutas que los animales siguen en sus migraciones, los posibles dialectos en los sonidos que emiten, las diferencias en la coloración (especialmente en la parte dorsal de las aletas) y la morfometría, o con el empleo de marcas; pueden utilizarse también análisis bioquímicos, a lo largo de un período más prolongado de tiempo. En el noroeste del Atlántico hay dos pesquerías de subsistencia de ballena jorobada. En Bequia, en las Antillas Menores, se capturan anualmente alrededor de cuatro animales, sobre todo hembras paridas y cachorros, frente a una población quizás inferior a 100 animales. Quizás estas capturas impidan que aumente el volumen de la población. Para describir esta pesquería puede utilizarse un modelo estocástico de Schaefer, en el que el aumento de la abundancia es exponencial y el índice de reproducción y la mortalidad natural son constantes (derivados de otras poblaciones). Frente a la costa occidental de Groenlandia se capturan todos los años hasta 10 ballenas jorobadas. Es preciso describir las características ecológicas de los criaderos tropicales de ballenas jorobadas y su relación con el volumen original de la población y con su volumen actual. Es preciso estudiar también la energética de la especie, particularmente en relación con la abundancia, el comportamiento social y los movimientos migratorios. Los estudios históricos pueden ser importantes en varios sectores de las investigaciones sobre esta especie. La interferencia del hombre con estas ballenas y su explotación aumentarán probablemente en el futuro, especialmente a medida que aumente el número de ballenas. Es preciso preparar a tiempo planes para explotarlas racionalmente y establecer santuarios.

H.E. Winn
Graduate School of Oceanography, University of Rhode Island, Kingston, RI 02881, USA

G.P. Scott
Graduate School of Oceanography, University of Rhode Island, Kingston, RI 02881, USA

Introduction

Confrontation between man and the humpback whale (*Megaptera novaeangliae*) may be expected to increase as humpback numbers grow and as man's activity on the continental shelf also increases. Certain activities affect humpbacks directly, such as high speed boats crossing the mating grounds in Hawaii, oil wells being drilled off Angola and at least 3 subsistence level fisheries (Bequia, Tonga, Greenland). Other effects will be indirect such as pollution and the harvesting of whale's food.

The international regime is changing, as many nations claim their 200 mile economic zones, within which all important humpback calving areas are likely to fall. While the International Whaling Commission (IWC) has some control over humpback take by member nations, a number of countries whose activities affect humpbacks do not belong to the IWC. Clearly, international control of all humpback populations is the only eventual answer for the conservation of this species.

It is wise to be prepared to predict the numbers of animals that could be taken at levels that would allow the most efficient sustainable yields while allowing for other factors such as increased incidental mortalities due to man's activities. The possibility exists that the humpback may be exploited again, as the more reduced populations hopefully increase. Certainly there are pressures for reopening humpback exploitation in some nations.

A more critical definition of stock structuring is required if, in the future, we are to manage this species better. Closer examination and correlative analysis of existing data such as density distribution records, stranding information and historical catches will aid in such definition. Institution of research programmes into topics such as dialectic analysis of humpback song types, dorsal flipper and ventral fluke colouration patterns (plus other "visual tag" studies), chemical tracer studies of blubber pesticide levels, radio tagging, and the modelling and analysis of subsistence fisheries may also prove valuable. We will also have to consider the ecological requirements on both the tropical calving grounds and Nearctic feeding grounds and the effects of man's past, present and probable future activities in these areas if intelligent management is to be accomplished.

Distribution of humpbacks

Townsend (1935), in one of the most valuable contributions to humpback whale biology, published the distribution of humpbacks around the world based on logbook records of whales, mostly from the nineteenth century. Although most of the information content was lost when the data were placed on maps, particularly information on specific geographic locations, his study still serves as the basis for most of our present day distributional knowledge. An analysis of Townsend's maps reveals 11 prime world populations of humpbacks (Fig. 1).

In addition to the distributional information available in Townsend's (1935) study, we may be able to calculate supplementary and/or new estimates of the "virgin" levels of the prime populations using his raw data and other data sources such as import-export duty records associated with shore stations for humpbacks (after Adams, 1971, 1975). More recent catch and effort data from the Committee on Whaling Statistics and the Yearbook of Fishery Statistics may also be used to calculate additional estimates in order to identify more precisely the trend in abundance relative to man's activities.

The value of Townsend's (1935) work has been identified; however, the information published is only general in nature. A closer examination of distributional data is required if we are to define more precisely the stock structuring within the 11 prime populations.

173

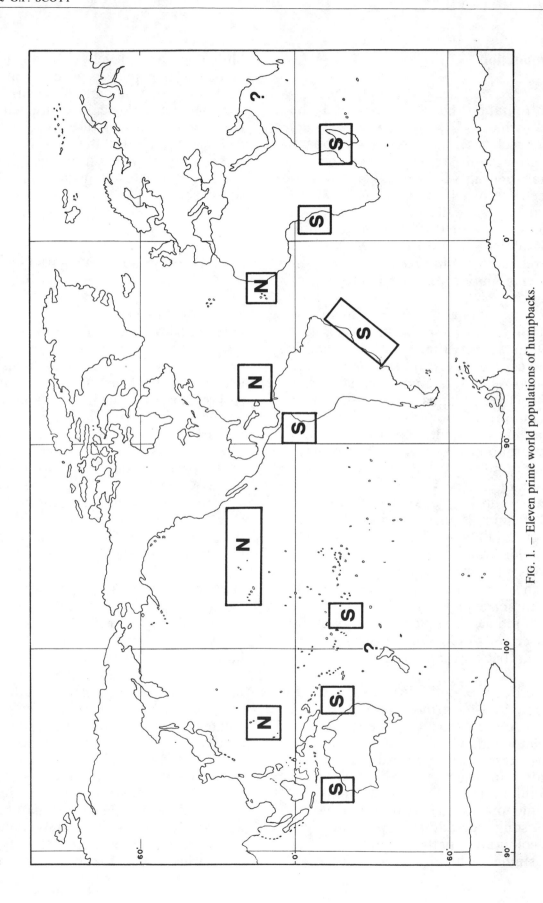

Fig. 1. — Eleven prime world populations of humpbacks.

For example, in the western North Atlantic, Winn et al. (1975) found different humpback densities in different geographic locations in the Caribbean breeding areas. More recently, Winn and Scott (in press) interpreted the density differences to result from at least 3 stocks in the Caribbean, each with different histories of exploitation.

Migratory route studies based on at-sea observations, tag and recapture studies, historical catches and, to some extent, strandings also provide information about the stock structuring of the prime populations. The generalized routes as shown by Kellogg (1929) and MacIntosh (1965) have often been quoted as "well known" because of this species' shallow water tendencies. However, there is some evidence (Townsend, 1935; miscellaneous unpublished observations) for a trans-oceanic migratory route in the western North Atlantic and in the Southern Ocean (Gaskin, 1976). Anomalies such as these in the generally accepted patterns indicate that closer study is needed.

In addition to the standard methods mentioned above, there are 3 potentially available to aid in migratory route studies. One, radio tagging, involves direct observation of the animal's movements while the others, chemical tracer analyses and fluke pattern recognition, examines similarities in the end points of the migratory route. Although still in need of some technological improvements, the radio tag developed by Watkins and Schevill (1977) shows special promise for the direct observation of daily and seasonal movements of humpbacks. To date, the radio tag has been successfully tested several times including one tag in a free-ranging Alaskan humpback for 6 successive days (Tillman and Johnson, 1976). A review of the advances in this method is given by Leatherwood and Evans (in press).

Chemical tracer studies similar to the iodine analyses of Lund (1950, 1950a, 1951) would also prove useful in humpback stock analyses. As an example, Winn and Scott (in press) have interpreted differing pesticide levels in humpback blubber samples as support for a 3 stock model for humpbacks in the western North Atlantic. They hypothesized (after Taruski, Olney and Winn, 1975) that animals from different stocks feed in different geographic areas with characteristic pollutant levels, which would be reflected in animal blubber levels. Winn and Scott (in press) found little evidence of metabolic or age differences in pesticide levels to dispute this hypothesis. This method may prove to be especially useful since blubber samples may be obtained from living animals without harming them using a modified version of the biopsy sampler described by Winn, Bichoff and Taruski (1973). Thus, with a concentrated effort, large numbers of blubber samples may be obtained and thereby statistically compared.

A visual tag or biological marker useful in stock identification, migratory route analyses and local movement studies, is the ventral fluke pattern and dorsal fin combination. In the western North Atlantic this method of individual identification has been described by Katona et al. (in press). Individual humpbacks have highly characteristic patterns of black and white on the ventral side of the flukes and, at times, in the shape of the dorsal fin. Additionally, the behavioural traits of "humping" up and then throwing flukes for a long dive allows one to photograph and subsequently catalogue and analyse these patterns for matches. To date, 4 individuals have been identified in both the northern and southern ends of their migratory range in the western North Atlantic (see Winn and Scott, in press) and a similar programme has been initiated in the eastern North Pacific.

Songs

It is well known that humpbacks produce characteristic songs while on their breeding grounds (Payne and McVay, 1971; Winn, Perkins and Poulter, 1970). It is clear that the songs produced by isolated popula-

175

tions (Gulf of California, West Indies) are different (Winn and Winn, 1978; Winn et al., work in progress) and it is possible that stocks in the same ocean basin have distinct dialects that can be recognized (Winn, 1975; Winn and Winn, 1978; Winn et al., work in progress). Unfortunately, there is at present some confusion in defining dialects for the northwestern Atlantic stocks, arising, perhaps, from crossover from high to low density Carribbean stocks. This may be caused by intermingling at high latitudes as described for some southern stocks of humpbacks (Chittleborough, 1959).

Colour patterns

From observations of humpbacks in Alaska, southern California, Hawaii and the Atlantic, it is probable that some groups exhibit stock-specific coloration patterns, particularly on the dorsal surface of the flipper. Wolman and Juraz (1977) have used this character difference to define the Hawaiian stock (mostly white flippers dorsally) as being distinct from that stock which summers in southeastern Alaska (mostly dark flippers dorsally). If this coloration is consistent, we would expect differences between other groups of humpbacks as well.

Feeding and energetics

Humpbacks generally migrate annually from their feeding grounds in the high latitudes to their mating and calving grounds in the lower latitudes (Townsend, 1935; Dawbin, 1964; Chittleborough, 1965; MacIntosh, 1965). It is likely that animals migrating to traditional mating and calving grounds require less energy than those that are forced into marginal habitats (Winn and Winn, 1978),

leaving themselves more energy for migration and feeding. This problem would be more critical for calves and yearlings because of their higher surface area: volume ratio.

Recent considerations of whale population energetics include those of Brodie (1975) Kawamura (1975) and Lockyer (1976). Kawamura (1975) particularly addresses the problem of the cost of seasonal migrations in terms of energy consumption and potential energy sources as related to the general distribution of whales. His calculations indicate that it is likely that fin whales and other large rorquals including humpbacks, must feed to some extent during migration. Certainly, evidence of humpbacks feeding on migration has been previously documented (Dawbin, 1966).

It thus seems possible that food source levels are quite related to population levels of humpbacks. The indirect influence of man on humpback populations through exploitation of this species' food sources may be of much greater significance than previously considered. There is some evidence (historical reports of starved or scrog whales, examination of photographs) that some individuals in the western North Atlantic and perhaps in other populations as well may be exhibiting symptoms of starvation. We hypothesize that the species has evolved specialized feeding strategies for certain food types and cannot adapt with 100 % efficiency to new food sources as man's exploitation of traditional sources continues. Thus, stocks may never recover to their former levels.

We also know that certain groups of humpbacks stay in areas 6 months out of the normal cyclic phase, both in the tropics and higher latitude feeding areas (e.g., Barents Sea, some tropical regions; Tomilin, 1957; Townsend, 1935). The extended stay may be a response to areas of high productivity and large food sources, especially in the high latitudes, whereas the extended stay in the tropics might be better explained by a late migration (Chapman, 1974) than a response to food. For certain tropical areas (regions of intense upwelling), the food source hypothesis certainly remains viable (e.g., Cape Verde Islands, Gulf of Panama).

Northwest Atlantic humpbacks and subsistence fisheries

In the northwestern Atlantic the humpback has been protected by the IWC since 1955. However, subsistence level exploitation has continued at Bequia and exploitation has continued at Bequia and in Greenland. Nearly 75 % of the estimated 1 013 animals in the West Indies were found to overwinter on Silver Bank by Winn, Edel and Taruski (1975). More recently Winn and Winn (1977) proposed that this area be considered a sanctuary.

The density of animals found by Winn, Edel and Taruski (1975) in the northern Caribbean may be high enough to cause movement into the lower stock as defined by Winn and Scott (in press). It is also possible that reverse movement may occur, caused by fishing activities and man's possible use of and encroachment on the humpback habitat.

Sergeant (1966) estimated the number of animals around Newfoundland for the end of the nineteenth century using regression methods based on Newfoundland catches from 1898 to 1915. The estimate of 1 500 animals does not represent the original level of humpbacks in the western North Atlantic since no allowance was made for catches in the tropics or in other regions during this time. Additionally, the humpback was hunted by both Yankee whalemen and natives in the tropics prior to 1898 (Adams, 1970). The level of this exploitation and to what extent it pre-dates the twentieth century is only vaguely known.

An example of a fishery targeted mainly at humpbacks and dating back to the 1870s is the Bequia fishery. The fishery stems from the American whaling operations which flourished near the Grenadine Islands during the 1860s. This fishery is, as were other shore-based and vessel-based operations for humpbacks during this period, directed primarily at the cow-calf component of the population (Sinckler, 1913; Fenger, 1918; Clark, 1887). Thus, the major impact of the fisheries was on the reproductive heart of the populations. The Bequia fishery has had a long history of success and failure (Adams, 1970) and to date is the last of numerous shore stations in the Caribbean. The average annual catch for the past 21 years of operations is 1.5 (s = 1.2; Winn and Scott, in press), however, the average annual kill may be on the order of 1.3-3 times this value (after Adams, 1970; Fenger, 1918; Winn and Scott, in press).

Winn, Edel and Taruski (1975) estimated only 49 ± 19 (95 % C.I.) humpbacks from Martinique to Venezuela (Bequia stock). Historical evidence suggests that the level of humpbacks in this area was once much higher (Winn and Scott, in press). We have modelled the possible response of the Bequia stock to protection and continued exploitation by the Bequians using an exponential model with fishing mortality terms for the cow, calf and yearling components. We used vital rate parameters reported in the literature for humpbacks (Chittleborough, 1965) and other species (Ottestad, 1956; Perrin, Holtz and Miller, 1975). Assuming an annual kill of 4 distributed among the calf, yearling and cow components, but weighted toward the mother-calf combination, we found that this level of exploitation is sufficient to maintain the stock at the low level observed for a period of 10 years or more.

Our analysis assumed no effect on the sex ratio of the stock. This assumption does not seem valid, considering that the main target of the fishery is the female (up to 75 % of the kill is female). Thus the fishery may have equal or even intensified effects on the population size at kill levels much lower than the assumed value of 4.

In addition to Bequia there is another fishery directed at humpbacks in the western

177

North Atlantic. Mitchell (1973) reported that Greenland natives along the west coast of that island take up to 10 animals per year. In fact, the average catch by this fishery from 1958 through 1976 was 3.1 humpbacks (s = 3.1) with peak catches of 9 occurring in 1973, 1974 and 1975 (Winn and Scott, in press). We have hypothesized that the Bequia animals migrate to the east of Greenland and Iceland to feed.

The first order analysis of the Bequia fishery indicates that the present catch may be suppressing the population response to protection. Similar modelling and analysis of the other known subsistence level fisheries using, perhaps, a stochastic birth-death process approach in order to more closely approximate actual events, should prove useful in planning proper management schemes.

Conclusions

In conclusion, the present state of knowledge about humpback populations worldwide is perhaps greater than that for all other rorquals. However, there are still many areas in which we have little or no information. In this paper we have identified some research methods that may aid in generating some of the information we need for *one* population of humpbacks. However, these methods need not be restricted for usage in the western North Atlantic since hopefully they are generalized enough to have application to all humpback stocks.

Acknowledgements

We greatly appreciate the editorial assistance provided by Ms J. Gordon Clark, without whose aid this paper would not have been completed. Dr W. Palm provided advice on the modelling and energetics sections; Mr T. Thompson supplied editorial comments on style and content, and Ms M. Nigrelli typed various drafts of the manuscript. The second author completed parts of this paper with support of a fellowship from the National Wildlife Federation and research funds from the Lerner Fund for Marine Research. The major development of this paper was the result of research support by the Office of Naval Research No. N00014-76-C-0226 to the first author.

References

ADAMS, J.E., Marine industries of the St. Vincent Gre-
1970   nadines, West Indies. Chapter 5. Whaling. Chapter 7. Minor fishing activities. University of Minnesota Ph.D. Thesis.

—, Historical geography of whaling in Bequia Island,
1971   West Indies. *Caribb. Stud.*, 11:55-74.

—, Primitive whaling in the West Indies. *Sea Frontiers*,
1975   21:303-13.

BRODIE, P.B., Cetacean energetics, an overview of in-
1975   traspecific size variation. *Ecology*, 56:152-61.

CHAPMAN, D.G., Status of Antarctic rorqual stocks. *In*
1974   The whale problem, edited by W. Schevill. Cambridge, Mass., Harvard University Press, pp. 218-38.

CHITTLEBOROUGH, R.G., Intermingling of two popula-
1959   tions of humpback whales. *Norsk Hvalfangst-tid.*, 48(10):510-21.

—, Dynamics of two populations of the humpback
1965   whale *Megaptera novaeangliae* (Borowski). *Aust. J. Mar. Freshwat. Res.*, 16(1):33-128.

CLARK, A.H., The whale fishery. 1. History and present
1887   conditions of the fishery. *In* The fishery industries of the United States, edited by G.B.

Goode. Washington, D.C., Government Printer, Section V, Vol. 2:3-293.

DAWBIN, W.H., Movements of humpback whales
1964   marked in the southwest Pacific Ocean, 1952 to 1962. *Norsk Hvalfangsttid.*, 3:68-78.

—, The seasonal migratory cycle of humpback whales.
1966   *In* Whales, dolphins and porpoises, edited by K.S. Norris. Berkeley, University of California Press, pp. 145-70.

FENGER, F.A. Longshore whaling in the Grenadines.
1918   *Outing*, 61:664-79.

GASKIN, D.E., The evolution, zoogeography, and ecolo-
1976   gy of Cetacea. *Oceanogr. Mar. Biol.*, 14:247-346.

KATONA, S. *et al.*, Identification of humpback whales by fluke photographs. *In* Behaviour of marine animals, Vol. 3. The natural history of cetaceans, edited by H.E. Winn and B. Olla. New York, Plenum Press, (in press).

KAWAMURA, A., A consideration of an available source
1975   of energy and its cost for locomotion in fin whales with special reference to the seasonal migrations. *Sci. Rep. Whales Res. Inst., Tokyo,* (27):61-79.

KELLOGG, R., What is known of the migrations of some
1929   of the whalebone whales. *Rep. Smithson. Inst.,* (1928):467-94.

LEATHERWOOD, S. and W.E. EVANS, Some recent uses and potentials of radio-telemetry in field studies of cetaceans. *In* Behaviour of marine animals, Volume 3. The natural history of cetaceans, edited by B.E. Winn and B. Olla. New York, Plenum Press, (in press).

LOCKYER, C., Growth and energy budgets of large ba-
1976   leen whales from the Southern Hemisphere. Paper presented to the Scientific Consultation on the Conservation and Management of Marine Mammals and their Environment, Bergen, Norway, 31 August-9 September 1976. Rome, FAO ACMRR/MM/SC/41:179 p.

LUND, J., Charting of whale stocks in the Antarctic on
1950   the basis of iodine values. *Norsk Hvalfangsttid.*, 39(2):53:60.

—, Charting of whale stocks in the Antarctic in the

1950a   season 1949/50 on the basis of iodine value. *Norsk Hvalfangsttid.*, 39(7):298-305.

—, Charting the whale stocks in the Antarctic in the
1951   season 1950/51 on the basis of iodine values. *Norsk Hvalfangsttid.*, 40(8):124-46.

MACINTOSH, N.A., The stocks of whales. London, Fish-
1965   ing News (Books) Ltd., 232 p.

MITCHELL, E.D., Draft report on humpback whales
1973   taken under special permit by eastern Canadian land stations 1969-71. *Rep. IWC,* 23 (App. IV, Annex M): 138-54.

OTTESTAD, P., On the size of the stock of Antarctic fin
1956   whales relative to the size of the catch. *Norsk Hvalfangsttid.*, 45(6):298-308.

PAYNE, R. and S. McVAY, Songs of humpback whales.
1971   *Science, Wash.*, 173:585-97.

PERRIN, W.F., D.B. HOLTZ and R.B. MILLER, Prelimina-
1975   ry estimates of some parameters of growth and reproduction of the eastern spinner porpoise, *Stenella longirostris* subsp. *Admin. Rep. Southwest Fish. Cent.*, (LJ-75-66):33 p.

SERGEANT, D.G., Populations of large whale species in
1966   the western North Atlantic with special reference to the fin whale. *Circ. Fish. Res. Board Can. Arctic Biol. Stn.*, (9): 13 p.

SINCKLER, E.G., The Barbados handbook. London,
1913   Duckworth and Co., 277 p.

TARUSKI, A.G., C.E. OLNEY and H.E. WINN, Chlorinated
1975   hydrocarbons in cetaceans. *J. Fish. Res. Board Can.*, 32(11):2205-9.

TILLMANN, M.F. and J.H. JOHNSON, Radiotagging of
1976   humpback whales. Report to the National Marine Fisheries Service, Permit No. 136:4 p. (mimeo).

TOMILIN, A.G., Cetacea. *In* Mammals of the USSR and
1967   adjacent countries. Jerusalem, Israel Program for Scientific Translations, IPST Catal. No. 1124.

TOWNSEND, C.H., The distribution of certain whales as
1935   shown by logbook records of American whaleships. *Zoologica, N.Y.*, 19(1):50 p.

WATKINS, W.A. and W.E. SCHEVILL, The development
1977   and testing of a radio whale tag. *ONR Tech. Rep.*, (WHOI-77-58):38 p.

WINN, H.E., Dialects and social organization of hump-
1975   back whales. Paper presented at the Congress
        on the Biology and Conservation of Marine
        Mammals, University of California, Santa
        Cruz, Dec. 4-7, 1975.

WINN, H.E. and G.P. SCOTT, A model for stock struc-
        turing of humpback whales (*Megaptera no-
        vaeangliae*) in the western North Atlantic. *Rep.
        IWC*, (in press).

WINN, H.E. and L.K. Winn, Status of the Humpback
1977   whale (*Megaptera novaeangliae*) (synonym *M.
        boops, M. nodosa*) in the area of Silver, Navi-
        dad and Mouchoir Banks, West Indies, and
        recommendations on measures to ensure their
        conservation. Report to the Marine Mammals
        Commission, 19 Feb., 1977, 17 p.

—, The song of the humpback whale in the West Indies.
1978   *Mar. Biol.*, 47:97-114.

WINN, H.E., W.L. BISCHOFF and A.G. TARUSKI, Cytolo-
1973   gical sexing of cetacea. *Mar. Biol.*, 23:343-6.

WINN, H.E., R.K. EDEL and A.G. TARUSKI, Population
1975   estimate of the humpback whale (*Megaptera
        novaeangliae*) in the West Indies by visual and
        acoustic techniques. *J. Fish. Res. Board Can.*,
        32(4):499-509.

WINN, H.E., P.J. PERKINS and T.C. POULTER, Sounds of
1970   the humpback whale. *In* Proceedings of the
        Seventh Annual Conference on Biological
        Sonar and Diving Mammals, 23-24 October
        1970. Menlo Park, Stanford University, Bio-
        logical Sonar Laboratory, pp. 39-52.

WOLMAN, A.A. and C.M. JURAZ, Humpback whales in
1977   Hawaii: vessel census. *Mar. Fish. Rev.*,
        39(7):1-5.

STATUS OF THE EASTERN PACIFIC (CALIFORNIA) STOCK OF THE GRAY WHALE

D.W. RICE

Abstract

The gray whale, *Eschrichtius robustus*, survives only in the North Pacific Ocean, where 2 geographically isolated populations have been distinguished; the eastern Pacific or "California" population, summering in the northern Bering Sea and adjacent seas to the north and wintering in calving grounds along the west coast of Mexico, and the western Pacific or "Korean" population, the continued existence of which is uncertain, summering in the northern Sea of Okhotsk and wintering off southern Korea and probably also in the Seto Inland Sea. Basic information on reproduction and mortality of California gray whales is given. Prior to commercial exploitation (old style whaling, 1846 to about 1900; modern whaling, 1905 to 1946), the California population numbered no more than 15 000 animals, being reduced to perhaps half its original size by 1900. Thereafter followed a population increase, especially rapid from 1947 until about 1960; during the past 10 years, the population has remained stable or increased slightly in size, and now includes about 11 000 animals (± 2 000). Siberian Eskimos continue their traditional use of gray whales, with the average annual catch in recent years at 170 whales (all now taken by a modern catcher boat).

Commercial whaling for the gray whale was banned under the International Convention for the Regulation of Whaling in 1946. The greatest present threat to the California gray whale population is increasing industrial development and vessel traffic, including "whale-watching" boats, in the calving lagoons.

Résumé

La baleine grise, *Eschrichtius robustus*, survit uniquement dans le Pacifique nord, où l'on a distingué deux stocks isolés géographiquement: la population du Pacifique oriental ou "de Californie" qui passe l'été dans la mer de Béring et les eaux boréales limitrophes et qui passe l'hiver sur des terrains de reproduction situés le long de la côte ouest du Mexique; et la population du Pacifique ouest, ou "coréenne" dont la poursuite de l'existence semble problématique actuellement, passe l'été dans le nord de la mer d'Okhotsk et l'hiver dans les eaux de la Corée du sud et probablement aussi dans la mer intérieure de Seto. L'auteur donne des informations de base sur la reproduction et la mortalité des baleines grises de Californie. Avant l'exploitation commerciale (ancienne pêche baleinière, 1846 à 1900 environ; pêche baleinière moderne, 1905 à 1946), la population de Californie ne comptait pas plus de 15 000 animaux et était réduite à peut-être la moitié de sa taille initiale vers 1900. La population a augmenté ensuite, à une cadence particulièrement rapide entre 1947 et 1960 environ. Ces dix dernières années, la population est restée stable ou a faiblement augmenté et elle compte actuellement environ 11 000 animaux (± 2 000). Les Esquimaux de Sibérie continuent à pratiquer la chasse traditionnelle des baleines grises, avec un chiffre moyen annuel de 170 animaux au cours de ces dernières annés (toutes ces captures étant maintenant réalisées par un baleinier moderne).

La pêche commerciale de la baleine grise a été interdite par la Convention internationale baleinière de 1946. Le plus grand danger menaçant la population de baleines grises de Californie est représenté par l'expansion industrielle et l'activité maritime, y compris les croisières touristiques dans les lagunes de reproduction.

Extracto

La ballena gris (*Eschrichtius robustus*) sobrevive sólo en el norte del Pacífico, donde se han distinguido dos poblaciones geográficamente aisladas: la población del este del Pacífico o de California, que transcurre el verano en el mar de Bering y mares septentrionales adyacentes y el invierno en las zonas de cría a lo largo de la costa occidental de México; y la población del oeste del Pacífico o coreana, de la que no se sabe si aún subsiste, que transcurre el verano en la parte septentrional del mar de Ojotsk y el invierno frente al sur de Corea y probablemente también en el mar interior de Seto. Se dan algunos datos básicos sobre la reproducción y la mortalidad de la ballena gris de California. Antes de su explotación comercial (cacería de tipo tradicional, 1846 a 1900 aproximadamente; cacería de tipo moderno, 1905 a 1946), la población de California contaba no más de 15 000 ballenas, que para 1900 se habían reducido tal vez a la mitad de la cifra original. Posteriormente la población aumentó, sobre todo entre 1947 y 1960, aproximadamente, y durante los últimos diez años se ha mantenido estable o ha aumentado ligeramente, contando en la actualidad unos 11 000 animales ($\pm$ 2 000). Los esquimales siberianos prosiguen su cacería tradicional de ballena gris, con una captura media en los últimos años de 170 animales (todos los cuales se capturan hoy con una embarcación moderna).

La caza comercial de ballena gris se prohibió en 1946 en virtud del Convenio Internacional para la Reglamentación de la Caza de la Ballena. En la actualidad, las principales amenazas para la población de ballena gris de California es el desarrollo industrial y el tráfico marítimo, incluido el de las embarcaciones desde las que los turistas observan a las ballenas en las lagunas costeras donde estos animales se reproducen.

D.W. Rice
NMFS, Northwest and Alaska Fisheries Center, Marine Mammals Division, 7600 Sand Point Way, Bldg 32, Seattle, WA 98115, USA

The gray whale *Eschrichtius robustus* is of considerable biological interest, being the only member of the primitive baleen whale family Eschrichtiidae. The eastern Pacific population of this species is also of great interest to conservationists since it is the only population of great whales which, following severe depletion, has recovered to near virginal size.

Distribution and separation of populations

Subfossil and early historical records (Fraser, 1970) bear witness to the former occurrence of the gray whale *Eschrichtius robustus* in the North Atlantic, but it has been extinct there for several centuries. The species now survives only in the North Pacific, where two geographically isolated populations have been distinguished: the eastern Pacific or "California" population, and the western Pacific or "Korean" population (Rice and Wolman, 1971).

The California whales spend the summer mostly in the northern Bering Sea, the southern Chukchi Sea, and the southwestern Beaufort Sea, with scattered individuals occurring along the west coast of North America as far south as northern California. They migrate down the North American coast to winter along the west coast of Mexco. The main calving grounds are:

— Laguna Ojo de Liebre and the adjacent Laguna Guerrero Negro, Laguna San Ignacio, and Bahia Magdalena and confluent waterways, all in Baja California;

— Yavaros, Sonora, and Bahia Reforma, Sinaloa (Gilmore, 1960).

Contrary to many published statements, there is no evidence that San Diego Bay, California, was ever a calving area (Henderson, 1972).

The Korean stock spent the summer in the northern Sea of Okhotsk. It migrated down the Asian coast to calving grounds off the south coast of Korea (Andrews, 1914). Another migration route probably led down the eastern side of Japan to calving grounds in the Seto Inland Sea (Omura, 1974).

According to Mizue (1951), the last gray whales reported on the eastern side of Japan were killed in 1914, and the last reported in Korea were killed in 1933. The gray whale was alleged to be unknown to present day South Korean whalers (Bowen, 1974). However, recent evidence raises some doubts. Brownell and Chun (1977) report that a few were killed as recently as 1966. In the Sea of Okhotsk there have been two recent sightings — four animals in 1967 and one in 1974 (Berzin, 1974; Berzin and Kuz'min, 1975; Kuz'min and Berzin, 1975).

In the northern Kuril Islands one gray whale was killed at Otomae (Shiashkotan) in 1942 (Mizue, 1951); one was seen off the southeast coast of Honshu in 1959, and one died in the same area in 1968 (Nishiwaki and Kusuya, 1970).

Nishiwaki and Kasuya (1970) and Bowen (1974) hypothesized that these three recent records involved vagrants from the California stock, rather than survivors of the Korean stock. Since gray whales mate during their autumn migration, such rare vagrants would make interbreeding between the California and Korean population possible — a possibility that would, however, be greatly reduced if as seems likely, most vagrants are immature animals as in the case of the 1968 whale. It appears equally likely that all three animals were members of the Korean stock.

Reproduction and mortality

The following data are from samples of gray whales collected mostly from 1964 to 1969 (Rice and Wolman, 1971).

Sexual maturity is attained at a mean

183

age of 8 years (range: 5-11 years). The mean annual ovulation rate is 0.52, and the mean pregnancy rate is 0.46.

Annual mortality estimates for animals over 8 years, calculated from the age composition as determined by ear-plug readings, were 0.095 for females and 0.081 for males. A similar estimate for sexually mature females, based on ovarian corpora counts, was 0.082. These estimates are probably higher than the true rate because the population was increasing during the preceding years. The ratio of males to females is about equal.

Exploitation

The Eskimos living on the shores of the northern Bering Sea and the Chukchi Sea have hunted whales since time immemorial. On the Alaskan side the catch is mostly bowhead whales (*Balaena mysticetus*), and very few gray whales are taken. However, on the Siberian side the catch is almost entirely gray whales. All the gray whales used by the Siberian Eskimos are now taken by one modern style catcher boat (Berzin, pers. comm.). Catch statistics are available only for recent years, in which period the catch has averaged about 170 gray whales per year (Table 1).

Several Indian tribes on Vancouver Island and in the State of Washington traditionally hunted gray whales, but have not done so since 1928.

From 1846 until about 1900, American whalers exploited gray whales, mostly on their wintering grounds, taking a few in northern waters during the summer. On the basis of available historical records, Henderson (1972) estimated that the total catch from 1846 to 1874 was about 8 100; this estimate was intentionally on the high side. During the peak whaling period of 1855 to 1865 the annual catch averaged 474 whales (Table 2). Catches during the three winter seasons from 1883/84 to 1885/86 were 58, 68, and 41 respectively (Townsend, 1887).

Table 1. Aboriginal catches of gray whales in Bering and Chukchi Seas: 1965-1976 (blank spaces indicate no data available)

Year	Number of whales caught		
	USSR [1]	USA [2]	Total [2]
1965	175		
1966	194		
1967	125		
1968	135		
1969	139		
1970	146	5	151
1971	150	3	153
1972	181	1	182
1973	173	1	174
1974	181	1	182
1975	171	7 [2]	178
1976	165	2 [2]	167

[1] Data from All-Union Research Institute of Marine Fisheries and Oceanography, Moscow.
[2] Species identification probable, but not certain.

Modern style whaling began on the west coast of North America in 1905. A few gray whales were taken in the winter off Baja California and California, mostly between 1925 and 1929. Factory ships took an average of 52 gray whales per year in the Bering Sea from 1933 to 1946 (Table 3), when the International Convention for the Regulation of Whaling was signed. One of the provisions of the Schedule to the Convention was the banning of commercial whaling for gray whales. Under Special Scientific Permits allowed by this Convention, 316 gray whales were killed off California between 1959 and 1969 (Table 4). From 1966 to 1969 the combined scientific (Table 4) and Siberian eskimo (Table 1) catches averaged 221 per year.

Other factors

Considerable harassment is caused by commercial cruise boats which take people

Table 2. Estimated catches of gray whales by old-style whaling: 1846-1874 [1]

Period	Number of whales killed		
	California and Mexico	Bering and Chukchi Seas	Total
1846-1854	716 [2]	55	771
1855-1865	4 938	275	5 213
1866-1874	2 007	110	2 117

[1] Data summarized from Henderson (1972).
[2] Midpoint; Henderson estimated 661-771.

Table 3. Catches of California gray whales by modern style whaling, 1905-1947 [1]

Year	Baja California	California	Washington	Alaska[2]	Bering and Chukchi Seas[3]	Total
1913				1		1
1914	19					19
1920	2					2
1921	1					1
1922	5					5
1924			1			1
1925	100				33	133
1926	41	1				42
1927	29	3				32
1928	9	1		2		12
1929	2					2
1933				2	2	4
1934					54	54
1935					34	34
1936					102	102
1937					14	14
1938					54	54
1939					29	29
1940					105	105
1941					57	57
1942					101	101
1943					99	99
1945					30	30
1947					1	1

[1] Data summarized from Rice and Wolman (1971), except that the figures for 1943, 1946, and 1947 have been changed to agree with those in Kleinenberg and Makarova 1955.
[2] Gulf of Alaska (shore stations).
[3] Pelagic whaling.

Table 4. Catch of gray whales off California under Special Scientific Permits, 1958/59 to 1968/69

Season	Number of whales caught
1958/59	2
1961/62	4
1963/64	20
1965/66	26
1966/67	125
1967/68	66
1968/69	73

into the calving lagoons to see the whales, and by small pleasure craft brought overland down the new Baja California highway. There is also a possibility that exploratory drilling for petroleum will soon begin in or near some of the lagoons. The government of Mexico has declared the most important calving area, Laguna Ojo de Liebre ("Scammon's Lagoon"), a whale sanctuary, and boat traffic is banned during the calving season.

Population size

Scammon (1874) estimated that the California gray whale population was probably not over 30 000 from 1853 to 1856 and that by 1874 the number did not exceed 8 000 or 10 000. After a careful analysis of the historical data, however, Henderson (1972) concluded that the population did not exceed 15 000 prior to the initation of exploitation in 1846.

In 1885/86, Townsend (1887) estimated that only 160 gray whales migrated south past San Simeon, California. Andrews (1916) wrote that "For over 20 years [preceding 1910] the species had been lost to science and naturalists believed it to be extinct". Howell and Huey (1930) said it was "doubtful whether more than a few dozen individuals survive". However, K.W. Kenyon (pers. comm.) says that he commonly observed gray whales migrating past La Jolla, California, during the thirties.

Systematic shore counts of the south-ward migration were initiated at San Diego, California, in 1952/53, and continued intermittently until 1967/77 (Gilmore, 1960; Rice, 1961). These counts indicated a steadily increasing population until 1959/60 (Table 5), but by the mid-1960's it became apparent that geographical features caused the majority of the migrating whales to pass too far offshore to be seen from land (Rice, 1965). A marked decrease in the 1967/68 and 1968/69 counts suggested that increasing boat traffic was causing a still greater proportion to pass far offshore, but the next counts in 1975/76 and 1976/77 were the highest on record.

From 1967/68 to 1973/74, a shore count was made every winter at Yankee Point near Monterey, California, where 95% of the whales pass within 2 km of the shore, and boat traffic is at a minimum. From 1974/75 to 1976/77 the count was made at Granite Canyon, 7 km south of Yankee Point, because of real estate development at the latter site. These counts (Table 5) indicate a stable or slightly increasing population. When the counts are extrapolated to allow for whales missed during periods of poor visibility, and for whales passing at night (Rice and Wolman, 1971), they indicate a total population of about 11 000 (± 2 000). A detailed report on the past 10 years' censuses is in preparation.

Between 1952 and 1973, a number of aerial surveys of the gray whale wintering grounds were conducted by several observers (Hubbs and Hubbs, 1967; Gard, 1974). The aerial counts agree with the shore counts in that they indicate that the population was increasing during the 1950's, and has remained stable since about 1960.

In the summer of 1968, Zimushko (1970) conducted an aerial survey of gray whales in the coastal waters off the Chukotski Peninsula. On the basis of 124 whales actually sighted, he calculated the total population in the area to be about 5 000. In the summer of 1973, Berzin (1974) conducted another aerial survey. On the basis of 290 whales sighted, he calculated a total population of about 7 700.

Table 5. Counts of southward-migrating gray whales, 1952/53 to 1976/77 [1]

Season	Number of whales counted	
	Point Loma [2]	Yankee Point/Granite Canyon [3]
1952/53	982	
1954/55	1 646	
1956/57	1 839	
1959/60	2 344	
1967/68	1 324	3 120
1968/69	1 154	3 081
1969/70		3 064
1970/71		3 034
1971/72		2 588
1972/73		3 304
1973/74		3 492
1974/75		3 348
1975/76	2 822	3 797
1976/77	3 648	4 054

[1] Data for 1952/53 to 1956/57 from Gilmore (1960).
[2] Total counts; not quite strictly comparable from year to year, since beginning and ending dates varied slightly.
[3] Figures include only those whales which passed the point between 07000 and 1700 hours, from 18 December to 4 February (excluding 25 December and 1 January). A few additional whales were observed outside these time limits, but such observations were not strictly comparable from year to year.

Conclusions

Prior to exploitation, which began in 1946, the California stock of the gray whale numbered no more that 15 000. By the year 1900 it was reduced to perhaps about half its original numbers. It probably increased somewhat during the first half of the 20th century, and appeared to grow rather rapidly from 1947 (when the International Convention for the Regulation of Whaling came into effect) until about 1960. During the past ten years it has remained stable or has increased slightly, and now numbers about 11 000 (± 2 000), with an average annual Eskimo catch of about 170. The greatest threat to the California gray whale population at present is the increasing industrial development and vessel traffic in the calving lagoons.

References

ANDREWS, R.C., Monographs of the Pacific Cetacea. 1.
1914 The California gray whale (*Rhachianectes glaucus* Cope). *Mem. Am. Mus. Nat. Hist. (New Ser.)*, 1(5):227-86.

—, Whale hunting with gun and camera. New York, D.
1916 Appleton and Co., 333 p.

BERZIN, A.A., Aktual'nye problemy izucheniya kitoo-
1974 braznykh. *Zool. Pozvonochn.*, 6:159-89.

BERZIN, A.A. and A.A. KUZ'MIN, Seryi i gladkie kity
1975 Okhotskogo morya. *In* Morskie mlekopitayushchie, Materialy VI Vsesoyuznogo So-
veshchaniya, edited by G.B. Agarkov *et al.*
Kiev, Akademiia Nauk SSSR, pp. 30-2.

BOWEN, S.L., Probable extinction of the Korean stock of
1974 the gray whale (*Eschrichtius robustus*). *J.
Mammal.*, 55(1):208-9.

BROWNELL, R.L. and CHAN-IL CHUN, Probable existence
1977 of the Korea stock of the gray whale (*Eschrichtius robustus*). *J. Mammal.*, 58(2):237-9.

FRASER, F.C., An early 17th century record of the Cali-
1970 fornia gray whale in Icelandic waters. *In*
Investigations on Cetacea, vol. 2:13-20.

GARD, R., Aerial census of gray whales in Baja Califor-
1974 nia lagoons, 1970 and 1973, with notes on behavior, mortality, and conservation. *Calif.
Fish Game*, 60(3):132-43.

GILMORE, R.M., A census of the California gray whale.
1960 *Spec. Sci. Rep. USFWS (Fish)*, (342):30 p.

HENDERSON, D.A., Men and whales at Scammon's la-
1972 goon. Los Angeles, Dawson's Book Shop, 313
p.

HOWELL, A.B. and L.M. HUEY, Food of the gray and
1930 other whales. *J. Mammal.*, 11(3):321-2.

HUBBS, C.L. and L.C. HUBBS, Gray whale censuses by
1967 airplane in Mexico. *Calif. Fish Game*,
53(1):23-7.

KLEINENBERG, S.E. and T.I. MAKAROVA (eds.), Kitoboi-

1955 nyi promysel Sovetskogo Soyuza. Moscow,
Izdatel'stvo Zhurnala "Rybnoe Khozyaistvo",
119 p.

KUZ'MIN, A.A. and A.A. BERZIN, Raspredelenie i sovre-
1975 mennoe sostayanie chislennosti gladkikh i serykh kitov v dal' nevostochnykh moryakh. *In*
Biologicheskie resursy morei dal' nego vosto-
ka, edited by A.A. Berzin *et al.* TINRO, Vladivostok, pp. 121-2.

MIZUE, K., Grey whales in the East Sea area of Korea.
1951 *Sci. Rep. Whales Res. Inst.*, (5):71-9.

NISHIWAKI, M. and T. KASUYA, Recent record of gray
1970 whale in the adjacent waters of Japan and a
consideration on its migration. *Sci. Rep. Whales Res. Inst.*, (22):29-38.

OMURA, H., Possible migration route of the gray whale
1974 on the coast of Japan. *Sci. Rep. Whales Res.
Inst.*, (26):1-13.

RICE, D.W., Census of the California gray whale. *Norsk
1961 Hvalfangsttid.*, 50(6):219-25.

—, Offshore southward migration of gray whales off
1965 southern California. *J. Mammal.*, 46(3):500-1.

RICE, D.W. and A.A. WOLMAN, The life history and
1971 ecology of the gray whale (*Eschrichtius robustus*). *Spec. Publ. Am. Soc. Mammal.*,
(3):1-142.

SCAMMON, C.M., The marine mammals of the northwe-
1974 stern coast of North America. San Francisco,
John H. Carmany and Co., 319 p.

TOWNSEND, C.H., Present condition of the California
1887 whale fishery. *Bull. U.S. Fish. Comm.*,
(6):340-50.

ZIMUSHKO, V.V., Aerovizual'nyi uchet chislennosti i na-
1970 blyudeniya za raspredeleniem serykh kitov v
pribrezhnykh vodakh Chukotki. *Izv. Tikhookean. Nauchno-Issled. Inst. Rybn. Khoz.
Okeanogr.*, 71:289-94.

NEW TECHNIQUES FOR ASSESSING POPULATIONS OF RIGHT WHALES WITHOUT KILLING THEM

H. WHITEHEAD **and** R. PAYNE

Abstract

Aerial photographs of the population of southern right whales (*Eubalaena* sp.) off Peninsula Valdez, Argentina, were used to recognize and measure individual whales. Individuals could be recognized by examination of photographs showing the unique pattern of light-coloured head markings, or callosities. Between 1971 and 1976, 1 648 identifications of 484 different whales were made. The lengths of whales were measured directly by comparing the image of the whale on film with a disk of known diameter carried on a fast boat next to the whale when it surfaced for breath; 29 animals were measured in this way. The lengths of 4 whales were found by comparing them to whales of known length swimming parallel to them. Similarly, ratios of the lengths of mothers and their calves were used to estimate the initial growth rate of calves and provide evidence for birth dates and lengths at birth. Results indicated that most calves seen at Peninsula Valdez are born in or near August, are about 5.5 m at birth and grow about 35 mm each day, for the first few weeks of life. A ratio of snout-to-blowhole length to overall length was found for 202 known whales photographed clearly but without the disk being present for size comparison. Formulae relating this ratio to length and describing changes in them over time, were used to estimate the growth rate of older animals and produce an approximate age-length key. Based on these results, it was estimated that females in this population become sexually mature at about 3 to 4 years of age. These new techniques provide estimates of parameters used in population models of large whales without killing them and, in some cases, do so in a more direct way than do traditional methods dependent on a fishery. Some of them can probably be applied to other species of large whales.

Résumé

Des photographies aériennes de populations de baleines franches australes (*Eubalaena* sp.), se trouvant au large de la péninsule Valdez, en Argentine, ont été utilisées pour reconnaître et mesurer les individus. Les sujets ont pu être reconnus par l'examen de photographies montrant la disposition unique de marques claires, ou callosités, de la tête. De 1971 à 1976, on a effectué 1 648 identifications de 484 baleines différentes. La longueur des baleines a été mesurée directement en comparant l'image de la baleine avec un disque de diamètre connu porté par un bateau rapide proche de la baleine quand elle émergeait pour respirer. On a mesuré 29 animaux de cette façon. On a trouvé la longueur de 4 baleines en les comparant à des sujets de longueur connue qui leur étaient parallèles. De même, le rapport entre la longueur des mères et de leurs petits a été employé pour estimer le taux de croissance initial des jeunes; il donne des indications sur la date des naissances et la longueur à la naissance. Les résultats ont indiqué que la plupart des jeunes observés à la péninsule Valdez sont nés en août ou à une date voisine, mesurent environ 5,5 m à la naissance et grandissent d'environ 35 mm chaque jour pendant les premières semaines de leur vie. Le rapport entre la distance museau-trou d'évent et la longueur totale a été déterminé pour 202 baleines connues, photographiées clairement mais sans le disque permettant de comparer la taille. Des formules établissant une liaison entre cette distance et la longueur, et décrivant les changements qui les affectent avec le temps, ont été utilisées pour estimer le taux de croissance des animaux plus âgés et établir un tableau approximatif âge-longueur. Sur la base de ces résultats, on a estimé que les femelles de cette population atteignent la maturité sexuelle vers 3 ou 4 ans. Ces nouvelles techniques ont fourni des estimations de paramètres employés dans les modèles de populations de grands cétacés sans qu'il soit nécessaire de les tuer et, dans certains cas, elles le font de façon plus directe que les méthodes traditionnelles, dépendant de la pêche. Certaines peuvent être probablement appliquées à d'autres espèces de grandes baleines.

Extracto

Gracias a fotografías aéreas de la población de ballena franca austral (*Eubalaena* sp.) que se encuentra frente a la Península Valdés (Argentina) ha sido posible reconocer y medir algunos individuos. El reconocimiento de los individuos se consiguió con fotografías que mostraban las marcas características (o callosidades) de coloración clara que presentan esas ballenas. Entre 1971 y 1976 se hicieron 1 648 identificaciones de 484 ballenas distintas. La longitud de las ballenas se midió directamente, comparando la imagen de la ballena en la película con un disco de diámetro conocido que se desplazaba en una lancha rápida cerca de la ballena cuando ésta salía a la superficie para respirar; con este sistema se midieron 29 animales. La longitud de cuatro ballenas se halló comparándolas con ballenas de longitud conocida que nadaban paralelas a ellas. De igual forma, se utilizó la relación entre la longitud de la madre y los ballenatos para estimar el índice inicial de crecimiento de éstos y obtener datos sobre las fechas de nacimiento y su longitud en ese momento. Los resultados indican que la mayoría de los ballenatos observados en la Península Valdés habían nacido en agosto o poco antes o después de ese mes, tenían en el momento de nacer 5,5 m de longitud y crecían unos 35 mm al día durante las primeras semanas de vida. En 202 ballenas que se habían fotografiado con claridad, pero sin la presencia del disco necesario para determinar su talla, se halló la relación entre la distancia desde el hocico al orificio nasal y la longitud total. Mediante fórmulas que relacionan esa proporción con la longitud y describen los cambios que se producen con el pasar del tiempo, se estimó el índice de crecimiento de los animales adultos y se obtuvo una clave aproximada de la relación edad-longitud. Sobre la base de esos resultados se estimó que las hembras de esta población alcanzan la madurez sexual entre los 3 y los 4 años de edad. Estas nuevas técnicas permiten estimar algunos parámetros utilizados en los modelos de población de grandes ballenas sin necesidad de matarlas y, en algunos casos, permiten hacerlo en forma más directa que los métodos tradicionales, que dependen de la pesquería. Probablemente algunas de estas técnicas podrán aplicarse también a otras especies de ballenas de gran talla.

H. Whitehead
Cambridge University, Cambridge, England

R. Payne
Center for Field Conservation and Biology, New York Zoological Society, Weston Road, Lincoln, MA 01773, USA

Introduction

In 1975 the International Whaling Commission adopted what was called the New Management Procedure (NMP). The NMP requires full protection for any stock which falls below 10 % of the maximum sustainable yield level.

However, once a stock is severely enough depleted to require full protection, it is difficult to monitor its recovery. This is because, with the exception of aerial and shipboard censuses (which are often ambiguous), the data used in traditional techniques for estimating whale populations have come from corpses provided by the whaling industry. It is frequently argued that without these corpses one cannot calculate ages (by counting growth layers in teeth, baleen or plugs of wax filling the ear canal), determine sex, or find out calving rates (by examining ovaries).

Migration studies, as well, will suffer because they have traditionally been based on finding numbered steel darts in the cooker after a whale is melted down for oil: these darts were fired into the same individual at some other time and place.

It is clear that if we are to monitor the recovery of depleted stocks, we must develop new methods for determining such parameters as length, age, individual identity, sex, reproductive history, and migration routes that do not rely on killing the whales. In this paper we introduce several new techniques for determining these parameters which we have developed over the last 8 years while studying a small population of right whales (*Eubalaena* sp.) off the coast of Argentina in the protected waters of Peninsula Valdez. As will be explained, these techniques should be applicable, with minor modifications, to most if not all other species of baleen whales.

We also give length frequencies within the Peninsula Valdez herds, calculate length at birth, and determine the growth rates for newborn calves and later stages.

Methods

Between 1970 and 1974 one of the authors, working from the coast of Peninsula Valdez in southern Argentina, censused and photographed right whales from a single engine plane [1]. We attempted to obtain several photos of every whale using a 35-mm single lens reflex camera and 300-mm f/2.8 lens, hand-held, shooting from the window of the plane. In all, we had 52 flights of an average duration of 3.2 h, which repeatedly covered the waters in the vicinity of the peninsula. Between 1971 and 1976 we obtained a total of 19 590 photographs of whales from which we have been able to make 1 648 identifications of our total group of 484 recognizable individuals. Sometimes there are several identified whales per picture (though this is infrequent) and sometimes several pictures per whale (the usual case). The technique of identifying individuals from aerial photographs is described in another paper (Payne *et al.*, in prep.) but, briefly, is as follows: the head of every right whale is naturally adorned on top and sides with a series of raised thickened patches of cornified epidermis called callosities. The pattern of these callosities is different in every whale. Changes after birth are insignificant and do not interfere with individual recognition. Because most callosities appear to be lighter then the surrounding normal skin, they show prominently and their pattern can be used to distinguish individuals (we currently recognize 454 adult right whales in the Peninsula Valdez population). The callosity pattern on top of the head is best seen from above and can be photographed from a plane circling overhead whenever a whale surfaces to breath. Most right whales can be photographed well enough for identification in less than 6 minu-

---

[1] These flights were kindly continued by B. Würsig (1975) and Clark (1976-77).

tes. Their identity is determined by comparison of the resulting photographs with a "head catalogue" which contains clear pictures of every head we have seen since the project started. A "known" whale as referred to below is a whale we can see well enough in our photographs to identify.

From 1971 to 1973, 2 or 3 flights were made each year to explore a new technique (see Appendix 1) for measuring whales by photogrammetry, using an object of known size, a 1-m disc, next to the whale. This is accomplished by photographing both whale and object in the same frame from a plane circling overhead. The photographer is in radio contact with the driver of the boat and tells him where to steer so as to be next to the selected whale when it surfaces for air. In the resulting photographs the whale is measured by using the maximum diameter of the disc to represent 1 m. The reason one uses the maximum diameter is that no matter what angle one is at in relation to a circle, his line of sight to the centre of the circle will always be perpendicular to at least one diameter of the circle, and that diameter will always appear to be the maximum diameter. Thus, by choosing the maximum diameter of the circle shown in the photograph, one has a scale with which to measure with useful precision any dimension of the whale that is perpendicular to the observer's line of sight and which lies in approximately the same plane as the circle.

In an improved version of this technique, the circle is held level, allowing any dimension of a whale to be measured, provided that the whale is lying parallel to the water's surface. The scale chosen is that diameter of the '1-m disc which is parallel to the desired dimension on the whale. As long as a disc is held level, each diameter in the resulting photograph will be correctly foreshortened for use in measuring any line in the plane of the disc that lies at the same orientation to the observer's line of sight as the diameter selected.

When the whale was not beside the disc, we applied a correction (see Appendix 1 for measurement techniques). The standard error

in measurements of the same whale, measured from different photographs taken during the same year, is 0.32 m (using 23 pairs of measurements). This error includes all qualities of data, for much of which the end points of the whale were not clearly visible. For measurements where these points were distinct, the errors approximated 0.216 m. An error of 0.32 m for a 12-m whale would be 2.7 % (1.8 % for clear pictures). However, only 2 pairs of measurements of total length were used in the calculation of the standard error (the other 21 were of shorter dimensions such as snout-to-blowhole) and thus the actual errors in measuring length may have been somewhat greater.

Results

With the 1-m disc we obtained length measurements of 18 known whales, and 16 unidentified whales. We also made measurements of total and of snout-to-blowhole lengths in 202 known whales which we expressed only as a ratio (no scale being available in these photos). In 4 additional cases we used whales as rulers to measure each other, in photographs in which a whale of unknown length was parallel to one of known length. We built up a catalogue of these absolute and comparative measurements, on known and unknown whales. In some cases measurements not actually made could be inferred (e.g., the length, L, of a whale could be calculated even when the only measurement made by comparison with the circular disc was α, snout-to-blowhole, as long as there was another frame of the same individual taken in the same year from which the ratio α/L could be estimated).

Each measurement was graded from 1 to 6 (good to bad) depending on how accurately the measurement was judged to have been made (for a discussion of grading, see Appendix 1). A summary of the notations used is given before the appendixes.

192

LENGTH FREQUENCIES

Fig. 1 gives histograms for frequencies of 3 different measurements on the Peninsular Valdez right whales. It shows, at 1-m increments, frequencies of measured absolute length, L. It gives frequencies of snout-to-blowhole length, α, for every 0.25 m. Fig. 1 also shows the most accurate determinations (grades 1 to 4, with the whale perpendicular to camera) of z (z is α as a percentage of L) for each 1 %.

Fig. 1 indicates that known mothers are the longest whales in our population, followed by whales whose sex was undetermined (we

suspect here a high percentage of males), which are in turn larger than calves accompanied by mothers.

The shape of the histogram for z and its similarity to that for L suggest that z might be an indicator for L. Thus we would be able to determine length of a whale from any photograph in which the ratio of α (snout-to-blowhole distance) to overall length could be calculated without the necessity of having a reference object in the picture.

We made most of our direct measuring flights in November (a few were made in October) which is almost certainly an important factor in the relative abundance of different lengths and sexes that we observed. There are 2 reasons for this: (i) more males than females had left the area by this time; (ii) most small calves are seen earlier in the season.

GROWTH IN CALVES

There were photographs of 87 calves swimming parallel and right next to their mothers. Using these pictures we could compare the calf's dimensions with those of its mother's, and we could calculate ratios such as L_c/L_m (length of calf/length of mother).

Growth by month

Fig. 2 is a series of histograms L_c/L_m for each calendar month in which aerial census flights were made. It is apparent that from September through November the calves are growing. The best data, grades 1 to 4, are shaded. They seem representative of the rest of the data but not noticeably better grouped than it.

In order to quantify this growth we have plotted in Fig. 3 the mean value of L_c/L_m for each month together with approximate 95 % confidence limits. The growth of the calves is shown clearly in Fig. 3.

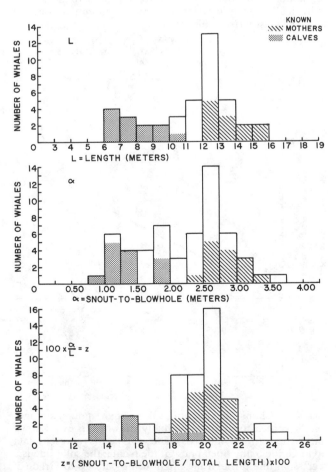

FIG. 1. – Histograms of length (L) and snout-to-blowhole length (α) for the measured whales, and of z (α as a percentage of L) for the most accurate determinations (Grades 1-4). The similarity of the histograms for z and L should be noted.

193

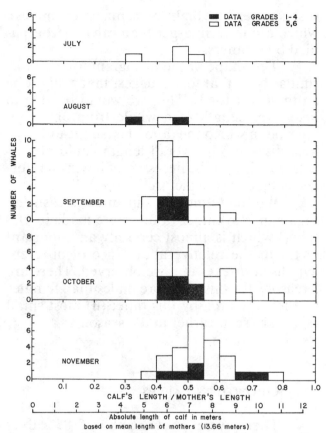

FIG. 2. – Histograms of calf's length/mother's length for each month. The best data (Grades 1-4) are blackened. A scale is included to show the absolute length of the calf if it had a mother of "mean length" (13.66 m). The calves appear to be growing and a length at birth of 5 to 5.5 m is suggested.

Growth for individuals

For 20 known mother and calf pairs, L_c/L_m could be estimated from photographs taken on 2 or more days in the same year. In Fig. 4 the change in L_c/L_m is plotted against the interval (in days) between measurements. There is much scatter in the plots, due mainly to measurement inaccuracies, but significant growth is shown (significant at a 5 % level). This growth was numerically estimated by a regression line (see Appendix 2), which gives a rate of growth for the calf of .00253 of mother's length/day. Our mean of length measurements for 12 known mothers was 13.66 m. If calves are growing at .00253 of this length per

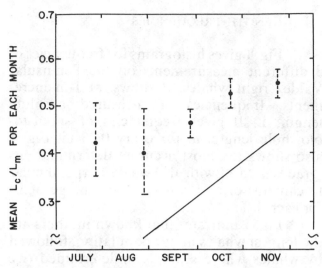

FIG. 3. – Mean calf's length/mother's length for each month, together with 95 % confidence limits. The slope of the line at the bottom right-hand corner gives the ratio of growth of individually known calves that were seen more than once in the same season (from Fig. 4).

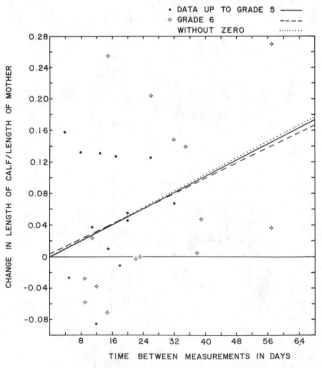

FIG. 4. – Change in the ratio, calf's length/mother's length (on the vertical axis) for individually known pairs over known time periods (horizontal axis). The solid points represent data up to Grade 5, the open ones Grade 6 data, the solid line is a regression line through the origin using data up to Grade 5, the dotted line a regression line through the origin using all the data. The dashed line a regression line, not artificially made to pass through the origin, using all the data. The slopes of these lines give the approximate initial rate of growth of calves.

day, it would mean that the initial growth rate of right whale calves is 34.6 mm/day. This figure is in remarkably close agreement with Klumov (1962) who found the rate of growth of the North Pacific right whale foetus just prior to birth to be 30-33 mm/day.

To compare the rates of growth from the monthly histograms and from the known mother and calf pairs seen more than once, a line with slope corresponding to .00253 mother's length/day is drawn in Fig. 3. The slope of this line seems in good agreement with the placing of the points for August, September and October, but in November the rate at which the overall length of the calf is increasing seems to have slowed.

Date of birth

The agreement in the rates of growth for August through October indicates that only a small proportion of the observed calves were born in the latter part of this period. For if many new small calves were being added to the population, the rate of growth of the mean of the population would be slower than the average rate of growth of known individuals, and it is about the same.

Young calves are usually close to their mothers, but sometimes either the mother or the calf is photographed alone (with the other being just out of the frame). In order to determine when calves are born, we have analysed pictures containing known mothers noting what months they were seen with their calves. We have then calculated the percentage of known mothers with calves to known mothers without calves for each month June through December (see Table 1).

This indicates that few mothers are photographed before their calves are born. There are several plausible explanations for this: (i) mothers rarely come to Valdez before their calves are born; (ii) they do come but are hard to photograph; (iii) most of the calves are born before September when the bulk of our photographing begins. It seems likely that the

Table 1. Numbers and proportions of known mothers seen each month

	June	July	Aug.	Sep.	Oct.	Nov.	Dec.
No. known mothers seen	3	13	20	102	73	30	6
No. known mothers seen with calves	0	12	15	93	67	28	6
% known mothers seen with calves	0	92	75	91	92	93	100

figures above reflect a combination of these 3 explanations. However, the impression given by the evidence in Table 1 and by the growth of calves is that the majority of calves are born before the beginning of September. Since some small calves are seen in October (e.g., 6.35 m) and they are growing at approximately 1.00 m per month, these calves cannot have been born long before.

We are aware of measurements for only a few, small, southern right whale calves and foetuses in the literature. These are given in Table 2.

It seems from the table that the majority

Table 2. Small southern right whale calves and foetuses

Source	Date of Measurement	Length Ft	Length M	Calf/ Foetus
Committee for International Whaling Statistics, 1938	12 July 1937	19'4"	5.89	Foetus
	22 July 1937	19'9"	6.02	Foetus
	28 July 1937	17'2"	5.23	Foetus
	21 Sep. 1937	1'0"	0.30	Foetus
Matthews, 1938	26 Aug. 1926	–	6.5	Calf
Lonneborg, 1906	12 May 1905	–	4.19	Foetus
D'Oyly, 1968	23 June 1832	20'0"	6.10	Foetus
Our data [1]	20 Sep. 1972	13'6"	4.11	Calf

[1] Local residents reported it had beached "in August". The corpse was in good condition indicating a relatively short time in the water.

of calves are born in or near August. This can be compared with Klumov (1962) who gives the birth date of North Pacific right whales to be December or January.

Length at birth

The minimum lengths of calves, and the data on foetuses in Table 2, indicate that there seems to be considerable variation in length at birth. The dead calf we found on a beach in Argentina and measured in September may have been premature. Other very small dead calves were seen beached in Argentina. The 2 foetuses greater than 6 m could be distorted – 1 comes from the caption to a picture in the sketchbook of a travelling artist which was drawn in 1832. As the measurement given is as an even 20 ft it raises doubts as to how carefully it was made. It also seems possible that this measurement of length might have been taken from the hind margin of the flukes rather than the fluke notch as is usual. (The calf from Lonneborg, 1906, was measured incorrectly – to the hind margin – thus exaggerating the length of the whale.)

In Fig. 2 we have noted on the abcissa the lengths the calves would have had, had they all had average sized mothers (13.66 m). The result is very few calves smaller than 5.5 m, approximately 17 % of the calves between 5.5 and 6.0 m, and all the other calves greater than 6.0 m, all of which suggests a possible length at birth of 5.5 m.

As further evidence for a length at birth of approximately 5.5 m we have 3 calves whose mothers were photographed alone early in the season and with their calves later that same season (see Table 3).

It must be borne in mind that the calf could have hidden on the date that the mother was seen "alone" and so the ages when calves were measured are estimates. However, using the mean value of L_c/L_m for these whales (0.427) and the mean length of mothers (13.66 m) we have an estimated average length for the calves when first photographed of 5.83 m.

Table 3. Mothers seen alone and then with calves [1]

Date measured	L_c/L_m	Date mother seen alone	Date calf first seen	Age of calf when measured (days) Min	Max
20 Oct.	.432	8 Sep.	26 Sep.	24	42
21 Oct.	.478	15 Oct.	21 Oct.	0	6
4 Oct.	.372	29 Sep.	4 Oct.	0	5

[1] Calves could have been born on the days when mothers were seen "alone".

We therefore conclude that while the length at birth is variable for individuals, its mean value is approximately 5.5 m, and almost certainly lies between 4 and 6.2 m. This is in agreement with Klumov (1962) who put the length at birth between 5.0 and 6.0 m for the North Pacific right whale.

Initial growth curve

Using a length at birth of 5.5 m, and the calculated growth rates, an initial growth curve was constructed and is shown in Fig. 5. It is divided into 2 parts, the division occurring at an age of 2 months. If we assume that southern right whales are born in August, it can be seen from Fig. 3 that at an age of about 2 months (October) the growth slows down. The actual initial growth curve is probably more like the smooth dashed line in Fig. 5. As a check, the mean length of the 9 calves measured in November is also shown in Fig. 5 when they would be around 3 months old (again assuming an August birth date). We have also indicated the 95 % confidence limits for this mean and it can be seen that the growth curve lies well within them.

GROWTH DURING LATER STAGES

Unfortunately, this growth curve cannot at present be continued in the same fashion for

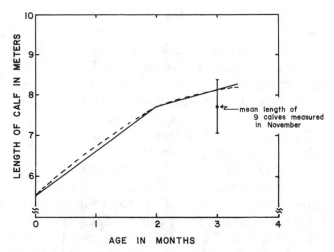

FIG. 5. — Initial growth of calves using data summarized in Figs. 3 and 4. The dashed line is a smoothed version of the 2 straight line portions. The mean length of 9 calves measured in November together with 95 % confidence limits for this mean is marked at 3 months of age (assuming calves born in August).

older whales, as this would require measurements of known whales over several years. Because our measuring flights were primarily intended to determine the feasibility of our new measuring technique, we could not afford specifically to seek out individuals measured in previous years for remeasurement. As a result, we have only one known whale (No. 156), a male, which was measured directly in more than 1 year (we now plan to extend our measuring programme, concentrating on remeasuring previously measured individuals). However, we have been able, using a less direct method, to estimate the later growth stages.

From Fig. 1 it appears that z (snout-to-blowhole distance, α, as a percentage of length L) is related to length. By using the regression of Log α on Log L, the following relationship is obtained:

$$z = 7.36 \times L^{.376\%} \tag{1}$$

(see Appendix 3 for details)

Because z is a ratio, its determination does not require having an object of known size in the same frame with the whale, thus we are able to measure z from any clear, full-length photo-

graph of a whale. Using this method, we were able to measure 46 known whales in 2 or more years. Using the most accurate sets of measurements, we have expressed z as a function of the growth in z per year ($\delta z/\delta t$):

$$z = -4.138 \, (\delta z/\delta t) + 21.66$$
(see Appendix 1 for details)   .... (2)

On integration this gives a growth curve for z:

$$z = 21.66 \, (1 - e^{-0.2417(t-t_k)}) \tag{3}$$

(where t is the age in years and t_k a constant of integration).

Combining equations (1) and (3) we can create a growth curve for L:

$$L = 17.04 \, (1 - e^{-0.2417(t-t_k)})^{2.66} \tag{4}$$

To fix t_k, one further piece of information was used: the measured lengths of whale No. 156 (7.77 m on 17 November 1972 and 9.9 m on 21 November 1973). Using the initial growth curve (Fig. 5) this would make No. 156 2-3 months old on 17 November 1972, with a birth date in the first half of August of that year. He was first seen by us on 8 September 1972 (fortunately after his theoretical birth date!). On 24 November 1973 he would be 1.2 years old. Putting t = 1.2, L = 9.9 in equation (4) we find $t_k = -5.79$ years and the formulae for growth in z and L become:

$$z = 21.66 \, (1 - e^{-0.2417(t+5.79)}) \tag{5}$$

$$L = 17.04 \, (1 - e^{-0.2417(t+5.79)})^{2.66} \tag{6}$$

Fig. 6 shows the growth curve given by equation (6) added to the curve of initial growth from Fig. 5. It is gratifying that the curves appear to join smoothly.

In Fig. 7, the growth curve for z is plotted for whales older than 1 year, using formula (5). Also plotted are the 8 smallest whales for which we have measured L (age, t, was estimated from Fig. 6). There seems to be rapid

197

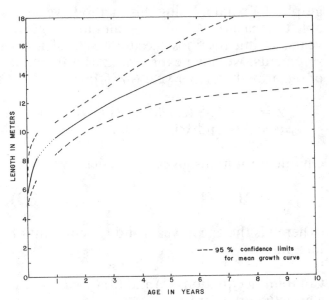

FIG. 6. – Approximate mean growth curve for the southern right whale. Dashed line indicates 95 % confidence limits for this mean (see Appendix 4).

variation in the growth of z in the first year of life and we feel there is not sufficient data currently available to estimate this growth.

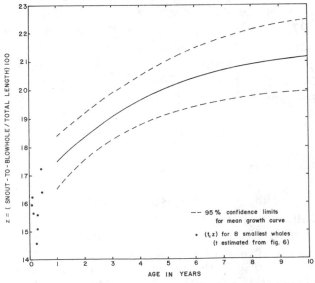

FIG. 7. – Approximate mean growth curve for z (snout-to-blowhole) as a percentage of total length. Dashed line indicates 95 % confidence limits for this mean curve (see Appendix 4). Points are plotted for the eight smallest whales measured, their ages being estimated from Fig. 6.

Sexual maturity of females

The smallest mother accompanied by a calf that we have measured was 12.44 m (we have measured 5 mothers between 12 and 13 m (see Fig. 1). It seems reasonable to assume, therefore, that an average female becomes sexually mature at between 12.5 and 13 m or, using our age-length key (Fig. 6), between 2.5 and 6 years (average 3 or 4 years). We now have 15 calves of unknown sex seen at age 4 or older, of which 9 were seen at age 5 or older. In no case has one of these calves appeared with a calf of its own. This strongly suggests that the age of sexual maturity for females is considerably older than indicated by our age-length key.

Further applications or our methods

With the amount of data analysed to date our age-length key can be no more than tentative but as more information becomes available the accuracy can be increased and perhaps someday we will be able to draw separate curves for each sex. Even with our present confidence limits it may be possible to use the curves, in conjunction with the measured lengths and z's to obtain estimates of mortality.

In the traditional methods of constructing age-length keys, an intermediate indicator of age (such as dentinal layers or ear plug laminae) is used. The validity of these indicators has been questioned particularly with baleen whales. Because our method of measuring whales does not require a dead whale, whales can be measured (by L or z) several times in their life, and growth can be determined directly.

Some of the ideas presented here can probably be applied to other large whale species: comparisons between measurements on mothers and calves, when photographed swimming parallel to one another, should be possible; the estimation of length and thus age from z could be particularly important in estimating gross length or age frequencies of other species. All that is needed is one aerial photograph per whale, preferably with the whale's long axis approximately perpendicular to the photographer's line of sight, showing the snout, blowhole and flukes. The position of the dorsal fin might also be an important point to measure. The important factor in the success of this method is the slope of the regression line in the plot of Log α against Log L. The further it is from unity the better the method works, indicating that head and body grow at quite distinct rates. Shown below are some values of this slope for different species in the adult (defined as post-weaning) growth stages, based on measurements reported in the literature.

Southern right whale	– this paper –	1.320
Balaenoptera	– Ohsumi, 1960	1.41
Megaptera	– Ohsumi, 1960	∞ 1.41

These figures indicate that our technique should have broad applicability within the mysticetes.

It is frequently said that some form of fishery is essential to obtain valid population data on whales. Our work in analysing measurements from aerial photographs provided estimates of some of the parameters (such as age at sexual maturity) on which population models are based. The time span between calvings of known females is being obtained from direct observation, and stock estimates are being made by noting the ratio of known to unknown whales in successive airflights over the same population. Because this technique uses natural patterns of individual whales (of which we currently recognize 484) we can also study migration paths and destinations by searching for known right whales elsewhere in their range. This work shows, we think, that it is feasible to determine parameters for population models of large endangered whale species without killing them, and that, at the very least, the approach is worth investigating further.

We intend to improve on our data on eight whales to increase the validity and applicability of our conclusions, but we already feel that other scientists should consider these methods before they sanction further reduction of endangered whale species on grounds that no other technique can be used to estimate stocks.

Summary

(1) A new technique is described for measuring whales by photogrammetry to an accuracy of approximately 3 %. It employs a disc of known diameter driven by a boat into the camera's field of view when a surfacing whale is being photographed by a hand-held camera from a light plane.

(2) The results indicate that:

– the initial growth rate of the southern right whale is approximately 35 mm/day, for the first few weeks of life;

– the majority of calves seen at Peninsula Valdez are born in or near August;

– southern right whales are approximately 5.5 m at birth.

(3) We have constructed an age-length key for the southern right whale, together with confidence limits for it.

(4) We have given a curve of z (snout-to-blowhole length as a percentage of total length) against time. It is pointed out that measurements of z from single photographs of other baleen whales could very probably be used successfully as an indication of age.

199

Acknowledgements

The authors acknowledge with gratitude the many contributions of others to this study, particularly Oliver Brazier, Nancy Davis, Lydia Leon, Judith Perkins and Victoria Rowntree who did most of the work of identifying individual whales from aerial photographs. We are also grateful to Hugo Callejas who served as pilot during most flights, to Nancy Davis for typing the many versions of the manuscript, to Judith Perkins for analysing and collating data on known mothers with calves, to Victoria Rowntree and Katharine Payne for drawing the figures, and to Cheng-yuan Shao of the Smithsonian Astrophysical Observatory who made measuring equipment available to us for calibrating our camera lenses.

Appendix 1

Measurement Techniques

Live whales were measured by photogrammetry using an object of known size next to the whale photographed by a 35-mm, single-lens reflex, hand-held, motor-drive camera, through an f/2.8 300 mm lens. On most occasions the object photographed was a white disc, 1 m in diameter, painted on a flat board; it was carried next to the whale on the bow of an outboard motor boat, which was in radio contact with an aeroplane circling overhead.

Pictures containing the disc, in which the whale was lying perpendicular to a line joining the whale and the camera, were analysed. The judgement that the whale was perpendicular was made by eye, with a probable error of less than 5°. Usually a sequence of frames was taken with a motor drive camera from an aircraft circling close to the whale so that there was at least 1 frame in which the whale appeared perpendicular to the line joining it to the camera.

The length and other dimensions of the whale were calculated by comparing their measured lengths on the negative with the maximum diameter of the disc.

A correction was applied in cases where the distances camera-to-whale and camera-to-disc were different so that the length of the whale (L) was given by:

$$L = \frac{l}{u} \left[1 \pm \frac{h}{v} \sqrt{\left(\frac{u}{s}\right)^2 - 1} \right] \text{ metres}$$

This was done using a + (plus) sign if the whale was further from the camera than the disc and a − (minus) sign if the whale was closer to the camera than the disc.

l mm = measured length of whale on negative
u mm = measured maximum diameter of disc on negative
h mm = measured distance from centre of whale to centre of disc on negative
s mm = measured diameter of disc which when extended passes through centre of whale
v mm = "image distance" of camera lens. This was calculated by photographing a series of objects of known size at known distances and measuring their image lengths on the resulting negatives using a photogrammetry measuring stage accurate to 1 micron. The resulting measurements were used to calculate image distances for our camera lenses according to the formula:

$$v = \text{image distance} = \frac{\text{object distance x image size}}{\text{object size}}$$

In our final year (1974) we developed a new technique — holding the disc horizontal during measurement. When a whale was parallel with the surface (e.g., while breathing) any dimension of the whale could

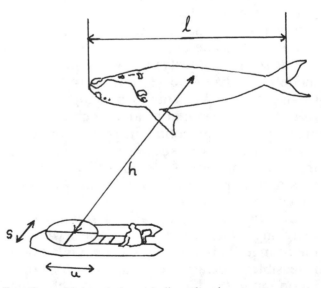

FIG. 8. — Right whale and disc, showing measurements made from each negative and used in calculations of length.

Notation

Letter	Units	Definition
α	m	Snout tip to centre of blowholes
a		Constant used in allometry formula
β	degrees	Angle of yaw of whale (between midline of whale and a line lying on the water's surface which is perpendicular to the line-of-sight from camera to whale)
b		Constant used in allometry formula
d	m	Depth of whale
e_1	m or %	Errors in growth curve due to errors in regression of $\bar{z}$ on $\delta z/\delta t$
e_2	m or %	Errors in growth curve due to errors in regression of $\log \alpha$ on $\log L$
e_3	m or %	Errors in growth curve due to errors in estimation of t_k
h	mm	Measured distance from centre of whale to centre of disc on negative
l	mm	Measured length of whale on negative
L	m	Length of whale — fluke notch to snout tip
L_c	m	Length of calf
L_m	m	Length of mother
L^1_m	m	True length of whale (see Appendix 1)
μ		Refractive index of water
m	m	Height of camera above water surface
n		Number of data points when applying regression error formulae
$\Phi(k)$	%	Indicator of how accurately measurement (k) is made — defined in Appendix 1
$\Psi(k)$		Indicator of how much measurement (k) is related to length — defined in Appendix 1
ρ		Standard error in height of regression line
r		Correlation coefficient
s	mm	Measured diameter of disc which, when extended, passes through centre of whale
τ	degrees	Inclination angle (pitch), between midline of whale and water's surface
θ	degrees	Angle between camera-to-disc line of sight and line parallel to water surface passing through centre of disc and centre of whale
t	years	Age of whale
t_1	years	Time when z_1 is measured
t_2	years	Time when z_2 is measured
t_k	years	Constant of integration
u	mm	Measured maximum diameter of disc on negative
v	mm	Image distance of camera lens
(x, y) (x_i, y_i)		Used in allometry and regression error formulae
$(\bar{x}, \bar{y})$		Mean values of x_i, y_i
z	%	$\alpha/L \times 100$
z_1	%	Value of z at time t_1
z_2	%	Value of z at time t_2
$\bar{z}$	%	$(z_1 + z_2)/2$
$\delta z/\delta t$	% yr^{-1}	$(z_1 - z_2)/t$

be measured by comparing it with whatever diameter of the horizontal disc was parallel to the dimension in question. Regardless of the point of view from which a circle is observed, the circle always shows a correctly foreshortened diameter for measuring any line parallel to that diameter in the same plane as the circle.

For each identified whale we recorded (i) all absolute length measurements made either (a) directly from the disc, or (b) through an intermediate whale or whales; (ii) all comparative ratios made between the identified whale and other whales, known or unknown, that were photographed parallel to it; (iii) all ratios of measurements of the identified whale (e.g., z).

Each measurement or ratio was graded from 1 to 6 (good to bad) depending on how accurately we thought the measurement was determined. Factors influencing the choice of grade included: (a) the clarity of the endpoints of the measurements, (b) certainty of identity of the whale (any uncertainty automatically gave the measurement grade 6), (c) the straightness of the whale, (d) how parallel 2 whales being compared really were, and (e) how many steps were involved in calculations of lengths from parallel whales. The orientation of the whale to the line between camera and whale was also recorded when recording ratios such as z.

As an example of grading: a measurement made from a well identified whale with its body held straight and perpendicular to the camera-whale line whose endpoints were clearly visible would be graded 1; one which had a small failing in any of these categories might be graded 3, while one which required visual extrapolation to find the endpoints would be graded 6.

Theoretical errors in our method of measuring photographs taken from the air

We considered 5 different sources of errors: (i) errors due to the whale not being beside the disc, (ii) errors due to the long axis of the whale failing to perpendicular to the line joining whale and camera, (iii) errors due to inclination of the whale, (iv) distortion of dimensions of whale due to refraction of light by water, (v) measurement errors. We will consider these in order.

Errors due to whale not being beside the disc

$$\cos ec\ \theta = \frac{u}{s} \quad \text{and} \quad \sqrt{(\frac{u}{s})^2 - 1} = \cot \theta$$

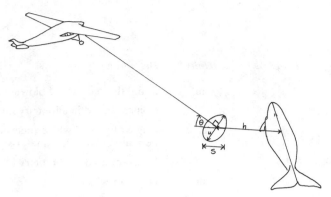

FIG. 9. – Diagram of a plane through camera, disc and whale. The angle between the camera-to-disc line and the water surface is θ.

In our photography $\theta > 30°$, $h < 20$ mm (as 35-mm film was used and the camera was held with the long axis of the frame horizontal) and $v \approx 300$ mm.

Therefore the correction due to the whale not being beside the disc,

$$\left(\pm \frac{h}{v} \sqrt{(\frac{u}{s})^2 - 1} \right)_1$$

was less than

$$\frac{24}{300} \times \sqrt{3} \approx .139$$

The errors in this correction were principally due to the whale not lying in the plane of the disc. For often either the disc was tilted when the boat rode over waves, or the whale would be lying beneath the surface. These errors were, we feel, less than 5°. Using $\theta = 35°$ and $25°$ we have corections of .114 and .172. Thus in the extreme case the error in the calculated length of the whale would be less than about .03 (.172 − .139 = .033) of the true length, or 3 %.

Errors due to yaw of whale

If the whale is yawed in the horizontal plane, and not perpendicular to the line joining the whale and the camera, the true length of the whale (L^1) is given by:

$$L^1 = L \sqrt{\cos^2 \beta + \sin^2\beta \ \sin^2 \theta} \ \text{metres}$$

Where L metres is the calculated length of the whale by the formula in the text, $\beta°$ is the yaw of the whale in the horizontal plane.

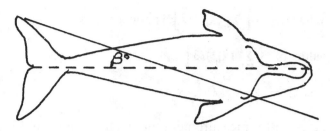

FIG. 10. – Bird's eye view of whale showing angle of yaw. Solid line lies on water surface and is perpendicular to line between camera and the whale.

From this we find that to produce a given error δ % in the calculated whale length, the yaw (β°) needed is given by:

$$\beta = \cos^{-1}\left(\sqrt{\frac{(1 - \delta/100)^2 - \sin^2\theta}{\cos\theta}}\right)$$

Table 4. Critical angles for α

	$\theta = 80^\circ$	$\theta = 60^\circ$	$\theta = 45^\circ$	$\theta = 30^\circ$	$\theta = 0^\circ$
$\delta = 1.0\%$	$\beta = 54^\circ$	$\beta = 16.0^\circ$	$\beta = 11.5^\circ$	$\beta = 9.4^\circ$	$\beta = 8.1^\circ$
$\delta = 0.5\%$	$\beta = 35^\circ$	$\beta = 11.5^\circ$	$\beta = 8.1^\circ$	$\beta = 6.6^\circ$	$\beta = 5.7^\circ$

Thus, if $\theta = 60^\circ$ and 6 pictures are taken regularly as the aeroplane flies in a semicircle around the surfacing whale, the error due to the whale not being perpendicular to the line between the whale and the camera in the final result, calculated from the "best" photograph, is less than 1 % (the change in β between pictures is

$$\frac{180^\circ}{6} = 30^\circ$$

and so for the "best" picture β is less than

$$\frac{30^\circ}{2} = 15^\circ.$$

Errors due to inclination of whale

Exactly the same theory can be used to calculate the errors due to the inclination of the whale to the water surface in the vertical plane τ, but θ must be replaced by $90 - \theta$.

Table 5. Inclinations of whales needed to produce given errors in length measurements

	$\theta = 90^\circ$	$\theta = 60^\circ$	$\theta = 45^\circ$	$\theta = 30^\circ$
$\delta = 1.0\%$	$\tau = 8.1^\circ$	$\tau = 9.4^\circ$	$\tau = 11.5^\circ$	$\tau = 16.0^\circ$
$\delta = 0.5\%$	$\tau = 5.7^\circ$	$\tau = 6.6^\circ$	$\tau = 8.1^\circ$	$\tau = 11.5^\circ$

The inclinations (τ) needed to produce given errors (δ) are shown in Table 5.

Errors due to refraction

Finally we studied the errors due to refraction, as often the extremities of the whale, between which the measurements are made, will be beneath the water as the photograph is taken.

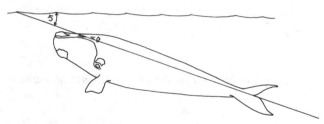

FIG. 11. – Inclination angle of whale to water surface.

The error (δL) due to refraction, in the calculated length of the whale, δ metres below the surface, photographed from m metres above the surface, is given by:

$$\delta L \approx \frac{L\delta}{m}\left(1 - \sqrt{\frac{\sin\theta}{\mu^2 - \cos^2\theta}}\right) \text{ metres,}$$

where μ is the refractive index of water.

The percentage error in the final length of the whale

$$\frac{\delta L}{L} \times 100$$

is given by:

$$\frac{\delta L}{L} \times 100 \approx 100\frac{}{m}\left(1 - \sqrt{\frac{\sin\theta}{\mu^2 - \cos^2\theta}}\right)\%$$

203

Table 6. Relation of angle of view (θ) to errors in calculated length due to refraction by sea water

% error $(\frac{\delta L}{L} \times 100)$	$\theta = 90°$	$\theta = 60°$	$\theta = 45°$	$\theta = 30$
	0.25 %	0.30 %	0.37 %	0.50 %

Using $\delta = 1.0$ metres, $m = 100$ metres, $\mu = 1.332$, we get the values indicated in Table 6.

This formula is linear in δ and m, so the percent error for the above values of θ can be calculated with different depths (d) and camera heights (m). For instance, if $\theta = 45°$, d = 2 metres, m = 50 metres, the error in the final length of the whale is approximately

$$.37 \times \frac{2}{1} \times \frac{100}{50} \approx 1.5 \%.$$

In our situation $\theta > 30°$ d < 2 metres, m $\approx$ 100 metres, and the errors due to refraction are less than 1 %.

Errors in longitudinal aerial measurements of southern right whales

Measurement (our notation) (k)	$\Phi(k)$	$\Psi(k)$
[1]	0.2 %	—
[4]	0.99 %	0.102
[5]	0.82 %	0.091
[6]	2.72 %	0.125
[7]	1.14 %	0.085
[8]	0.85 %	0.094
[9]	20.40 %	0.520

Twenty different whales were used (i = 1 . . . 20). Six different head measurements were considered (k = 4, 5, 6, 7, 8, 9) plus the total length of the whale. Each measurement was made 3 times (j = 1, 2, 3) for each whale, x_{ij}^k is the jth time the kth measurement was made on whale i:

$$\Phi(k) = \sum_{i=1}^{20} \sqrt{\sum_{j=1}^{3} (x_{ij}^k)^2 - \left(\sum_{j=1}^{3} (x_{ij}^k)\right)^{2/3}} / 20 \; \underset{i}{\text{Sup}} \left[\sum_{j=1}^{3} (x_{ij}^k)/3\right]$$

Where $\underset{i}{\text{Sup}} \left[\sum_{j=1}^{3} (x_{ij}^k)/3\right]$ is the greatest value of $\left[\sum_{j=1}^{3} (x_{ij}^k)/3\right]$

Let $y_i^k = \sum_{j=1}^{3} x_{ij}^k / \sum_{j=1}^{3} x_{ij}'$

(x_{ij}' is j'th measurement of length-snout tip to fluke notch) (Fig. 12). Then:

$$\Psi(k) = \left[\sqrt{\sum_{i=1}^{20} (y_i^k)^2 - \left(\sum_{i=1}^{20} (y_i^k)\right)^{2/20}}\right] / \sum_{j=1}^{20} y_i^k$$

$\Phi(k)$ gives an indication of how accurately the measurements are made. For instance, when making measurement 7 we can expect a measurement error of the order of 1.14 % in each measurement from the true value. A high value of $\Phi(k)$ means that the measurement is hard to make, a low value that it is easy. $\Psi(k)$ gives an indication of how closely related a particular measurement is to length. High values of $\Phi(k)$ indicate poor correlation, low values a good correlation.

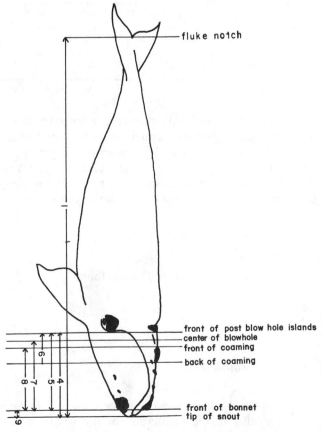

FIG. 12. — Measurements made on aerial photographs of southern right whales.

The α (3 in "Discovery" notation) we use is measurement [7] + measurement [9] and while not directly measured in this survey probably would have Φ and Ψ values close to those of measurement [7], with perhaps Φ a little higher, and Ψ a little lower.

Summary

In the normal case where a circling aeroplane is taking pictures almost continuously, the total error, if $\theta = 30°$, due to errors in correction for the whale not being beside the disc ($< 3\%$), the yaw of the whale ($< 1\%$), the inclination of the whale ($< 0.5\%$) and refraction ($< 1\%$) is less than 5.5% ($3 + 1 + 0.5 + 1$) and probably around 3% as these effects will rarely all produce their maximum errors in the same direction at the same time (errors decrease, in general, as θ increases).

Errors due to mismeasurement were reduced by making 3 separate attempts at each measurement and using their mean.

Appendix 2

Regression Lines for Individual Growth Rates

The initial rate of growth of calves was estimated by fitting a regression line which passed through the origin to the points in Fig. 4 (as no calf would grow an appreciable amount instantaneously). Lines using the best data (with both endpoints Grade 5 or better), the worst data (one or both endpoints Grade 6), and all the data, agree very closely — the slopes of these lines are respectively .00255, .00253 and .00253 of the mothers' length per day.

As a check against bias a regression line was fitted to the points but not artificially made to pass through the origin. The intercept was .0033 on the y axis and the slope .00242, a very close agreement with the lines which were made to pass through zero (see Fig. 4).

In Fig. 4 it can be seen that 20 whales appear to be growing and 9 shrinking; that is 20 points lie above the line $L_c/L_m = 0$, and 9 below. The probability of 20 or more points, out of 29, being above the line if they were equally likely to fall above or below it (a situation of no growth) is approximately .02, i.e., the probability of obtaining this or more significant data when the whales were not growing is about .02. We therefore conclude that significant growth is shown (at the 5 % level).

Appendix 3

Methods Used in Obtaining Growth Curve

Huxley (1932) determined experimentally that 2 different dimensions of animals (x and y) tend to be related as:

$$y = bx^a \tag{7}$$

Ohsumi (1960) found that for balaenopterids the constants a and b are, in general, different for the different growth stages (such as birth to weaning, or weaning to physical maturity). Plotting Log x against Log y, Huxley's formula [7] is a straight line, of slope a, and an intercept Log b. This relationship is called by Ohsumi the "allometry formula".

We considered the allometry formula of α and L for those whales where both α and L had been measured. Where just one of them was present we combined it with a calculation of z from another frame taken in the same year. In this category we had a total of 17 known whales and 24 unknown whales (one whale, No. 156, is plotted in 2 different years).

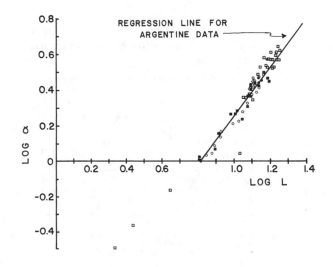

FIG. 13. — Allometry α (snout-to-blowhole length) and L (total length), Log α plotted against Log L. A regression line of Log α on Log L for the Argentine data is shown.

In Fig. 13 we have plotted Log α against Log L as Ohsumi (1960) did for fin whales. Also plotted are measurements of right whales given in the literature. There seem to be only 5 measurements of α and L for the southern right whale in the literature (all are from Matthews, 1938). We have, therefore, added in values for North Pacific right whales, of which we know of 27, and North Atlantic right whales, of which we know of 2. There seems to be little variation in the value of L for any given α between the different species (or subspecies as some would have it), or the sexes. We feel that the validity of our technique of determining α and L from photographs is strongly supported by the good agreement between our values and those from the literature.

One pair of measurements from Klumov (1962) (L = 10.75 m, α = 1.1 m) appear to be mismeasured.

The points for non-foetal whales appear to lie on a straight line, which, for the Argentine data, we have estimated to be (using linear regression):

$$\text{Log } \alpha = 1.376, \quad \text{Log L} = -1.127$$

Which gives:

$$z = 7.46 \times L^{0.376} \% \quad (z = \alpha/L \times 100 \%)$$

Correlation coefficients (r) were calculated for different sets of data, as shown in Table 7:

Table 7. Correlation coefficient (r) for different sets of data, for Log α and Log L

	No. points	r
Argentine data (all grades)	40	.970
Argentine data (grades 1-4)	18	.973
Argentine data (grades 5-6)	22	.962
Southern Oceans (literature)	5	.987
All literature [1]	28	.964

[1] Note including data on foetuses or Klumov's doubtful measurement.
α Snout-to-blowhole length.
L Total length.

The best measurements (grades 1-4) are only slightly better correlated than the worst measurements (grades 5-6) indicating that the deviation from the regression line is mainly due to variation between individuals and not to measurement errors.

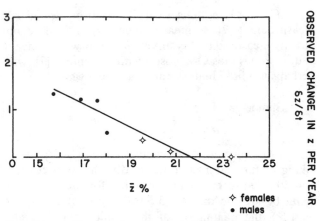

FIG. 14. — Changes in z for A quality data. From a known whale two estimates of z are made, z_1 % in year t_1 and z_2 % in year t_2 and then

$$\bar{z} = \frac{z_1 + z_2}{2}, \quad \frac{\delta z}{\delta t} = \frac{z_1 - z_2}{t_1 - t_2}$$

Points representing $\bar{z}$ and $\delta z/\delta t$ for 7 pairs of measurements are given and a regression line is drawn through the points.

In Fig. 14 we have plotted the change in z per year

$$\left(\frac{\delta z}{\delta t}\right)$$

against the mean value of z in the intervening period ($\bar{z}$)

$$\bar{z} = \frac{z_1 + z_2}{2}$$

Where a whale was measured as z_1 at time t_1 and as z_2 at time t_2 then:

$$\frac{\delta z}{\delta t} = \frac{z_2 - z_1}{t_2 - t_1} \% \text{ per year.}$$

The points entered in Fig. 14 were graded as follows:

Grade A — both measurements of z better than grade 5 and made with the whale perpendicular to a line between the camera and the whale

Grade B — each measurement of z either better than grade 5 and within 15° of being perpendicular to the line between camera and whale (angle estimated by eye) or grade 5 or 6 and perpendicular to the line between the camera and the whale

Grade C — other pairs of measurements.

Plots of Grade C are so widely scattered as to be meaningless.

Regression techniques are not theoretically applicable since errors in measurement of either z_1 or z_2 will produce an error in both $\bar{z}$ and $\delta z/\delta t$.

However, in Fig. 14 we have attempted to fit a line to these points as follows: we used only the most accurate measurements — Grade A — to minimize the effect of measurement errors.

The variation of the plots due to individual differences between whales can be looked at in 2 ways:

Case 1 — different whales with the same value of z may have this ratio (z) growing at different rates;

Case 2 — different whales with the same rate of growth in z ($\delta z/\delta t$) may have different values of z.

If we interpret Fig. 14 as Case 1 we would have to use regression of $\delta z/\delta t$ on $\bar{z}$, while if we interpret it as Case 2 we would use the regression of $\bar{z}$ on $\delta z/\delta t$.

Included in data of Grade A there is a whale with $\bar{z} = 23.23$ %, a value which most whales probably never

Table 8. Selecting between Case 1 and Case 2

Case	Slope	($\delta z/\delta t = 0$)	($\delta z/\delta t = 1$)	r
Case 1 ($\delta z/\delta t$) on $\bar{z}$	-0.1980	22.28	17.23	0.905
Case 2 $\bar{z}$ on ($\delta z/\delta t$)	-4.138	21.66	17.53	0.905

attain and so it seems more reasonable to use Case 2, i.e., the regression of $\bar{z}$ on $\delta z/\delta t$.

The maximum value of $\bar{z}$ is when $\delta z/\delta t = 0$ and is given in the table above. We can convert these into maximum theoretical lengths using the regression line obtained from our data in Fig. 13:

$$z = 7.46 \times L^{.376} \% \qquad . \ (1)$$

In the case of the regression line for $\bar{z}$ on $\delta z/\delta t$ the maximum length ($\delta z/\delta t = 0$) would be 18.36 m while for the regression of $\bar{z}$ on $\delta z/\delta t$ the maximum length would be 17.04 m. This latter seems more reasonable on the basis of the greatest lengths of southern right whales given in Table 9, and the line for Case 2 gives equation (2).

Unfortunately at present there is not enough data to draw separate regression lines that are meaningful for each sex.

Table 9. A summary of maximum and minimum lengths of right whales (m)

Source	Total whales measured	Known whales measured	Maximum male length	Known females measured	Maximum female length	Calves measured	Min. calf. length measured	No. of foetuses measured	Largest foetus measured
SOUTHERN OCEANS									
Our data - alive	41	3	12.37	12	15.56	12	6.35		
Our data - dead						1	4.11		
Lonneborg, 1906	3	2	15.21					1	4.19
Committee for Whaling Statistics, 1938				4	15.55			4	6,02
Matthews, 1938	5	2	13.54	2	15.23	1	6.54	2	15.23
NORTH ATLANTIC									
Various sources summarized in Omura, 1958	12	2	12.93	8	16.46	2	8.48		
NORTH PACIFIC									
Omura, 1958	4	2	13.60	2	17.80				
Omura et al., 1969	13	6	17.10	5	16.10			2	2.70
Klumov, 1962	10	5	17.06	5	18.30			1	4.40

Appendix 4

Confidence Limits for Growth Curves

For early growth (0-0.5 year) there are 2 possible sources of error: (1) our estimates of the length at birth and (2) the estimate of the initial rate of growth. We have estimated 95 % confidence limits for the length at birth to be 5.0-6.0 m or ± 0.5 m. 95 % confidence limits for the initial growth rate can be found from the data in Fig. 4 and the standard formulae for the confidence limits in the slope of a regression line which is made to pass through the origin:

$$\pm\, t_{n-1}\, (0.05) \sqrt{\frac{\sum_1^n y_i^2 - (\sum_1^n x_i y_i)^2 / \Sigma x_i^2}{\sum_1^n x_i^2\, (n-1)}}$$

Where n is the number of data points.

(x_i, y_i) are the data points

t_{n-1} (0.05) is the 2-sided, 0.05 significance point of the student's t distribution with n −1 degrees of freedom.

These limits were ± 8.19 m/year in this case. So our confidence limits for the initial growth curve become:

$$\pm \sqrt{(0.5)^2 + (8.19\, t)^2}\ \text{metres,}$$

where t is the age in years. These are the confidence limits shown in Fig. 6 for the initial growth.

As there are checks for the calculated initial growth curve (the growth by month, and the mean length of calves measured in November), which were not taken into account in these calculations, the confidence limits are almost certainly conservative.

For the later growth stages, the errors in the formula for growth in L and z (equations (5) and (6)) were broken down into 3 possible sources: errors in the regression of z on $\delta z/\delta t$, errors in the regression of Log α on Log L, and errors in the estimation of the constant t_k.

The standard error, ρ, in the height of the regression line of y on x at x is given by:

Where n is the number of data points,

(x_i, y_i) are the data points,

x and y are the means of x_i and y_i.

For the regression of z on $\delta z/\delta t$, the maximum value of ρ in the range $\delta z/\delta t$ we are considering, was 0.743.

Thus z = − 4.138 $\delta z/\delta t$ + 21.66 ± .743 (from (2)) for 1 standard error in this regression. For these limits new growth formulae for z and L were constructed and the errors in z and L for different values of t found. These errors are shown in Table 10.

Similarly we found the errors in the regression of Log α on Log. L.

t_k was fixed by the measurement of whale No. 156 at 9.9 m on 21 November 1973, when he was assumed to be 1.2 years old. This procedure introduces 3 sources of error:

(1)   errors due to whale No. 156 not being of average size;

(2)   errors in measuring whale No. 156 on 21 November 1973;

(3)   errors due to whale No. 156 not being 1.2 years old on 21 November 1973.

The standard errors from these 3 sources were estimated at (1) 0.5 m (at a length L of 9.9 m), (2) 0.22 m (L = 9.9 m) and (3) 0.1 years (when t = 1.2 years). The errors in length can be converted into errors in time from equation (4) and the 3 errors can be combined to make a total standard error in t_k of ± .39 years.

The corresponding errors in L were calculated and are shown in Table 10.

If the standard errors in the regression of $\bar{z}$ on $\delta z/\delta t$, Log α on Log L, and the estimation of t_k are denoted by e_1, e_2, and e_3 respectively, 95 % confidence limits for the errors in the growth curves (equations (5) and (6)) were approximated by:

$$\pm\, 1.96 \sqrt{e_1^2 + e_2^2 + e_3^2}$$

These values are shown in Table 10 and also as the confidence limits in Figs. 6 and 7.

$$\rho = \sqrt{\frac{\left[\sum_1^n (y_i - \bar{y})^2 - \left(\sum_1^n (y_i - \bar{y})\ (x_i - \bar{x})\right)^2 \Big/ \left(\sum_1^n (x_i - \bar{x})^2\right)\right]}{n-2} \left(\frac{1}{n} + \frac{(x-\bar{x})^2}{\sum_1^n (x_i - \bar{x})^2}\right)}$$

Table 10. Summary of errors in growth curves

Age (years)	Regression of $\bar{z}$ on $(\delta z/\delta t)$ (e_1)	Error arising from: Regression of Log α on Log L (e_2)	Estimation of t_k (e_3)	95 % confidence limits for errors $1.96\sqrt{e_1^2 + e_2^2 + e_3^2}$
	In length curve (Fig. 6)			
	± (m)	± (m)	± (m)	± (m)
1	.05	.04	.57	1.12
2	.22	.06	.49	1.06
3	.45	.15	.41	1.23
4	.66	.28	.34	1.55
5	.84	.45	.27	1.94
7	1.10	.71	.17	2.59
8	1.31	.99	.09	3.29
	In z curve (Fig. 7)			
	± %	± %	±	± %
1	.04	.28	.39	.94
2	.14	.22	.31	.79
3	.27	.17	.24	.78
4	.36	.13	.18	.83
5	.45	.11	.15	.95
7	.55	.06	.09	1.10
10	.66	.03	.05	1.30

References

ALLEN, G.M., The whalebone whales of New England.
1916 *Mem. Boston Soc. Nat. Hist.*, 8(2):114-5.

ANDREWS, R.C., Notes upon the external and internal
1908 anatomy of *Balaena glacialis* (Bonn.). *Bull. Am. Mus. Nat. Hist.*, 24(10):171-82.

Committee for Whaling Statistics, International whal-
1938 ing statistics. 11. *Int. Whaling Stat.*, (11).

D'OYLY, C., The Cape sketchbooks of Sir Charles
1968 D'Oyly, 1832-33 depicting Cape Town. Cape Town, A.A. Balkema.

HUXLEY, J.S., Problems of relative growth. London,
1932 Methuen.

KLUMOV, S.K., Right whales (Japanese) of the Pacific
1962 Ocean. *Tr. Inst. Okeanol.*, 58:202-97 (in Russian).

LONNEBORG, E., Contributions to the fauna of South
1906 Georgia. 1. Taxonomic and biological notes on vertebrates. *K. Sven. Vetenskapsakad. Handl.*, 40 (5):104 p.

MATSUURA, Y. and K. MAEDA, Biological investigations
1942 of whales from the northern Pacific. *Semi-kujira Hogei-shiryo*, 9(1):44-5 (in Japanese).

MATTHEWS, L.H., Notes on the southern right whale
1938 *Eubalaena australis. Discovery Rep.*, (17):169-82.

OMURA, H., North Pacific right whale. *Sci. Rep. Whales
1958 Res. Inst., Tokyo*, (13):52 p.

OMURA, H. *et al.*, Black right whales in the North Pacific.
1969 *Sci. Rep. Whales Res. Inst., Tokyo*, (21):78 p.

REVIEW OF PYGMY BLUE WHALE
STOCK IN THE ANTARCTIC

T. ICHIHARA

Abstract

Pygmy blue whales were not identified until 1961 as a population separate from ordinary blue whales, *Balaenoptera musculus,* differing from the latter in the following characteristics: body proportion, shape of baleen plate, growth (smaller lengths at sexual and physical maturity and when full grown) and migratory range and sub-population. The 3 former characteristics together with additional distinctive osteological information, indicate that the pygmy blue whale is a new sub-species, *B. brevicauda,* as proposed by Ichihara in 1963 (published 1966). A selective harvest of pygmy blue whales is practically impossible due to an absence of distinguishing features observable at sea, providing good reason for the subspecies' inclusion in completely protected stocks since 1966 in order to allow recovery of ordinary blue whales from earlier overexploitation.

Except for a provisional assessment of population size (10 000 whales in sub-Antarctic regions from 0 to 80°E prior to the 1960/61 season) and of maximum sustainable yield by Ichihara and Doi (1964), there is no other information on the state of pygmy blue whales, and inadequate data prevents calculation of additional unknown reproductive and mortality parameters.

Résumé

Il a fallu attendre 1961 pour reconnaître que les baleines bleues naines forment une population distincte des populations de baleines bleues ordinaires, *Balaenoptera musculus,* dont elles diffèrent par les caractères suivants: proportions du corps, forme des fanons, croissance (taille plus petite à l'âge de la maturité sexuelle et physique et au stade du plein développement), champ migratoire et sous-population. Les trois premières caractéristiques, ainsi que d'autres traits ostéologiques distinctifs, montrent que la baleine bleue naine est une sous-espèce nouvelle, *B. brevicauda,* comme Ichihara l'avait suggéré en 1963 (texte publié en 1966). La pêche sélective des baleines bleues naines est pratiquement impossible en l'absence de caractères distinctifs observables en mer, ce qui justifie le fait qu'on ait inclus cette sous-espèce, depuis 1966, dans les stocks complètement protégés pour permettre la reprise des stocks de baleines bleues ordinaires antérieurement surexploités.

A part une évaluation provisoire de la taille de la population (10 000 animaux dans les régions subantarctiques comprises entre 0 et 80° Est avant la campagne de 1960/61) et du rendement maximal soutenu, effectuée par Ichihara et Doi (1964), on ne possède aucune information sur l'état des stocks, et l'imperfection des données empêche de calculer d'autres paramètres de reproduction et de mortalité, qui restent inconnus.

Extracto

La ballena azul enana no se identificó hasta 1961 como población independiente de la de ballena azul común (*Balaenoptera musculus*), de la que difiere por las características siguientes: proporción del cuerpo, forma de la placa de barbas, crecimiento (talla menor en el momento de la madurez sexual y física y cuando está completamente desarrollada), amplitud de los movimientos migratorios y subpoblación. Las tres primeras características, junto con otros datos osteológicos distintivos, indican que la ballena azul enana constituye una nueva subspecies, *B. brevicauda,* tal como propuso Ichihara en 1963 (publicado en 1966). La captura selectiva de ballenas azules enanas es prácticamente imposible debido a la falta de características distintivas importantes claramente apreciables mientras la ballena está en el agua, lo que constituye una buena razón para incluir esta subspecie entre las poblaciones totalmente protegidas desde 1966 para permitir la recuperación de la ballena azul ordinaria, antes explotada en exceso.

Si se exceptúa una evaluación provisional de la población (10 000 ballenas en las regiones subantárticas, entre los 0 y los 80° de latitud Este, antes de la campaña de 1960/61) y la evaluación del rendimiento máximo sostenible hecha por Ichihara y Doi (1964), no se dispone de otros datos sobre la situación de la ballena azul enana y, como los datos biológicos son aún incompletos, no es posible calcular otros parámetros, como los de reproducción y mortalidad.

T. Ichihara
Department of Fisheries, Faculty of Marine Science and Technology, Tokai University, 1000 Orido, Shimizu, Shizuoka-ken 424, Japan

Stock Identity

CHARACTERISTICS OF PYGMY BLUE WHALES

Pygmy blue whales which were probably taken by the old Antarctic whaling in the first half of this century were not identified until 1961 (Ichihara, 1961). The biological research on 311 blue whales taken by the Japanese pelagic whalers around Kerguelen Island in the 1959/60 season provided the indication that these whales might be distinct from ordinary blue whales which were described by Mackintosh and Wheeler (1929). Pygmy blue whales differ from ordinary blue whales in the following ways:

Body proportion

In the seasons 1959/60 through 1961/62, Japanese whaling expeditions took a total of 1 826 pygmy blue whales in the waters around Kerguelen, Marion and Crozet Islands which are in the Antarctic far north of the ice edge. A study of the morphometry of 195 whales of both sexes between 21 m (69 ft) and 23 m (75 ft) long revealed that the tail region of the pygmy is significantly smaller than that of the ordinary blue gion (Ichihara, 1963, 1966). Omura, Ichihara and Kasuya (1970) present additional data on the body proportions of 3 pygmy blue whales, which support the suggestion that the tail region in full-grown male pygmy blue whales is significantly smaller than in ordinary blue whales, and that full-grown pygmy blue wha-

212

les have a longer central body region than ordinary blue whales. Ichihara (1966) finds that this peculiarly long central body region is formed in the foetal stage of pygmy blue whales.

From examination of the whole skeleton from a pygmy blue whale, 18.6 m (61 ft) long, which was taken at 42°08'S and 44°09'E in the sub-Antarctic in 1966 under a special permit of the International Whaling Commission, Omura, Ichihara and Kasuya (1970) propose an additional good reason to separate the pygmy blue from the ordinary blue whale as a sub-species, i.e., *Balaenoptera musculus brevicauda*. From this physically mature whale, they found distinctive characters in the shape of the skull and other bone structures.

Baleen plate

The food trapping horny baleen plates hang from the upper jaw but their tips begin to wear off at a comparatively early stage of growth. The colour, shape and morphological characteristics of the fringes of baleen plates are species-specific. Ichihara (1966) compared the shape of baleen plate from both types of blue whales, measuring the length and the breadth for the longest plates in which wear of tips had already taken place, and found that baleen plates of pygmy blue whales are significantly shorter relative to breadth than those of ordinary blue whales.

Growth

From an examination of reproductive organs, Ichihara (1966) estimates that female pygmy blue whales attain sexual maturity at 19.2 m (63 ft) in length, shorter by over 10 ft than ordinary blue whales. Male pygmy blue whales become mature below 19.2 m in length. About the same number of growth layers accumulate in the ear plug by sexual maturity for

both sexes in pygmy blue whales as well as in ordinary blue whales. From the number of growth layers in the ear plug and the number of corpora in ovaries, Ichihara (1966) concludes that female pygmy blue whales attain physical maturity at a length of about 21.6-21.9 m (71-72 ft). The longest female pygmy blue whale taken by the pelagic investigation was 24.1 m (79 ft) long.

In ordinary blue whales from the Antarctic, on the other hand, body length at sexual maturity is 22.6 m (74.2 ft) for males and 23.7 m (77.8 ft) for females (Mackintosh, 1942; Brinkmann, 1948; and Ichihara, 1961). Female ordinary blue whales attain physical maturity at 25.9 m (85 ft) and above (Laurie, 1937; Brinkmann, 1948; Nishiwaki and Hayashi, 1950). The difference in the full-grown length between both types of blue whales exceeds 3.0 m in the Antarctic.

Migratory range and sub-population

Catch statistics indicate that the major herd of pygmy blue whales inhabit the waters north of 54°S and from 0° to 80°E during the summer season in the Antarctic, while ordinary blue whales are widely distributed in seas of high latitude. It is difficult to find a geographical separation between the two types of blue whales, however, because the marking of a pygmy blue whale, and subsequent recovery of the mark showed that the pygmy whales occasionally reach far south in the course of the feeding migration in early summer.

According to personal communications, a small number of pygmy blue whales were caught by coastal whaling in Australia and most of them were taken along the west coast, which faces the Indian Ocean (R.G. Chittleborough, 1961 to 1963). Gambell (1964) reports that a female pygmy blue whale, 20.1 m (66 ft) in length, was taken off Durban, South Africa, in September 1963. Examining the external characters of this whale, he states that there is no

213

evidence of disease or deformity which might have inhibited normal growth.

Many small parasites were found in whitish-yellow spots on the baleen plates of pygmy blue whales from the Antarctic. The same commensal harpacticoid copepod, *Balaenophilus unisetus,* also occurred on the baleen plates of a pygmy blue whale taken off Durban (Gambell, 1964). Vervoort and Tranter (1961) found numerous individuals of *Balaenophilus unisetus* on the baleen plates of a pygmy blue whale collected in 1959 at the whaling station of Western Australia. Bannister and Grindley (1966) state that an immature pygmy blue whale taken off Saldanha, South Africa, was infested with *Balaenophilus.*

Bannister and Grindley (1966) examined many baleen plates of balaenopterid whales killed in South Georgia and pelagic Antarctic whaling, and in South African whaling operations, *Balaenophilus* occurred most frequently in sei whales. The ratio of infection was about 80 % for sei whales, 1.0 % for blue whales, 0.8 % for fin whales and 0.0 % for humpback whales. Blue whales taken off South Africa in 1930, 1962 and 1963, show a high infection rate of 12.0 %. On the other hand, the occurrence of *Balaenophilus* is very rare in ordinary blue and fin whales taken in the Antarctic. Bannister and Grindley suggest that heavy infestations of *Balaenophilus* survive best in those species which are more warm water or sub-Antarctic in habit. Pygmy blue whales mostly live in warmer waters in the Antarctic summer than do ordinary blue whales. It seems that the degree of incidence of *Balaenophilus* is closely related to the ambient water temperature.

From the viewpoint of population genetics, Fujino (1962) observed intra-population differences in the distribution of blood types among pygmy blue whales in the Antarctic although more study is needed to examine the racial differences. His findings suggest that there may be sub-populations of the pygmy blue whales which spend the summer in the Antarctic.

Pygmy blue whales have also been taken in other waters of the Southern Hemisphere.

Aguayo (1974) reports that 10 pygmy blue whales were identified among 168 blue whales caught off Quintay (33°11'S, 71°42'W) in Chile from the 1965/66 through the 1966/67 season. It is of behavioural importance that pygmy blue whales off Chile formed separate pods from ordinary blue whales and the 2 species did not intermingle in the feeding area (A. Aguayo, personal communication in 1974). His paper strongly suggests that there may be a few sub-populations of pygmy blue whales in the Southern Hemisphere which do not migrate to the Antarctic in the summer. The major herds of pygmy blue whales which live in the sub-Antarctic region adjacent to Kerguelen Island in the summer do not appear to migrate north to the waters off Chile in the winter.

Conclusions

Body colour does not provide a definite criterion for discriminating pygmy from ordinary blue whales, since its expression varies from observer to observer. Whale marking has not given information of much value on the migratory range of pygmy blue whales, because only one Discovery type of whale mark has been recovered. Gunners with long experience of Antarctic whaling informed me that pygmy blue whales look like big ordinary blue whales as the long anterior region of the body surfaces, but surprise them by their unexpectedly short tail which shows as they begin to dive. This observation coincides with the result of proportional measurements for pygmy blue whales, and is the only way at present to distinguish the live pygmy from the live ordinary blue whales in the Antarctic. Observations at sea also suggest that a selective catch of pygmy blue whales is practically impossible, and provided a good reason for protecting pygmy blue whales completely since 1966 in order to save the stock of ordinary blue whales suffering from the intensive damage done by whaling.

After killing, however, the combination of characters — body proportion, shape of baleen plate and stage of maturity — help to identify pygmy blue whales. On the basis of several specified characters, Ichihara proposed a new sub-species name, *Balaenoptera musculus brevicauda* for pygmy blue whales at the First International Symposium on Cetacean Research which was held in Washington, D.C., in 1963. This proposal was not published until 1966 because of the delay in editing the papers submitted to the Symposium. Meanwhile, Zemsky and Boronin (1964) published the denomination *brevicauda* without calling it a new sub-species and without crediting Ichihara (Rice and Scheffer, 1968). Priority of the scientific name for the pygmy blue whale must be decided to eliminate future confusion.

Unitil now there have been no other papers on the problem of the taxonomic position of pygmy blue whale. From the above biological evidence, I disagree with Small's opinion (1971) that the pygmy blue whale was a fraud, used as an excuse to continue killing blue whales in a portion of the Antarctic where a few could still be found. His denial of the existence of pygmy blue whales appears to have no basis in science and may be explained by an incomplete understanding of whale biology.

Population size

Some biological data and statistics from Antarctic whaling suggest that there may have been a catch of pygmy blue whales in old days. Ichihara (1961) comments on old records found in Laurie (1937), Budker (1954), and in the International Whaling Statistics, but the catch in the old days was not large enough to have adversely affected the size of the stock, even if all these records were of pygmy blue whales.

Ichihara and Doi (1964) estimated the initial population size of pygmy blue whales in the Antarctic, examining the data available collected from the Japanese expedition from the 1959/61 through the 1962/63 season. Unfortunately, no information on the commercial catch of pygmy blue whales by other countries in the same period was published, therefore catch data from other sources is not included in their population assessment.

Ichihara and Doi used two methods to estimate the size of stock of pygmy blue whales, prior to the 1960/61 season in which the catch effort was more concentrated for this whale than in the 1959/60 season. One estimate was an application of De Lury's method. The value of catch-per-unit effort, which was a relative index of the stock size, was revised by season to estimate the number of adult whales. Revision was made on the basis of the sea conditions and the maturity ratio of whales in the catch. Then, the virginal stock size of mature whales was estimated to be 4 350 pygmy whales prior to the 1960/61 season. The other estimate was calculated from 3 parameters of population: catch-per-unit effort, natural mortality rate and rate of fishing mortality. The natural morality rate was estimated for both sexes through the age composition of pygmy blue whales taken in the 1960/61 season.

Of whales taken, 40.3 % provided ear plugs with readable growth layers. The coefficient of natural mortality for pygmy blue whales was 0.11 assuming a biannual formation of growth layer, and consequently 0.05 assuming an annual formation of one growth layer. The annual accumulation rate of growth layer in the ear plug is still not determined accurately for both pygmy and ordinary blue whales. At any rate, the whales were fully recruited at 20 growth layers in the 1960/61 season. An equation to estimate the fishing mortality rate is

$$(1 - f) e^{-M} = \sum_{i=11} \varphi_i (t + 1) / \sum_{i=10} \varphi_i (t)$$

where

f   = rate of fishing mortality

215

M = coefficient of natural mortality

$\varphi_i(t)$ = index for the amount of i-year class in the season t

$\varphi_i(t)$ is calculated from the catch-per-unit effort. Thus, the abundance of whales in each year-class is estimated from known values of M and f for the stock prior to the 1960/61 season. Considering the age of maturity, the number of mature animals was estimated to be about 5 000 whales, which is very close to another estimate from De Lury's method.

Finally, Ichihara and Doi (1964) consider that the virginal population size of pygmy blue whales which inhabit the sub-Antarctic region from 0° to 80°E was about 10 000 whales prior to the 1960/61 season. The sighting of whales in the Antarctic has been continued after all blue whales were completely protected from Antarctic whaling (Masaki and Fukuda, 1975). No useful sightings data on pygmy blue whales can be obtained unless planned sighting surveys are regularly carried out, because of their limited area of distribution in the summer.

Maximum sustainable yield

As pygmy blue whales were only taken in 4 seasons, the parameters needed to construct population models were only tentatively available. Ichihara and Doi (1964) apply one method to estimate the maximum sustainable yield from the stock of pygmy blue whales. The known number of mature females and recruits provides several reproduction curves for assumed ages of recruitment. The respective reproduction curve indicates the number of recruits for each level of mature female whales. The amount of stock is shown as follows, when 10 year old whales are recruited.

Year-class	Amount of stock
5	Re^{5M}
6	Re^{4M}
7	Re^{3M}
8	Re^{2M}
9	Re^{M}
10	R
11	Rs
12	Rs^2
—	—
—	—

Then, the usable stock is

$$N = R + Rs + Rs^2 + ----- = \frac{R}{1-s}$$

where

R = number of recruits

M = coefficient of natural mortality

s = rate of survival

N = exploitable stock (numbers)

As the stock of 5 year-class and above, assuming biannual formation of growth layers, is composed of mature whales, the relation of recruits to the size of the mature stock is

$$\frac{2S}{R} = \frac{1 + e^M + e^{2M} + e^{3M} + e^{4M} + e^{5M}}{1-s}$$

S is the size of the mature female stock and hence 2S shows mature whales of both sexes. Consequently s is obtainable through the calculation. The fishing mortality rate f is obtained from the equation

$$s = (1 - f) e^{-M}$$

sustainable yield C is

$$C = N f$$

When this graphic method is applied to recruits of 9, 8, 7, 6 and 5 year-classes, the sustainable yield corresponding to each recruitment is estimated. From this procedure the maximum sustainable yield is 268 whales, which are produced at the full recruitment of the 9 year-class if 2 growth layers are formed per-year. On the other hand, it is 101 whales at the full recruitment of 9 year-class if 1 growth layer is formed per year. The body length at this age of recruitment is from about 18.9 m (62 ft) to about 20.7 m (68 ft) for pygmy blue whales. From the above procedures, Ichihara and Doi (1964) estimate that the maximum sustainable yield from pygmy blue whales is 200 whales or less.

With regard to the growth layer in the ear plug, it is reasonably well accepted that 1 growth layer is accumulated per year in fin whales. Assuming a similar rate of accumulation, the maximum sustainable yield from pygmy blue whales is possibly lower than the estimate by Ichihara and Doi (1964). Other unknown parameters are related to the imperfect biological knowledge of pygmy blue whales. The natural mortality rate for whales younger than the age of recruitment cannot be estimated from the data available. The abnormally low pregnancy rate which is 35.6 % for pygmy blue whales cannot be explained because of a lack of biological data. Except for a provisional assessment of population by Ichihara and Doi (1964), there is no other information on the status of pygmy blue whales.

References

AGUAYO, Baleen whales off continental Chile. *In* The
1974   whale problem, a status report, edited by W.E. Schevill. Cambridge, Mass., Harvard University Press, pp. 209-17.

BANNISTER, J.L. and J.R. GRINDLEY, Notes on *Balae-*
1966   *nophilus unisetus* P.O.C. Aurivillius, 1879 and its occurrence in the southern hemisphere (Copepoda, Harpacticoida). *Crustaceana*, 1966(3):296-302.

BRINKMANN, A., Studies on female fin and blue whales,
1948   report on investigations carried out in the Antarctic during the season 1939-40. *Hvalraad. Skr.*, 31:1-38.

BUDKER, P., Whaling in French overseas territories.
1954   *Norsk Hvalfangsttid.*, 43:320-6.

FUJINO, K., Blood types of some species of Antarctic
1962   whales. *Am. Nat.*, 96:205-10.

GAMBELL, R., A pygmy blue whale at Durban. *Norsk*
1964   *Hvalfangsttid.*, 3:66-8.

ICHIHARA, T., Blue whales in the waters around Ker-
1961   guelen Island. *Norsk Hvalfangsttid.*, 1:1-20.

–, Identification of pygmy blue whales in the Antarctic.
1963   *Norsk Hvalfangsttid.*, 5:128-30.

–, The pygmy blue whale, *Balaenoptera musculus bre-*
1966   *vicauda*, a new subspecies from the Antarctic. *In* Whales, dolphins and porpoises, edited by K.S. Norris. Berkeley, University of California Press, pp. 79-113.

ICHIHARA, T. and T. DOI, Stock assessment of pygmy
1964   blue whales in the Antarctic. *Norsk Hvalfangsttid.*, 6:145-67.

LAURIE, A.H., The age of female blue whales and the
1937   effect of whaling on the stock. *Discovery Rep.*, 15:223-84.

MACKINTOSH, N.A., The southern stocks of whalebone
1942   whales. *Discovery Rep.*, 22:197-300.

MACKINTOSH, N.A. and J.F.G. WHEEELER, Southern
1929   blue and fin whales. *Discovery Rep.*, 1:257-540.

MASAKI, Y. and Y. FUKUDA, Japanese pelagic whaling
1975   and sighting in the Antarctic, 1973/74. *Rep. IWC*, (25):106-28.

NISHIWAKI, M: and K. HAYASHI, Biological survey of fin
1950   and blue whales taken in the Antarctic season 1947-48 by the Japanese fleet. *Sci. Rep. Whales Res. Inst., Tokyo*, (3):132-90.

OMURA, H., T. ICHIHARA and T. KASUYA, Osteology of
1970   pygmy blue whale with additional informa-
        tion on external and other characteristics. *Sci.
        Rep. Whales Res. Inst.,* Tokyo, (22):1-27.

RICE, D.W. and V.B. SCHEFFER, A list of the marine
1968   mammals of the world. *Spec. Sci. Rep. Fish.
        NMFS/NOAA,* (579):16 p.

SMALL, G.L., *The blue whale.* New York, Columbia
1971   University Press, 248 p.

VERVOORT, W. and D.J. TRANTER, *Balaenophilus unisetus*
1961   P.O.C. Aurivillius (Copepoda Harpacticoida)
        from the southern hemisphere. *Crustaceana,*
        1961(1):70-84.

ZEMSKY, V.A. and V.A. BORONIN, On the question of the
1964   pygmy blue whale taxonomic position. *Norsk
        Hvalfangsttid.,* 11:306-11.

A NOTE ON THE ABUNDANCE
OF ANTARCTIC BLUE WHALES

J.A. GULLAND

Abstract

The methods used to estimate the relative or absolute abundance of Antarctic blue whales are discussed. These include catches per unit effort, sightings and marking. None of these are entirely free from bias: for example, the c.p.u.e. data underestimate the relative abundance during the last years of commercial catching of blue whales because of the increased preference for fin whales rather than blue whales. While no single method is satisfactory, there is a broad agreement between different methods, suggesting a total exploitable stock in 1963 of around 4 000 blue whales. Possible changes since the cessation of commercial hunting in 1963 are discussed. The only information comes from sightings from survey vessels, which are not sufficient to detect small rates of increase. Information on the likely values of the natural parameters suggest that the population may be increasing at 4-5 percent per year. The total population of blue whales in the southern ocean (including juveniles and pygmy blue whales) may be around 10 000 animals.

Résumé

L'ouvrage expose les méthodes qui sont utilisées pour estimer l'abondance absolue et relative des baleines bleues dans l'Antarctique. Ces méthodes incluent les prises par unité d'effort (c.p.u.e.), les repérages (à vue), et les marquages. Aucune de ces méthodes n'est totalement exempte de distorsion: c'est ainsi par exemple que les données concernant les c.p.u.e. sous-estiment l'abondance relative se rapportant aux dernières années de chasse commerciale des baleines bleues, et cela en raison d'une préférence croissante des baleiniers pour les rorquals communs par rapport aux baleines bleues. Si aucune méthode n'est en soi satisfaisante, il existe une large concordance entre les différentes méthodes, qui laissait apparaître un stock total exploitable d'environ 4 000 baleines bleues en 1963. La note traite également de l'évolution possible de ce chiffre depuis 1963, date où la chasse commerciale a cessé. Les seules informations disponibles proviennent des repérages à vue effectués par les navires de surveillance, qui sont en nombre insuffisant pour déceler de faibles taux d'augmentation. Les informations sur les valeurs plausibles des paramètres naturels laissent penser que la population peut être actuellement en augmentation de 4 à 5 pour cent par an. La population totale de baleines bleues dans l'océan Austral (y compris les juvéniles et les baleines bleues naines) peut être de l'ordre d'environ 10 000 individus.

Extracto

Se examinan los distintos métodos utilizados para estimar la abundancia relativa o absoluta de la ballena azul del Antártico, en especial examen de las capturas por unidad de esfuerzo, avistamientos y marcado. Ninguno de estos métodos está totalmente exento de defectos: por ejemplo, los datos relativos a las capturas por unidad de esfuerzo subestiman la abundancia relativa durante los últimos años de la captura comercial de ballenas azules debido a la mayor preferencia por el rorcual de aleta que por la ballena azul. Aunque ninguno de los métodos, aplicado individualmente, resulte satisfactorio, los distintos métodos concuerdan en general, indicando una población explotable en 1963 de unas 4 000 ballenas azules. Se examinan las modificaciones que pueden haber tenido lugar desde el cese de la captura comercial en 1963. Las únicas informaciones proceden de los avistamientos de los buques de reconocimiento, que no permiten detectar las tasas pequeñas de aumento. Los datos sobre los valores probables de los parámetros naturales indican que la población puede estar aumentando a razón del 4-5 por ciento anual. La población total de ballenas azules (incluidos ballenatos y ballenas azules enanas) se cifra en unos 10 000 animales.

J.A. Gulland
Chief, Marine Resources Service, Fishery Resources and Environment Division, Department of Fisheries, FAO, Via delle Terme di Caracalla, 00100 Rome, Italy

Introduction

This note was prepared in response to a request for a general review of the state of the stocks of blue whales in the Antarctic. As there have been virtually no commercial catches of blue whales since 1964, few new data (other than for sightings) have become available since the preparation of the Report of the IWC Committee of Three (Chapman, 1964). The present note is, therefore, concerned mostly with some comments on the results of the Committee of Three, including the use of catch-per-unit effort as a measure of abundance, with the estimates of stock size and also with a brief analysis of the sighting data.

Catch-per-unit effort as an index of abundance

The catch-per-unit effort of the commercial whaling fleets is by far the best measure currently available to study historical changes in the abundance of whales. However,

it is only a valid index if corrected for any changes in the efficiency of the whalers, and for changes in their preferences between species. Efficiency changes have been dealt with (probably not completely, but to a useful first approximation) by taking account of the tonnage or horsepower of catchers.

The effects of changes in species preference on indices for fin whales have long been recognized; for instance, the increase in catch-per-day between 1945/46 and 1955/56 was not evidence of an increase in stock, but of increased attention being paid to fin whales (plus some increase in catcher efficiency). Presumably this was not without cost, and must have involved some decreased attention to blue whales (otherwise the whalers would not have had any incentive in the earlier years to pay less than full attention to fin whales). At the tactical level there may not have been much change, e.g., the whalers would always prefer to pursue a blue whale, once seen, than a fin whale, but at the strategic level, e.g., the choice of area, there have been changes.

One large-scale change has been a shift of whaling effort northward. This was allowed for by the Committee of Three by using the catch-per-day south of 60°S as an index of blue whale population. However, even within a latitudinal band there are considerable differences in species composition. For example, Table 1 shows the catch-per-day in 2 zones of series B (60-70°S) in the South Atlantic during February in the 1949/50 and 1950/51 seasons.

This shows marked differences in species composition, with relatively more blue whales being found further east. The combined catch in blue whale units (BWUs) is roughly the same in both sectors, which is not surprising since otherwise the whalers would have presumably kept in the more profitable areas. If blue whales had been much scarcer, presumably the whalers would have kept in the eastern parts, and the catch-per-day of these whales would have been much lower.

Unfortunately, the table also shows that even within this comparatively detailed time/area breakdown, the species composition is not consistent. Between 1949/50 and 1950/51 there appeared to be a substantial increase in blue whales in both areas, and a roughly corresponding drop in fin whales. This could have been due to changes in migration patterns, but since the areas are still very large (roughly 600 x 1 200 miles), it could also be due to changes in the sections searched within each area, and greater preference for blue whales in the second season.

Another index of the degree to which selection can take place between different species of whales is given in Table 2. This shows in the first columns the ratio of blue to fin whales in the catches of different national fleets. Over the period there was a decline in

the ratio for all fleets, but considerable variation between fleets. This ratio represents a combination of 2 factors — the actual ratio of numbers in the sea, and the relative preference between the species expressed by each fleet. Some indication of the latter is expressed in the last columns, which give the preference of each national fleet for blue whales, relative to the average preference of the Antarctic pelagic expeditions as a whole (including fleets other than those of Japan, Norway and the USSR). Mathematically, the relative selection by fleet A, P_A is calculated as

$$P_A = \frac{B_A/F_A}{B_S/F_S}$$

where B_A, B_S are the numbers of blue whales caught by Fleet A, and all fleets respectively, and similarly F_A, F_S are the numbers of fin whales.

As well as considerable year-to-year changes, Table 2 shows some clear features. Japanese whalers usually had a marked relative preference for blue whales, except for 3 odd years (1950, 1951 and 1953); Norwegian preference for blue whales declined fairly steadily, particularly after 1959, while that of the USSR increased. Unfortunately, while this analysis shows that the preferences of individual national fleets have changed relative to the average of all fleets, it does not show what the absolute changes in preference between species have been, i.e., it does not allow corrections to be made to the observed catch-per-day of blue whales so as to arrive at unbiased indices of blue whale abundance. It is virtually certain, from the decline in relative performance, that the decline in catch-per-unit effort of Norwegian vessels overestimates the true decline in abundance, but not at all clear that the increased relative preference by Soviet vessels means that between 1947 and 1959 they actually concentrated more on blue whales (i.e., their catch-per-unit effort underestimated the decline), or merely that they turned to fin whales less than other countries (i.e., their

Table 1. Catch-per-day in the S. Atlantic

	60°W-20°W Catch-per-catcher day			20°W-20°W Catch-per-catcher day		
	Blue	Fin	BWU	Blue	Fin	BWU
1949/50	0.05	1.39	0.75	0.19	0.92	0.65
1950/51	0.22	1.03	0.71	0.46	0.66	0.79

Table 2. Indices of selection between fin and blue whales by different national fleets

	Ratio of blue : fin whales in the catches				Relative selection		
	Japan	Norway	USSR	All Pelagic	Japan	Norway	USSR
1946/47	1.45	.846	.681	.688			
1947	1.17	.641	.312	.361	3.24	1.77	.86
1948	.624	.250	.299	.430	1.45	.58	.70
1949	.773	.342	.216	.342	2.26	1.00	.63
1950	.132	.245	.401	.399	.33	.61	1.01
1951	.089	.343	.098	.250	.36	1.37	.39
1952	.273	.216	.058	.182	1.50	1.18	.32
1953/54	.061	.122	.054	.107	.58	1.14	.50
1954	.140	.095	.013	.084	1.67	1.14	.16
1955	.084	.064	.076	.064	1.32	1.00	1.19
1956	.104	.060	.032	.059	1.78	1.02	.55
1957	.090	.056	.086	.067	1.35	.84	1.29
1958	.053	.058	.144	.046	1.15	1.26	3.12
1959	.048	.037	.148	.046	1.03	.79	3.18
1960	.128	.013	.069	.064	2.00	.20	1.08
1961	.041	.014	.083	.042	.97	.33	.99
1962	.069	.019	.047	.051	1.36	.37	.93
1963	.005	.002	.027	.008	.62	.25	3.34
1964/65	—	—	.011				

catch-per-unit effort still overestimated the decline, but less than the total catch-per-unit effort).

One semi-quantitative deduction from Table 2 that is significant is the degree of selection that can take place. National differences consistent over several years are commonly of a factor of 2 or 3 and as much as 10 exceptionally in single years (e.g., 1954 and 1960). The overall changes in selectivity are likely to be at least as great, i.e., the changes in catch-per-unit effort between 1946 and 1964 may overestimate the actual decline in blue whale abundance manyfold.

In principle, this difficulty may be tackled by using detailed statistics of catch and effort, to take account of the relative distributions of whales and hunting. In practice, the most detailed available areas ("squares" of 10° latitude and longitude) are both too large (so that there is significant patchiness within a square) and too small (so that there are many squares without observations). For the present we probably have to accept that no analysis of catch and effort data alone will give an entirely reliable index of abundance of species of whales that are not at the time a major objective. With specific reference to blue whales, we can also conclude that the observed decline in catch-per-unit effort up to 1962 will tend to overestimate the true decline in abundance.

Antarctic stock size during exploitation

The Committee of Three used 3 independent approaches to estimating stock size (sightings, marking and the De Lury method).

Sightings

The sightings were based on the data of Mackintosh and Brown (1956) for the period 1933/34 to 1938/39. The estimate obtained was 15 % of 220 000 ± 50 %, i.e., 16 500-49 500 with a central value of 33 000. The Committee of Three noted that this figure tended to be an underestimate to the extent that (a) it excluded whales outside the Antarctic Convergence, and (b) whales, particularly older whales, avoided the survey vessel, or were not seen. On the other hand, the count included young whales not normally included in the estimates of the exploited population. There seems little to add to these comments by the Committee.

Marking

Because a large but unknown proportion of marks in captured whales are not reported, the normal methods of analysis of marking data cannot be used directly to give sensible estimates, though it may be noted that (discarding returns during the season of marking because of possible uneven distribution) 5.2 % of the marks placed in blue whales in the post-war period were returned in the first full season after marking. This gives a lower limit to the annual rate of exploitation. Annual catches were around 7 000 animals, giving an upper limit to the exploited population in the immediate post-war period of 140 000 whales.

An estimate that is less vulnerable to some types of bias can be obtained from the fact that of marks placed before the war, 122 out of 3 243 (= 3.76 %) fin whale marks, and 7 out of 693 (= 1.01 %) blue whale marks were returned in the post-war period. Disregarding sampling variances, this difference could be due to differences in marking success with species, the probability of capture or return of marks, or to higher mortality of blue whales between marking (around 1937) and recapture (around 1947). Differences in loss of mark or in reporting seem unlikely. If differences in probability of capture can be ignored, the difference is due to mortality rates, and using the formulation of the Committee of Three, we have the following equation

$$^{1.01}/_{3.76} = e^{-(Z_b - A_f)t}$$

The average time between marking and recovery was 10 years and included the war period, so that the fishing mortality over the 10 years was therefore believed to be equivalent to about 5 years "normal". We therefore have

$$-\ln\left(\frac{1.01}{3.76}\right) = -\ln(0.269) = 1.131$$
$$= 10(M_b - M_f) - 5(F_b - F_f)$$

or, assuming, as did the Committee, $M_b = M_f$ $F_b = F_f + 0.26$ (the Committee of Three, rounding off differently, used 0.28).

Other marking data, for fin whales only, gave an estimated total mortality of fin whales during the same period of 0.11. This gives estimates of F_b, the fishing mortality on blue whales, between 0.26 (if $F_f = 0$, $M = 0.11$) and 0.37 (if $F_f = 0.11$, $M = 0$), with a more probable value (for $F_f = 0.06$, $M = 0.05$) of 0.32.

The Committee of Three did not comment on this value. It would, however, imply that the fishing mortality on blue whales was 5 times that on fin whales, at a time when the catches, in number, of fin whales exceeded that of blue whales. This seems surprising. One possible explanation lies in the small number of marked blue whales recovered in the post-war period. If only 8 more (15 in all) had been recovered, the estimate of F_b corresponding to $F_f = 0.06$ would be 0.17, and if 4 less (3 in all), $F_b = 0.46$.

Another possible reason is that by 1947 the whalers had swung away from blue whales to a positive preference (in terms of general strategy at least) for fin whales, i.e., that the fishing mortality in 1947 (and hence the probability of a marked whale being caught) was greater for fin whales than blue whales. If this difference was big enough it could account for all the difference in return rate, and completely invalidate the method. Indeed, the method and the estimate above, are only valid if (for marks returned in 1947)

$F_b = 0.32$ $\quad$ $F_f = 0.06$ $\quad$ 1937-46
$F_b = F_f$ $\qquad\qquad\qquad$ 1947 (actual value is irrelevant)

223

In fact, even ignoring effects of chance variation, the observed marking data are consistent with a wide range of values of fishing mortality on blue and fin whales.

De Lury estimate

This was the basic method used by the Committee of Three. Between 1953/54 and 1960/61 the catch-per-catcher day, south of 60°S, fell by 68 % (from .256 to .081), while the cumulative catch was 11 650 whales. If the population also declined by 68 %, and if the only source of change was the removal of 11 650 animals, the initial population was 17 132, leaving 5 482 at the end of the period.

As the Committee of Three pointed out, neither assumption is fully justified. The population numbers also change because of gains from recruitment, and losses from natural mortality.[1] The effect of these net gains is that the population declines more slowly than expected from the numbers caught, and this fact leads to the population being overestimated. This can be in part allowed for by using only the decline in the last part of the period. A more accurate method of allowance has been used in the fin whale studies. Making allowance for this net recruitment, the Committee of Three considered figures of 9 000-10 000 whales in 1953/54 as being the best available estimate.

Allowance should also be made for the probable decrease in the attention paid to blue whales, as discussed in section 2. This decrease would tend to overestimate the reduction in blue whale stock. It is not possible to estimate the amount of this bias, though the differences between countries in Table 2 indicate its possible extent. A change in attention that altered the ratio of catch-per-unit effort to abundance by a factor of 2 is clearly possible, so that the 1960/61 population could be as much as 64%, rather than 32 %, of the 1953/54 population. This would imply, using the simple De Lury estimate, that the 1953/54 population was 11

650/.32 = 32 360, and the 1960/61 population 20 790.

Recent data on stock size: sightings

The most extensive series of observations of blue whales in the Antarctic since commercial catching ceased are those carried out by Japanese scouting vessels. The most recent report on these sightings is that submitted by Masaki and Yamamura (1977) to the 1977 IWC Meeting. These authors conclude (Table 20 of their paper) that the average number of blue whales in the Antarctic (south of 30°S) in 1975 was 11 840. Of these, 6 450 were in zones E and D of areas III and IV, in the southern Indian Ocean, and were probably pygmy blue whales. Subtracting these, the estimate for the 1975 blue whale population proper is 5 390. This is probably the best single estimate of the absolute members of Antarctic blue whales, but is subject to considerable variance and bias. A negative bias could occur because whales are not seen, or if they react to the presence of the vessels, and move away before the vessel comes within sighting range. Bias in either direction will occur insofar as the area covered by the vessels does not have the same density of blue whales as the unsurveyed. This has been somewhat reduced by considering sightings by zones and areas, but some bias probably remains.

To the extent that these sources of bias (especially the fraction not seen) are likely to be the same from year to year, the annual sightings could be used as indices of abundance. The year-to-year changes, particularly over a period, could then be used to determine whether the abundance was changing. However, the high year-to-year variability, added to the fact the annual values, as tabulated in Table 19 of Masaki and Yamamura, do not appear to be wholly independent, make it difficult to apply the data as they stand. Further analysis of the original observations would probably be valuable.

Sustainable yield

It is a commonplace of a simple population theory that if a population is reduced by exploitation it will generally tend to increase toward its original level of abundance. The relation of this tendency to population abundance and other factors determines the nature and form of the sustainable yield curve, e.g., whether sustainable yield is a simple, single-valued, parabolic function of abundance or something much more complicated.

A tendency to increase, if it occurs, must be caused by changes in 1 or more of the natural parameters (growth, mortality, or reproduction). The form of the sustainable yield curve can therefore be studied in 2 ways — either directly and empirically by observing the net changes in population (i.e., natural increase, if any, less catches) and relating these to population abundance or other factors, or, more analytically, by observing changes in the parameters. The first approach was attempted by the Committee of Three.

Figure 6 of their report shows their best estimate of the relation between stock abundance and sustainable yield and suggests very clearly as the points have been plotted, that for stock sizes of between 50 000 and 150 000 animals a significant annual yield, of 5 000 or more, with a maximum at a population somewhat in excess of 100 000, can be taken without decreasing the stock. The data points are, however, taken from Table 13 of the Committee's report. This table examines different relations between catch per catcher ton day and abundance, and the value chosen is that which gives the most reasonable result. To some extent their argument was circular and the figure is more an illustration of the results of the catch and effort analyses than an independent demonstration of the existence and shape of a sustainable yield curve. Until improved estimates of changes in abundance are available the nature of the sustainable yield curve may be better studied from examination of changes in some of the natural parameters.

Large-scale catching of blue whales

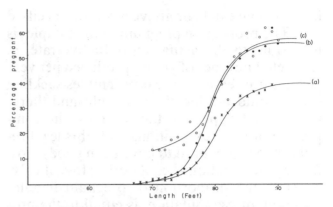

FIG. 1. – The percentage of blue whales in the catches that were found to be pregnant, as a function of length, for three periods - 1932/33-1936/37 seasons (curve (a), crosses); 1947/48-1951/52 (curve (b), closed circles) and 1957/58-1961-62 (curve (c), open circles). Note changes to a higher asymptotic pregnancy rate.

ended before the development of age determination from ear-plugs, so good age samples are not available, and direct estimates of mortality, recruitment rate, and other parameters cannot be obtained. However, the data do allow estimates of changes in birth rates to be obtained. Fig. 1 shows the percentage in the catches of females that were pregnant at each length during 3 periods – 1932/33-1936/37, 1947/48-1951/52 and 1957/58-1961/62. These all show the same typical sigmoid shape, but with important differences. The 1957/58-1961/62 period has relatively high values among very small whales. These are almost certainly the "pygmy blue whales" which were only caught during this period, and whether or not genetically distinct certainly mature at a smaller size. The curve for the first period rises to a lower asymptote (36 % pregnant). There is also a slight difference in the length at which the percentage pregnant is half the asymptotic value (79.5 ft against 79 ft). The latter would suggest that in the more recent periods, the females were maturing at a slightly smaller size, and hence also probably at a lower age, especially if there has been any increase in growth rate. This analysis of the pregnancy data, taking account of the sizes of whales, seems to avoid the chances of bias involved in looking at the percentage pregnant

225

in the whole catch, or above a certain specified size. The proportion pregnant in the samples is not equal to the actual reproductive rate, n, (expected number of young produced per year per mature female), because females suckling calves could not legally be caught and therefore do not appear in the samples (the compliance of most expeditions with this legal requirement is believed to have been good). To a first approximation (i.e., ignoring loss of calves, etc.), the proportion of pregnant females in the mature population, is equal to the proportion of suckling calves (i.e., those pregnant in the previous season) so that for the 1932/33-1936/37 data

$$.36 (1 - n) = n \quad \text{i.e.,} \quad n = \frac{.36}{1.36} = .265$$

and similarly, for the later periods

$$r = \frac{.55}{1.55} = .355$$

A tabulation can now be prepared following that of Gulland (1971) for fin whales, as given in Table 3, setting out the net rate of natural increase (r − M), as a function of the parameters of reproduction and mortality in the population.

In this tabulation, the differences between column A (1) and column B show only the changes that have been independently observed, i.e., an increase in pregnancy rate, an earlier maturity (taken to be 1 year), and an increase in females, because of the greater hunting pressure on males. The value of M′ was chosen so as to make the value of r in the unexploited stock equal to the value of M, that is, to assume that the stock is not changing. It is also possible that other changes, e.g., lower natural mortality among juveniles, may occur, which would tend to increase the net rate of increase (equal to the sustainable rate of harvest). A possible set of values, which would imply a sustainable harvesting rate not far short of the 0.1 implied in the figure of the Committee of Three's report is given in column A (2).

In the above tabulation too much significance should not be attached to the individual figures. Several, particularly the natural mortality rates, are not well known. The im-

Table 3. Possible values of natural parameters and net rate of natural increase of whale stocks during (A), and before (B), exploitation

| | Fin Whales (from Gulland, 1971) | | Blue Whales Mature Stock | | |
	A	B	A(1)	B	A(2)
P	0.6	0.5	0.6	0.5	0.6
q	0.9	0.9	1.0	1.0	1.0
n	0.4	0.35	0.355	0.265	0.355
M	0.15	0.15	0.133	0.133	0.10
T	4	9	8	9	8
r	0.119	0.041	0.074	0.040	.096
M	0.04	0.04	0.040	0.040	0.03
r-M	0.079	0.001	0.034	0.0	0.066

In this table:
- p = proportion of females
- q = proportion of females that are mature
- n = average number of young produced per mature female per year
- M′ = average natural mortality rate among immature animals
- T = mean age at maturity
- r = recruitment rate = $pqn\, e^{-M'T}$
- M = natural mortality rate among mature animals

portance of the table is the light it can shed on whether the blue whale is increasing, and if so, roughly at what rate. The 2 columns corresponding to the exploited stock suggest that taking account only of observed changes in basic parameters, the stock is increasing at about 3 % per year; if other, quite feasible, changes are occurring, it could be increasing at some 6 %.

The supposition that it is increasing will be untrue only if 1 or more of the following is true:

(i) the *changes* in sex ratio, pregnancy rate, and age at first capture have been badly overestimated;

(ii) the pre-exploitation population was not in approximate equilibrium, with a net rate of increase (r − M′) about zero, but was in fact decreasing;

(iii) there have been changes in other parameters (essentially M or M′, since the other parameters are reasonably well monitored), in a direction (i.e., increased mortality), and with a magnitude that would cancel the tendency to increase arising from the observed.

None of these seem likely; we may, therefore, be justified in expecting that the blue whale stock is increasing possibly at a rate rather better than 3 %, though until more accurate estimates of abundance are available over a period it is not possible to confirm or disprove this expectation.

Discussion

The preceding sections suggest that no single method can at present be used to give a reliable quantitative estimate of the size of the Antarctic stock of blue whales. We might, therefore, conclude that all that can definitely be said is that there are quite a number of blue whales about (as shown by the sightings discussed above, as well as other sighting records), and that there are fewer than there were (catch-per-unit effort data) and that it is more likely than not that they are increasing (sustainable yield analysis, and changes in natural parameters). This is perhaps a useful conclusion, in view of some of the more alarmist statements that have been published in recent years. However, some more quantitative estimates can be suggested if all the data are considered together. The sightings provide an estimate of an average of 5 000 animals between 1965 and 1977. To compare with this, we have the 2 estimates (for the Antarctic stock less pygmy blue whales) from the De Lury analysis of 6 000 and 9 000-10 000 animals in 1953/54 which, due to bias in the catch-per-unit effort, are likely to be biased downward. Taken together, and allowing for the catches between 1953 and 1963, a likely figure for the catchable stock of Antarctic blue whales is about 4 000 whales in 1963, increasing at about 4-5 % per year, to perhaps 7 000-8 000 by 1975 if the pygmy blue whale stock, which may be somewhat larger than the blue whale stock, is included, an estimate of rather over 10 000 blue and pygmy blue whales is arrived at for the whole of the Southern Ocean.

References

CHAPMAN, D.G., Reports of the Committee of Three
1964   Scientists on the special scientific investigations of the Antarctic whale stocks. *Rep. IWC*, (14):32-106.

GULLAND, J.A., The effect of exploitation on the numbers of marine animals. *In* Proceedings of
1970   Advanced Study Institute on Dynamics of numbers in populations, edited by P.J. den

Boer and G.R. Gradwell. Wageningen, Centre for Agriculture Publication and Documentation, pp. 450-68.

MACKINTOSH, N.A. and S.G. BROWN, Preliminary esti-
1956   mates of the southern populations of the larger baleen whales. *Norsk Hvalfangsttid.*, 45:467-80.

OHSUMI, S. and Y. MASAKI, Status of whale stock in the
1974   Antarctic 1972/73. *Rep. IWC*, (24):102-13.

POPULATION ASSESSMENTS OF ANTARCTIC FIN WHALES

R. GAMBELL

Abstract

Methods for determining the movements and stock units of fin whales, *Balaenoptera physalus,* in the Southern Hemisphere are examined. Analyses suggest a number of separate breeding stocks either in each southern ocean or subdivided in each ocean, remaining in the warm northern breeding grounds in the winter and migrating to Antarctic feeding grounds during the summer, where limited dispersal and intermingling of stocks takes place.

The methods for estimating the population size of *B. physalus* in the Southern Hemisphere and associated problems and modifications are also reviewed. Pre-season estimates of the total Antarctic fin whale populations were agreed upon in 1970 by the Scientific Committee of the International Whaling Commission (IWC) as 171 800 animals in 1957/58 and 100 500 animals in 1961/62. A number of methods used in the analyses depend upon catch and effort data; precise quantification of the latter has proven difficult, especially in a multispecies fishery, and deserves additional study. Since 1962, whaling interest has shifted from fin whales to successively smaller balaenopterid species, eliminating population assessments which rely heavily on catch-per-unit of effort data, as above; subsequent estimates since 1970 have depended upon some form of interative process, allowing room for cumulative errors, with the major problem being the determination of the rate of recruitment. Based on such methods, the IWC Scientific Committee has agreed on a current population estimate of about 83 000-84 000 for Antarctic fin whales. Recent sharp declines in fin whale stock indices in Antarctic Areas IV and III are not consistent with the model being used and suggest a need to revise the estimates. In classifying southern fin whales under the IWC's new management policy, the size of the Antarctic population before whaling was given to be 400 000 animals, reduced to a current size of 80 500; the maximum sustainable yield was taken to be 8 080, produced from a population of 227 000 whales.

Résumé

L'auteur décrit les méthodes utilisées pour étudier les déplacements et les stocks unitaires de rorquals communs, *Balaenoptera physalus* ("fin"), dans l'hémisphère austral. Les analyses donnent à penser qu'il existe dans chaque océan austral un ou plusieurs stocks reproducteurs distincts qui passent l'hiver dans les eaux chaudes des basses latitudes et gagnent en été les zones d'alimentation de l'Antarctique où se produisent une certaine dispersion et un certain mélange des stocks.

L'auteur précise également les méthodes employées pour estimer la taille des populations de *B. physalus* dans l'hémisphère austral, les problèmes qui se posent et les modifications introduites. Le Comité scientifique de l'IWC a estimé en 1970 qu'à l'ouverture des campagnes, la population antarctique de rorquals communs s'élevait au total à 171 800 animaux en 1957/58 et à 100 500 en 1961/62. Plusieurs méthodes d'analyse reposent sur le calcul des prises et de l'effort de pêche. La quantification précise des secondes données est difficile, surtout dans une pêche multi-spécifique, et elle demande un supplément d'étude. Depuis 1962, l'intérêt des baleiniers s'est détourné des rorquals communs vers des espèces de balcinoptères de plus en plus petites, de sorte qu'il n'est plus possible d'estimer la population par les méthodes qui reposent essentiellement sur la c.p.u.e. Les estimations postérieures à 1970 dépendent de procédés itératifs qui comportent un risque d'erreurs cumulatives. Le problème principal a été le calcul du taux de recrutement. En se fondant sur ces méthodes, le Comité scientifique de l'IWC a estimé la population actuelle de rorquals communs de l'Antarctique à environ 83 000-84 000 animaux. Les fléchissements considérables observés récemment dans les indices des stocks des zones IV et III de l'Antarctique sont en désaccord avec le modèle utilisé et donnent à penser qu'il faudrait réviser les estimations. En plaçant les rorquals communs de l'Antarctique dans le cadre de la nouvelle politique d'aménagement de l'IWC, on a indiqué que la taille de la population de l'Antarctique était de 400 000 animaux avant le début des opérations de chasse baleinière et qu'elle a été réduite au chiffre actuel de 80 500. On a considéré que le rendement maximal soutenu était de 8 080, à partir d'une population de 227 000 animaux.

Extracto

Se examinan los métodos utilizados para determinar los movimientos y las unidades de población de las ballenas de aleta (*Balaenoptera physalus*) en el Hemisferio Sur. Los análisis hechos indican la presencia de varias poblaciones de reproductores, bien en cada uno de los mares australes o subdivididas dentro de ellos, que permanecen en los criaderos septentrionales más cálidos durante el invierno y emigran a las zonas de alimentación del Antártico durante el verano, cuando se produce una dispersión limitada y una intermezcla de las poblaciones.

Se examinan también los métodos empleados para la estimación del volumen de las poblaciones de *B. physalus* en el Hemisferio Sur, los problemas que se plantean y las modificaciones que a veces son necesarias. El Comité Científico de la Comisión Ballenera Internacional (IWC) consiguió llegar a un acuerdo, en 1970, sobre las cifras de las poblaciones totales de ballena de aleta del Antártico, estimándolas en 171 800 animales en 1957/58 y 100 500 animales en 1961/62, antes del comienzo de las respectivas campañas de caza. Varios de los métodos utilizados para el análisis se basan en los datos de captura y esfuerzo; la cuantificación exacta de estos últimos resulta difícil cuando se trata de una pesquería en la que se capturan varias especies, y merece nuevos estudios. Desde 1962, el interés de los cazadores de ballenas se ha desplazado de las ballenas de aleta hacia especies de balenoptéridos cada vez menores, eliminando así la posibilidad de hacer evaluaciones de la población a partir de los datos de captura por unidad de esfuerzo, en la forma indicada más arriba. Las estimaciones hechas a partir de 1970 se han basado en una u otra forma de proceso iterativo, teniendo en cuenta posibles errores cumulativos, y el principal problema ha sido la determinación del índice de reclutamiento. Basándose en esos métodos, el Comité Científico de la IWC ha llegado a un acuerdo sobre una estimación de la población de ballenas de aleta del Antártico del orden de 83 000-84 000 animales. La notable disminución que se ha registrado recientemente en los índices de población de las ballenas de aleta en las Areas IV y III del Antártico no está de acuerdo con el modelo utilizado y sugiere que es necesario revisar las estimaciones. Al clasificar las ballenas de aleta australes, en la nueva política de ordenación de la IWC, el volumen de la población del Antártico antes de que comenzara su explotación se estimó en 400 000 animales, mientras que en la actualidad su número se ha reducido a 80 500. Como rendimiento máximo sostenible se ha dado la cifra de 8 080 ballenas, producidas a partir de una población de 227 000 animales.

R. Gambell
International Whaling Commission, The Red House, Station Road, Histon, Cambridge CB4 4ND, England

Distribution and migration

The movements of the Southern Hemisphere fin whales between the coldwater feeding grounds occupied in the summer months and the warm water which they frequent for breeding in the winter is known from such evidence as the times and places of whaling, direct observations of their seasonal population densities, and the recoveries of marks.

An analysis of whale sightings made by RRS *Discovery II* in the Antarctic during the period 1933-39 was presented by Mackintosh and Brown (1956). This was one of the first attempts to quantify the size of the southern baleen whale stocks. In addition, it demonstrated the seasonal variation in the numbers of these whales present in ice-free Antarctic waters. Some 75 % of the whales seen were reckoned to be fin whales. By plotting the numbers of sightings in good visibility, wind less than force 6, on a monthly basis the gradual rise in numbers from about September to February and the decline from then until July was clearly demonstrated. This was taken as evidence of a long stream of arrivals into the feeding grounds during the former period and departures in the latter.

A rather similar analysis of aircraft sightings data and catch records in the warm waters of the Durban whaling grounds by Bannister and Gambell (1965) during the winter months showed the reverse cycle. Fin whales appeared in this area in April, increased to a peak density in July and then declined to the end of the whaling season in October.

Stock units

The fin whale populations in the Southern Hemisphere are considered to be composed of a number of separate stocks. The identification and delineation of these stocks is by no means complete. In general, the schemes most favoured in published analyses suggest separate breeding stocks either in each southern ocean, or subdivided in each ocean. The existence of a central Indian Ocean stock has also been postulated (see Table 1).

All of the breeding stocks migrate to the Antarctic feeding grounds during the summer months, where they overlap and intermingle to a limited extent before returning to the winter breeding grounds in the equatorial waters.

For practical purposes, most population analyses using sub-division of the total Antarctic zone have been based on the traditional six areas (I, 120°W-60°W; II, 60°W-0°; III, 0°-70°E; IV, 70°E-130°W; V, 130°E-170°W; VI, 170°W-120°W) utilized by the Bureau of International Whaling Statistics. These areas were based primarily on the feeding concentrations of blue whales, but they seem to have some relevance to the fin whale stock units suggested above.

Whale marking

The most direct evidence for the migrations of fin whales comes from recoveries of

Table 1. A suggested division of the Southern Hemisphere fin whale populations (after Ivashin, 1969)

Stock	Wintering area	Summer distribution
Chile-Peruvian	West of N. Chile and Peru	100/110°W-60°W
South Georgian	East of Brazil	60°W-25°W
West African	West coast of Africa	25°W-0°
East African	East coast of Africa and Madagascar	0°-40°E
Crozet-Kerguelen	East of Madagascar	40°E-80°E
West Australian	Northwest of W. Australia	80°E-110/120°E
East Australian	Coral Sea	140°E-170°E
New Zealand	Fiji Sea and adjacent waters	170°E-145°W

Discovery-type internal marks. These can only be found when a carcass is processed at a whaling factory, and so there are no recoveries outside the whaling grounds. However, the returns to date suggest that there is in Southern Hemisphere fin whales a general pattern of segregation of the populations in the breeding areas, and of their dispersal and association on the feeding grounds (Brown, 1962, 1962a, 1970).

Fifteen recoveries under the International Marking Scheme demonstrate the north-south movement (Table 2). Fin whales marked off both coasts of South America have been caught in the western half of Area II, and whales caught off both coasts of South Africa have been marked in Area III. One whale marked off the east coast of South Africa has been caught further south and east in Area III (Brown, 1972), and one mark from the western part of Area IV has been recovered off the east coast of South Africa. East-west movements of whales marked and recaptured in the Antarctic indicate that of 466 marks recovered up to 1970, only 82 (17.6 %) were found in an area

different from that of marking (Brown, 1970). In all cases the area of recovery was next to that of marking. Movement eastward has taken place between all of the areas, and movement westward in all except Areas VI and V. It appears that movements between Areas II and III, III and IV and I and VI are about equal. There may be rather more movement between Areas I and II, at least in an eastward direction.

A study by Brown (1954) showed that the maximum range of dispersal appears to be limited within 50° east or west of the position of marking, and that dispersal is not necessarily in itself progressive. Within this limit, however, the proportion of fin whales dispersing with lapse of time does tend to increase. 34 % of the whales recovered within two years of marking had moved more than 10° east or west compared with 60 % of the longer term recoveries from the 208 records considered. Brown suggested that dispersal may take place among the younger whales of the population rather than among older whales.

Similar results can be derived from the

Table 2. **Marks returned from fin whales illustrating north-south migration**

Mark No.	Date marked	Position marked	Date recovered	Position recovered
Migration northward				
3482	24/02/35	64°52′S, 22°30′E	30/06/37	33°04′S, 17°50′ E
11389	10/12/57	52°12′S, 19°45′E	19/07/58	33°03′S, 15°45′ E
11428	16/12/57	59°40′S, 35°00′E	11/05/59	30°30′S, 32°30′E
12432	28/11/54	54°07′S, 09°47′E	3/07/62	Durban [1]
12634	24/11/54	53°16′S, 19°06′E	26/07/59	30°30′S, 32°00′E
13284	15/11/55	54°07′S, 07°03′E	20/06/63	Durban
13360	18/11/57	59°09′S, 33°35′E	9/68	Durban
13523	22/11/54	52°57′S, 24°02′E	29/07/59	30°30′S, 30°30′E
22285	12/12/62	56°33′S, 46°55′E	27/08/63	Durban
Migration southward				
9340	11/10/37	28°03′S, 46°17′W	7/01/49	52°55′S, 38°42′W
18405/9	4/11/58	33°40′S, 74°55′W	1/02/61	61°14′S, 58°32′W
18450	22/10/58	33°10′S, 74°40′W	13/12/61	59°01′S, 42°27′W
20173	4/11/58	33°48′S, 74°53′W	31/03/61	60°02′S, 56°23′W
20182	4/11/58	33°40′S, 74°55′W	8/02/60	62°04′S, 45°43′W
16335/18288	29/08/69	30°39′S, 33°06′E	25/02/71	43°32′S, 53°16′E

[1] Durban (29°52′S, 31°00′E).

Soviet marking data presented by Ivashin (1973). These indicate that of 43 fin whales recovered, 8 (18.6 %) had moved into an adjacent area (2 records appear to be in error in the listing as printed).

It should be emphasized that in many cases included in both the International and the Soviet Marking Schemes, the location of firing and finding of the marks was very close, with only a few degrees of longitude separating the positions even when these fell on different sides of an area boundary. It seems that many fin whales return to the same feeding place year after year.

Iodine values of fin-whale oil

During the feeding period, whales build up reserves of fat. The composition of the constituent oils varies according to the food supply, migration pattern and environmental and other factors, and these are taken to be reflected in the iodine value of the oil. Norwegian work has shown some evidence of regional segregation of the fin whales in the Antarctic using iodine values as a means of comparison between areas (Lund, 1950, 1950a, 1951).

Samples were taken on board factory ships working around the polar region, and at the South Georgian land stations. The iodine figures fell into 4 distinct groups, which were taken to indicate that there are more or less discrete groups of fin whales feeding in the areas 50°W-0°, 0°-70°E, 70°E-130°E and 160°-170°W. The average values of the samples are shown in Table 3.

The iodine values rose uniformly in all areas during the course of the whaling season by some 8-11 units.

These results suggest that there may be little interchange of whales between the areas indicated during the whaling season. However, the oil samples analysed were not entirely homogenous; although they were derived predominantly from the fin-whale catch, some blue-whale oil would also be included (a cor-

Table 3. Average iodine values for whale-oil samples

Whaling ground	1948/49	1949/50	1950/51
South Georgia	125	125	130
'2'	123	125	130
'3'	118	122	124
'4'	114	116	116
'5'	(107)	112	111

rection was applied to compensate for higher values due to any humpback catches taken). In addition, the rather small variations demonstrated in the iodine values make it impossible to determine to what extent whales may move both between areas outside the Antarctic and within the Antarctic Zone outside the whaling season.

Length composition

Length distributions of the catches between 50°-60°S and 60°-70°S over a long period of years both pre- and post-war were plotted by Laws (1960) by 10° sectors of longitude. Both male and female material showed a similar pattern in both zones, with a striking size segregation by longitude. There appeared to be four main size groups represented, roughly corresponding to South Atlantic, South Indian, West Pacific and East Pacific Oceans.

The size distributions were fairly closely correlated with the distribution catches. Within each group the largest animals were correlated with the greatest intensity of catching and there was a progressive decline in body size east and west of the modes. The largest animals and the greatest intensity of whaling occurred on the Greenwich meridian at the boundary between Areas II and III.

More recently, Ohsumi and Shimadzu (1970) have looked at mean length data by areas for fin whales aged 25 years and over, roughly corresponding to the size at physical maturity in this species. They found that the

233

largest mean length was recorded in Area II, decreasing eastwards through Areas III, IV, V and VI. The value for Area I was intermediate between those of Areas II and VI. The different lengths for neighbouring areas were not statistically significantly different, and while they cannot be used to differentiate between stock units, they do suggest that more than one stock may be present.

The pattern of sizes found by Laws, Ohsumi and Shimadzu is similar, but caution must be exercised when using such data because the mean lengths of whales may be affected by the status of the stocks after having experienced varying histories of exploitation.

Serological studies

Three major blood groups in Antarctic fin whales were identified by Fujino (1964) and they were considered to be produced by 2 major and equal allelic systems. Non-random geographic distributions of the incidence of the blood types suggested that 4 different breeding stocks enter the Antarctic Areas II, III (north of 50°S), III (south of 50°S) and IV. Adjacent stocks overlap to a certain degree on the feeding grounds, and intermingle to the extent of some 10 % between Areas II and III.

Variations associated with the blood types in the school size, pregnancy rate, age composition and ovarian corpora accumulation rate were also suggested. These could well account for the observed variations in these features around the Antarctic zone.

Unfortunately, these blood group analyses and other comparable biochemical studies have never progressed much beyond the initial investigation stage. The results look interesting and probably significant, but they have never been followed up sufficiently to make them conclusive. In particular, identification of breeding stocks by sampling whales on the feeding grounds is clearly not the most direct method, and it would be much more convincing if samples from warm waters could form the basis of stock identifications.

Methods of estimating population size

The International Whaling Commission set up the Committee of Three (later Four) Scientists to analyse the status of the Antarctic whale stock and advise on catch levels in 1960. This Committee introduced the techniques and methodology then developed in fisheries research into the field of whale investigations (Chapman, Allen and Holt, 1964). These methods were further extended and modified by the Committee and subsequent workers to give the range of assessment models now commonly employed.

A number of the methods of analysis depend to a greater or lesser extent on catch and effort data. The whale catch records are reasonably reliable, given the long series of data on numbers, lengths and sex of each species caught, which is held by the Bureau of International Whaling Statistics. The effort statistics are much less firmly based because the measurement of whaling effort in mixed species whaling in the Antarctic is not easy. Over the years whale catcher boats have become more powerful; they have been fitted with more sophisticated equipment for tracking whales, finding their own and the whales' positions, communicating with the rest of the whaling fleet, and overcoming the difficulties and dangers of poor weather conditions and ice. In addition, the skill of the gunners varies, and the different national fleets have differing modes of operation which affect their relative efficiencies. All of these factors are hard to quantify precisely, but Holt and Gulland (1964) suggested that many or all of them are reflected in a tonnage correction which had been shown to give a better adjustment than horsepower (Hylen, 1964).

The unit of effort, catcher-day's-work, is a variable which has not been standardized so far. Day length varies with latitude and time of year. Stop-catch days, when the whaling factory can process no more carcasses, shorten that day's catching effort. The use of scouting vessels and aircraft will also affect the value of a catching day's effort.

In multispecies whaling, there are very considerable problems of allocating the catching effort employed for various species. Again, this is a difficulty which has not been adequately studied. Rather, the problem has been circumvented to a greater or lesser extent by examining in detail only data from years or parts of years of more or less constant interest in one species; by looking at areas frequented more by one species than another; or by dealing with areas and periods of more or less equal whaling intensity.

Although a correction for weather had been calculated (Gulland and Kesteven, 1964), which incidentally showed considerable variations between expeditions, such corrections have not been utilized in any assessments. It has been generally assumed that weather is a random factor over several seasons, but the published analyses do not suggest this. With the changes in whaling grounds and species interest which have occurred over the years, it would seem rather important to take this factor into account.

Most methods of population assessment based on catch and effort data give estimates of the "exploitable population" size. This term is used to define the whales which may be killed within the restrictions imposed by size, area, time and other limits. The estimates given refer to this section of the population and appropriate additions must be made to obtain the total population sizes.

De Lury method

The size of a declining population can be estimated from catch and effort data alone using the De Lury (1947) method. In this, the accumulated catch is plotted against the catch-per-unit-of-effort in sucessive years. An extrapolation to the intercept with the zero catch-per-unit-of-effort axis gives an estimate of the stock size in the first year of the series. This simple model assumes that the population under study is a closed one. In practice,

modifications have to be introduced to take account of the fact that recruitment and mortality occur within a whale population during the time period covered by the data.

One modification developed by Chapman (1970) uses the relationship:

$$X_i = k [N_0 (1 + (r - M) i) - cc] \ldots\ldots (1)$$

where

X_i = catch-per-unit-effort in season i
N_0 = population size at start of study period
r-M = net recruitment rate (gross recruitment - natural mortality)
cc = cumulative catch to midpoint of season i

This is an approximation, since gross recruitment is derived from the parent population, while natural mortality applies to the actual population.

If $cc = di + f$ then equation (1) can be rewritten as

$$X_i = [N_0 (1 + (r - M)i) - di - f]$$
$$= (N_0 - f) + k [N_0 (r - M) - di]$$

d and f can be calculated from cumulative catch data, and the regression of catch-per-unit-of-effort on season derived using:

$$X_i = a + bi$$

Then with $k (N_0 - f) = a$
and $k (N_0 (r - M) - d) = b$

$$\frac{N_0 - f}{N_0 (r - M) - d} = \frac{a}{b} = h$$

or $$N_0 = \frac{f - dh}{1 - (r - M) h}$$

Clearly this procedure relies on an accurate independent estimate of $r - M$ being

235

available, and the assumption that it is constant during the study period.

Least squares method

This method, developed by Allen (1966, 1968), is based on the approximation:

$$N_{t+1} = (N_t - C_t) e^{-M} + R_t \quad \ldots \ldots \quad (2)$$

where

C_t = catch in season t
N_t = population at the beginning of season t
R_t = recruitment in year t

This assumes that the natural mortality (M) occurs after the fishing season but since the Antarctic whaling season is only some 3 or 4 months in duration, this is probably not too serious.

Estimates of recruitment are obtained from age composition data. The ratio of the catches from a partially recruited year-class to a fully recruited year-class is compared in successive seasons:

$$\frac{C_{it}}{C_{i} - 1, t} = r_i \frac{N_{it}}{N_{i} - 1, t}$$

where r_i = fraction of the year-class i which is recruited in season t.

It is assumed that the year-class $i - 1$, one year older, is fully recruited, so that for the catches in the next season:

$$\frac{C_{i, t+1}}{C_{i-1, t+1}} = \frac{r_i N_{it} e^{-Zt} + R_i, t + 1}{N_{i-1, t} e^{-Zt}}$$

where $R_{i, t+1}$ = recruits from year-class i in season $t + 1$.
z = total mortality coefficient (natural + fishing mortalities).

With

$$A_{i, t+1} = \frac{R_{i, t+1}}{r_i N_{it} e^{-Zt} + R_{i, t+1}}$$

the proportion of new recruits in the population $N_{i, t+1}$ (and also in the catch $C_{i, t+1}$ if this is now a fully recruited year-class).

Then

$$\frac{C_{i, t+1}}{C_{i-1, t+1}} = \frac{r_i N_{it}}{N_{i-1, t}} \cdot \frac{1}{1 - A_{i, t+1}} \cdot$$

With

$$B_{i, t+1} = \frac{C_{it}}{C_{i-1, t}} \cdot \frac{C_{i-1, t+1}}{C_{i, t+1}}$$

then

$$A_{i, t+1} = 1 - B_{i, t+1}.$$

This estimates the proportion of new recruits in the year-class fully recruited for the first time, and by working backwards in time it can be extended to estimate the proportions of new recruits in all the catches.

Using this calculated recruitment, the population size is estimated from equation (2) by computing the value for the population which minimizes the sum of the squares of the difference between actual catches and the catches expected from the population.

Mortality and catchability coefficient methods

The basic relationship assumed in fisheries analysis is:

$$C = qf\bar{N} \quad \ldots \ldots \ldots \ldots \ldots \ldots \quad (3)$$

where

C = catch
q = catchability coefficient
f = unit of effort
N = mean population during the fishing season.

The size of a year-class in year t is related to its size a year later $t + 1$ by

$$N_{i,t+1} = N_{it}e^{-Zt}$$

where

N_{it} = size of year-class i at beginning of season t

$N_{i,t+1}$ = size of year-class i at beginning of season t + 1

Zt = total instantaneous mortality coefficient

The total mortality coefficient can be derived from the ratio of the catch-per-unit-of-effort (c/f) in 2 successive years of a fully recruited year-class, thus:

$$Z_t = \log \frac{(C/f)_{i,t+1}}{(C/f)_{it}}$$

Estimates are usually calculated by month, comparing similar months a year apart, and by sex and suitable sub-areas. In this way a number of estimates can be derived which may be pooled to give an average for a particular season, sex and area. Even so, the estimates often differ quite widely, due to variable migration patterns of the whales, fluctuations in catching effort from place to place and season to season, and variations in the catches resulting from weather and other environmental factors.

In an effort to reduce this variability, the Committee of Three had averaged their total mortality estimates over 4 seasons and these smoothed results were considered satisfactory (Chapman, Allen and Holt, 1964). Subsequently, Doi *et al.* (1970) developed another smoothing technique. After calculating the total mortality coefficients by area and season, they estimated the corresponding natural mortality coefficients from the numbers of the year-classes are represented in recent catches as the oldest whales and obviously the method depends on having good samples of accurately aged whales. The difference between the abundance of successive year-classes of these old animals recruited before exploitation and which have subsequently experienced the same history of exploitation is a measure of their natural mortality rate.

Now since $z_t = M + F_t = M + qf_t$

where

M = natural mortality coefficient

F_t = fishing mortality coefficient in season t

f_t = fishing effort in season t

the fishing mortality coefficient can be calculated by simple subtraction of the natural mortality coefficient from the total mortality coefficient. This makes the assumption that the natural mortality remains unchanged throughout the period, and this may not be true when there have been very large changes in stock size.

For each estimate of the fishing mortality coefficient F_t there is a corresponding value for f_t, the fishing effort applied and for q, the catchability coefficient. The estimates of q are combined to give a smoothed value, particularly for periods of more or less uniform effort, and the smoothed estimate can then be applied in formula (3) to find the population size.

It should be noted that this method gives an estimate of the population size during the whaling season, whereas most of the other methods employed give estimates for the start of the season.

Recruitment curve method

Doi *et al.* (1970) have also argued that since whales are only caught during a fairly short period of time each year, the equation (2) model is a reasonable approximation:

$$N_{t+1} = (N_t - C_t)e^{-M} + R_{t+1}$$

where

N = catchable stock size

C = catch

R = recruits

237

The recruitment appeared to be calculated from the number of mature females in the population. The mature females themselves were determined by multiplying the total stock by a factor which declined from 1930 to 1960 and then remained constant. The reason for this, and the precise manner for determining the recruitment were not made clear.

Four types of reproduction curve were deduced which could possibly apply to the fin whale situation. These fell within the maximum and minimum values calculated for initial and grossly depleted populations, but again, the basic data and reasoning supporting the erection of these reproductive relationships were not clearly indicated.

Whale marking methods

Because the numbers of baleen whales marked and recovered in the Antarctic are relatively small, the marking data have not generally been used directly for estimating the population sizes. Instead, there have been various analyses in which the data have been manipulated to yield estimates of the total, natural and hence fishing mortalities, leading on to population estimates from the catch data in certain periods of the year.

Table 4. Efficiency of returns of pre-war marked fin whales

Seasons	% return
1945/46-1948/49	9
1949/50-1953/54	10
1954/55-1958/59	20
1959/60-1963/64	46

Any analyses should take account of the fact that recovery and reporting of whale marks is not fully efficient. Calculations based on the recoveries of pre-war placed fin whale

Table 5. Returns of fin whale marks by different countries

Country	No. of marks returned	No. of whales caught	Marks per 1 000 whales	Efficiency cf Japan
Japan	234	79 992	2.925	100
UK	29	22 099	1.312	45
Norway	68	57 402	1.185	41 [1]
Netherlands	9	10 626	0.847	29

[1] Original gives 48 (International Whaling Commission, 1967, Table 8, p. 36).

marks in the post-war years, allowing for mortality, changes in whaling effort and species interest, have shown a remarkable increase in the efficiency of returns (see Table 4 - International Whaling Commission, 1967). This probably results from the increasing use of carcasses for meat rather than oil, so that the marks are more likely to be found when the whale is cut up. Also, Japanese fleets took a larger proportion of the catch in the later years, and they have the highest mark recovery rate of the pelagic whaling nations, as shown in Table 5.

Even the Japanese fleets are not fully efficient. Doi *et al.* (1970) showed that of 101 marks fired into baleen whale carcasses during pelagic operations, only 70 (70 %) were found during cutting up, and only 15 (65 %) of 23 marks thrown into sperm-whale cookers were later recovered.

An alternative approach is based on Seber's (1962) method of analysis. This makes a comparison of the number of marks recovered in the same season from different marking experiments. The Committee of Three's original analysis was similar, and the method eliminates the problems of varying efficiencies of recovery between expeditions and years, since these factors are common to the different experiments. The ratio of the proportion of recoveries from the various marking experiments in a period after the last is a measure of the total mortality during the period between the experiments.

Population estimates up to 1970

The starting point for most current assessments of Antarctic fin whale populations are the estimates adopted at the IWC Scientific Committee's special meeting held in March 1970 in Honolulu (International Whaling Commission, 1971). This meeting reviewed all the available estimates and found that they agreed reasonably closely over the period from 1954 to 1962. At this time about 80 % of the Antarctic catch was composed of fin whales, so that any bias in catch-per-unit-of-effort data introduced by the catch of other species was not very large. The main stock estimates included in the Scientific Committee's synthesis were as follows:

De Lury estimates

Chapman (1970) presented modified De Lury estimates of 1957/58 initial fin whale populations by areas calculated from catch-per-unit-of-effort data from 1957/58 to 1961/62 (1959/60 - 1963/64 for Area I because of its different history of exploitation). Effort was adjusted for tonnage: (catcher days x tonnage)/1 000. Results employing a range of values for r − M were given, and also those from a further correction to the effort data. This correction prorated the effort by the proportion of fin whales in the total catch, as an extreme measure of the division of effort directed toward fin whales. The results are summarized in Table 6.

Least squares estimates

Allen (1970) made population estimates for the whole Antarctic fin whale stocks using his least squares technique (Allen, 1966, 1968). They were based on catch data for the period 1954-64, and were extrapolated back to 1932 and forward to 1969. All Southern Hemisphere catches were allowed for, but the comparison of actual and expected catches was for pelagic whaling only.

Values of M of 0.03 and 0.04 were adopted and recruitment was calculated from age data derived from (i) annual age compositions; (ii) a smoothed age-length key, both provided by Japanese scientists, and (iii) the original age-length key data from many nations provided for the Committee of Three. The estimates of population size resulting from these analyses are summarized in Table 7.

Table 6. Modified DeLury estimates ($\times 10^{-3}$) of 1957/58 initial fin whale populations by area calculated from catch-per-unit-of-effort data from 1957/58 to 1961/62

	Area						Total
	I[1]	II	III	IV	V	VI	
r-M = 0.03	9.2	43.6	72.5	32.1	17.5	11.8	186.7
r-M = 0.04	8.9	41.0	68.0	30.0	16.5	11.3	175.7
using prorated effort data							II-VI
r-M = 0.03		50.7	67.0	31.1	14.9	11.7	175.4
r-M = 0.04		47.1	63.2	29.3	14.3	11.1	165.0

[1] A separate analysis of the Area I data was carried out, because this Area was essentially unfished until 1955/56. A modified DeLury analysis was based on the decline in catch-per-unit-of-effort from 1955/56 to 1957/58 and 1959/60 to 1963/64. There was no catch in the Area in 1958/59. The estimates for the 2 periods were consistent with one another, and the 1957/58 value was obtained by extrapolation back from the second period. This analysis also indicated that r-M = 0.03 gave the best agreement between the two periods; but it was subsequently decided from other evidence that the average value for the whole Antarctic from 1958 to 1962 was about 0.04.

Table 7. Least squares estimates of total Antarctic fin whale stocks (× 10⁻³) using 3 sets of age data

Age data	(1)		(2)		(3)	
M =	0.03	0.04	0.03	0.04	0.03	0.04
1931/32	288	338	330	280	275	323
1957/58	164	171	195	186	177	185
1961/62	90	93	112	107	103	107
1967/68	44	46	61	58	56	59

Mortality and catchability coefficient estimates

No new estimates of this type were available at the Honolulu meeting, but an analysis produced the year before (Doi *et al.*, 1970) was reviewed. Key data are summarized in Table 8.

In this analysis, the total mortality coefficient was calculated by year, area, month, sex and age, and the mean values from ages 7 to 49 years were used as the coefficients for each area.

The natural mortality was derived from the oldest age groups by year, area and sex, and an average figure of 0.0415 used in subsequent calculations.

The fishing mortality coefficient was obtained by subtracting the mean natural mortality figure from the estimated total mortality values.

The catchability coefficient for each year was calculated by dividing the fishing mortal-

Table 8. Mortality and catchability coefficients for Antarctic fin whales, and resulting population estimates

Area	I	II	III	IV	V	VI	Whole	I-VI
Z total mortality coefficient								
Period 1954/57	0.420	0.106	0.081	0.073	0.040	0.137	0.095	0.143
M natural mortality coefficient								
January	0.036	0.023	0.043	♂♂ 0.042	0.035	0.033	0.035	0.035
All months	0.021	0.024	0.043	0.042	0.040	0.033	0.034	0.034
January	0.035	0.053	0.065	♀♀ 0.064	0.036	0.034	0.057	0.048
All months	0.043	0.053	0.066	0.061	0.037	0.035	0.057	0.049
F fishing mortality coefficient								
1954/57	0.379	0.065	0.040	0.032	-0.002	0.096	0.054	0.102
q catchability coefficient								
1954/58	0.405	0.026	0.014	0.045	-0.003	0.100	0.007	0.098
Estimated population sizes (× 10⁻³)								
1957/58	14.8	44.3	66.7	25.0	18.0	11.4	168.7	180.2
1961/62	9.1	25.1	39.7	11.9	7.0	3.7	91.5	96.5

ity coefficient by the fishing effort, the latter expressed as catcher-day's-work corrected for tonnage.

Finally, the mid-season population estimates were determined, apparently using the relationships:

$$N = \frac{C}{E}$$

$$E = \frac{F}{tM + F} (1 - e^{-(tM + F)}) 1 - e^{-qX}$$

where

N = mid-season population size
C = catch
t = length of season
X = corrected fishing effort

It is not clear if individual values of q for season and area were used, or smoothed values for combined groups of years, in this final step.

Recruitment curve estimates

Using an iterative procedure from an assumed starting population size, the known annual catches and estimated recruitment each year were deducted and added on respectively to calculate the history of the fin whale population by Doi *et al.* (1970).

Four reproductive curves were considered, giving the results summarized in Table 9. The bases for the calculations are not known.

Table 9. Estimated populations of fin whales ($\times 10^{-3}$) based on 4 reproductive relationships

Case	1	2	3	4
Season				
1929/30	283.7	378.5	378.5	378.5
1957/58	107.7	147.9	123.1	144.4
1961/62	58.4	92.0	59.4	89.2
1969/70	57.7	95.8	44.6	89.2

Synthesis of results in 1970

The results from the main estimates outlined above were combined to provide agreed estimates of the Antarctic fin whale populations, as shown in Table 10.

It should be noted that other analyses were taken into account in evaluating this result. For example, Chapman (1970) had updated the Committee of Three's estimates of total mortality to correct for the new understanding of age determination in fin whales. The original work had been based on the assumption that 2 ear plug layers were formed a year. The catches-per-unit-of-effort of "four" year olds in one year were compared with the "five" year olds in the next. The work of Roe (1967, 1967a) on the histology of the ear plug showed that only 1 layer was formed a year. The original coefficients of mortality would then be half as large, to allow for the doubling of age, as shown in Table 11.

Natural mortality had been estimated by the Committee of Three from catch curves for 1931/32, a season following relatively small fin whale catches. Heincke's method was used for "5, 6, 7 year old whales and older", because it is insensitive to inaccuracies in the estimation of the older age groups. The average mortality coefficient for males was 0.047 and for females 0.055, so that a mean of 0.05 was adopted as the best estimate of natural mortality. Again this value should be halved when the ages were

Table 10. Estimates of total exploitable Antarctic fin whale populations (pre-season values $\times 10^{-3}$)

Method	1957/58	1961/62
Modified DeLury (M = 0.04)	176	108
Least squares (M = 0.04, annual keys)	171	93
Catchability coefficient (adjusted from mid-season)	194	110
Recruitment curve (average of caos 2 and 4)	146	91
Average	171.8	100.5

Table 11. Original and corrected estimates of the total mortality coefficients of fin whales

Area	Original		Corrected	
	1953/54-1956/57	1957/58-1960/61	1953/54-1956/61	1957/58-1960/61
I	0.656	0.280	0.328	0.140
II	0.263	0.268	0.132	0.134
III	0.201	0.386	0.100	0.193
IV	0.286	0.532	0.143	0.266
V	0.121	0.541	0.060	0.270
VI	0.164	0.490	0.082	0.245
Average	0.282	0.416	0.141	0.208

doubled, although this result, 0.025, is rather low when compared with the Japanese estimate of 0.0415.

Whale marking estimates

The results of marking analyses were also considered at Honolulu. The Committee of Three had compared the mark returns in 1957/58 to 1960/61 from both pre-war and post-war marking experiments. The ratio of the percentage returns is equal to e^{-zt}, with t about 20 years, the time between the mean dates of pre-and post-war marking, 1935 and 1955. The results are shown in Table 12.

The results of post-war marking experiments in each season from 1953/54 to 1959/60 were analysed by the Seber (1962) method (International Whaling Commission, 1967). Zero year recoveries were disregarded because they may introduce a bias due to incomplete

Table 12. Estimates of total mortality (z) from pre- and post-war marking

	Experiments			
	Area II	Area III	Other Areas	Total
z	0.071	0.065	0.147	0.110

mixing. The mortality rates were calculated from the formula:

$$1 - \frac{t_1 S_{02}}{t_0 (S_{12} + 1)}$$

where

t_0 = number marked less zero year recoveries

t_1 = number marked in following season, less zero year recoveries

S_{02} = number of marks recovered in second season and later

S_{12} = number of marks recovered from following season's marking in first or later seasons

The estimates were very variable, being based on only a little over 1 000 marked whales but gave an average value for the 7 seasons of z = 0.27. This calculation was updated by Chapman (1970) taking additional mark recoveries into account, and gave an estimate of z = 0.026.

A further analysis, making appropriate corrections for the efficiency of recoveries outlined in Table 4 was carried out to estimate the actual number of marked whales recaptured each season. This overcomes the problem of non-finding or reporting of marks, although a value of M = 0.05 was used in the calculations. A Seber-type analysis was then performed giving an average figure for z = 0.040.

The Honolulu meeting decided that the marking results generally supported those from the other methods of analysis which were utilized, but because of the variability of the marking data they could not be considered very conclusive by themselves.

Population estimates since 1970

Since 1962 there has been a shift of interest, in the Antarctic fishery, from fin to sei, and latterly to minke-whales. This has meant that it is no longer possible to estimate the size of the fin whale populations by methods relying heavily on catch-per-unit-of-effort data, since this is no longer an accurate measure of the fin whale abundance. The fishing grounds also have moved to lower latitudes where the sei whales are found in greater concentrations. The net result has been that all the estimates of fin whale population size in recent years have depended on some form of iterative process incorporating the known catches and the assumed or calculated recruitment, starting from the agreed populations in 1957/58 and 1961/62. There is clearly scope here for cumulative errors to build up, but the major problem has been in determining the actual rate of recruitment.

Two main methods of estimating recruitment have been employed; one based on mathematical models developed by Japanese scientists, i.e., a theoretical recruitment curve technique; the other developed by Allen on his least squares approach, using actual estimated recruitment from age data.

Theoretical recruitment rate

Using mathematical models, Japanese scientists have used the best available estimates of the relevant parameters to calculate population size and sustainable yields in a formulation submitted by Ohsumi (1972, revised 1973). The fluctuations in the key parameters of pregnancy rate, age at sexual maturity in females, age at recruitment and natural mortality rate after recruitment were examined in the context of relative stock size compared with the initial size, using the assessment of Doi et al. (1970) as the standard. Pregnancy rates (p) were obtained from the Bureau of International Whaling Statistics data. These show an overall increase since the nineteen thirties. They were modified in 2 ways to compensate for the absence of lactating females in the catches. One correction was a smoothing to reduce the maximum value to

50 %, a figure agreed at the Honolulu meeting on biological grounds. The second correction was a calculation due to Shimadzu (1970). This assumed that half the lactating females weaned their calves during the whaling season and so became exploitable, the number of lactating animals being based on the number pregnant in the preceding season. This gave a maximum value of just under 40 %.

Age at sexual maturity in females (t_m) has declined over the years. It has been estimated from the age at which 50 % of the catch sample is mature (Ohsumi, 1972, 1973) and by the ear plug layer counts to the transition phase (Lockyer, 1970, updated 1972). Unpublished Japanese data comparable with the latter were also plotted and 2 rather arbitrary curves fitted decreasing from 11 to 5 and 9 to 4.5 years with decreasing stock size.

Age at recruitment (t_c) was not estimated directly because no age data from ear plugs are available before 1957. It is not correct to apply recent age-length keys to these earlier years because of the changes in growth rates which may have occurred. Instead, the average age at recruitment (knife-edge) was calculated from the following equations:

$$S = \frac{R_m}{2(1-s)}$$

$$R_0 = \frac{R_m}{2(1-s)} \, p$$

$$R_0 = R_0 e^{-tcM}$$

$$R_c = z \, \frac{R_m}{e^{-(tm-tc)Z}}$$

$$R_m = R_c e^{-(tm-tc)Z}$$

$$N = \frac{R_c}{1-s}$$

$$\frac{ZS}{N} = e^{-(tm-tc)Z}$$

243

Where

R_c = number at age of first recruitment to exploitable population

R_m = number at mean age of sexual maturity

R_0 = number at birth

S = population size of mature females

s = survival rate

N = exploitable populutaion size

p = pregnancy rate

t_m = age at sexual maturity, females

t_c = age at recruitment

M' = natural mortality before recruitment

Z = total morality

z was calculated from the fishing rates for 1957/58 and 1961/62, using the agreed stock sizes, and assuming natural mortality of 0.04. Values of t_c for these 2 years were therefore obtained, and also for the initial stock, with z = M = 0.055 and 0.04.

For each set of estimates of the age at sexual maturity there were thus 2 sets of estimates for the age at recruitment in 1958, 1962 and the initial stock level, to which straight lines were fitted.

The natural mortality coefficient after recruitment (M) was given 2 arbitrary sets of values in relation to stock size, because of the lack of more precise information. One was of a constant value of 0.04 and the other decreased from 0.055 to 0.035 over the full range of stock size.

The natural mortality coefficient before recruitment (M') is also not known directly, but can be calculated from the equation:

$$1 - e^{-M} = \frac{p}{2} e^{-t_c M'} e^{-(t_m - t_c)}$$

Using the 2 sets of values for each of M, p, t_c and t_m already decided, 8 estimates of M' were obtained, ranging from 0.0820 to 0.1972. Although it is possible that the value of this coefficient changes with stock size, it was treated as constant in further calculations.

The rate of recruitment was estimated by the following parameters:

$$R = p\, S\, e^{-t_c M'}$$

$$s = e^{-(M+F)}$$

$$K = \frac{R}{2S} = \frac{1-s}{s^{(t_m - t_c)}}$$

$$E = \frac{F}{F + TM} (1 - e^{-(F+TM)})$$

$$M = \frac{R}{1-s}$$

$$C = EN$$

With p, t_c, t_m, M and M' determined at each stock level, the above equation can be solved for the corresponding R, K, s, F, E, N and C. (T = length of whaling season).

The rate of recruitment (r) is then calculated as:

$$r = \frac{R}{N}$$

The averaged results obtained using the 8 sets of combinations indicated that as the population level decreases, the net recruitment rate (r − M) increases to a maximum of 0.0555 at 20 % of the initial population level. The replacement yield reaches its maximum (MSY) at 53 % of the initial population level and is 1.99 % of that initial level.

Population assessments were then carried out using the calculated parameter values and an initial population size which conformed with the agreed 1957/58 and 1961/62 population estimates. This gave the results shown in Table 13, from the updated calculations of Ohsumi and Masaki (1974).

This analysis seems to be particularly sensitive to the value of the natural mortality coefficient (M) which is employed. While it is agreed that the present figure is about 0.04

Table 13. Recruitment curve model estimates of populations and yields ($\times 10^{-3}$)

Population level	Population size	Replacement yield (RY)
Average combination of parameters		
Initial	395.0	0
MSY	209.4	7.68
1973/74	97.1	4.89
Lowest combination of parameters		
Initial	425.0	0
MSY	221.0	6.84
1973/74	83.0	3.38

with the stock at 20-25 % of its initial level, the values at larger stock sizes are unknown. If the natural mortality rate is not so much bigger at higher stock levels, the effect is to move the MSY level to a larger stock size, and also to increase the maximum sustainable yield.

Least squares estimates of recruitment rate

Allen (1971, 1972 and 1973) has carried out further analyses using his least squares method to investigate the rate of recruitment and related aspects of the fin whale populations in recent years.

The model was revised to take account of the fairly steady reduction in the proportion of males in the catches from about 55 % to about 45 % since the thirties. It was found (Allen, 1972) that a combination of values for natural mortality of M = 0.035 for males and 0.045 for females with a protection factor for females of 0.2 would account very largely for the observed changes in the sex ratio of the catches. The protection factor makes allowance for the reduced whaling mortality of females due to the protection of those accompanied by a calf. Shimadzu (1970) had calculated that the proportion of females protected in this

way is 0.2, on the observed migrations and weaning behaviour.

With the revised model, recruitment was calculated from Japanese age data from the late fifties onward applied to length compositions covering a much earlier period of years. This introduces the problem of possible changes in growth rates during the period considered and hence biased estimates of recruitment in the earlier years. Allen's (1971, 1972) conclusions were of relatively high and consistent recruitment rates (around 0.06) to about 1952, and that the rate then declined quite rapidly to a stable level (of about 0.05) which extended to at least 1962. Thus, although there appears to be evidence of an increase in the pregnancy rate and an advance in the age at sexual maturity, direct measurement of the recruitment rate gave no evidence of increase with declining population size.

However, more recently Allen (1973) has found some evidence of an increase in the recruitment rate in the last few years. For this analysis Japanese age-length keys from 1967 to 1971 were available. The component of the exploitable population consisting of the 1960 and earlier year-classes was estimated by the least squares method for the period 1954-63 and extrapolated to 1971. This assumed that no recruitment occurred after the 1960 year-class was fully recruited and that recruitment is very small in the years preceding this. The total exploitable population was then found by adding on the percentages in the catches of the appropriate ages in the years from 1960 onward.

The catch in 1 year was then subtracted from the estimated population at the beginning of that year. Natural mortality (at 4 %) was applied and the estimated population at the beginning of the following year was subtracted to give the estimated number of recruits. These recruits were allocated to their parent years by an averaged composition schedule. The sum of the recruits derived from each parent year-class was then expressed as a percentage of the parent total exploitable stock. From 1957 to 1962/63 the recruitment rate was about 6 % as in the earlier analyses, but in the last 3 years of the series to 1966 there was a sharp increase to 10-15 %. This jump seems to be too large to be entirely natural, and it may be that really there has been a more gradual but undetected rise in the recruitment rate.

The overall conclusion from the least squares analyses during most of the years studied is of a gross recruitment rate of 0.05-0.06. With natural mortality taken as 0.04, the net recruitment is then 0.01-0.02. This clearly leads to lower estimates of the sustainable yield than the Japanese method and, until the Allen (1973) analysis, lower estimates also of the current population size by extrapolation.

However, Allen's estimates by slightly differing methods are very variable for a Schaefer-type analysis (Allen, 1972) gives much higher estimates of recruitment. In this the potential increases in the population were estimated as the sum of catch and population change, using the estimated population to provide the latter parameter. The results were smoothed by 5-year periods to reduce fluctuations due to the intensity of whaling. The results showed that the net recruitment rate remained very constant over a large part of the population range with values of 0.030-0.036. It appeared to decline rapidly for populations above 300 000-350 000, which is, therefore, the MSY level with a yield of about 10 000-12 000.

The main results from the least squares analyses are summarized in Table 14.

Conclusions

The IWC Scientific Committee has agreed on a current population estimate of about 83 000-84 000 for the Antarctic fin whale. This is based largely on the lowest of the Japanese recruitment model estimates, together with analyses by the least squares method and recalculations due to Chapman (Ohsumi and Masaki, 1974). The Japanese model makes assumptions about changes in key parameters which have not yet been demonstrated resulting in the problem of the recruitment rate discussed above.

In addition, it has been noted (International Whaling Commission, 1975) that there have been sharp but erratic declines in the fin whale stock indices in Areas III and IV, and also a sharp decline in the total catch-per-unit-of-effort in Area V, although the data are inadequate to make firm inferences regarding changes in stock size. The declines in Area IV and to a lesser extent in Area III are not consistent with the Japanese model and suggest the need to revise the estimates.

In reviewing the available estimates for the purpose of classifying Southern Hemisphere fin whales in terms of the IWC's new management procedure, Chapman (1976)

Table 14. Least squares method estimates of Antarctic fin whale population size ($\times 10^{-3}$)

	1955	1960	1965	1970
From Allen, 1972				
II	62.0	38.5	14.4	16.4
III	93.4	66.6	33.2	34.5
IV	22.4	16.7	8.2	7.7
V	14.2	12.1	3.7	2.9
VI	20.6	9.6	4.9	9.6
II-VI	212.6	143.6	69.4	71.1
Total	253.7	159.4	72.8	66.4
From Allen, 1973				
Total	242.8	148.8	72.5	85.4

Table 15. Estimated stock sizes and yields ($\times 10^{-3}$) of Southern Hemisphere fin whales

Area	Original	Current	MSY level	Ratio of current to MSY level	MSY
I	12	7.0	7	1.00	0.24
II	124	17.5	70	0.25	2.51
III	152	34.7	86	0.40	3.07
IV	60	7.4	34	0.22	1.21
V	28	2.7	16	0.17	0.57
VI	24	11.2	14	0.80	0.48

drew up a set of figures for each of the 6 statistical areas. The estimated original population size in the whole Antarctic of 400 000 was based on the average results of 3 calculations by Ohsumi and Masaki (1974), including those shown in Table 13, the results of Allen (1972), and a quite separate analysis of density changes, catch and effort data, and assumed recruitment made by Jones (1973). This total was subdivided into the separate areas according to the allocation suggested by Gulland (1974), based on the total catches b, based on the total catches by areas in 1931-68.

The current population sizes were taken from a synthesis of sightings (Ohsumi and Masaki, 1974) and catch and effort data

(Allen, 1972) for the separate areas reviewed by Fox et al. (1974). The sum of the area figures selected by Chapman came to 80 500 fin whales.

The MSY level was taken as 227 000 from the average of the 3 models for the whole Antarctic given by Ohsumi and Masaki (1974) and this was subdivided into areas using the Gulland allocation. The total estimated MSY of 8 080 was calculated from the same sources as those used for the original population size and subdivided into areas using the proportions employed for the current population estimates.

The full set of figures developed are shown in Table 15.

References

ALLEN, K.R., Some methods for estimating exploit-
1966   ed populations. *J. Fish. Res. Board Can.*,
         23(10):1553-74.

–, Simplification of a method of computing recruit-
1968   ment rates. *J. Fish. Res. Board Can.*,
         25(12):2701-2.

–, Estimates of total population size and recruitment
1970   for the fin whale stocks of the Antarctic. Paper
         presented to the IWC Special Meeting on
         Antarctic fin whale Stock Assessment, Hono-
         lulu, Hawaii, 13-25th March 1970, Doc. F/3
         (mimeo).

–, Notes on the assessment of Antarctic fin whale
1971   stocks. *Rep. IWC*, (21):58-63.

–, Further notes on the assessment of Antarctic fin
1972   whale stocks. *Rep. IWC*, (22):43-53.

–, Revised estimates of Antarctic fin whale popula-
1973   tions. *Rep. IWC*, (23):99-114.

BANNISTER, J.L. and R. GAMBELL, The succession and
1965   abundance of fin, sei and other whales off
         Durban. *Norsk Hvalfangsttid.*, 54 (3):45-60.

BROWN, S.G., A note on migration in fin whales. *Norsk
1962   Hvalfangsttid.*, 51(1):13-6.

–, The movement of fin and blue whales within the
1962a   Antarctic zone. *Discovery Rep.*, (33):1-54.

–, A note on the migrations and movements of fin
1970   whales in the Southern Hemisphere as re-
         vealed by whale mark recoveries. Paper pres-
         ented to the IWC Special Meeting on Ant-
         arctic fin whale Stock Assessment, Honolulu,
         Hawaii, 13-25th March 1970, Doc. F/1
         (mimeo).

–, Whale marking – progress report, 1971. *Rep. IWC*,
1972   (22):37-40.

CHAPMAN, D.G., Re-analysis of Antarctic fin whale pop-
1970   ulation data. *Rep. IWC*, (20):54-9.

–, Review of analysis of southern baleen whale stocks.
1970a   Paper presented to the IWC Special Meeting
         on Antarctic fin whale Stock Assessment, Ho-
         nolulu, Hawaii, 13-25th March 1970, Doc. F/4
         (mimeo).

–, Estimates of stocks (original, current, MSY level and
1976   MSY) as revised at Scientific Committee
         Meeting, June 1975. *Rep. Pap. Sci. Comm.
         IWC*, (1975):44-7.

CHAPMAN, D.G., K.R. ALLEN and S.J. HOLT, Reports of
1964   the Committee of Three Scientists on the spe-
         cial scientific investigation of the Antarctic
         whale stocks. *Rep. IWC*, (14):32-106.

DELURY, D.B., On the estimation of biological popula-
1947   tions. *Biometrics*, 3:145-67.

Doi, T. *et al.*, Advanced assessment of the fin whale
1970   stock in the Antarctic. *Rep. IWC*, (20):60-87.

Fox, W. *et al.*, Report of Sub-committee on sub-division
1974   of Antarctic baleen whale quota. *Rep. IWC*,
       (24):69-71.

Fujino, K., Fin whale sub-populations in the Antarctic
1964   whaling Areas II, III and IV. *Sci. Rep. Whales
       Res. Inst., Tokyo*, (18):1-27.

Gulland, J.A., Distribution and abundance of whales
1974   in relation to basic productivity. *In* The whale
       problem: a status report, edited by W.E.
       Schevill. Cambridge, Mass., Harvard Uni-
       versity Press, pp. 28-52.

Gulland, J.A. and G.L. Kesteven, The effect of wea-
1964   ther on catches of whales. *Rep. IWC*,
       (14):87-91.

Holt, S.J. and J.A. Gulland, Measures of abundance
1964   of Antarctic whale stocks. *Rapp. P.-V. Réun.
       CIEM*, 155(27):147-51.

Hylen, A., Catcher efficiency. *Rep. IWC*, (14):92.
1964

Ivashin, M.V., O local'nosti nekotorykh promyslovykh
1969   vidov kitov v iuzhnom polusharii. *Rybň Khoz.*,
       45(10):11-3.

—, Marking of whales in the Southern Hemisphere
1973   (Soviet materials). *Rep. IWC*, (23):174-91.

IWC, Report of IWC/FAO Joint Working Party on
1967   whale stock assessment held from 26th Ja-
       nuary to 2nd February 1966, in Seattle. *Rep.
       IWC*, (17):27-47.

—, Report of the special meeting on Antarctic fin whale
1971   stock assessment. *Rep. IWC*, (25):34-9.

—, Report of the Scientific Committee. *Rep. IWC*,
1975   (25):62-77.

Jones, R., Population assessments of Antarctic fin and
1973   sei whales. *Rep. IWC*, (23):215-59.

Laws, R.M., Problems of whale conservation. *Trans. N.
1960   AM. Wildl. Conf.*, 304-19.

Lockyer, C., A new method of estimating age at sexual
1970   maturity in southern fin whales. Paper pre-
       sented to the Scientific Committee of the
       International Whaling Commission, London,
       June 1970, (mimeo).

—, The age at sexual maturity of the southern fin whale
1972   (*Balaenoptera physalus*) using annual layer
       counts in the ear plug. *J. Cons. CIEM*,
       34(2):276-94.

Lund, J., Charting of whale stocks in the Antarctic on
1950   the basis of iodine values. *Norsk Hvalfang-
       sttid.*, 39(2):53-60.

—, Charting of whale stocks in the Antarctic in the
1950a   season 1949/50 on the basis of iodine values.
       *Norsk   Hvalfang-
       sttid.*, 39(7):298-305.

—, Charting of whale stocks in the Antarctic 1950/51 on
1951   the basis of iodine values. *Norsk Hvalfangst-
       tid.*, 40(8):384-6.

Mackintosh, N.A. and S.G. Brown, Preliminary esti-
1956   mates of the southern populations of the lar-
       ger baleen whales. *Norsk Hvalfangsttid.*,
       45(9):469-80.

Ohsumi, S., Examination of the recruitment rate of the
1972   Antarctic fin whale stock by use of mathema-
       tic models. *Rep. IWC*, (22):69-90.

Ohsumi, S., Revised estimates of recruitment rate in the
1973   Antarctic fin whales. *Rep. IWC*, (23):192-9.

Ohsumi, S. and Y. Masaki, Status of whale stocks in the
1974   Antarctic, 1972/73. Rep. IWC, (24):102-13.

Ohsumi, S. and Y. Shimadzu, Comparison of growth of
1970   fin whales among various areas of the Ant-
       arctic Ocean. Contributed paper to the IWC
       Special Meeting on Antarctic fin whale Stock
       Assessment, Honolulu, Hawaii, 13-25th
       March 1970. Doc. F/7 (mimeo).

Roe, H.S.J., The rate of lamina formation in the ear
-1967   plug of the fin whale. *Norsk Hvalfangsttid.*,
       56(2):41-3.

—, Seasonal formation of laminae in the ear plug of the
1967a   fin whale. *Discovery Rep.*, 35:1-30.

Saber, G.A.F., The multisample single recapture cen-
1962   sus. *Biometrika*, 49:339-50.

Shimadzu, Y., Problems and some consideration on re-
1970   production relationship of fin whales in the
       Antarctic. Paper presented to the IWC Special
       Meeting on Antarctic fin whale Stock Assess-
       ment, Honolulu, Hawaii, 13-25th March 1970,
       Doc. F/8 (mimeo).

A REVIEW OF POPULATION
ASSESSMENTS OF SOUTHERN MINKE WHALES

R. GAMBELL

Abstract

The Antarctic fishery for minke whales, *Balaenoptera acutorostrata,* is a newly developed enterprise providing little information on which firm population assessments can be
based. Catches in the Antarctic totalled no more than 2 000 animals up to 1970/71, while in
the 4 following seasons catches were 3 054, 5 754, 7 713 and 7 000, respectively.

Size estimates of the Antarctic population of this species made in the early seventies and
associated population models and figured maximum sustainable yields (MSY) are reviewed,
with assessments ranging from an initial population of 299 000 with an MSY of 12 230 whales
at a population size of 158 500 (53 percent of initial level), to an initial mature population of
about 204 000 with an MSY of 5 000 at a population size of 122 400. Earlier, much smaller
estimates are also given; however, none of the assessments made up to 1973, based mainly on
sightings data and employing a very limited range of relevant biological information, can be
verified. Another assessment, made in 1975 under the new management policy of the International Whaling Commission, estimated the size of the minke whale population in the
Southern Hemisphere in 1971 (initial) and 1975 at 139 700 and 122 000, respectively. These
estimates may be too low, being based primarily on a Japanese analysis of the number of
whales in Antarctic Area IV which takes inadequate account of factors affecting catching
effort. It is also possible that southern minke whales were increasing in number when
exploitation of them began, because of the effects of earlier reductions of larger balaenopterids.

Résumé

La chasse au petit rorqual, *Balaenoptera acutorostrata,* est dans l'Antarctique une
activité récente qui n'a apporté que peu d'informations sur lesquelles on puisse fonder des
évaluations fiables de la population. Les captures réalisées dans l'Antarctique n'ont pas
pas représenté au total plus de 2 000 animaux jusqu'en 1970/71, alors qu'au cours des quatre
campagnes suivantes, les captures ont été respectivement de 3 054, 5 754, 7 713 et 7 000.

L'auteur passe en revue les estimations de la population antarctique de cette espèce
effectuées au début des années 70, ainsi que les modèles de population correspondants et les
chiffres du rendement maximal soutenu (MSY). Les estimations vont d'une population
initiale de 299 000 animaux avec un MSY de 12 230 pour une taille de population de 158 500
(53 pour cent du niveau initial) à une population initiale de quelque 204 000 animaux matures
avec un MSY de 5 000 pour une taille de population de 122 400 animaux. L'auteur présente
aussi des estimations antérieures, beaucoup plus faibles. Cependant, aucune des estimations
faites jusqu'en 1973, fondées surtout sur des repérages à vue et employant une gamme très
limitée d'informations biologiques pertinentes, ne peut être vérifiée. Une autre estimation,

effectuée en 1975 dans le cadre de la nouvelle politique d'aménagement de l'IWC, considère que la taille de la population de petits rorquals de l'hémisphère austral en 1971 (initiale) et en 1975 était respectivement de 139 700 et 122 000. Ces estimations peuvent être trop faibles, car elles se fondent surtout sur une analyse japonaise du nombre de baleines dans la zone antarctique IV, qui ne tient pas suffisamment compte des facteurs affectant l'effort de capture. Il est aussi possible que le petit rorqual austral ait été en augmentation quand son exploitation a débuté, en raison de l'effet des réductions antérieures des grands baleinoptères.

Extracto

La pesquería de ballena enana (*Balaenoptera acutorostrata*) del Antártico es muy reciente, y se dispone de poca información que pueda servir de base para evaluaciones bien fundadas de la población. Hasta 1970/71 las capturas en el Antártico no ascendían a más de 2 000 animales, mientras en las cuatro campañas siguientes se capturaron, respectivamente, 3 054, 5 754, 7 713 y 7 000 ballenas.

Se examinan las estimaciones del volumen de la población antártica de esta especie hechas a principios de los años setenta y los modelos correspondientes de población, así como las cifras de rendimiento máximo sostenible (RMS). Las estimaciones van de una población inicial de 299 000 ballenas, con un RMS de 12 230 animales y un volumen de población de 158 500 (53 por ciento del nivel inicial), hasta una población madura inicial de unos 204 000 cetáceos, con un RMS de 5 000 animales y un volumen de población de 122 400. Se dan también algunas estimaciones hechas anteriormente, todas ellas mucho menores. Pero ninguna de las evaluaciones hechas hasta 1973, basadas principalmente en datos de avistamientos y en información biológica muy limitada, pueden verificarse. Otra evaluación, hecha en 1975 en virtud de la nueva política de ordenación de la Comisión Ballenera Internacional, ha estimado el volumen de la población de ballena enana del Hemisferio Sur en 1971 (inicial) y 1975 en 139 700 y 122 000 cetáceos, respectivamente. Quizás estas estimaciones sean demasiado bajas, ya que se basan ante todo en un análisis japonés del número de ballenas presentes en el Area IV del Antártico, en el que no se tienen suficientemente en cuenta los factores que influyen en el esfuerzo de captura. Es también posible que la ballena enana austral estuviera aumentando en número cuando comenzó su explotación, como consecuencia de la reducción de las poblaciones de balenoptéridos de mayor talla.

R. Gambell
International Whaling Commission, The Red House, Station Road, Histon, Cambridge CB4 4ND, England

Introduction

Catches of minke whales in the Antarctic totalled no more than 2 000 animals in the seasons up to 1970/71. In the 4 following seasons the catches were of 3 054, 5 745, 7 713 and 7 000 minke whales respectively (Committee for Whaling Statistics, 1961-75).

The Antarctic minke whale fishery is thus such a newly developed enterprise that very little information has been gathered to date on which firm population assessments can be based. Whale sighting is the main method by which the population size has been estimated, and the maximum sustainable yield (MSY) has been calculated from a population model employing a very limited range of data on the relevant biological parameters.

Estimates of population size

Preliminary estimate

The first attempt to estimate the minke whale population size from sightings by the Japanese expedition scouting boats in the Antarctic was made by Ohsumi, Masaki and Kawamura (1970). They used the formula:

$$N = \frac{A \cdot n}{p \cdot S \cdot L}$$

where

N = population number
A = size of area
n = number of whales seen
p = sighting rate
S = search area
L = scouting distance

The search area was taken as 11.1 miles and the sighting rate 0.344 at a finding efficiency of 0.7. These values were derived from theoretical considerations by Nasu and Shimadzu (1970), based on sighting data specially collected in the 1966/67 season by research vessels and scouting boats.

The coverage of the Antarctic zone was very limited, but estimates were made using the formula given for the sectors (10° latitudinal zones within areas) shown in Table 1.

From these admittedly scanty data it was suggested that the population size in Area IV was about 23 000, that the area covered contained some 41 000 minke whales, and that the total Antarctic population might number about 70 000.

Revised estimates, 1971 and 1972

A new theoretical model for converting sightings into population estimates was developed by Doi (1974). The main difference was in calculating the average sighting rate (p), to take account of the duration of dive and the visual angle of observation. From new data collected in the 1969/70 season, $p = 0.112$ was derived for minke whales.

Table 1. Estimated minke whale population sizes from sightings in the Antarctic, 1965/66-1968/69 ($\times 10^{-3}$)

Zone/Area	II	III	IV	V	VI	I
40°-50°S	0.27	0.93	0.81	0.22	0.04	?
50°-60°S	?	5.37	11.41	3.40	0.59	?
60°-70°S	?	?	10.94	6.61	0.75	?
70°-80°S	?	—	—	?	?	?

Ohsumi and Masaki (MS) used this new value for $\bar{p}$, and an estimate of the sighting range $X_1 = 4.9$ miles calculated by Doi (1974) in the equation:

$$N = \frac{n \cdot A}{2\,p \cdot X_1 \cdot L}$$

With this formula, the abundance of minke whales was calculated by season and sector. The sighting data did not cover the whole region south of 30°S, in particular Area I (120°W-60°W) was not sampled, nor three other sectors. There were also large seasonal variations in the same sectors. Averaging all the results obtained from 1966/67 to 1970/71 gave an estimated population in the sectors sampled of 127 000 minke whales. Extrapolation to the unsampled sectors on the basis of average densities by zones gave the results shown in Table 2.

The extrapolated total was 217 000 minke whales, but the population suggested for Area I looks rather large by comparison with the relative abundance in this area of the larger baleen whales, which were less numerous here than elsewhere in the Antarctic.

The following year the actual sightings estimate was revised slightly with the addition of the 1971/72 observations. The new population figure for Area II-VI was 150 000, mainly due to the addition of a value for sector III B (Masaki, 1973).

Area III estimates

The Area III population in this revised estimate was 31 200 minke whales. Another estimate for this area was provided by Best (1973, 1974). He compared the minke and fin whale densities off Durban derived from catches and aircraft sightings per unit of effort from 1968 to 1971. The Area III fin whale population was taken from Chapman (1971), and the minke whale estimates obtained are shown in Table 3.

Best (1973) also used sightings data from a catcher in August and September 1971, the peak of minke whale abundance off Durban, to estimate their numbers. Using the Doi (1974) formula, this gave a figure of 58 400 minke whales for sector III E, which may represent most of the mature animals in Area III at this time of the year.

The mean of all these estimates is around 60 000, which suggests that the Japanese figure for Area III and perhaps the rest of the Antarctic, may be too low. However, the Table 3 Durban estimate depends on an equal intensity of catching and sighting effort being applied for fin and minke whales, and the minke/fin ratio has increased considerably in the four seasons. It also supposes that the total abundance of the whales in Area III is properly reflected in the coastal fishery off Durban. Both these factors add a note of caution to the conclusion suggested.

Table 2. Sightings estimates (and extrapolations) of minke whales, 1966/67-1970/71 ($\times 10^{-3}$)

Zone/Area	I	II	III	IV	V	VI	Total
E (30°-40°S)	(2.3)	2.0	3.7	2.8	1.3	(1.9)	9.8 (14.0)
D (40°-50°S)	(1.9)	1.0	3.3	2.2	3.4	0.2	10.1 (12.0)
A (50°-60°S)	(9.3)	3.8	11.8	26.0	4.2	2.1	47.9 (57.2)
B (60°-70°S)	(28.8)	(17.7)	(27.8)	34.3	18.6	6.6	59.5 (133.8)
Total	— (42.3)	6.8 (24.5)	18.8 (46.6)	65.3 65.3	27.5 27.5	8.9 (10.8)	127.3 (217.0)

Table 3. Area III minke whale estimates from comparison with fin whale abundance off Durban

		Minke/fin ratio	Fin pop. Area III	Minke pop. Area III
1968	catch	1.6	29 400	47 040
	sightings	1.3		38 200
1969	catch	0.7	29 600	20 720
	sightings	1.1		32 560
1970	catch	3.6	29 000	104 400
	sightings	3.1		89 900
1971	catch	2.6	29 000	74 500
	sightings	3.2		92 800
	Average			62 500

Estimates in 1973

As a result of the sightings obtained in the 1972/73 season, which covered the highest latitudinal zone B, 60°-70°S more extensively than before, the Japanese estimates of minke whale abundance were dramatically increased (Ohsumi and Masaki, 1974), as shown in Table 4.

The extrapolation of the very high values in zone B to the unsampled sectors of Areas I and II increased the estimated population size from a figure for the surveyed sectors of about 220 000 to nearly 300 000 for the whole region south of 20°S. Considering the very recent catching history of this species, these figures could be considered as representing the initial, unexploited population size of the southern minke whales.

Population model and yield

The only estimate of the MSY of the southern minke whales is that made using the model proposed by Ohsumi and Masaki (MS), based on the fin whale model developed by Ohsumi (1972) and using the minke whale data reported by Ohsumi, Masaki and Kawamura (1970).

This model (Gambell, 1976) assumed that the observed values of age at recruitment (t_c) and natural mortality (M) will decline linearly as the population is reduced. The value of M was varied from 0.10 in the initial stock to an assumed figure of 0.07 at low population levels. The recruitment rate (r) is the same as natural mortality in the initial population, and was calculated with a pregnancy rate increasing from 0.70 to 0.95 as the population decreased to 30 % and then falling again at still lower population levels, using the equation:

$$r = \frac{pe^{-t_cM}}{2}$$

Table 4. Sightings estimates (and extrapolations) of Antarctic minke whales, 1966/67-1972/73 ($\times 10^{-3}$)

Zone/Area	I	II	III	IV	V	VI	Total
E	(1.3)	1.8	2.7	2.1	0.7	0	7.3 (8.6)
D	0	1.0	4.0	2.5	2.6	0	10.1
A	0	3.9	11.4	23.2	1.7	2.0	42.2
B	(48.6)	(29.9)	85.3	43.4	20.2	11.0	159.9 (238.4)
Total	0 (49.9)	6.7 (36.6)	103.4	71.2	25.2	13.0	219.5 (299.3)

The age at recruitment (t_c) was varied between 7 and 4 years with declining population size, although this is a particularly difficult value to set because there is no minimum size limit for the species.

With these parameters, the natural mortality in the immature whales was calculated as declining from 0.172 to 0.140 with decreasing population size.

The model with these parameters indicates that the population level providing the MSY is about 53 % of the initial population level, and the MSY is about 4.1 % of the initial population size.

Applying these proportions to an initial population of 299 000 minke whales gives an MSY of 12 230 whales at a population size of 158 500.

Chapman (MS) thought that this estimate of MSY was much too large. He argued that the Area I population extrapolated in Table 4 is unrealistic since this area has been very unproductive for baleen whales. He suggested, on the assumption that 90 % of the zone B population and 50 % in the other zones were sexually mature, that the mature population might be about 204 000 animals.

No whale species has been shown to have a net recruitment greater than 4 % at MSY levels, and Chapman suggested that the MSY is more likely to lie around 60 % of the initial population size. He therefore calculated the MSY as:

$$204\,000 \times 0.060 \times 0.04 \simeq 5\,000 \text{ minke whales.}$$

Assessments in 1975

Unfortunately, our knowledge of minke whale biology and likely changes in the basic parameters are so sketchy that neither of the two rather arbitrary assessments above can be verified at present.

Besides the problems of knowing the true calving rate (a calf every one or two years?), there have been no observable changes in the key parameters. It is not even certain that they will change in response to a reduction in the population size; the changes may already have occurred, as in the sei whale, before exploitation (Gambell, 1975).

The initial population size itself is still in doubt, and has only been assessed by the one method of sightings analysis. Whilst it is possible that the present population is at the initial level, it could also be greater, through occupation of the niche previously filled by a now reduced larger baleen whale species.

However, the adoption by the IWC of its new management policy in 1975 required that the southern minke whale stocks should be assessed and classified as best they could in order to properly regulate their further harvesting. To this end, the Scientific Committee set up a small sub-committee to examine the latest evidence (Best et al., 1976).

Area IV Analysis

This group agreed that the best estimate of exploitable stock size was that for Area IV, which had been derived from corrected catch per unit effort data by Ohsumi (1976). This was a preliminary attempt to take into account some factors which make the use of crude catch per catcher days work (CDW) unreliable as estimates of whale density. Using data from a Japanese expedition which caught only minke whales, Ohsumi suggested that catcher hours work (CHW) was a more realistic measure of effort. The daily catching period varied according to the freezing capacity of the expedition and other operational controls resulting from the optimisation of the product yield, so this seems a reasonable assumption.

Catching efficiency, especially of minke whales, is very dependent on the weather conditions. Ohsumi calculated a correction for each wind force based on the catch per CHW. He then used the sum of the corrections times the number of days with a given wind force as a proportion of the number of days in each season for the wind correction coefficient in each season.

Because the abundance of minke whales varies during the whaling season, reaching a maximum in early January, the timing of the catching season is also an important factor. Ohsumi calculated the catch per CHW for 10-day periods in the 1971/72 and 1972/73 seasons, and used the relative abundance indices to determine a correction coefficient for each whaling season, as in the wind correction above.

A number of other important modifying factors could well be added to this analysis. They include the effect of visibility as well as wind in the weather correction; the year-to-year variability of the migration and abundance patterns in the seasonal timing correction; the latitude of the catching operations depending on the northward extent of the pack-ice, with the associated minke whale density variations; and the distribution of catching hours during the whaling season, associated with the seasonal density changes of the whales.

Whilst recognising these deficiencies, and also the inadequacy of the corrections calculated on the basis of only four seasons' catches, Ohsumi estimated the population sizes in Area IV using these corrections for comparison with the results employing uncorrected catch per CDW effort data. The initial population size was calculated using Chapman's (1974) modification of the DeLury (1947) method. This takes into account the

known catches and an assumed natural mortality coefficient (M) of 0.125 from Ohsumi and Masaki (1975). The latter figure is an average value for both sexes based on the age composition of the catches in 1971/72 and 1972/73, as determined by earplug age readings of males and females, and ovarian corpora counts in the females. The later population sizes were calculated in two ways as:

1) $N_i = N_0 \cdot I_i/I_0$

where

N_i = population size in year i
N_0 = population size in initial year
I_i = corrected cpue* in year i
I_0 = corrected cpue* in initial year

2) $N_i + 1 = (N_i + C_i)e^{-M} + N_0 M$

where

C_i = catch in year i.

The coefficients and resulting population estimates are shown in Table 5.

Extension to the whole Antarctic

The average sightings indices of minke whales in the six Antarctic areas from 1965/66 to 1974/75 (Masaki, 1976) were taken by the sub-committee (Best et al., 1976) and normalized relative to the index for Area IV. It was

Table 5. Minke whale catches, effort, correction coefficients and population estimates in Area IV

Season	Catch		Effort		Corr. coeff.		Pop. (× 10⁻³)	
	Total	Japan	CDW	CHW	Wind	Season	(1)	(2)
1971/72	2 659	2 659	285	2 815	0.705	0.973	23.7	23.7
1972/73	4 559	2 091	192	2 216	0.674	1.025	23.5	21.5
1973/74	4 569	2 043	282	2 948	0.696	0.933	18.4	17.9
1974/75	2 231	841	225	2 100	0.548	0.910	13.8	14.8

Table 6. Sightings indices and stock size estimates ($\times 10^{-3}$) for Southern Hemisphere minke whales

Area	Relative index	Exploitable stock	Initial stock (1971)	Current stock (1975)	MSY level (60%)	Ratio 1975: MSY level
I	0.86	20.4	20.4	18.1	12.2	1.48
II	1.42	33.7	33.7	31.5	20.2	1.56
III	1.65	39.1	39.1	36.1	23.5	1.54
IV	1.00	23.7	23.7	14.0	14.2	0.99
V	0.32	7.6	13.5	13.0	8.1	1.60
VI	0.22	5.2	9.3	9.3	5.6	1.66
Total	—	—	139.7	122.0	—	—

assumed that these average figures applied to the mid point of the period, which is the start of 1971. Estimates of the exploitable stock size in the other areas were then obtained by multiplying the normalized index for each by the estimate of 23.7 x 10³ for Area IV (from Ohsumi, 1976, see above). The results are shown in Table 6, but it appears that this procedure underestimates the stocks in Areas V and VI.

Estimates of the stock sizes in Areas I and V were available from Soviet sighting data (Pokrovsky and Emelyanov, 1976), of 96.5 and 64.0 x 10³ respectively. Unfortunately, neither the data nor the precise method of analysis used in this work were specified, although it seems likely that a modified form of the Doi (1974) model was used. These high figures were therefore taken as indices of abundance, and their ratio applied to the previous estimate for Area I, giving an adjusted estimate for Area V. A new estimate for Area VI was in turn calculated by multiplying the adjusted estimate for Area V by the relative indices for Areas V and VI. The final figures were then taken as the estimates of the initial stock size, and are also shown in Table 6.

Lastly, the 1975 stock sizes were forward calculated using equation (2) above, with M = 0.125. Brazilian catches were included with those from Area II, and Natal catches with those from Area III, so the stock figures shown in Table 6 represent estimates for the whole Southern Hemisphere. The MSY level was assumed to be 60% of the exploitable stock, in the absence of firm information.

This analysis depends very heavily on the accuracy of the original assessment of the stock in Area IV. It may well need to be revised in the light of new effort corrections which take fuller account of the operating factors affecting the catching of minke whales. However, the trend of such revisions is likely to be to increase the estimates of stock size, so that the present results may be underestimates of the true number of whales. Another important factor which may modify these results is the probability that the southern minke whale stocks were not in equilibrium at the start of exploitation but were increasing as a result of the reduction of the blue and fin whale stocks, a situation comparable to that of the southern sei whales (Gambell, 1976a).

References

BEST, P.B., Status of whale stocks off South Africa, 1971.
1973 *Rep. IWC*, (23):115-26.

—, Status of the whale populations off the west coast of
1974 South Africa, and current research. *In* The

whale problem: a status report, edited by W.E. Schevill. Harvard University Press, pp. 53-81.

BEST, P.B. *et al.*, Report of the Sub-Committee to esti-
1976   mate stock sizes of Southern Hemisphere minke whales. *Rep. Pap. Sci. Comm. IWC*, (1975):52-3.

CHAPMAN, D.G., Analysis of 1969/70 catch and effort
1971   data for Antarctic baleen whale stocks. *Rep. IWC*, (21):67-75.

–, Maximum sustainable yield of minke whales in the
1973   Antarctic. Contributed paper to the meeting of the Scientific Committee of the International Whaling Commission, London, June 1973, MS.

–, Estimation of population size and sustainable yield
1974   of sei whales in the Antarctic. *Rep. IWC*, (24):82-90.

Committee for Whaling Statistics, *Int. Whaling Stat*, 1961-75 (1961-74).

DeLURY, D.B., On the estimation of biological popula-
1947   tions. *Biometrics*, 3:145-67.

DOI, T., Further development of whale sighting theory. *In* The whale problem: a status report, edited by W.E. Schevill. Harvard University Press, pp. 359-68.

GAMBELL, R., Variations in reproduction parameters
1975   associated with whale stock sizes. *Rep. IWC*, (25):182-9.

–, A review of population assessment of Antarctic fin
1976   whales. Paper presented to the Scientific Consultation on the Conservation and Management of Marine Mammals and their Environment, Bergen, Norway, 31 August - 9 September 1976. Rome, FAO, ACMRR/MM/SC/9, 23 p.

–, Population biology and the management of whales.
1976a   *Appl. Biol.*, 1:247-343.

MASAKI, Y., Estimation of abundance of whales by
1973   means of whale sightings in the Antarctic. *Rep. IWC*, (23):155-63.

–, Japanese pelagic whaling and sighting in the An-
1976   tarctic, 1974/75. *Rep. Pap. Sci. Comm. IWC*, (1975):333-49.

NASU, K. and Y. SHIMADZU, A method of estimating
1970   whale population by sighting observation. *Rep. IWC*, (20):114-29.

OHSUMI, S., Examination of the recruitment rate of the
1972   Antarctic fin whale stock by use of mathematical models. *Rep. IWC*, (22):69-90.

–, An attempt to standardize fishing efforts as applied
1976   to the stock assessment of the minke whale in the Antarctic Area IV. *Rep. Pap. Sci. Comm. IWC*, (1975):404-8.

OHSUMI, S. and Y. MASAKI, Revised estimates of popu-
1971   lation size and MSY of the Antarctic minke whale. Contributed paper to the meeting of the Scientific Committee of the International Whaling Commission, Washington, D.C., June 1971, MS.

–, Status of whale stocks in the Antarctic, *Rep. IWC*,
1974   (24):102-13.

–, Biological parameters of the Antarctic minke whale
1975   at the virginal population level. *J. Fish. Res. Board Can.*, 32(7):995-1004.

OHSUMI, S., Y. MASAKI and A. KAWAMURA, Stock of the
1970   Antarctic minke whale. *Sci. Rep. Whales Res. Inst., Tokyo*, 22:75-126.

POKROVSKY, B.I. and S.V. EMELYANOV, Minke whale
1975   population estimation based upon observation data. Paper presented to the 27th Meeting of the International Whaling Commission, London, June, 1975, Doc. No. SC/27/Doc. 20.

INVESTIGATIONS INTO THE POPULATION MORPHOLOGY OF SPERM WHALES, *PHYSETER MACROCEPHALUS* L. 1758, OF THE PACIFIC OCEAN

A.A. BERZIN and G.M. VEINGER

Abstract

Sperm whales show marked variation in several morphological characteristics, such as the form of bifurcation of the tail lobes, the structure of the first finger of the hand, a light spot on the umbilicus, the number and insertion of the sternal ribs, the shape of the spleen and the liver, etc. By investigating pregnant females caught by the whaling fleet, it was observed that for various of these features there was a rather high degree of occurrence of the same characteristics in female and foetus. This suggested a genetic basis for these characteristics and strengthened the hypothesis that these characteristics can be used to distinguish separate sperm whale populations.

A large number of whales from the western, the central and the eastern North Pacific, and from the southwestern Pacific was investigated for the occurrence of a number of characteristics, and a table is given of the percentage occurrence of the different types of form of the caudal fin, the shape of the percentage occurrence of the differences in the observed frequencies indicate the existence of separate populations, and that the method used can be an important tool for distinguishing between separate populations of the same whale species. More research is needed, and further data should be collected not only on sperm whales, but also on baleen whales. This requires international cooperation.

Résumé

On constate chez les cachalots des variations marquées de plusieurs caractères morphologiques, tels que la bifurcation des lobes caudaux, la structure du premier doigt de la main, une légère tache sur l'ombilic, le nombre et l'insertion des vraies côtes, la forme de la rate et du foie, etc. Lors de travaux portant sur les femelles enceintes capturées par la flottille baleinière, on a observé que pour certains de ces traits l'apparition de caractères identiques se retrouvait à un degré élevé chez la femelle et le fœtus. Cela laissait supposer qu'ils avaient une origine génétique, tout en consolidant l'hypothèse selon laquelle lesdits caractères peuvent servir à distinguer diverses populations de cachalots.

On a étudié un grand nombre de cachalots provenant des secteurs occidental, central et oriental du Pacifique nord, ainsi que du Pacifique sud-ouest, pour déceler la fréquence d'un certain nombre de caractères, en indiquant dans un tableau le pourcentage de fréquence des diverses formes de la nageoire caudale et de la rate, du mode de pigmentation et d'autres traits morphologiques qui se retrouvent dans chacun des secteurs. On a conclu que les écarts de fréquence observés indiquent l'existence de populations distinctes et que la méthode utilisée pourrait constituer un critère important pour distinguer entre des populations séparées d'une même espèce de baleines. Il faudra poursuivre les recherches et recueillir de nouvelles données non seulement sur les cachalots, mais aussi sur les mysticètes. Cela suppose une coopération internationale.

Extracto

Los cachalotes presentan diferencias marcadas en algunas características morfológicas, como la forma de bifurcación de los lóbulos de la cola, la estructura del primer dedo de la mano, una mancha clara en el ombligo, el número e inserción de las costillas verdaderas, la forma del bazo y el hígado, etc. Mediante el examen de hembras preñadas capturadas por las flotas balleneras, se observó que algunos de esos rasgos presentaban, con una frecuencia bastante elevada, las mismas características en las hembras y los fetos. Esta constatación sugirió una base genética para esos caracteres y dio fuerza a la hipótesis de que esas características podrían utilizarse para diferenciar distintas poblaciones de cachalotes.

Al objeto de verificar la frecuencia de algunas de esas características se procedió a examinar gran número de cachalotes del Pacífico Norte (oeste, centro y este) y del Pacífico Sudoccidental. En un cuadro se muestra el porcentaje de frecuencias de las diferentes formas de aleta caudal, configuración del bazo, pigmentación y otros rasgos morfológicos en cada una de esas zonas marítimas. Se llega a la conclusión de que las diferencias en las frecuencias observadas indican la existencia de poblaciones distintas, y de que el método utilizado puede constituir un instrumento importante para diferenciar poblaciones distintas de la misma especie de cachalote. Hace falta más investigación, y deberán recogerse nuevos datos no solamente sobre los cachalotes sino también de las ballenas de barbas. Para ello se requiere cooperación internacional.

A.A. Berzin
Pacific Research Institute, Fisheries and Oceanography (TINRO), Shevchenko Al. 4, 690613 Vladivostok, USSR

G.M. Veinger
Pacific Research Institute, Fisheries and Oceanography (TINRO), Shevchenko Al. 4, 690613 Vladivostok, USSR

Introduction

Identification of the populations of exploited animals permits different types of theoretical research and the resolution of problems caused by man's activity in a strictly scientific way. It seems that the seventies will see the end of large-scale and irrational whaling. It is now very important to preserve the largest possible populations of whales, and to make every effort to prevent the disappearance of some populations.

To do this we need to know both population structure and numbers. In the last years of whaling, the catch quotas for killing should not be set by area or section[1] but according to the place of occurrence of discrete breeding populations. Much research is needed into population identity, if possible, necessary and practical ways of managing whaling on a strict scientific basis are to be found.

It is generally impracticable to carry out genetic population research for large, long-lived wild animals, and particularly for whales. All feasible methods have therefore to be used.

Lately it has been shown that one of the most reliable and exact methods of population investigation based on genetic (phenetic) principles is the study of qualificative and quantitative variability of morphological features (Yablokov, 1966; Shvarts, Smirnov and Dobrynsky, 1968; Timofeev-Ressovsky, Yablokov and Glotov, 1973, and others).

Use of morphological features

Although variability occurs between individuals of the same population, it may also be used to distinguish between populations (Simpson, 1948; Yablokov, 1966). All features and properties of animals are subject to variation (Shvarts, 1974). Data have been accumulated on morphological variations in whales since the end of the 19th century, but referred only to variation within the species and did not focus on identifying discrete populations. One may cite, for example, the discovery of different numbers of vertebrae in spine sections of different individuals and other meristic variations in whales' skeletons (Flower, 1867; Beddard, 1900), or the two forms of sternum structure which were found to occur in the white whale, *Delphinapterus leucas* (Smirnov, 1935). Some papers bring together similar data showing clearly the variability of features (Kleinenberg, 1956; Yablokov, 1964; Berzin, 1971). Using all collected data, Yablokov (1964) was able to extend Smirnov's work of 1935 and show that there were 13 types of sternum structure in the white whale. The wide variability in pigmentation is a good subject of study for population identification – see Yablokov (1966) and his analysis of its variation in dolphins (*Delphinus delphis*) and killer whales (*Orcinus orca*). Naturally, these data alone are not sufficient to say that qualificative characteristics of groups of cetaceans are different.

Yablokov (1966) summarized different kinds of variability and established theoretical bases and principles for selecting features. Investigations by Lounde (1951, in Ivashin, 1969) could be related to morphological studies of populations; using iodine content indices in whale oil he distinguished different areas inhabited by separate groups of whales[2] in Antarctic waters; similar investigations by Fujino (1963) showed the differences in the blood composition of sperm whales inhabiting Japanese coastal waters and waters off the Aleutians.

---

[1] Currently, quotas for species are set by geographical areas; it is not known, however, whether the whales occurring in each area all belong to one discrete population.

[2] While there is clearly a possibility of distinguishing between whale populations in this way, it is also possible that differences in this feature may be caused by abiotic factors which can overlap whatever different whale populations exist.

Of prime importance in population research was the proof that the polymorphic features of the white whale flipper depended on genetic factors (Bel'kovich and Yablokov, 1965). Various anatomical differences in the fingers of these animals determined the difference between northern and southern populations in the Sea of Okhotsk (Yablokov, 1966).

Large whale population divisions

Large whales were divided into sub-species between Northern and Southern Hemispheres, generally on the basis of differences in the size structure of the catches. This division was later confirmed by morphological features (Berzin, 1971; Kuzmin, 1972). Southern Hemisphere whale species were divided into populations by doubtful methods: differences in age at maturity, duration of pregnancy, lactation and even milk fat content[3]. Population investigations of morphological variability in sperm whale teeth dentine layers in different areas of the Southern Hemisphere were made by Klevezal and Tormosov (1971); Clarke, Aguayo and Paliza (1968), and Veinger (1974) did it on the basis of the frequency of distinguished colour types of these whales.

While determining the main problems in cetacean study, we have summarized the results of population study and have drawn the attention of scientists to the morphopopulation method of whale study; we also suggested for uniformity of terminology using the term "population" instead of "stock" when dealing with large independent structural groups of whales (Berzin, 1974).

To summarize our current knowledge of Pacific sperm whale populations: in the North Pacific 3 populations are suggested (western, central and eastern), and 2 others are suggested in the warm ocean areas — Fiji and the Galapagos Islands (Berzin, 1974). The most recent data allows us to distinguish one more population — the central South Pacific. There is no doubt that some 2 or 3 populations exist in the South Pacific.

Even for North Pacific sperm whales, however, the information is limited, partly contradictory and the area boundaries are tentative. Summer intermixing, rate of isolation and therefore the rate of exchange of genetic information are not at all clear. There is still less information on sperm whale populations in the South Pacific.

Analysis of whale population structure

Fig. 1 shows the sub-species and the principal theoretical inter- and intra-population relationships, illustrating the different rates of exchange of genetic units and genetic information for whales in general, and in particular for whales of the Pacific, including sperm whales.

Family

The smallest unit is the family, more clearly defined in the monogamous baleen whales, where it consists of 2 to 3 animals (male and female, or male, female and calf). In the polygamous sperm whales this is less clear — family groups consist of 4 to 10 animals or more; in addition there are more classes of sexually mature animals. In Fig. 1 all types of grouping have been reduced to family or male groups.

Populations

In the feeding areas these groups form larger aggregations in which the genetic relationships between individuals are close. These

---

[3] See footnote No. 1.

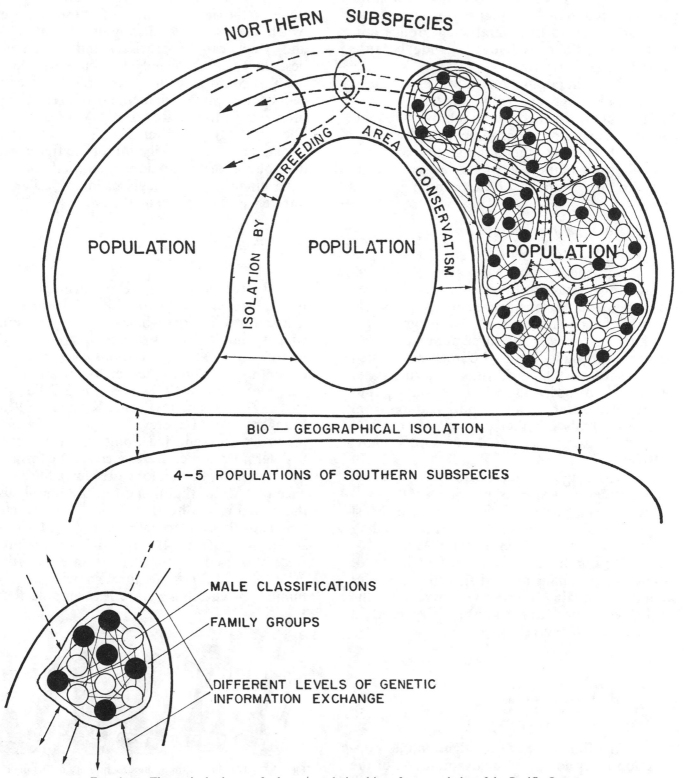

FIG. 1. – Theoretical scheme of subspecies relationships of sperm whales of the Pacific Ocean.

aggregations, however, are not permanent and only occupy a given biocenoses in one or other limited area in one or other season. A population is made up of several such groups and within it genetic exchange is wide because there are no barriers.

Between populations, however, isolation occurs, although no visible barriers exist between them. (In theory, the relative homogeneity of the marine environment, its absence of physical barriers, the adaptability and homeothermic nature of whales all create unlimited opportunities for intermingling.)

Isolation mechanisms

Sub-species of all whales living in different hemispheres are isolated by the half-year difference in seasons caused by geographical factors. This we have called a bio-geographical isolation. It causes the appearance of stable morpho-physiological differences, which should follow the '75 %' rule. This rule states that 75 % of individuals in the group (popution) must differ by one or more features from all individuals of other groups (Mair, 1974; Timofeev-Ressovsky, Yablokov and Glotov, 1973).

Breeding-area conservatism, seen in highly migratory species, is a cause of the existence of separate populations of whales. Most of the population returns to the same breeding grounds (Naumov, 1963; Rovnin, 1970).

Within a sub-species, many kinds of isolating mechanisms exist, which prevent complete mixing between members of the sub-species and cause morpho-physiological differences to evolve.

Method of work

The first stage of investigation into population morphology is the examination of the variability of features within a species, which Yablokov (1968) has described as "study of intra-population variability". The second stage is the identification of those features which may serve to distinguish populations and which can be precisely and objectively described. The selected characters should if possible conform with the criteria of the concept of the "pheno-elementary discrete" features (see Timofeev-Ressovsky, Yablokov and Glotov, 1973). The final stage is the collection of data and the investigation of the frequency distribution of the features identified and the determination of the level of interpopulation differences by various methods.

Type of feature used

In recent years much information has been collected on the variability of morphological features in large whales, especially sperm whales, and thus it was easy to determine both interior and exterior morphological features exhibiting variation (Clarke, Aguayo and Paliza, 1968; Berzin, 1971; Klevezal and Tormosov, 1971; Urmanov and Kuzmin, 1973; Veinger, 1974; Lagerev, 1975, etc.). Twenty features were identified. Some were clear, easy to determine and discrete, such as: the form of bifurcation on the tail lobe (Table 2), the presence of a white (light) spot on the umbilicus, the structure of the first finger (Fig. 2), etc. Others dealt with sternal ribs (Fig. 3), form of spleen, liver (Fig. 4), etc. Meristic features were also used — number of vertebrae ribs, etc. (Table 2). Those features showing a large number of variations were less suitable than

FIG. 2. — Polymorphism in the structure of the first digit of sperm whale hand.

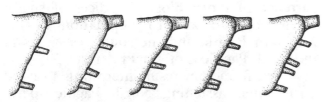

FIG. 3. – Number and pattern of connexion to sternum of sternal ribs of sperm whales.

those with less variables, both for the collection and the analysis of the information.

It is most important to check at all stages of research, particularly at the beginning, whether the features are genetically inherited (which can only be done through the female line. Between 24 and 50 mother-embryo pairs were examined and it was found that 9 features were reproduced from the female to the embryo in between 30 and 100 % of the observations (average 62 % – see Table 1).

Features of age, such as number of teeth, degree of grinding down in upper and lower jaws, and a number of features of the interior morphology of large whales were not included.

Collection of data

A special card used by factory ships facilitated the collection of data as whales were being processed. It was used in all regions but material from the southern North Pacific populations was processed first; it probably came from breeding areas, i.e., the centre of the areas, since samples from the periphery of the areas where superposition might have occurred were excluded. We have reason to suppose that our samples came from "pure" populations.

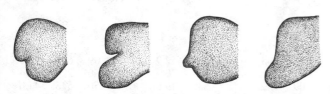

FIG. 4. – Types of sperm whale liver forms.

Table 1. The inheritance of morphological signs

Morphological signs	Pairs observed	Resemblance	
		Pairs	%
Form of caudal fin	49	27	55.1
Number of papillae on caudal fin	49	17	34.6
Head pigmentation:			
by simplified scheme	49	48	97.9
by complicated scheme	24	12	50.0
Body pigmentation:			
by simplified scheme	50	50	100.0
by complicated scheme	41	23	56.0
Dark spots on palate	47	27	57.4
Texture of the skin:			
by simplified scheme	50	50	100.0
in detailed analysis	42	25	59.2
Lateral split of head	38	22	57.9
Jaw wrinkles	50	15	30.0
Form of the spleen	6	3	50.0

Methodological defects revealed while processing the 1974 materials reduced the number of usable features to 6, but even that confirmed our suggestion that there are local populations of sperm whales in the North Pacific (Berzin, Isakov, Lagerev, 1975).

Analysis of data

In 1975 in the North Pacific data were collected from more than 2 200 sperm whales and analyses were made on more than 12 features. Altogether in the North Pacific population data were collected from 3 000 sperm whales. 1 200 sperm whales were studied in the South Pacific, for a lesser number of features, however.

Recognition of the occurrence of many identical features by year in certain regions was important from the point of view of methods.

The analysis of the frequency of occurrence of some features illustrates clearly the morphological differences between populations. In other features these difference do not show so clearly (Table 2). The greatest differences, as one would expect on theoretical grounds, are found between populations that are remotest from each other — western and eastern — in the following features: numbers of papillae on the caudal peduncle, pigmentation of the body, texture of the skin, shape of spleen and frequency of damage to the caudal fin. This latter (traumatic) feature, though not genetically determined, is reliable enough and is used as a criterion for belonging to a population (Timofeev-Ressovsky, Yablokov and Glotov, 1973; Mair, 1974). The occurrence of these differences shows the different conditions under which the populations live and their lack of mixing.

The great similarity in frequency of occurrence of morphological indices of sperm whales of the western population in the Northern Pacific and the southwest areas of the South Pacific is of great interest.

We found one result interesting: to avoid error when assigning North Pacific sperm whales occurring between 160°E and 170°E to one area or another where western and central populations meet, they were analysed separately (over 400 individuals). The frequency of occurrence of most features showed that whales of this area were identical to the western population. This, we believe, confirms the effectiveness of our method for determining which populations sperm whales belong to.

Problems of sample size

Table 2 shows that data on different populations are not identical. For instance, 1 092

Table 2. **Frequency of morphological features in different sperm whale populations**

Population	Damage of caudal fin		Form of caudal fin n				Types of form of spleen n				Body pigmentation (without spots)	
	n	%		%	%	%		%	%	%	n	%
Western N. Pacific	146	60.3	838	52.6	17.6	38.1	72	22.2	48.1	18.1	824	2.7
Central N. Pacific	205	70.2	259	43.2	10.8	45.9	113	27.4	53.1	13.3	246	4.2
Eastern N. Pacific	685	20.3	680	36.9	10.7	52.2	479	33.4	49.1	7.9	685	7.2
Southwestern Pacific	—	—	1 183	64.3	3.5	32.2	—	—	—	—	1 003	16.6

Population	Head pigmentation (without spots)		Lack of dark spots on palate		No. of papillae on caudal fin (average)		No. of jaw wrinkles (average)		Lateral splits on head		Smooth surface of body		Lumbar vertebrae (average)		No. of sternal ribs (average)	
	n	%	n	%	n	%	n	%	n	%	n	%	n	%	n	%
Western N. Pacific	843	6.4	704	8.4	838	3.00	846	2.42	143	44.1	788	2.9	657	9.09	693	3.86
Central N. Pacific	274	15.3	263	12.9	257	3.26	262	3.23	200	46.5	263	20.9	110	8.90	224	4.04
Eastern N. Pacific	687	12.8	649	22.9	683	4.90	676	3.09	642	42.1	689	19.0	409	8.50	546	4.00

sperm whales in the western population and 441 in the central population of the North Pacific were investigated. In order to check the results obtained from unequal samples, the larger sample was divided into smaller ones of approximately the same size as the smaller sample, and the frequency of the various feature types in the sub-samples was calculated. Comparison of the results has shown that the variations of the indices of these sub-groups from the mean indices of the total sample are insignificant, and are much less than the differences between the two samples.

Mathematical processing

There are two uses for mathematical processing of data: elaboration of a programme for the construction of a logical model for diagnosing sperm whale populations on electronic computers, and the determination of the populations to which sperm whales from two regions of whaling belong. The Diagnostic Laboratory of the Institute of Automatics and Processes of Control of the Far Eastern Scientific Centre, Academy of Sciences, USSR has started this work.

Fifteen morphological features were put into tables for mathematical processing. In working out the programme they were reduced to 9, other features being discarded because they occurred only in some sperm whales. Of 773 cards (thus 773 lines) in the analysis tables only those which had a full set of 9 signs were used. Using accepted methods of analysis of a complex of features, it was shown that 80 % of the sperm whales will be assigned to the correct population.

Future needs

More thorough population-genetic (phenetic) investigations, using a wider spectrum of features (phenes) according to different forms of variability, further accumulation of data using evaluation of the distribution of the variable characters and the modification and use of other methods will help to clarify the main population parameters.

It is necessary to extend investigations of the intra-species variability of features and to start accumulation of similar data on baleen whales.

During these present years of intensive whaling it is especially necessary to unite and concentrate all the efforts of biologists in different fields and countries on developing coordinated methods of choosing the same features for the collection of compatible data from different areas, and on the investigation of baleen whale and sperm whale population identity from many different areas. It is also necessary to use simple, uniform and clearly defined terminology, especially for sub-specific divisions.

Acknowledgements

We thank the many scientists of the Laboratory for Cetacean Research of TINRO and specialists from the scientific groups of the Far Eastern Whaling Flotilla, who participated in the elaboration of methods at different stages of this work and in the collection of data: A.A. Kuzmin, V.V. Melnikov, V.I. Privalikhin, G.I. Kucheruk and others; we are grateful to them all.

References

Barabash-Nikiforov, I.I., Fauna kitoobraznych Cernovo morja, jejo sostav i proischodjenie (Fauna of the Black Sea cetacea; its composition and origin). Voronezh, Izdatelstvo Voronezhskogo Gos. Universiteta, 85 p. (in Russian).
1940

BEDDARD, F.E., A book of whales. London, 320 p.
1900

BEL'KOVICH, V.M. and A.V. YABLOKOV, The structure of
1965   a stock of toothed whales. *In* Morskie mleko-
       pitayushchie. Moscow, Izdatelstvo Nauka, pp.
       265-9 (in Russian).

BERZIN, A.A., The sperm whale. (Kashalot). Edited by
1972   A.V. Yablokov. Jerusalem, Israel Program for
       Scientific Translations, IPST Cat. No. 60070
       7:367 p. Translated from the Russian, pu-
       blished in Moscow by Izdatel'stvo 'Pishche-
       veya Promyshlennost' (1971).

—, Actual problems of cetacea investigations. *In* Itogi
1974   nauki i tekhniki, Zoologia pozvonochnykh.
       Moscow, vol. 6:158-89 (in Russian).

BERZIN, A.A., and G.I. ISAKOV, S.I. LAGEREV, Data on
1975   sperm whale morphology in the Pacific
       Ocean. *In* Marine mammals. Kiev, Naukova
       Dumka, pp. 28-30 (in Russian).

CLARKE, R., A. AGUAYO and O. PALIZA, Sperm whales of
1968   the southeast Pacific. Part 1. Introduction.
       Part 2. Size range, external characters and
       teeth. *Hvalraad. Skr.*, (51):80 p.

FLOWER, W.N., On the osteology of the Cachalot
1867   or sperm whale. *Trans. Zool. Soc. Lond.*,
       6(2):309-78.

FUJINO, K., Identification of breeding sub-populations
1963   of sperm whales in waters adjacent to Japan
       and around the Aleutian Islands, by means of
       blood typing investigations. *Bull. Japan. Soc.
       Sci. Fish.*, 29(12):1057-63 (in Japanese).

IVASHIN, M.V., On locality of some commercial whales
1969   in the Southern Hemisphere. *Rybn. Khoz.,
       Mosk.*, (10):11-3 (in Russian).

KLEINENBERG, S.E., Mammals of the Black Sea and Sea
1956   of Azov. Moscow, Akademiia Nauka, 288 p.
       (in Russian).

KLEVEZAL, G.A. and D.D. TORMOSOV, On distinguishing
1971   of local classifications of sperm whales on
       character of layers in the tooth dentine. *Tr.
       Atl. Nauchno-Issled. Inst. Rybn. Khoz. Okea-
       nogr.*, 39:35-43 (in Russian).

KUZMIN, A.A., Correlation of osseous elements of the
1972   skeleton of the sperm whale body. *In* Tezisy
       dokladov i vsesoyuznogo soveshchaniya po
       izucheniyu mlekopitayushchie, Pt. 2, Mak-
       hachkala, pp. 119-21 (in Russian).

LAGEREV, S.I., Some data on spleen morphology of
1975   toothed and baleen whales. *Izv. Tikhookean.
       Nauchno-Issled. Inst. Rybn. Khoz. Okeanogr.*,
       96:253-7 (in Russian).

MAIR, E., Populations, species and evolution. Moscow,
1974   MIR, 447 p. (in Russian).

NAUMOV, N.P., The ecology of animals. Moscow, Vys-
1963   shaya shkola, 618 p. (in Russian)

ROVNIN, A.A., Whale distribution in the north and cen-
1970   tral Pacific. *Avtoref. Diss. Soisk. Uch. Stepeni
       Kand. Biol. Nauk. Vladivost.*, 1970:23 p. (in
       Russian).

SHVARTS, S.S., Inside-species variability and formation
1974   of species: evolution and genetic aspects of
       problems. *In* Pervy mezhdunarodny teriolo-
       gichesky kongress. *Ref. Dokl.*, 2:328-31 (in
       Russian).

SHVARTS, S.S., V.S. SMIRNOV and V.S. DOBRYNSKY, Me-
1968   thod of morphological indicators in the eco-
       logy of terrestrial invertebrates. *Tr. Inst. Ekol.
       Rast. Zhivotn., Sverdl.*, 58:387 (in Russian).

SIMPSON, J.I., Rates and forms of evolution. 358 p.
1948

SMIRNOV, N.A., (ed.), Animals of the Arctic seas. Le-
1935   ningrad, Glavsevmorputi (in Russian).

TIMOFEEV-RESSOVSKY, N.V., A.V. YABLOKOV and N.V.
1973   GLOTOV, Article on population study. Mos-
       cow, Nauka, 267 p. (in Russian).

URMANOV, M.I. and A.A. KUZMIN, On the liver mor-
1973   phology of large whales. *Izv. Tikhookean.
       Nauchno-Issled. Inst. Rybn. Khoz. Okeanogr.*,
       87:205-15 (in Russian).

VEINGER, G.M., Revealing of population structure of
1974   sperm whales on the base of pigmentation
       analysis. *Invest. Fish. Biol. Fish. Oceanogr.*,
       5:153-65 (in Russian).

YABLOKOV, A.V., Skeleton and skeleton musculature. *In*
1964   Belukha, pp. 73-103 (in Russian).

—, Izmenchivost mlekopitayushchie (Variability of
1966   mammals). Moscow, Akademiia Nauk, 364 p.
       (in Russian).

—, Population morphology of animals. *Zool. Zh.*,
1968   47(8):1749-65 (in Russian).

REVIEW OF BALAENOPTERIDS
IN THE NORTH ATLANTIC OCEAN

C.J. RØRVIK and Å. JONSGÅRD

Abstract

Present knowledge of the geographic ranges, stock identity, migration and abundance of balaenopterids in the North Atlantic Ocean and the methods used for determining them are reviewed. Most research to date has been centred upon the fin whale, *Balaenoptera physalus,* indicating that the population may be divided into 6 independent groups: (1) Arctic eastern North Atlantic and off north Norway, migrating northward and southward seasonally through the Denmark Strait; (2) off east Greenland and west Iceland; (3) off western Norway and the Faroes; (4) off the Scottish islands, migrating seasonally to the coasts of Spain and Portugal; (5) off west Greenland, and (6) Northwest Atlantic off Nova Scotia, Newfoundland and Labrador, perhaps composed of 2 or more stocks. Blue whales, *B. musculus,* on each side of the North Atlantic seem to be composed of several stocks with migration possibly taking place between the eastern and western sides in high northern latitudes. Population estimates by area are given for both species. Little information is available on the identity and size of stocks of sei whales, *B. borealis,* and minke whales, *B. acutorastrata,* in the North Atlantic, though a recent estimate for minke whale populations is given, along with a tentative division into stocks.

The exploitation of balaenopterids in the North Atlantic is presently limited to one Icelandic whaling station, 3 shore stations in Spain and annual catches of a few fin, sei and minke whales by aborigines in west Greenland (current to 1974). Pelagic whaling for baleen whales in the Atlantic and its dependent waters north of 40°S has been forbidden under international convention since 1937; blue whales have been completely protected under international convention since 1960.

Résumé

Les auteurs font le point des connaissances sur le domaine géographique, l'identité des stocks, les migrations et l'abondance des baleinoptères de l'Atlantique nord et décrivent les méthodes utilisées pour déterminer ces facteurs. Jusqu'ici, les recherches ont surtout porté sur le rorqual commun, *Balaenoptera physalus*. Elles ont fait apparaître que la population peut se diviser en six groupes indépendants: (1) secteur arctique de l'Atlantique nord-est et nord de la Norvège, avec migration saisonnière vers le nord et vers le sud à travers le détroit du Danemark; (2) eaux du Groenland oriental et de l'Islande occidentale; (3) eaux de la Norvège occidentale et des Féroés; (4) eaux des îles écossaises avec migrations saisonnières vers les côtes de l'Espagne et du Portugal; (5) eaux du Groenland occidental; (6) Atlantique nord-ouest, au large de la Nouvelle-Ecosse, de Terre-Neuve et du Labrador, avec peut-être deux ou plusieurs stocks. Les populations de baleines bleues, *B. musculus,* rencontrées de part et d'autre de l'Atlantique nord paraissent comprendre plusieurs stocks. Il y a peut-être des

269

migrations entre le secteur est et le secteur ouest aux hautes latitudes boréales. Les auteurs donnent des estimations de population ventilées par zone géographique pour chacune de ces deux espèces. On possède peu de données sur l'identité et la taille des stocks de rorquals de Rudolf, *B. borealis* et de petits rorquals *B. acutorostrata,* dans l'Atlantique nord, bien qu'une estimation récente des populations de petits rorquals soit indiquée, avec une répartition préliminaire en stocks.

L'exploitation des baleinoptères de l'Atlantique nord est actuellement limitée aux opérations d'une station baleinière islandaise et de trois stations terrestres espagnoles et aux pêches traditionnelles des indigènes du Groenland occidental qui ont capturé chaque année quelques *B. physalus, B. borealis* et *B. acutorostrata* (jusqu'en 1974). La pêche pélagique des mysticètes dans l'Atlantique et dans les eaux limitrophes au nord de 40°S est interdite par convention internationale depuis 1937. Les baleines bleues sont intégralement protégées par convention internationale depuis 1960.

Extracto

Se examinan los conocimientos actuales sobre la distribución geográfica, la identidad de las poblaciones, los movimientos migratorios y la abundancia de balenoptéridos en el norte del Atlántico y los métodos utilizados para obtener esos datos. La mayoría de las investigaciones realizadas hasta la fecha se han centrado en la ballena de aleta (*Balaenoptera physalus*) e indican que quizás la población esté dividida en seis grupos independientes: (1) nordeste del Atlántico ártico y norte de Noruega, que emigra hacia el norte y hacia el sur estacionalmente a través del Estrecho de Dinamarca; (2) este de Groenlandia y oeste de Islandia; (3) oeste de Noruega y Feroes; (4) Islas Escocesas, con migraciones estacionales hacia las costas de España y Portugal; (5) oeste de Groenlandia, y (6) noroeste del Atlántico, frente a Nueva Escocia, Terranova y Labrador, quizás compuesto de dos o más poblaciones. La ballena azul (*B. musculus*) que se encuentra a ambos lados del norte del Atlántico parece estar dividida en varias poblaciones, que es posible realicen movimientos migratorios de este a oeste y viceversa, a latitudes septentrionales muy elevadas. Se dan estimaciones de la población, por zonas, de ambas especies. Se dispone de poca información sobre la identidad y volumen de las poblaciones de ballena boba (*B. borealis*) y ballena enana (*B. acutorostrata*) del norte del Atlántico, aunque se da una estimación reciente de las poblaciones de ballena enana y una división provisional en poblaciones.

La explotación de los balenoptéridos en el norte del Atlántico se limita en la actualidad a una estación ballenera de Islandia, tres estaciones en tierra en España y algunas capturas de ballenas de aleta, bobas y enanas, por los aborígenes, en el oeste de Groenlandia (hasta 1974). La caza pelágica de mistacocetos en el Atlántico y en sus aguas adyacentes al norte de 40°S se ha prohibido desde 1937 por un convenio internacional. La ballena azul está totalmente protegida por un convenio internacional desde 1960.

C.J. Rørvik
Institute of Marine Research, Nordnesparken 2, 5011 Bergen-Nordnes, Norway

Å. Jonsgård
Institute of Marine Biology and Limnology, University of Oslo, Postboks 1064, Blindern, Oslo 3, Norway

Introduction

Modern whaling began in 1864 when Svend Føyn carried out his whaling experiments at a shore station in Finmark in north Norway.

By 1974, the only "great" whaling station [1] operating in the northern North Atlantic was Icelandic. The last Norwegian "great" whaling station was shut down after the 1971 season. In Canada, whaling for large rorquals has been illegal since December 1972. A few fin, *Balaenoptera physalus* (L.), sei *B. borealis* Lesson, humpbacks, *Megaptera novaeangliae*, and some minke whales, *B. acutorostrata* Lacépède, are caught annually off West Greenland by Eskimos. In addition, balaenopterid species are caught from 3 shore stations in Spain.

This study was undertaken in 1974 in order to review what is known about different stocks of balaenopterids in the North Atlantic, their distribution and size. Most of the research has focused on the fin whale — an important economic species. There is little information about different population of blue whales, *B. musculus* (L.) and even less on sei and minke whales.

Regulations

The International Whaling Conference held in London in 1937 agreed to forbid the use of a factory ship or a whale catcher attached thereto for the purpose of taking or treating baleen whales in the Atlantic Ocean and its dependent waters north of 40° South Latitude. It was forbidden to take or kill calves, suckling whales or female whales which were accompanied by calves or suckling whales. It was forbidden to take or kill blue whales below 70 ft and fin whales below 55 ft (Committee for International Whaling Statistics, 1947).

The International Whaling Conference held in Washington in 1946 reaffirmed these regulations and also set minimum lengths of 40 ft on sei whales. However, it was agreed that baleen whales not more than 5 ft below minimum length may be taken for delivery to land stations provided that the meat of such whales is to be used for local consumption as human or animal food (Committee for International Whaling Statistics, 1948). No minimum lengths were or are set for the minke whale.

The hunting of blue whales in the North Atlantic and adjacent Arctic waters was prohibited by the International Whaling Commission from the beginning of the 1955 season for a period of 5 years. Denmark and Iceland objected to this decision and therefore continued hunting blue whales up to and including the 1959 season. They did not, however, renew their objection when the prohibition was extended for a further 5 years from the beginning of the 1960 season. The blue whale has been completely protected since that date in the North Atlantic and Arctic, although a few blue whales have been killed by accident. On a national level governments have regulated the number of stations, the number of catcher boats permitted to operate, and the length of the seasons to conform with IWC regulations. Quotas by weight for minke whales were set for each of the Norwegian vessels in 1973 and the following seasons. National regulations have in some cases been of great importance in the management of the whale stocks in the North Atlantic.

Identification of breeding populations

Effective management of whaling requires that the distribution, size and dynamics

---

[1] *Editor's note:* Great whales are usually taken to be the sperm whales and the baleen whales (blue, fin, humpback, right, sei, Bryde's, gray) with the frequent addition of the minke whale — at 30 ft the smallest baleen whale. The authors here no doubt exclude the minke whales since these are currently being taken by Norway. The Spanish whaling station referred to also takes great whales — mostly fin and sperm but also a few humpbacks.

of the populations being exploited be known. It seems clear that the effective biological unit in most species of whales is not the species itself, but the individual breeding populations. Thus, unequivocal identification of populations is of the greatest importance. A practical definition of a breeding population is as follows: "a relatively homogeneous and self-contained population whose losses by emigration and accessions by immigration, if any, are negligible in relation to the rates of growth and mortality" (ICNAF/ICES/FAO, 1960).

METHODS

Marking

Table 1 shows the results of whale marking of balaenopterid species in the North Atlantic. Most of the marking has been done under a Canadian marking programme; information on the marking positions is only specified by areas (Mitchell, 1974b).

The advantage of the old method of marking whales using inconspicuous (internal) marks recovered after death, has decreased, because several land stations have been closed. A "streamer" mark with monofilament line was developed by Ruud. The main aim of this mark was to increase the recovery rate of marks during "flensing" (Ruud, Clarke and Jonsgård, 1953). Another "streamer" mark has been developed by Mitchell and Kozicki (1975). According to Mitchell (1975) it is visible at sea at ranges of 1/2 mile and more. Kozicki and Mitchell (1974) have also developed a drug mark which contains quinacrine to mark growth zones in the ear plugs for absolute age determination. In 1973, 10 fin whales and 3 blue whales were marked with combined drug and visual tags in the Gulf of St. Lawrence (Mitchell, 1975).

Patterns of density and migration

Discontinuity in distribution of a species of whales does not necessarily indicate different populations, though groups of whales inhabiting different feeding grounds and following different migration routes are likely to belong to different populations. Had more information been available about the localities of the different breeding grounds, stock identification would have been much easier. The great majority of the balaenopterids in the North Atlantic have been caught from shore stations. The density of catch positions is, however, to a large extent dependent on the distance from the stations, as well as on the distribution of the whales. Altogether, 16 pelagic whaling expeditions were made in the North Atlantic in the period 1929-37. Logbooks from 9 of these were examined by Jonsgård (1966).

Sightings from merchant ships (Brown, 1958) research boats (Mitchell, 1975) and catchers (Brown, 1972) have been used.

Different trends in catch-per-unit effort

If catch-per-unit effort in different areas shows different trends, this indicates that the areas are used by different stocks.

Table 1. Balaenopterids marked in the North Atlantic

	Blue	Fin	Sei	Minke
Canada (1960/73)	20	287	30	12
France (1965/69)	–	2 [1]	–	–
Iceland (1965/72)	–	17	–	–
Norway (1954/75)	–	29	–	207
United Kingdom (1950/55)		9	–	–
Total number marked	20	344	30	219
Recoveries to 1974	–	46	3	4 [2]

[1] Marked in the Mediterranean.
[2] Including 1 recovery in 1975.

Biological parameters

Differences in biological parameters in the catches from different areas indicate different breeding populations. However, such differences may also occur through segregation within a single breeding population. Such biological parameters are:

— Mean length of the catches
— Rate of ovulation
— Mean age and length at sexual and physical maturity
— Relation between number of corpora lutea and length
— Relation between laminations in the earplug and length
— Differences in the total mortality rate
— Food and feeding habits

Other methods

— Morphological differences
— Biochemical differences
— Pattern and shape of earplug

RESULTS

Blue whale

Geographic range

The northern limit of blue whales is the pack ice or the edge of the ice in the Arctic regions of the North Atlantic Ocean (Jonsgård, 1966a). On the east side of the North Atlantic, 12 blue whales were observed as far south as the part of the ocean between Cape Verde Islands and the coast of Africa in March 1911 by Ingebrigtsen (1929) who had extensive experience as a whaler. According to Moore (1953), blue whales are apparently not known off Florida. Harmer (1923) reports that a large blue whale entered the harbour at Cristobal in the Panama Canal zone, in January 1922.

Discrete breeding populations

Studies of the blue whale populations in the North Atlantic have mainly been based on catch statistics in seasons before the Second World War when they were severely overexploited in almost all whaling areas.

On the basis of observations and statistics obtained from the stations in Iceland, the Faroes, the Hebrides and Ireland, Ingebrigtsen (1929) concludes that one population of blue whales migrates along East Greenland and passes northwest Iceland to the northeastern Atlantic Ocean and the Arctic. Another population migrates past Ireland, the Hebrides and the Faroes, and then proceeds toward Iceland and mixes with the first population.

Risting (1922) mentions that blue whales migrate along the coast of Newfoundland as early as February, and he assumes that animals belonging to these western stocks occur off Iceland in the last part of March. Hjort, Jahn and Ottestad (1933) adopted the hypothesis that the baleen whales off Iceland could be treated as one discrete breeding population, for management purposes. His reasons were that whaling was being carried on from north Norway, the Shetlands, the Hebrides, Ireland, Svalbard, the Murmal Coast and Newfoundland concurrent with Icelandic whaling from 1883-1915, and in all the former areas it became unremunerative, doubtless because of the near extermination of the whales there (Hjort et al., 1933).

Hjort (1902) shows drawings of 2 small harpoons, one of which was found in a blue whale caught in 1888 and the other in a blue whale caught in 1898. Both whales were caught off Finmark. According to Hjort these types of harpoons were never used off Finmark, but were in use off the North American east coast. Also, in a blue whale captured off Iceland in the summer of 1886, a portion of an explosive harpoon was found with which the whale had been hit the previous year off Finmark (Collett, 1911).

Jonsgård (1955) analysed catch statistics from North Atlantic land stations and pelagic expeditions and concluded that it is possible that the blue whales which stay for a time in the northern part of the Barents Sea in the course of their migration for food, are hunted by land stations from Iceland, the Faroes, Norway and to some extent from Newfoundland. According to Jonsgård, there can be no doubt that a substantial number of the blue whales which are found off Newfoundland continue their migration northward through the Davis Strait to waters west of Greenland. However, blue whales in the Gulf of St. Lawrence may not migrate to the Davis Strait (Mitchell, 1975).

Jonsgård (1955) also points out that a separate breeding population of blue whales may occur off the Hebrides. This was substantiated in part by the fact that, in the period from 1927 onward, the catch of blue whales at stations in the Faroes and Norway did not change though catches diminished off the Hebrides.

The northern waters off north Norway and Spitsbergen also seem to have their local breeding populations because it would otherwise seem hardly possible that blue whales in some areas could be nearly exterminated while, at the same time, the numbers in adjacent areas remained relatively high (Jonsgård, 1955).

It may be concluded that blue whales on each side of the North Atlantic seem to be composed of several discrete breeding populations and that migration may take place between the eastern and western side in high northern latitudes.

Fin whale

Geographic range

Although fin whales no doubt migrate north and south during spring and autumn, they have been caught simultaneously in the summer months from the coasts of Portugal and Spain and from the edge of the ice in high northern latitudes (Jonsgård, 1966a). A single specimen was reportedly seen by D.M. Lindsay as far north as Melville Bay, West Greenland (Allen, 1916). A pelagic whaling expedition in the North Atlantic made their northernmost catches of fin whales at 80°42'N, 11°E in June 1930 (Jonsgård, 1966).

In May 1948, Kirpichnikov (1950) sighted an unspecified number of fin whales between the Canary Islands and the coast of Africa, and reported that 4 fin whales were seen off the coast of West Africa at about 20°50'N, 18°10'W.

In 1946, Brimely (1946) reported that the southernmost record to date, on the western side, was from North Carolina. Moore (1953) gives the information that a living fin whale was stranded in Florida in May 1950, at 29°17'N, 81°4'W.

Discrete breeding populations

As mentioned above, Hjort et al. (1933) assumed that fin whales off Iceland constitute a distinct population for management purposes. Jonsson (1965) suggests on the basis of catch statistics that fin whales off northern Norway and western Iceland belong to the same breeding population, but that the Icelandic catches are taken from different breeding populations from those inhabiting the waters off the Faroes and western Norway. This hypothesis is supported by Jonsgård (1966). Jonsgård (1958, 1966) also concluded that there seems to be a connexion between the breeding populations off western Norway and the Faroes because of the concurrent drastic decrease in number of fin whales caught which took place in these 2 areas in post-War seasons. Jonsgård (1966) shows a constant and similar trend in catch per boat from Iceland and from north Norway since 1948, and he points out that the higher catch per boat figure for Icelandic whaling may indicate that fin whales migrating through the rather narrow waters off the Denmark Strait disperse in Arctic waters further north.

Off the west coast of Iceland, between 1965 and 1971, 14 fin whales were marked under the Icelandic programme and 3 fin whales under the Norwegian programme. In the same period off east Greenland, 16 fin whales were marked under the Norwegian programme. Of these 33 marks, 5 had been recovered by 1971 from the catches of 1 770 fin whales taken by the Icelanders since 1965. But none of these marks were recovered in the catches of fin whales off north Norway in the same period. However, since only 346 fin whales were caught off north Norway from 1965 to 1971, which is only 1/5 of the numbers in the Icelandic catch, the marking data neither support nor disprove the hypothesis that fin whales off northern Norway and Iceland belong to the same breeding population.

Fin whales occur off east Greenland (Jonsgård, 1966; Jonsgård and Christensen, 1968; Christensen, 1974). Of the 27 fin whale marked off east Greenland, 2 have been recovered off Iceland. Rørvik *et al.* (1975) conclude that the same discrete breeding population of fin whales which occurs off the west coast of Iceland, probably also occurs regularly off east Greenland and possibly off north Norway.

Jonsgård (1966) examined 170 and 40 sexually mature female fin whales from west Norway and north Norway, respectively. Assuming that the corpora lutea persist and accumulate in the ovaries at the same rate in both areas, Jonsgård concludes that females found off north Norway have a faster growth and attain a greater length than those found off western Norway. The hypothesis that fin whales off western Norway and north Norway belong to different breeding populations is further supported by the fact that fin whales are very seldom seen in the waters between these two whaling areas (Jonsgård, 1966).

Jonsgård (1966) also finds that the catch statistics indicate a connexion between the fin whales occurring in the Gibraltar area (which is a possible breeding ground) and those found off the Scottish islands. According to Jonsgård

this view is supported by the previous observations that when whaling was introduced off the Shetland Islands, the fin whales behaved as if they had never been chased before, and also that they were different in colour and size from those caught off the Faroes (Risting, 1922).

Lund (1934) analysed whale oil bought for industrial use from various areas of whaling. In his opinion, the iodine value of the oil and to some degree also the saponification value characterized a whale stock. From a comparison of the iodine values of oil from whales caught on different North Atlantic whaling grounds, he states that distinct breeding populations of fin whales might be found in the following geographical areas; southern Spain, north Spain, the coast of Norway, the Scottish islands and the Arctic Ocean. Lund's method has not, to the authors' knowledge, been taken into consideration by other scientists although he has also published papers on the iodine values of oil from Antarctic whales (Lund, 1938, 1950, 1950a, 1951).

The catch statistics for fin whales caught off the west coast of Iceland, 1948-74, were examined by Rørvik *et al.* (1975). They found that on the average from 1951 to 1973, the mean length both for males and females is larger in fin whales caught in waters outside the continental shelf than in those caught over the shelf or the slope, especially in the early (May and June) and in the late (September) parts of the season. The proportion of the fin whales caught over the shelf was largest in July and August. This indicates in our opinion that it is mainly one stratified breeding population which inhabits the waters of the west coast of Iceland. The larger and sexually mature animals seem to dominate among those whales that arrive first on the grounds. The immatures occur off west Iceland mainly in July and August, and are more restricted to the waters over the shelf. Fin whales seldom occur in waters of less than 200 m depth. Rørvik *et al.* conclude that the segregation by length and its variations through the season is best explained by assuming that one breeding population in-

275

habits the waters off west Iceland.

It may be concluded that differences in biological parameters such as mean length do not imply different breeding groups, but may be caused by variations in one stratified population.

According to Mitchell (1974a) the Canadian marking of fin whales in the North Atlantic has not demonstrated any interchange between the fin whales in the west Greenland area and in the Labrador and Nova Scotia areas. The data also suggest that the fin whales in the Nova Scotia area are clearly separable as a breeding population from the fin whales being taken in the Denmark Strait by Iceland. Marking data indicate that there is some interconnexion, of the order of 10 %, between the population of the Nova Scotia area and that in the Labrador area during summer months. Mitchell also points out that males caught at Blandford Station, Nova Scotia, are sexually mature at a mean length which is 2 ft shorter than those caught further north at Dildo and Williamsport, Newfoundland. The fact that the average age of sexually mature males at Blandford (34.4 lamina, n = 47), is much greater than at Dildo (24.2 lamina, n = 61) and Williamsport (20.4 lamina, n = 57), indicates that they represent a different breeding group of whales or that there is a cline in this feature. Taking the presence of a single corpus luteum or a corpus albicans of an early stage as the definition of the attainment of sexual maturity in female fin whales, Mitchell finds the average length at sexual maturity to be 58 ft at Blandford (n = 12), 58 ft at Dildo (n = 11) and 60.48 ft at Williamsport (n = 23). Plotting of ear plug laminations against number of corpora in both ovaries for the years 1966 to 1969 gave 0.90 ovulations per 2 lamina in Blandford whales. In other words, there are 0.90 ovulations for a period of time represented in ear plugs by 2 light bands (or 2 dark bands). In Dildo and Williamsport whales the results were 0.82 and 0.68 ovulations, respectively.

Mitchell (1974a) also found a decrease in the mean length of fin whales sampled at the 3 whaling stations from the most northerly one (Williamsport) to the most southerly one (Blandford).

	Males (ft)	Females (ft)
Williamsport	57.56 (n = 425)	60.53 (n = 652)
Dildo	57.55 (n = 469)	60.08 (n = 462)
Blandford	54.72 (n = 589)	57.33 (n = 752)

According to Mitchell (1974a) the differences between mean lengths at each of the 3 different stations imply that, without considering gunner selection, there are either 3 discrete breeding populations being fished; or a number of isolated schools available to each station; or that within one latitudinally dispersed population, there may be a cline in such characters as mean length.

Mitchell and Kozicki (1974) examined 1 122 ear plugs, sampled off the east coast of Canada. They concluded that data concerning the external ear plug shape are not related to possible discrete breeding populations defined by other parameters.

Mitchell (1975a) notes that an external pigmentation scale is being divised in order to quantify small differences found on the Canadian east coast between individual whales and possibly between populations, and that body measurements from northwest Atlantic fin whales also will be analysed.

Electrophoretic analyses of fin whales from the west coast of Iceland are also being carried out (IWC, 1974).

We conclude that the fin whales in the North Atlantic may be divided into the following independent groups:

(1) The breeding population – or populations – in the Arctic eastern North Atlantic and off north Norway. These whales probably migrate northward and southward through the Denmark Strait and may have been subject to catch both off north Norway and off Iceland;

(2) The fin whales off east Greenland and west Iceland which probably belong to the same breeding population;

(3) The fin whales off the west coast of Norway which may belong to the same breeding population – or populations – as those found off the Faroes;

(4) A separate breeding population of fin whales may inhabit the waters off the Scottish islands during the feeding season, and spending the breeding season off Spain and Portugal;

(5) The fin whales off west Greenland;

(6) The northwest Atlantic fin whales off Nova Scotia, Newfoundland and Labrador, which may be composed of two or more breeding populations (Nova Scotia and Gulf of St. Lawrence versus Newfoundland/Labrador).

Sei whale

Geographic range

Sei whales have been caught as far north as 79°N and 100 nautical miles off west Spitsbergen (Ingebrigtsen, 1929). However, Ingebrigtsen concludes that the majority of this species in the northeastern part of the Atlantic Ocean and the Norwegian Sea does not proceed beyond 72°N. On the western side of the North Atlantic, sei whales have been observed in the Davis, Strait (Mitchell, 1974a). In the southern part of the North Atlantic sei whales may be confused with Bryde's whales when sighted. Catches of sei whales have been made in the winter months off Cape Blanco, West Africa (Ingebrigtsen, 1929). Sei whales have been marked off northern Venezuela (Mitchell, 1974a).

Discrete breeeding populations

Ingebrigtsen (1929) states that the regions where sei whales afford opportunities of remunerative whaling are the Norwegian Sea off the Shetlands, the Faroes and the west coast of Norway. According to him, sei whales are scarce off Iceland, the greatest number captured in the 1900 season by 30 boats being 20 whales. However, since 1948, sei whales have been caught regularly off west Iceland, in some years in great numbers. In 1971, 240 sei whales were caught by 4 boats, compared with only 3 sei whales in 1968. On the average for 1948 to 1974, 63 sei whales were caught per season by 4 boats.

With regard to the northwest Atlantic, Mitchell (1974a) states that geographic limits, migration and ranges are not known certainly for the population being fished there. The 3 recoveries of marks from sei whales in the Canadian fisheries indicate that sei whales caught from Blandford in the early and late summer are from the same breeding population (Mitchell, 1975a). However, it is possible that 2 discrete breeding populations of sei whales, a northern and southern one, occur off the east Canadian coast (Mitchell, 1974a).

It may be concluded that very little is known about the identity of sei whale breeding populations in the North Atlantic Ocean.

Minke whale

Geographic range

Like blue and fin whales, minke whales are limited in their northern range by the ice. On the eastern side of the North Atlantic minke whales have been caught by Norwegian whalers as far north as the edge of the ice (Jonsgård, 1966a). A plot of the localities where minke whales were taken by Norwegian whalers in recent years (1966-72) is given by Solvik and Jonsgård (1974).

Little is known of the southern limit of this species on the European side of the North Atlantic although it has been reported as far south as the Mediterranean (Collett, 1911).

The northwest Atlantic breeding population of minke whales apparently summers along the coast between Ungava Bay and Cape Cod, and winters offshore in waters south to Florida (Mitchell, 1974a).

Off the west coast of Greenland, concentrations of minke are observed as far north as off Svartenhuk (Christensen, 1974).

Discrete breeding populations

Jonsgård (1951, 1962) reports that minke whales during their migration in Norwegian and adjacent waters are segregated in time and area according to size and sex. O-group calves are mainly found in late summer from Stadt (western Norway) to Vesterålen with a concentration in the Lofoten area. On their northward migration, larger females often visit the fjords while larger males migrate further out. Mature females and males also segregate in the eastern Arctic waters. In all probability, the same breeding population of minke whales occurs off Norway and in the Barents Sea.

Sergeant (1963) concludes from length frequencies, that the Newfoundland catch consisted mainly of mature animals, and suggests that the immature whales are distributed further south in the summer months. Mitchell (1974) who examined 191 minke whales caught off Newfoundland and Nova Scotia between 1966 and 1972, found that 74.3 % were females. The length frequencies of males and females indicated no segregation by size.

Christensen (1974) reports on the sex distribution in the Norwegian catches of minke whales in the 1973 season. He found that females made up 70 % (n = 120) of the catch in the Barents Sea, 34.2 % (n = 38) off east Greenland and 85.2 % (n = 81) off west Greenland.

Christensen (1974) also found that the mean length of minke whales caught in the Barents Sea was smaller than that of those caught off west Greenland, and suggested that the differences in the length group frequencies and sex composition between these two areas indicate separate stocks of minke whales in the northeast and the northwest Atlantic.

While little is known about stock units of minke whales in North Atlantic waters, there are indications that different stocks may occur in northern North Atlantic waters during the feeding migration in the summer.

ADDENDUM

The stocks of minke whales in the North Atlantic were the subject of a working group of the Scientific Committee of the International Whaling Commission (IWC, 1977). This group concluded that the following areas probably contain fairly independent stocks of minke whales:

(1) The Canadian east coast;

(2) The west Greenland area;

(3) The east Greenland-Iceland-Jan Mayen area;

(4) The region from Svalbard, the Barents Sea, along the Norwegian coast, including Skagerak, the North Sea and other areas around the British isles.

Estimation of population size

METHODS

(1) Estimates from the old catch statistics

(2) Sightings

(3) Mark/recapture

(4)   Catch-per-unit effort, and De Lury method

(5)   Estimates from the total mortality rate

The methods of sightings and mark/recapture have been discussed by Best (1974).

Estimates from the old catch statistics

During the period from 1868 to about 1915, whaling took place in most parts of the northern North Atlantic. The catch statistics from this period as given in the International Whaling Statistics are incomplete for two reasons:

(1)   The statistics give only the number of whales brought to the stations and worked up (Tønnessen, 1967). Tønnessen writes that the number actually killed must have been considerably higher, because of inefficient catching equipment. According to Tønnessen, it has been suggested that in the first 20-25 years after 1867 the statistics should be doubled to get a correct estimate of the number of whales actually killed, and in the following 15-20 years the statistics should be increased by a factor of 50 %. Tønnessen comments that these suggested increases seem to be too high, but he is not able to give more correct data. We assume that the percentages should probably be different for different species.

(2)   The number of species brought to the stations is only partly specified in the statistics. Also, those which are specified may not be representative of the actual catch.

An estimate of the total number of whales killed may give an idea of the stock size of the species in question, especially when the stock is much reduced. Old catch statistics may provide such an estimate. The following notation is introduced:

C_{T_j} is the total number of whales caught in season j, of which

C_{S_j} is the number of whales recorded by species,

C_j is the number of the actual species which are caught and specified in the catch statistics.

Assuming that the statistics have to be increased by 25 % to compensate for those which were lost at sea, and that the number specified is representative of the total catch, the estimated number of whales of the actual species killed is:

$$1.25 \times \Sigma_j \ (C_j \times C_{T_j}/C_{S_j}) \ \ldots\ldots\ldots \quad (1)$$

An alternative formula is:

$$1.25 \times \Sigma_j \ C_j \times \Sigma_j \ C_{T_j}/ \ \Sigma_j \ C_{S_j} \ \ldots\ldots \quad (2)$$

Catch-per-unit effort and De Lury method

This method has also been discussed by Best (1974). However, the assumption used in De Lury's (1947) method that catch-per-unit effort is proportional to stock density and stock size should be further discussed. Most of the whaling in the North Atlantic has been carried out from shore stations. It is likely that an essential part of the time of the catcher boats on each trip is spent en route to and from the catching grounds. This part of the operation time is independent of the stock density. Therefore, the number of whales caught per catcher in the season is not likely to be proportional to the stock density (Rørvik *et al.*, 1975). Rørvik *et al.* conclude from theoretical considerations that a decrease in catch per catcher indicates a relatively larger decrease in stock size. Catch per hour used for searching and hunting seems to be a better measure of stock density than number of whales caught

279

per catcher. However, estimation of catch per hour used for searching and hunting requires often a tedious search of logbooks in which such data occurs.

The way in which catch per hour used for searching and hunting relates to stock density is also likely to depend on how the number of whales per school changes when the stock size changes.

Estimates from the mortality rate

The total instantaneous mortality rate, Z, can be determined from the age distribution as read from the earplugs. The instantaneous whaling mortality, F, can be estimated from Z when the instantaneous mortality rate, M, is known. From the catch, C, the stock size, N, can be estimated by:

$$\hat{N} = C/F = C/(Z - M)$$

However, estimates of Z may be biased if samples are taken from a stock which is segregated by body length. Such a segregation has been demonstrated for the fin whales off the west coast of Iceland (Rørvik et al., 1975). Fin whales caught on different grounds of the eastern coast of Canada also have different mean lengths (Mitchell, 1974a). Segregation may also influence the results from the markings. We cannot assume that whales marked in a small area after a given time are uniformly distributed in the area inhabited by the stock, if the stock is segregated.

RESULTS

Blue whale

North Norway

Off north Norway, 17 745 whales were reported caught from 1868 to 1904 and, of the 11 607 whales specified, 1 833 were recorded as blue whales (Committee for International Whaling Statistics, 1931). Using formula (2), this indicates that about 3 500 blue whales were killed in this area.

Denmark Strait

In 1915, the blue whale population off Iceland was very much reduced. Only 9 blue whales were caught by 4 boats in 1915 (Committee for International Whaling Statistics, 1931) compared with 219 blue whales caught by 5 boats from one station both in 1895 and 1896 (Risting, 1922). Using formula (1) and the catch statistics for the period from 1883 to 1915, we conclude that about 10 000 blue whales were killed in this period off Iceland. In this estimate, we have assumed that 25 % of the whales caught in the 1905 to 1910 seasons and in 1912, for which no specification is given, were blue whales. This estimate also includes blue whales caught off the east coast of Iceland, where whaling took place from 1900 to 1913; these blue whales may belong to a different breeding population than those caught off the western and northern coasts of Iceland.

Taking into consideration the recruitment of the breeding populations in the period 1883-1915, and a possible over-representation of blue whales in the specified part of the catch, it is likely that the virgin stock must have been less than 10 000. But it may be concluded that the virgin stock of blue whales off Iceland must have numbered several thousand.

Brown (1972) concludes from a preliminary analysis of sighting records from the Icelandic catcher boats, that the number of blue whales occurring on the whaling grounds off west Iceland during the whaling season has increased considerably during the whaling season since 1960, when the prohibition on catching blue whales came into force in Icelandic waters.

Using the data from Table 4 of Brown (1972) the mean abundance of this population is calculated to be minimum 30 and maximum

37 blue whales in the 1957-59 seasons; for 1969-71 the mean abundance is 138 minimum and 249 maximum. This would imply an average rate of increase of around 15.6 % per annum, which is so great as to be highly unlikely, given our knowledge of great whale biology. Thus, while Brown's data and our observations indicate an increase, its actual size is open to question. Sightings data may have been influenced by other factors than a change in numbers of blue whales.

Northwest Atlantic

Allen (1970) assumes that the abundance of fin whales did not change appreciably between 1903 and 1951, and uses the ratio of blue whale to fin whale catches as an index of blue whale stock size. He concludes that the initial stock size was slightly over 1 100 blue whales and that the maximum sustainable yield was probably not more than 100 blue whales. Mitchell (1974a) points out that since it is likely that all of these whales were on migration to or from Davis Strait, this estimate of 1 100 whales probably accounts for the entire northwest Atlantic population of blue whales.

As judged from cumulative sightings recorded by whale catchers, and from estimates based upon strip censuses, the population of blue whales in the northwest Atlantic can probably be numbered in the low hundreds (Mitchell, 1974a).

Fin whale

North Norway

No stock estimates of fin whales off north Norway have yet been made. However, if the breeding population of fin whales off north Norway is independent of that off Iceland, and if it can sustain a catch of 61 fin whales annually, which was the average from 1948 to 1971, then the recruited virgin stock must have been at least 2 034 animals, assuming 6 % yield at MSY level, where MSY level is 50 % of the virgin stock.

In the catch statistics from 1868 to 1904 (Committee for International Whaling Statistics, 1931), 11 607 out of 17 745 whales are specified, 5 864 of which are fin whales. Using formula (2), we arrive at a figure of 10 500 for fin whales killed from 1868 to 1904, mostly during the last 20 years of the period. It may, therefore, be concluded that the virgin breeding population or populations inhabiting the northern Northeast Atlantic waters must have numbered several thousand.

Western Norway and the Faroes

Jonsgård and Rørvik (1975) estimated the recruited stock in 1946 to be 2 700 fin whales using De Lury's (1947) method, with the modifications given by IWC (1964) which take natural mortality and recruitment into account. The stock size in 1963 was estimated to be 430. Since then, only 84 fin whales have been caught in these waters. However, the stock size probably decreased relatively more than the corresponding decrease in catch per catcher from 1946 to 1963. The present stock of fin whale off western Norway and the Faroes is, therefore, probably in the low hundreds. The virgin stock must have been more than 2 700 because over-exploitation of this stock (or stocks) seems to have taken place in between 1918 and 1939.

Spain, Portugal and the Scottish Islands

A total of 6 433 fin whales were caught off Spain and Portugal from 1921 to 1927. In this period, the stock was severely reduced (Hjort, Jahn and Ottestad, 1933). In 1903 whaling started off Scotland (Risting, 1922). 2 229 fin whales were caught in these waters from 1922 to 1929 and the catch per catcher decreased from 71.6 fin whales in 1924 to 18.3

in 1929 (Jonsgård, 1966). This gives a total of 8 662 fin whales caught from 1921 to 1929 off Scotland, Spain and Portugal.

It may be concluded that the virgin breeding population or populations which inhabit the waters off Spain, Portugal and Scotland (Jonsgård, 1966) must have numbered at least 5 000 animals.

Denmark Strait

In a preliminary analysis, Gambell, Jonsson and Jonsgård (1973) estimated the total instantaneous mortality rate, Z, from the age distribution as determined from 428 earplugs available from fin whales caught at Iceland. However, only 266 (62 %) were readable. They found Z to be 0.07 and, assuming that the natural instantaneous mortality rate, M, is 0.04, the instantaneous fishing mortality, F, is $0.07 - 0.04 = 0.03$. From the average annual catch of 241 fin whales (1948-72), the total exploitable breeding population was then estimated to be 8 033. Because of the low readable proportion of the earplugs, Gambell et al. (1973) doubted their reliability as a random sample of the population.

The sample may also be biased because of the segregation by length of fin whales off west Iceland.

Rørvik et al. (1975) estimate the stock size from marking results from 1965 to 1973 using the method of direct multiple sample census (Chapman, 1952).

Symbols,

T_j = number of marked fin whales alive at the beginning of the season j

C_j = number of fin whales caught in season j

t_j = number of marks recovered in season j

m_j = number of fin whales marked at the beginning of the season j

N = the number of fin whales

The method of direct multiple sample census (Chapman, 1952) gives,

$$\hat{N} = \Sigma_{C_j T_j}^{1973} / (\Sigma_{t_{j+1}}^{1973})$$

$$j = 1965 \qquad j = 1965$$

The method assumes that no recruitment or mortality occurs during the experiment — a condition which is not fulfilled in this case. Since catch-per-unit effort showed no significant decline from 1948 to 1973, the stock could not have changed much in the period 1965 to 1973.

The mortality rate of marked fin whales was taken into account by

$$T_{j+1} = (T_j - t_j) e^{-0.04} + m_{j+1}$$

The approximated Poisson distribution of t_j also allowed the calculation of a 95 % confidence interval of $\hat{N}$. Assuming 100 % recovery rate, the recovery of 6 fin whales out of 20 marked off the west coast of Iceland, gives the following estimate,

$$\hat{N} \quad = \quad 4\,265$$
$$N_{min} \quad = \quad 1\,791$$
$$N_{max} \quad = \quad 11\,584$$

Assuming that the same discrete breeding population of fin whales occurs in the waters off the west coast of Iceland, north Norway and east Greenland, and taking into account the 27 fin whales marked off east Greenland (2 recoveries at Iceland), the following estimate is obtained:

$$N \quad = \quad 7\,661$$
$$N_{min} \quad = \quad 3\,627$$
$$N_{max} \quad = \quad 17\,651$$

Northwest Atlantic

Mitchell (1974b) has made 3 estimates of stock size using different methods: from censuses a preliminary conservative figure of less than 4 182; from tag recapture, a preliminary figure of 6 310 of which 1 992 are referred to the Nova Scotia area; from catch/effort data a figure of at least 3 590 fin whales. From cumulative catch data Mitchell (1974b) estimates the upper limit of MSY to be 400.

Allen (1973) estimates from records of catch-per-unit effort that the fin whale population in 1972 was 320 animals off Blandford, Nova Scotia, 550 off Dildo, Newfoundland and 1 500 off Williamsport, Newfoundland, or a total of 2 370 fin whales. The maximum number of northwest Atlantic fin whales would then be at least 4 500 with a MSY of 193 animals (Allen, 1973).

No estimates of the number of fin whales occurring off west Greenland have been made.

Sei whale

Northeast Atlantic

From 1885 to 1904 15 326 whales were caught off north Norway, and of the 10 815 which are specified, 2 826 were sei whales (Committee for International Whaling Statistics, 1931). Allowing for 25 % lost at sea, and using formula (2) about 5 000 sei whales are estimated to have been killed off north Norway in this period.

According to Jonsgård (1974), sei whales are seldom seen so far north nowadays, which is confirmed by the fact that only 3 animals were caught from 1948 to 1972, when the shore station in north Norway was in operation. He also gives the information that sei whales were rather plentiful off western Norway during and after the Second World War, but since the sixties very few have been observed (Jonsgård, 1974). Jonsgård concludes that although this species is known for its irregular occurrence, it cannot be denied that its disappearance may be caused by over-exploitation. Jonsgård mentions that only 1 824 sei whales have been reported caught in the northeastern North Atlantic in the period from 1946 to 1970 (Norway – 220, Faroe Islands – 224, Iceland – 840), or an average of 51 animals per season. If catches of this order have harmed the stocks of sei whales, it could safely be concluded that the sei whale stocks just after the Second World War were very small (Jonsgård, 1974).

Another and more likely explanation, according to Jonsgård, may be that their route of migration has altered. This explanation is in accordance with reports of plentiful sei whales off the coast of western Norway during the Second World War (Jonsgård, 1974). It is also supported by the fact that sei whales have been caught much more frequently off Iceland since 1948, as compared with the period from 1883 to 1915 (see IDENTIFICATION OF BREEDING POPULATIONS – Sei Whale).

Northwest Atlantic

Mitchell (1974b) estimates the number of sei whales in the Nova Scotia fishery to be 1 856 in 1966, from marking and recapture. A census in 1969 gave 828 sei whales in the Labrador Sea (Mitchell, 1974b).

Minke whales

Estimates of minke populations from ship census data involves different difficulties from other mysticete species. Minke whales tend to approach ships, especially stationary vessels, while most other mysticetes tend to avoid or ignore ships at comparable ranges. This "ship-living" tendency introduces an important bias into resulting data that is related to the size, type and acoustic characteristics of the survey vessels, and to other features (Mitchell, 1974a).

Jonsgård (1974) writes that the annual total Norwegian catch of minke whales in the North Atlantic increased from a little less than 2 000 animals at the end of the Second World War to 4 338 in 1958, and thereafter decreased to 2 307 in 1970. He points out that this decrease is mainly due to the fact that the number of licences was gradually reduced by the Norwegian Government from 183 in 1960 to 118 in 1970, resulting in a slight increase in the catch per boat during the sixties. In the preliminary analysis by Solvik and Jonsgård (1974) where the catches for different "old" whaling areas in the Northeast Atlantic are shown, and catch per unit effort is corrected for weather conditions, it is concluded that the stock of minke whales in the eastern North Atlantic has not decreased significantly during the period 1960-72.

No stock estimates and MSY for the Northeast Atlantic minke whales are available up to now [1974], mainly because of the difficulties involved in having usable parameters for stock assessment.

A catch of about 35 minke whales per annum for 15 years has been taken off Newfoundland, but there are no good estimates of the stock size available (Mitchell, 1974a).

ADDENDUM

The Minke Whale Working Group of the Scientific Committee of the IWC (IWC, 1977) concluded that the best estimate of North Atlantic minke whale numbers, based on markings of the Svalbard-Norway — British Isles stock, is 50 952 animals, with a 95 % confidence interval of 26 895-100 172. A yearly quota of 1 790 (for 1977, 1978 and perhaps 1979) gives an exploitation rate of 3.5 % with a 95 % confidence interval ranging from 1.8 % to 6.7 % (Christensen and Rørvik, 1978).

References

ALLEN, G.M., The whalebone whales of New England.
1916   Mem. Boston Soc. Nat. Hist., 8:107-322.

—, A note on baleen whale stocks of the Northwest
1970   Atlantic. Rep. IWC, (20):112-3.

—, Catch per effort estimates of Northwest Atlantic fin
1973   whale stocks. Rep. IWC, (23):97-8.

BEST, P.B., Review of world sperm whale stocks. 58 p.
1974

BRIMLEY, C.S., The mammals of North Carolina. Carol.
1946   Tips, 7:2-30.

BROWN, S.G., Whales observed in the Atlantic Ocean.
1958   Mar. Obs., 28:142-6, 209-16.

—, Blue and humpback whales in Icelandic waters.
1972   ICES, C.M. 1972/N:4:5 p.

CHAPMAN, D.G., Inverse, multiple and sequential sam-
1952   ple census. Biometrics, 8:286-306.

CHRISTENSEN, I., Minke whale investigations in the Ba-
1974   rents Sea and off east and west Greenland in
        1973. Fisk. Gang, 60 (12):278-86.

CHRISTENSEN, I. and C.J. RØRVIK, Stock estimate of

1978   minke whales in the Svlabard, Norway —
        British Isles area. Paper presented to the
        Scientific Committee of the International
        Commission on Whaling, Doc. SC/30/60:4 p.

COLLETT, R., Norges pattedyr. Kristania, H. Aschehoug
1911   & Co. (W. Nygaard). 744 p.

Committee for Whaling Statistics (ed.), International
1931   whaling statistics, 2. Int. Whal. Stat., (2).

—, International whaling statistics, 17. Int. Whal. Stat.,
1947   (17).

—, International whaling statistics, 18. Int. Whal. Stat.,
1948   (18).

DeLury, D.G., Estimation of biological populations.
1947   Biometrics, 3:145-67.

GAMBELL, R., J. JONSSON and Å. JONSGÅRD, Preliminary
1973   report on analysis of the fin whales off Ice-
        land. Report presented to the Twenty-fifth
        Meeting of the International Commission on
        Whaling, Doc. SC/25/5 (Unpubl.).

HARMER, C.F., Cervical vertebrate of a giant blue
1923   whale from Panama. Proc. Zool. Soc. Lond.,
        1923:1085-9.

HJORT, J., Fiskeri og hvalfangst i det nordlige Norge.
1902   Bergen, John Griegs Forlag, 251 p.

—, Whales and whaling. *Hvalraad. Skr.,* 7:7-29.
1933

HJORT, J., G. JAHN and P. OTTESTAD, The optimum
1933   catch. *Hvalrqad. Skr.,* 7:92-127.

ICNAF/ICES/FAO, Proceedings of joint scientific
1960   meeting of ICNAF, ICES and FAO on fishing
         effort, the effect of fishing on resources and
         the selectivity of fishing gear. Vol. 1 Reports.
         Spec. Publ. ICNAF, (2):1-45.

IWC, Special committee of three scientists. Final re-
1964   port. *Rep. IWC,* (14):39-87.

—, Report of the meeting on age determination in ba-
1974   leen whales. *Rep. IWC,* (24):62-8.

—, Report of the Working Group on North Atlantic
1977   whales. *Rep. IWC,* (27):369-87.

INGEBRIGTSEN, A., Whales caught in the North Atlantic
1929   and other seas. *Rapp. P.-V. Réun. CIEM,*
         56:1-26.

JONSGÅRD, Å., Studies on the little piked whale or minke
1951   whale (*Balaenoptera acuto-rostrata* Lacépè-
         de). *Norsk Hvalfangsttid.,* 40:80-95.

—, The stocks of blue whales (*Balaenoptera musculus*) in
1955   the northern Atlantic Ocean and adjacent
         Arctic. *Norsk Hvalfangsttid.,* 44 (9):297-311.

—, Taxation of fin whales (*Balaenoptera physalus* (L)) at
1958   land stations on the Norwegian west coast.
         Norsk Hvalfangsttid., 47 (9):433-9.

—, Population studies on the minke whale *Balaenoptera*
1962   *acuto-rostrata* Lacépède. *In* The exploitation
         of natural animal populations, edited by E.O.
         LeCren and M.W. Holdgate. Oxford, Black-
         well Scientific Publications, pp. 159-67.

—, Biology of the North Atlantic fin whale *Balaenoptera*
1966   *physalus* (L). Taxonomy, distribution, migra-
         tion and food. *Hvalraad. Skr.,* 49:1-62.

—, The distribution of Balaenopteridae in the North
1966a   Atlantic Ocean. *In* Whale, dolphins and por-
         poises, edited by K.S. Norris. Berkeley and
         Los Angeles, University of California Press,
         114-24.

—, On whale exploitation in the eastern part of the

1974   North Atlantic Ocean. *In* The whale problem,
         edited by W.E. Schevill. Cambridge, Mass.,
         Harvard University Press, pp. 97-107.

JONSGÅRD, Å. and I. CHRISTENSEN, A preliminary report
1968   on the HARØYBUEN cruise in 1968. *Norsk
         Hvalfangsttid.,* 57:174-5.

JONSGÅRD, Å. and C.J. RØRVIK, A note on the stock of fin
1975   whales *Balaenoptera physalsus* (L) off the west
         coast of Norway and the Faroes. *Rep. IWC,*
         (25):166-73.

JONSSON, J., Whales and whaling in Icelandic waters.
1965   *Norsk Hvalfangsttid.,* 54:245-53.

KIRPICHNIKOV, A.A., Nablyudeniya nad raspredele-
1950   niyem krupnykh kito-obraznakh v Atlanti-
         cheskom okeane (Observations on the distri-
         bution of large cetaceans in the Atlantic
         Ocean). *Priroda, Leningr.,* 10:63-4.

KOZICKI, V.M. and E. MITCHELL, Permanent and selec-
1974   tive chemical marking of Mysticeti ear plug
         lamination with quinacrine. *Rep. IWC,*
         (24):124-49.

LUND, J., Hvalens stammer og vandringer i lys av oljen
1934   fra fangstfeltene. *Norsk Hvalfangsttid.,*
         23:89-96.

—, Whale tribes in the Antarctic. Iodine value determi-
1938   nations. *Norsk Hvalfangsttid.,* 27:251-61.

—, Charting of whale stocks in the Antarctic on the
1950   basis of iodine values. *Norsk Hvalfangsttid.,*
         39:53-60.

—, Charting of whale stocks in the Antarctic in the
1950a   season 1949/50 on the basis of iodine values.
         Norsk Hvalfangsttid., 39:298-305.

—, Charting of whale stocks in the Antarctic 1950/51 on
1951   the basis of iodine values. *Norsk Hvalfangst-
         tid.,* 40:384-6.

MITCHELL, E., Preliminary report on Newfoundland
1974   fishery for minke whales (*Balaenoptera acu-
         to-rostrata*). *Rep. IWC,* (24):150-76.

—, Present status of northwest Atlantic fin and other
1974a   whale stocks. *In* The whale problem, edited by
         W.E. Schevill. Cambridge, Massachusetts,
         Harvard University Press, pp. 108-69.

—, Progress report on whale research May 1973. *Rep.
1974b   IWC,* (24):196-213.

−, Progress report on whale research, May 1973 to May
1975   1974. *Rep. IWC*, (25):270-82.

−, Preliminary report on Nova Scotia fishery for sei
1975a   whales *(Balaenoptera borealis). Rep. IWC*,
(25):218-25.

MITCHELL, E. and V.M. KOZICKI, North West Atlantic
1974   fin whales (*Balaenoptera physalus*): the ear
plug sample. *Rep. IWC*, (25):236-9.

−, Prototype visual mark for large whales modi-
1975   fied from DISCOVERY tag. *Rep. IWC*,
(25):236-9.

MOORE, J.C., Distribution of marine mammals to Flo-
1953   rida waters. *Am. Midl. Nat.*, 49:117-58.

RISTING, J.T., Av hvalfangstens historie. Kristiania, J.
1922   Petlitz Boktrykkeri, 625 p.

RØRVIK, C.J. *et al.*, Fin whales, *Balaenoptera physalus*
1975   (L), off the west coast of Iceland. Distribution,
segregation by length and exploitation. Paper
presented to the Scientific Committee of the
International Commission on Whaling 1975,
Doc. SC/27/22. Issued also in *Rit. Fiskideild*.

RUUD, J.T., R. CLARKE and Å. JONSGÅRD, Whale mark-
1953   ing trials at Steinshamn, Norway. *Norsk
Hvalfangsttid.*, 42:293-305.

SERGEANT, D.E., Minke whales, *Balaenoptera acu-
1963   to-rostrata* Lacépède, of the western North
Atlantic. *J. Fish. Res. Board Can.*,
20(6):1489-504.

SOLVIK, O.V. and Å. JONSGÅRD, A preliminary report on
1974   analysis of minke whales (*Balaenoptera acuto-rostrata* Lacépède) in the eastern North
Atlantic Ocean. Paper presented to the Scientific Committee of the International Commission on Whaling, 14 p.

TØNNESSEN, J.N. Den moderne hvalfangsts historie.
1967   Sandefjord, Kommandør Chr. Christensens
Hvalfangstmuseum, vol. 2:619 p.

ÉCOLOGIE DES CÉTACÉS DE LA MÉDITERRANÉE OCCIDENTALE

D. VIALE

Abstract

Using direct (sight counts) or indirect (assessment of mortality) methods, this document attempts to evaluate the cetacean populations in the northern part of the western Mediterranean basin (300 000 km^2): *Balaenoptera physalus* − 1 200; *Physeter macrocephalus* − 600; *Tursiops truncatus* (neritic) − 3 500; *Delphinus delphis* − 70 000; *Stenella coeruleoalba* − 4 000; *Ziphius cavirostris* − 1 250; *Globicephala melaena* − 1 350; *Grampus griseus* − 7 500.

We shall show the effect of the ecological characteristics of the western Mediterranean on the cetaceans inhabiting it. The fish-eating delphinidae travel between the south of the basin in winter and the north in summer.

Two herds of common rorquals frequent the Mediterranean. One, arriving in springtime, uses the north of the basin as its feeding area, its diet being composed mainly of Euphausiaceae, the other comes to the area in winter and uses the northeastern part of the basin as a breeding area.

Rorquals are exploited in the Bay of Biscay (about 100 catches per year).

The number of cachalots is small. They were and continue to be heavily exploited off the Atlantic coasts of Spain (Vigo). The depletion of the Mediterranean cachalot stock, although all exploitation at Gibraltar has ceased, suggests a relation between the stock exploited at Vigo and that of the Mediterranean.

Ziphius are relatively abundant. The northwestern Mediterranean seems to be favourable to them. In winter they find satisfactory trophic conditions near the coast between 0 and 20 m; production is not higher in winter than in summer, but is distributed in a euphotic stratum ten times less thick, making a higher performance possible of the consuming trophic levels by decreasing the levels of energy expended in the search for food. The young are dropped in winter. In the summer the animals probably live on dermersal cephalopods in the central areas of the Mediterranean; in fact their stomach contents have revealed species which do not appear in the list of the more common species.

Stenella, which were considered infrequent until 1970, are predominant in the annual landings. Most of the carcasses found are those of suckling young.

The main interactions between the different competing species are those of competition between plankton-eating sharks, of small Euphausiaceae-eating bathyal selacians, and of man. An evaluation of fish and cephalopod catches by fisheries shows that man takes away the same average quantity of epipelagic fish per day as do fish-eating cetaceans (or 1/8 of daily fish production), and one half the quantity of cephalopods taken by cetaceans. The western Mediterranean seems favourable to cuttlefish-eating cetaceans, since the stock depleted by catches on the coast of Spain are allowed to reconstitute themselves. In the present state of our observations it seems that the *Balaenoptera physalus* stock shows a slight recovery since 1974; on the other hand, the cachalot remains the rarest of all Mediterranean cetaceans at present, whereas formerly it was common. The small Delphinidae have decreased greatly in number along the European coasts but seem to be more abundant in the past few years on the coasts of

Africa and southern Spain, which may point to a shift of population due to the pollution of the northern Mediterranean.

Résumé

Par des méthodes directes (comptage à vue) ou indirectes (évaluation de la mortalité), ce travail tente une évaluation des effectifs de cétacés dans la partie nord du bassin occidental méditerranéen (300 000 km^2): *Balaenoptera physalus* — 1 200; *Physeter macrocephalus* — 600; *Tursiops truncatus* (néritique) — 3 500; *Delphinus delphis* — 70 000; *Stenella coeruleoalba* — 4 000; *Ziphius cavirostris* — 1 250; *Globicephala melaena* — 1 350; *Grampus griseus* — 7 500.

Nous mettons en évidence l'action des caractéristiques écologiques de la Méditerranée occidentale sur les cétacés qui la fréquentent. Les Delphinidae ichtyophages se déplacent entre le sud du bassin en hiver et le nord du bassin en été.

Deux troupeaux de rorquals communs utilisent la Méditerranée: l'un arrivant au printemps utilise le nord du bassin comme zone alimentaire; la nourriture est alors composée essentiellement d'euphausiacés. L'autre est présent l'hiver et utilise la partie nord-est du bassin comme zone de reproduction.

Les rorquals sont exploités en Biscaye (une centaine de captures par an).

Les cachalots sont peu nombreux; ils ont été exploités intensément et continuent de l'être sur les côtes atlantiques d'Espagne (Vigo). La déplétion du troupeau de cachalots méditerranéens alors que toute exploitation a cessé à Gibraltar suggère une relation entre le troupeau exploité à Vigo et le troupeau méditerranéen.

Les *Ziphius* sont relativement abondants; la Méditerranée nord-occidentale semble leur être favorable. En effet, ils trouvent l'hiver des conditions trophiques satisfaisantes près des côtes entre 0 et 20 m; la production n'est pas supérieure l'hiver à celle de l'été, mais est répartie dans une couche euphotique dix fois moins épaisse, ce qui permet un rendement supérieur des niveaux trophiques consommateurs en diminuant les dépenses énergétiques liées à la recherche de la nourriture. Les mises bas se font en hiver. L'été les animaux vivent probablement dans les zones centrales de la Méditerranée, aux dépens des céphalopodes profonds; en effet leurs contenus stomacaux nous ont montré des espèces qui ne figurent pas dans la liste des espèces les plus communes.

Les *Stenella*, considérés rares jusqu'en 1970, deviennent prépondérants dans les échouages annuels. La majorité des cadavres sont des jeunes, encore allaités.

Les interactions majeures entre les diverses espèces concurrentes sont la compétition des requins planctonophages, celle des petits sélaciens bathyaux mangeurs d'euphausiacés, et celle de l'homme. Une évaluation des prises de poissons et de céphalopodes par les pêcheries indique que l'homme prélève par jour en moyenne la même quantité de poissons épipélagiques que les cétacés ichtyophages (soit 1/8 de la production journalière de poissons), et une quantité de céphalopodes moitié de celle prélevée par les cétacés. La Méditerranée occidentale semble favorable aux cétacés teuthophages, dans la mesure où on laisse se reconstituer les stocks déprimés par les ponctions sur les côtes d'Espagne. Dans l'état actuel de nos observations, il semble que le troupeau de *Balaenoptera physalus* montre une légère recrudescence depuis 1974; par contre le cachalot reste le plus rare de tous les cétacés méditerranéens à l'heure actuelle, alors qu'il a été commun. Les petits Delphinidae ont beaucoup diminué en nombre au large des côtes européennes, mais nous semblent plus abondants ces dernières années sur les côtes africaines et d'Espagne du sud, ce qui pourrait indiquer un transfert de population du fait de la pollution nord-méditerranéenne.

Extracto

En este trabajo se trata de realizar una evaluación, mediante métodos directos (recuento a ojo) o indirectos (evaluación de la mortalidad), de las poblaciones de cetáceos en la parte septentrional de la cuenca del Mediterráneo occidental (300 000 km²): *Balaenoptera physalus* — 1 200; *Physeter macrocephalus* — 600; *Tursiops truncatus* (nerítico) — 3 500; *Delphinus delphis* — 70 000; *Stenella coeruleoalba* — 4 000; *Ziphius cavirostris* — 1 250; *Globicephala melaena* — 1 350; *Grampus griseus* — 7 500.

Ponemos de relieve la acción de las características ecológicas del Mediterráneo occidental en los cetáceos que viven en sus aguas. Los delfines ictiófagos se mueven entre el sur de la cuenca en invierno y el norte en verano.

Dos rebaños de rorcuales comunes frecuentan el Mediterráneo: uno llega en primavera, y aprovecha la parte norte de la cuenca como zona alimentaria; su alimentación se compone esencialmente de eufausiáceos. El otro se presenta en invierno y utiliza la parte nordoriental de la cuenca como zona de reproducción.

La caza del rorcual se practica en Vizcaya (un centenar de capturas al año).

Los cachalotes no son muy abundantes; han sido y son explotados intensamente en las costas atlánticas de España (Vigo). La depauperación del rebaño de los chachalotes del Mediterráneo, habiendo cesado su explotación en Gibraltar, denota una relación entre el rebaño explotado en Vigo y el rebaño del Mediterráneo.

Los *Ziphius* son relativamente abundantes; parece ser que la parte nordoccidental del Mediterráneo les es favorable. En efecto, en invierno esta zona les ofrece condiciones tróficas satisfactorias cerca de la costa: entre 0 y 20 m; la producción obtenida en invierno no es superior a la del verano, pero se distribuye en una capa eufótica diez veces menos espesa, lo que hace posible un rendimiento superior de los niveles tróficos de consumo, disminuyendo los gastos de energía inherentes a la búsqueda del alimento. El desove se realiza en invierno. En verano los animales viven probablemente en las zonas centrales del Mediterráneo, a expensas de los cefalópodos de aguas profundas; de hecho, en sus contenidos estomacales se han detectado especies que no figuran entre las especies más comunes.

Los *Stenella*, considerados raros hasta 1970, llegan a ser preponderantes en los encallamientos anuales. La mayoría de los cadáveres son de animales jóvenes, todavía en fase de cría.

Las principales interacciones entre las diversas especies rivales son la competencia de los escualos planctófagos, la de los pequeños seláceos de aguas profundas devoradores de eufausiáceos, y la del hombre. La evaluación de las capturas de peces y de cefalópodos indica que el hombre captura al día, por término medio, la misma cantidad de peces epipelágicos que los cetáceos ictiófagos (es decir 1/8 de la producción diaria de peces), y una cantidad de cefalópodos igual a la mitad de la que devoran los cetáceos. El Mediterráneo occidental parece ser favorable a los cetáceos que se alimentan de calamares en la medida en que permite que se reconstituyan las poblaciones depauperadas debidas a la intensa explotación en las costas de España. En el estado actual de nuestras observaciones, parece que la población de *Balaenoptera physalus* experimenta una ligera recuperación desde 1974; por el contrario, el cachalote continúa siendo el más escaso de todos los cetáceos mediterráneos cuando en un tiempo fue una especie común. Ha disminuido mucho el número de los delfines jóvenes frente a las costas europeas, pero parece que son más abundantes en estos últimos años en las costas africanas y del sur de España, lo que parecería indicar un traslado de la población, debido a la contaminación de la parte septentrional del Mediterráneo.

D. Viale
Université P. et M. Curie, Station marine de Villefranche-sur-Mer, Station zoologique, 06230 Villefranche-sur-Mer, France

Facteurs écologiques du bassin nord-occidental méditerranéen

EXTENSION GÉOGRAPHIQUE DE L'ÉCO-SYSTÈME ÉTUDIÉ

La Méditerranée occidentale constitue un bassin relativement isolé, par des seuils, du bassin oriental et des régions atlantiques adjacentes (lusitanienne au nord, mauritanienne au sud). Ces seuils ne sont pas des barrières pour les cétacés, qui effectuent des déplacements entre ces régions atlantiques et la Méditerranée occidentale; les échanges avec le bassin oriental sont beaucoup moins certains et certaines espèces constituent des populations endémiques (*Delphinus* de mer Noire; *Phocaena* présent dans le bassin oriental et absent dans le bassin occidental).

L'individualité du bassin occidental apparaît dans son hydrologie et dans la dynamique de sa production organique. Il s'ensuit que toutes les espèces de cétacés ont leur biologie influencée par les conditions physico-chimiques et trophiques de ce bassin (cf. para. 1-3).

Notre étude porte sur la partie du bassin située au nord du 40ème parallèle, et plus spécialement sur les 2/3 nord-est de cette zone, comprenant la haute Tyrrhénienne, la mer Ligure, et le nord du bassin algéro-provençal. Pour la partie nord-ouest du bassin nous avons utilisé les travaux de la Commission de cétologie de Barcelone. Cependant cette limite latitudinale arbitraire ne reflète pas la structure de l'écosystème, et par suite ne permet pas d'expliquer les déplacements des cétacés; nous avons donc été amenés à considérer également les conditions thermiques et trophiques de la partie sud du bassin.

FACTEURS ABIOTIQUES

Structure hydrologique

Quatre masses d'eau différentes:

Eau atlantique: pénètre par le seuil de Gibraltar, en formant un courant superficiel est-ouest, le courant algérien, dont une branche « algéro-provençale » traverse le bassin vers la Sardaigne et vers la Corse au niveau de laquelle il se divise en trois branches (salinité inférieure à 38,4 ‰).

Eau orientale (38,5 à 39 ‰): pénètre par le seuil de Sicile et la Tyrrhénienne, et constitue *l'eau intermédiaire* de notre zone.

Eau continentale: d'origine fluviale, surtout importante au niveau du Rhône où elle forme un courant vers les Baléares et un contre-courant le long du Roussillon (caractère important pour la dispersion des polluants apportés par le Rhône); ces eaux se mêlent aux eaux méditerranéennes du courant catalan.

Eau profonde ($S = 37,5$ ‰; $T \pm 13°$): se forme par refroidissement de l'eau superficielle dans la partie septentrionale du bassin, et plonge sous les eaux intermédiaires. Température de 13° jusqu'au fond.

La structure hydrologique en conditions stables est donc à trois couches: eau superficielle, intermédiaire et profonde. Cette structure s'interrompt au niveau de quatre zones de divergence: ligure, provençale, catalane et nord-Baléares, créées par des courants cycloniques entraînant des remontées d'eau profonde, donc des enrichissements en sels nutritifs. Ces phénomènes d'upwellings se situent différemment suivant les saisons, et leur position précise détermine à chaque saison la localisation des cétacés.

En fin d'hiver les régions côtières sont favorables à la production phyto- et zooplanctonique, mais sont défavorisées par leur pauvreté en sels nutritifs. Au début du printemps les sels nutritifs remontent au niveau des divergences, et se répandent par force centrifuge vers les zones côtières; ils se trouvent alors dans des eaux assez stables pour permettre la production. En début d'été, la stabilité gagne ce centre du bassin, et l'épuisement des régions côtières et intermédiaires repousse les fortes productions vers les centres de divergence. En fin d'été, les dômes de productivité s'enfoncent à 50 m et la productivité maximale devient faible en raison du manque de lumière.

Conditions thermiques

Le golfe de Gênes est le centre d'une dépression cyclonique qui est à l'origine de vents violents, déterminants pour les courants de surface; la vallée du Rhône apporte des masses d'air froid suscitant la formation d'eau froide, mais insuffisantes pour créer des upwellings côtiers susceptibles d'enrichir le milieu: en effet, les eaux intermédiaires sont oligotrophes.

Les variations saisonnières de la température de l'eau ne se font pas selon un gradient latitudinal: en effet, la présence du golfe du Lion soumis à un refroidissement par les vents et par la partie montagneuse de la Catalogne entraîne une dissymétrie des parties est et ouest du bassin. Une diagonale joignant la côte marocaine à Gênes en passant par le sud des Baléares isole la partie nord-ouest du bassin, qui se réchauffe plus tardivement au printemps et se refroidit davantage en hiver (10 à 13°C) que la partie sud-est, qui reste douce en hiver (17 à 18°). Ces températures sont déterminantes pour les déplacements de *Delphinus delphis* et de *Tursiops truncatus*, qui se maintiennent dans des eaux comprises entre 18 et 23°, avec optimum à 19° (Viale, 1977). Les déplacements de ces Delphinidae suivent la progression des températures de l'eau superficielle au-dessus de 18° et au-dessous de 23°. Les animaux se trouvent dans le sud-est du bassin occidental en hiver, et dès février amorcent une remontée vers la Sardaigne et la Corse, qu'ils atteignent en mars, puis vers la haute Tyrrhénienne en avril; ils sont nombreux à partir de mai-juin dans le golfe de Gênes et sur la Côte d'Azur, puis en août sont plus nombreux dans les régions centrales au large de la Provence et du golfe du Lion. Entre fin septembre et novembre les températures superficielles inférieures à 18° apparaissent dans le nord du bassin, et les deux espèces migrent le long des côtes d'Espagne ou de Corse et de Sardaigne, vers le sud du bassin; leur densité fin septembre dans la mer de Malaga est 10 fois supérieure à celle des zones centrales du nord en août; l'eau est alors à 19°. Plus tard en hiver les températures inférieures à 17° gagnent le sud-ouest du bassin alors que la partie sud-orientale reste plus chaude: la région sicilo-tunisienne constitue alors un refuge pour les cétacés, particulièrement pour les espèces qui utilisent la Méditerranée comme zone de reproduction (*Balaenoptera physalus*, *Physeter macrocephalus*).

Ce déplacement saisonnier correspond également à un déplacement des conditions trophiques à l'intérieur du bassin, puisque les refroidissements et les enrichissements organiques sont liés.

FACTEURS BIOTIQUES

Conditions trophiques

En hiver les conditions de stabilité hydrologique nécessaires à la photosynthèse planctonique se trouvent près de la côte, et cela rend compte des observations de cétacés tels que *Ziphius cavirostris* près de la côte. Au début du printemps, alors que les régions côtières ont épuisé leurs sels nutritifs et deviennent oligotrophes, la stratification gagne le large et atteint les zones intermédiaires enrichies par les divergences et où se produisent de fortes poussées phytoplanctoniques (mars-avril); cela détermine en mai-juin un maximum de production zooplanctonique dans le golfe de Gênes, au large de la Corse, de la Sardaigne et de la Côte d'Azur. En été les plus fortes productions ont lieu aux centres des divergences, et elles s'enfoncent avec la thermocline jusqu'à 50 m.

On connaît encore mal ce qui se passe en fin d'été et début d'automne pour la production primaire et secondaire.

Minas (1968) a mesuré la variation annuelle de la productivité à 45 milles au large de

Marseille, et trouve 560 mg C/m²/jour en avril, 250 en juin, 111 en septembre, 108 en décembre. En juin, le maximum de productivité se trouve à 50 m de profondeur car la couche superficielle a été vidée de ses sels nutritifs. En septembre le maximum d'abondance planctonique se trouve entre 75 et 100 m. En décembre la productivité est la même qu'en septembre mais son maximum est remonté entre 0 et 10 m, et est limité au-dessous par le manque de lumière.

La productivité primaire printanière en Méditerranée nord-occidentale est importante mais la biomasse constituée se répartit dans une couche euphotique épaisse, et par suite la production secondaire est faible (Minas, 1968): une partie importante de l'énergie assimilée par le zooplancton sert à ses déplacements pour la recherche du phytoplancton. La biomasse zooplanctonique au printemps est de 500 à 1 000 mg/m² à la côte, 1 600 mg/m² dans les zones intermédiaires, 500 mg/m² au centre du bassin.

La forte diminution de l'épaisseur de la couche euphotique pendant l'hiver est intéressante pour les cétacés planctonophages. Le double effet de la localisation de la couche productive entre 0 et 20 m et du déplacement vers les côtes des conditions favorables à la production phytoplanctonique explique le rapprochement de la côte des calmars, que l'on pêche essentiellement en hiver, en surface, entre la Corse et la Toscane. En conséquence, les cétacés teuthophages sont également plus abondants près de la côte à cette époque (*Ziphius, Grampus*).

Les variations nycthémérales de la biomasse mésoplanctonique sont négligeables; elles sont faibles pour le macroplancton, et d'un ordre de 1 à 10 pour le micronecton (Razouls et Thiriot, 1973).

Utilisation de l'écosystème par les cétacés.
Leurs réseaux trophiques

Le déplacement saisonnier des cétacés en Méditerranée occidentale correspond à celui des conditions trophiques à l'intérieur du bassin. Le sud de ce dernier est constamment enrichi par l'arrivée d'eau océanique et les productivités primaire et secondaire y sont importantes toute l'année; cependant les Delphinidae en sont repoussés en été par les températures superficielles trop élevées, et se déplacent vers le nord ou vers la mer d'Alboran. Dans la partie nord du bassin la circulation géostrophique engendre des upwellings saisonniers, maximaux en fin de printemps et début d'été, qui fonctionnent tour à tour ainsi que nous l'avons exposé plus haut. Les cétacés utilisent tour à tour ces divergences. Le déplacement des cétacés n'a pas été mis en évidence aussi nettement pour les autres espèces que pour *Delphinus* et *Tursiops*. Cependant des missions d'observation en août situent au niveau de la divergence provençale 5 espèces sur les 6 fréquentant les eaux du large: *Delphinus, Stenella, Grampus, Globicephala, Balaenoptera*, alors que ces espèces s'observent essentiellement en mer Ligure en juillet. Les baleinoptères utilisent la partie nord-occidentale du bassin toute l'année, mais avec un troupeau d'hiver et un troupeau d'été distincts (cf. *Balaenoptera physalus*); ils consomment en été du zooplancton et du micronecton. Les échantillons de micronecton prélevés au filet se révèlent constitués de 12 % de poissons, 5 % de décapodes Natantia, 59 % d'euphausiacés, 24 % d'hypériens. Dans les échantillons d'euphausiacés prélevés en Méditerranée on trouve *Euphausia krohnii* (65 %), *Meganyctiphanes norvegica* (17 %), *Nematoscelis megalops* (14 %) et *N. atlantica* (4 %) (Goy et Thiriot, 1976). Casanova-Soulier (1968) a montré que *Meganyctiphanes norvegica* est beaucoup plus abondante au niveau des zones de divergence, qu'elle a la fertilité la plus forte mais que la reproduction cesse pendant les mois chauds. D'autre part, Goy et Thiriot (1976) ont montré que cette espèce est surtout présente dans le golfe de Gênes. Les analyses de contenus stomacaux de baleinoptères capturés accidentellement montrent essentiellement des *Meganyctiphanes*. Cette espèce donne lieu à des phénomènes d'essaims importants dont Thiriot (com. pers.) pense qu'ils sont liés à la re-production: 80 % de femelles. Cependant les densités maximales en mer Ligure, telles qu'estimées d'après les prélèvements au filet, sont de l'ordre de l'individu par litre, densité incompatible avec l'entretien d'un troupeau de baleinoptères tel qu'estimé par les observations directes. Ces animaux ne sont pas comme les baleines franches des filtreurs passifs, mais doivent s'alimenter par comportement actif aux dépens d'essaims tels qu'il en existe probablement au niveau des DSL.[1] La connaissance de ces dernières en Méditerranée apportera certainement des éléments de compréhension de l'écologie de ces cétacés. On connaît déjà la biomasse considérable constituée par les poissons micronectoniques dans le nord du bassin méditerranéen (Sardous, com. pers.); c'est probablement aux dépens de cette biomasse que les baleinoptères survivent l'hiver.

Les cétacés teuthophages (*Ziphius*, *Physeter*, *Grampus*, *Globicephala*) consomment des céphalopodes pélagiques (*Todorodes sagittatus*, *T. eblanae*, *Loligo vulgaris*) et benthiques. Le détail du réseau alimentaire apparaît à l'examen des becs de céphalopodes trouvés dans les estomacs de *Ziphius* et de *Grampus;* cet examen révèle des fréquences relatives d'espèces très différentes de celles habituellement fournies par les pêcheries (en particulier, l'espèce la plus consommée par *Ziphius* ne figure pas parmi les 18 espèces les plus communes de céphalopodes de Méditerranée d'après Mangold-Wirz). Cela implique que les cétacés capturent les céphalopodes dans une zone différente et/ou avec des moyens plus efficaces (cf. plus loin: évaluation de la prédation).

Les espèces ichtyophages (*Delphinus*, *Tursiops*, *Stenella*) consomment 75 % de poissons épi- ou bathypélagiques (*Sardina*, *Engraulis*, Sparidae, Myctophidae, etc.) et 25 % de céphalopodes.

---

[1] Deep Scattering Layer.

Les espèces de cétacés en Méditerranée occidentale: écologie et éthologie

Les espèces

Balaenoptera physalus

Se fondant sur l'effondrement des nombres de prises sur les côtes du nord de l'Ecosse après les années 1926-27, Jonsgård (1966) fait une relation avec les phénomènes de surexploitation au niveau de Gibraltar: il s'agirait d'un troupeau hivernant en Méditerranée et passant l'été au nord de l'Ecosse. Cette hypothèse est confirmée par l'étude des teneurs en iode de l'huile. Jonsgård émet donc le premier l'hypothèse que la Méditerranée est à la fois une zone de reproduction et une zone alimentaire.

J'ai analysé les échouages rapportés dans la littérature et observés par moi-même en fonction de la taille et de la saison. Les résultats statistiques indiquent la présence de deux cohortes (Fig. 1). L'une correspond aux naissances d'avril à juin; elle arrive dans les classes de tailles 13-18 m entre septembre et novembre. Un individu qui a 6 m en mai atteint en moyenne 13,50 m en octobre, 15 m en février suivant, soit une croissance de 9 m en 9 mois. A partir de novembre la cohorte se fond avec le contingent pérennant d'individus de 13 à 18 m.

L'autre cohorte est issue des naissances de septembre-novembre (en décembre et janvier: observations d'un mort-né et d'un mort à la naissance). Elle atteint 9 m en décembre (retardataire) et disparaît, ce qui confirme l'hypothèse de Jonsgård.

D'après les observations de baleiniers ayant pratiqué dans la zone de Gibraltar, il semble qu'il y ait une entrée de baleinoptères à Gibraltar vers avril-mai, arrivant entre juin et mi-juillet au niveau de la Corse. Ce troupeau estiverait en Méditerranée.

En été on trouve des baleinoptères essentiellement à mi-chemin des segments

Nice-Calvi et Baléares-Corse: ce sont les emplacements des centres de divergence où se concentrent à cette époque de l'année les fortes productions de phyto- et de zooplancton. Or l'autopsie d'un baleinoptère tué en juillet près de cette zone m'a permis de constater que le contenu stomacal est fait de la même bouillie planctonique que celle trouvée dans deux requins pèlerins planctonophages capturés sur les côtes corses en avril et juin. Les baleinoptères semblent donc planctonophages dans cette partie de la Méditerranée à cette époque. Or les zones où on les observe sont également celles où se rencontrent les *Meganyctiphanes norvegica*. Le problème est celui de la quantité disponible: les phénomènes d'essaims compacts ne sont pas fréquents. Les euphausiacés comme le zooplancton se trouvent éparpillés dans une grande épaisseur d'eau correspondant à la zone euphotique, particulièrement épaisse en Méditerranée. Il semble donc que l'énergie à dépenser, pour un baleinoptère cherchant sa nourriture, est importante. On doit prendre ceci en considération quand on veut élaborer des modèles d'exploitation.

En conclusion, les baleinoptères sont visibles en Méditerranée toute l'année. Le maximum d'observations en été, particulièrement en août, ne correspond-il pas à une forte densité d'observateurs dans cette région très touristique?

Des observations personnelles (Viale, 1977) faites avec M. Ridell (Marineland d'Antibes) et avec l'équipage du N.O. Korotnef (Station zoologique de Villefranche-sur-Mer), me permettent d'évaluer le troupeau de baleinoptères de Méditerranée nord-occidentale (= zone comprise entre les côtes européennes et le 42ème parallèle) à 400 individus en été.

L'estimation:

$$\frac{\text{nombre d'observations de baleinoptères}}{\text{surface explorée par le navire}} \times \text{surface totale}$$

semble bonne puisque les navires de ligne Nice-Corse en ont observé 440 en 1975 (Duguy, 1975). Il faudrait une étude identique pour évaluer le troupeau d'hiver mais cette technique n'est alors pas très utilisable en raison de l'état de la mer. On peut évaluer à environ 1 200 la population de baleinoptères entre le 40ème parallèle et les côtes européennes.

Physeter macrocephalus

Il est curieux de constater que cette espèce s'est inscrite dans les dialectes des pêcheurs italiens, corses et provençaux, qui désignent toujours un gros cétacé par le mot « cachalot ». Est-ce parce que l'espèce était jadis prépondérante dans cette zone? Ce n'est pas le cas aujourd'hui où les observations à la mer sont devenues beaucoup moins fréquentes que celles de baleinoptères. Les bêtes échouées ne dépassent pas 10 m depuis 1943. Les enquêtes auprès des pêcheurs corses et toscans révèlent une diminution des observations de cachalots.

Une estimation indirecte (comparaison du taux annuel d'échouage avec celui de cétacés dont les effectifs sont connus, et dans l'hypothèse d'une mortalité annuelle de 7 % pour toutes les espèces) aboutit à un effectif de 600 cachalots entre le 40ème parallèle et les côtes européennes.

Ziphius cavirostris

Il fréquente en troupeaux les eaux tyrrhéniennes et ligures, et se rapproche des côtes en hiver. Des échouages ont lieu en troupeaux parfois considérables, comme à Gênes en 1963 (Tortonese, 1963). J'ai pu étudier trois échouages en groupes, dont certaines bêtes avaient été mitraillées (Viale, 1973, 1974). Ils mettent bas en hiver; taille des femelles gravides: 5,50 m; taille du nouveau-né: 1,15 m. Ils sont strictement teuthophages. Nous évaluons à 1 250 la population entre le 40ème parallèle et les côtes européennes.

Globicephala melaena

Les observations à la mer sont peu fiables, en dehors de celles des spécialistes. Les données sont donc maigres. On connaît deux cas d'échouages collectifs: 72 individus s'échouent à Calvi (NO de la Corse) en 1872, et un troupeau important sur la Côte d'Azur en 1975. Quelques échouages isolés, plus nombreux depuis 1975. Population évaluée à 1 350.

Grampus griseus

Il s'observe en haute mer à partir de 20 milles des côtes. Peu de renseignements car les observations en mer sont peu fiables. En utilisant une observation du Marineland d'Antibes, on peut estimer la population globale sur la zone considérée à 7 500 individus. Quatre échouages, correspondant à des femelles de 2,20 à 3 m.

Tursiops truncatus

Fréquente presque toute l'année le plateau continental autour de la Corse et des îles toscanes, où il constitue l'espèce la plus fréquente, alors qu'il est devenu beaucoup moins fréquent sur les côtes de Provence: les pêcheurs de la région marseillaise n'ont plus à se plaindre des dégâts commis dans leurs filets, alors qu'en juin 1976 tous les filets des pêcheurs du cap Corse ont été systématiquement visités chaque nuit. Les effectifs sont difficiles à évaluer parce qu'on ne connaît pas le rayon d'action quotidien d'un groupe. L'analyse des observations faites depuis 1962 (Viale, 1977) montrent que les animaux apparaissent près de la côte en février et augmentent en nombre jusqu'à un maximum prononcé en mai (Fig. 2). Il nous apparaît que cette augmentation est corollaire de l'apparition à la côte des grands bancs de sardines, bogues, etc., poissons planctonophages tributaires de l'enrichisse-ment de printemps (cf. plus haut). Le nombre d'animaux observés diminue légèrement et montre un plateau jusqu'en août, puis s'effondre. Pas d'observation en janvier. Les groupes comprennent en moyenne 5,6 individus. Les petits naissent en début d'hiver: nombreux échouages d'individus de 1,50 m, quand les *Tursiops* arrivent dans le nord de la Méditerranée occidentale en début de printemps, ce qui implique des naissances antérieures.

Delphinus delphis

Ils apparaissent sur les côtes corses et toscanes au printemps, plus tardivement sur les côtes de Provence. Leur apparition à la côte est toujours associée à celle d'un banc de sardines (*Sardina pilchardus*), de bogues (*Boops boops*) ou d'anchois (*Engraulis encrassichola*). En été on les observe dans les zones de divergence ligure, provençale et catalane. Ils se déplacent au cours de l'été de la Corse et de la Ligurie vers la Provence et la Catalogne, et en début d'automne, ils quittent le nord du bassin et gagnent la Tyrrhénienne ou la mer de Malaga, où leur densité en octobre a été évaluée à 10 fois celle observée en août sur la divergence algéro-provençale. Il semble qu'ils passent l'hiver dans la partie sud du bassin (Viale, 1977).

Nous avons estimé leur population à 70 000 individus entre le 40ème parallèle et les côtes européennes en été.

Stenella coeruleoalba

Il vit en haute mer. Busnel (com. pers., 1974) l'a observé en troupeau mêlé avec des *Delphinus delphis*. Il est curieux de constater que le nombre des échouages sur les côtes de France, nul en 1971, augmente d'année en année: 3 en 1972, 4 en 1973, 7 en 1974, 10 en 1975, 18 en 1976. Les 50 % de ces échouages correspondent à des individus inférieurs à 1,70

m. Les études statistiques de Kasuya et Miyazaki (1975) montrent que la taille moyenne à la maturité est de 2,12 m pour les femelles, 2,19 m pour les mâles: si on applique ces normes à la Méditerranée, un seul des échouages observés est un échouage d'adultes. Peut-être doit-on envisager une acquisition de la maturité sexuelle à une taille inférieure pour le biotope méditerranéen. L'estomac d'un individu de 1,43 m, qui devrait en principe être allaité, contenait des becs de céphalopodes; à une taille plus avancée (1,80 m) il contient également des otolithes et des vertèbres de poissons, ce qui montre que l'espèce est ichtyophage et teuthophage. Des phénomènes pathologiques ont été constatés sur ces deux carcasses: concrétions de phosphate de calcium abondantes dans tous les tissus (Viale, 1975, 1976, 1977). On peut tenter une estimation de la population d'après des observations faites par le Marineland d'Antibes: moyenne par troupeau = 15 bêtes; 265 troupeaux pour la Méditerranée nord-occidentale, soit environ 4 000 individus.

Orcinus orca

Un seul échouage, en 1974, sur la côte ouest de la Corse. Le Marineland a effectué de nombreux quadrillages pour rechercher des orques, et en a observé peu.

Monodon monocera

Un individu échoué sur la côte est de la Corse en 1960.

Phocaena phocaena

Pas d'échouages ni d'observations personnelles, ce qui confirme l'absence de cette espèce en Méditerranée occidentale. Cependant Duguy (1975) rapporte des observations qui auraient été faites au large de la côte ouest de la Corse. Il convient d'être prudent, parce que systématiquement les Corses traduisent le mot « delphini » de leur langue par le mot français « marsouin » (dont il n'existe pas d'équivalent en corse).

INTERACTIONS ENTRE LES DIFFÉRENTES ESPÈCES DE MAMMIFÈRES MARINS

Elles sont très limitées en Méditerranée, où le phoque moine (Monachus monachus), seule espèce de phoque en Méditerranée, a disparu des côtes de Provence, et pratiquement de Corse.

INTERACTIONS AVEC DES RESSOURCES ANIMALES PRÉSENTES OU POTENTIELLES

Les prédations humaines créent la concurrence la plus importante en Méditerranée. Elles portent sur:

— Le poisson bleu, c'est-à-dire toutes les espèces pélagiques vivant en grands bancs, consommé par Delphinus, Stenella et Tursiops. Nous avons évalué les prises faites par les pêcheurs à une moyenne de 300 tonnes par jour dans cette zone.

— Les céphalopodes. A la suite des travaux de Mangold-Wirz (1963, 1973, 1973a) sur l'écologie et la biologie des céphalopodes méditerranéens, divers auteurs tentent actuellement d'apprécier les biomasses. L'impact de la prédation humaine est difficile à évaluer; en effet, l'importance de ce matériel sur les marchés méditerranéens comparée aux faibles rendements des chalutages prospectifs évoquerait une surexploitation importante des stocks. En fait, ces derniers sont sans doute très mal échantillonnés par les moyens classiques, et l'opinion actuelle est que la biomasse de céphalopodes est très supérieure à ce que l'on estimait jusqu'ici.

Nous évaluerons la prédation humaine en céphalopodes d'après les débarquements moyens sur les côtes de Toscane en 10 ans, et en extrapolant à toute la zone méditerranéenne nord-occidentale: soit 400 tonnes par jour.

La prédation totale par les cétacés dans cette zone peut être évaluée à partir des effectifs de chaque espèce, et de la ration alimentaire moyenne d'après la littérature (pour les grandes espèces, 2 % du poids du corps chaque jour d'après Lockyer, 1976, 1976a; pour les petites espèces, 2 à 10 % du poids du corps chaque jour d'après Sergeant, 1969 et Anderson, 1965).

Les cétacés ichtyophages consomment environ 300 tonnes par jour.

Les cétacés teuthophages consomment en moyenne 600 tonnes par jour de céphalopodes, et les ichtyo-teuthophages en prélèvent 100 tonnes par jour.

On voit donc que la prédation humaine par la pêche est du même ordre de grandeur que celle des cétacés en Méditerranée nord-occidentale, alors que pour les céphalopodes la prédation par les cétacés est supérieure (presque le double) à la pêche par l'homme: l'efficacité de détection et de capture par les cétacés est supérieure à celle de l'homme. On peut également tirer la conclusion que la Méditerranée nord-occidentale offre plus de possibilités trophiques pour les teuthophages que pour les ichtyophages. On conclut que l'homme et les cétacés prélèvent en Méditerranée nord-occidentale un quart environ de la production de poissons épipélagiques présumée (Margalef, 1967).

Exploitation

Balaenoptera physalus

En dehors des données concernant les captures faites à Gibraltar en 1926-27 rapportées dans les « Whaling Statistics », il y a peu de données nouvelles et précises. Les statistiques de 1933-39 donnent un chiffre global de prises pour l'Espagne et le Portugal. La capture, même clandestine, semble avoir cessé à Gibraltar après la seconde guerre mondiale, par contre elle se poursuit de 1968 à 1974 sur les côtes cantabriques (Viale et Ridell, 1975) où elle s'élève à 4 ou 500 têtes par an, mais depuis 1974 elle est devenue si aléatoire que certains baleiniers ont dû désarmer (Fig. 3).

Physeter macrocephalus

Ils ont été exploités avant et après la guerre de 1939-45. De 1921 à 1927 Jonsgård rapporte que 489 cachalots ont été interceptés à Gibraltar. En 1927, l'industrie baleinière s'effondre par épuisement des stocks de baleinoptères. Elle reprend en 1933 essentiellement à partir des cachalots et des petits cétacés en ce qui concerne la totalité de l'Espagne. On manque de chiffres pour la zone méditerranéenne: on sait seulement que pour l'année 1934 les prises ont été de 66 baleinoptères et 5 cachalots. L'exploitation a ensuite été interrompue pendant toute la durée de la guerre. Il semble que cela ait permis une certaine reconstitution du troupeau puisqu'une exploitation a pu se réinstaller vers les années 50 à 60: un baleinier nous a rapporté avoir capturé jusqu'à 11 cachalots par jour dans le détroit de Gibraltar. Viale et Ridell (1975) rapportent les prises d'un baleinier au large des côtes nord de l'Espagne: 536 cachalots ont été pris en 8 ans, soit en moyenne 67 captures par an pour ce navire; 5 baleiniers travaillent de façon identique dans cette zone [2]. En analysant le contenu des observations et des prises, on voit que 19 troupeaux seulement sont signalés sur 280 cas d'observations, et 7 seulement comportent des petits. D'autre part, sur 261 observations (668 cachalots), un tiers concerne des animaux

---

[2] *Note éditoriale*: à présent 2 baleiniers sont entendus être employés sur ces côtes, mais 3 bateaux plus modernes attendent d'être utilisés.

isolés, d'où pour les observations d'animaux non isolés une moyenne de 3,3 individus par groupe. La sex-ratio est de 71/476 soit 13 % de femelles. Toutes ces observations confirment l'impression donnée par la courbe d'exploitation (Fig. 3): alors que les prises annuelles ne montrent pas de diminution significative au cours des années de pêche, les prises effectuées à l'époque du maximum de passage sont régulièrement décroissantes: les points entourés d'un contour en tireté constituent un nuage d'allure décroissante. La corrélation entre l'année et le nombre maximal mensuel de prises est de − 0,78, valeur significative au seuil 2 %. Ces faits indiquent soit un étalement dans le temps des migrations (biologiquement improbable), soit un étalement dans le temps de l'effort de pêche: plus concrètement, un effort accru pendant les mois de moindre passage. L'effort pendant le maximum de passage est toujours maximal, ce qui nous autorise à interpréter la diminution des maximums de prises, à effort de pêche constant, comme un signe de diminution des effectifs.

Cette surexploitation s'ajoute à l'exploitation pratiquée au large des Açores et semble expliquer la raréfaction du cachalot en Méditerranée occidentale, dans l'hypothèse où les cachalots de Méditerranée se détachent des stocks atlantiques lors de la migration estivale vers les eaux tempérées.

Un autre indice de surexploitation apparaît dans la diminution de taille des cachalots échoués en Méditerranée occidentale: pas d'échouage supérieur à 10 m depuis 1943.

Autres effets de l'intervention humaine

DESTRUCTIONS OCCASIONNELLES

Certains mitraillages de cétacés se font au cours de manœuvres militaires. Destruction au fusil de chasse par les pêcheurs à cause des dégâts commis dans les filets (Viale, 1973).

DESTRUCTION INDIRECTE PAR L'EFFET DE LA POLLUTION

Nous renvoyons au texte adressé à la FAO pour la réunion de Bergen, « Relation entre les échouages de cétacés et la pollution chimique en mers Ligure et Tyrrhénienne ».

Conséquences pour la gestion des stocks

Dans l'état actuel des connaissances sur la population de *Balaenoptera physalus* l'exploitation de cette espèce ne doit pas être reprise en Méditerranée. Quand au cachalot, sa rareté actuelle interdit d'envisager une exploitation rentable. L'exploitation des petits cétacés ne peut être envisagée tant que des études sérieuses n'auront pas fourni des renseignements sur la dynamique de leurs populations (il est vrai que l'absence d'exploitation rend problématique une telle étude).

Les besoins de recherche

OBSERVATIONS AÉRIENNES

Pour l'évaluation des troupeaux de baleinoptères, cachalots, etc., actuellement un Centre d'étude des mammifères marins fonctionne à La Rochelle sous la direction de R. Duguy, et collationne les observations bénévoles le long des côtes de France. Des crédits pour des prospections aériennes ont été demandés.

MARQUAGES DES CACHALOTS ET BALEINOPTÈRES

ÉTUDE DES POPULATIONS DE DELPHINIDAE

SURVEILLANCE DES EFFETS DE LA POLLUTION

Sur ce point, je propose un programme de travail en collaboration avec le Laboratoire de surveillance des nuisances du CEA (Centre de Pierrelatte):

— polluants chimiques dans les cétacés;

— recherche des pesticides;

— mise en évidence d'éventuelles accumulations de métaux dans les organes;

— recherches sur les transferts d'organe à organe; devenir des polluants chimiques dans l'organisme; formes de stockage éventuelles.

Références

ANDERSON, S., L'alimentation du marsouin (*Phocoena phocoena* L.) en captivité. *Vie Milieu. (A Biol. Mar.*), 16(2):779-810.
1965

CASANOVA-SOULIER, B., Les euphausiacés en Méditerranée. *Rapp. P.-V. Réun. CIESM*, 20(3):435-7.
1971

DUGUY, R., Dans Colloque des Embiez sur les cétacés des côtes de France. CIESM, Comité des vertébrés marins et céphalopodes. 4 p. (mimeo).
1975

GOY, J. et A. THIRIOT, Conditions estivales dans la divergence de Méditerranée nord-occidentale. 2. Macroplancton et micronecton. Etude qualitative et estimation quantitative des cnidaires et des euphausiacés. *Ann. Inst. Oceanogr., Paris*, 52(1):33-44.
1976

JONSGÅRD, Å., Biology of the North Atlantic fin whale *Balaenoptera physalus*. *Hvalråd. Skr.*, (40): 62 p.
1966

KASUYA, T. et N. MIYAZAKI, The stock of *Stenella coeruleoalba* off the Pacific coast of Japan. Paper presented to the Scientific Consultation on the Conservation and Management of Marine Mammals and their Environment, Bergen, Norway, 31 August-9 September 1976. Rome, FAO, ACMRR/MM/SC 25:37 p.
1975

LOCKYER, C., Estimates of growth and energy budget for the sperm whale, *Physeter catodon*. Paper presented to the Scientific Consultation on the Conservation and Management of Marine Mammals and their Environment, Bergen, Norway, 31 August-9 September 1976. Rome, FAO, ACMM/MM/SC/38:33 p.
1976

–, Growth and energy budgets of large baleen whales from the Southern Hemisphere. Paper presented to the Scientific Consultation on the Conservation and Management of Marine Mammals and their Environment, Bergen, Norway, 31 August-9 September 1976. Rome, FAO, ACMM/MM/SC/41:179 p.
1967a

MANGOLD-WIRZ, K., Biologie des céphalopodes benthiques et nectoniques de la mer Catalane. *Vie Milieu*, (Suppl. 13):285 p.
1963

–, La faune teuthologique actuelle en Méditerranée et ses rapports avec les mers voisines. *Rapp. P.-V. Réun. CIESM*, 21(10):779-82.
1973

–, Les céphalopodes récoltés en Méditerranée par le *Jean Charcot*, campagnes Polymède I et II. *Rev. Trav. Inst. Pêches Marit., Nantes*, 37(3):391-5.
1973a

MARGALEF, R., El ecosistema. *Monogr. Fund. La Salle Sci. Nat. Caracas*, (14):377-453.
1967

MINAS, H.J., Recherche sur la production organique primaire dans le bassin méditerranéen nord-occidental. Rapports avec les phénomènes hydrologiques. Thèse. Université d'Aix-Marseille.
1968

–, Résultats de la campagne Mediprod I du *Jean Charcot* (1-14 mars 1969 et 3-17 avril 1969). *Cah. Océanogr.*, 23(Suppl.1):93-144.
1971

RAZOULS, C. et A. THIRIOT, Données quantitatives du mésoplancton en Méditerranée occidentale. *Vie Milieu (B Océanogr.)*, 23(2):209-41.
1973

SERGEANT, D.E., Feeding rates of Cetacea. *Fiskeridir.*
1969   *Skr. (Havunders.)*, (15): 246-58.

TORTONESE, E., Insolita comparsa dei cetacei *Ziphius*
1963   *cavirostris* nel Golfo di Genova. *Natura, Milano*, 54:120-2.

VIALE, D., Sightings and strandings of Cetacea around
1973   Corsica. IWC Doc. SC/25/31: 8 p. (mimeo)

—, Odontoceta around Corsica. Paper presented to the
1974   IWC 26th Annual Meeting, 10 p. (mimeo)

—, Whalebone whales and toothed whales in the wes-
1975   tern Mediterranean. Paper presented to the
         63rd Meeting of ICES, Montreal, 5 p. (mimeo)

—, Quelques exemples d'imprégnation du tégument et
1976   des viscères chez les cétacés de Méditerranée.
         Document présenté au congrès Centenaire de
         la Société zoologique de France, Paris, septembre, 1976.

—, Ecologie des cétacés en Méditerranée nord-occi-
1977   dentale: leur place dans l'écosystème, leur
         réaction à la pollution marine par les métaux.
         Thèse. Université Pierre-et-Marie Curie, Paris, 310 p. (MS)

VIALE D. et M. RIDELL, Reflections on the observations
1975   of a whaling captain in the North Atlantic.
         Paper presented to the 63rd Meeting of ICES,
         Montreal, 5 p. (mimeo)

EVIDENCE OF WHALING IN THE NORTH SEA AND ENGLISH CHANNEL DURING THE MIDDLE AGES

W.M.A. De Smet

Abstract

Although cetaceans are not now abundant in the North Sea, historical evidence indicates that they were common there and in the English Channel during the Middle Ages and earlier. Whales were probably hunted regularly in this area from at least the 9th century onward, mostly by Flemings and Normans. Biscayan whales (*Eubalaena glacialis glacialis*) and perhaps also gray whales (*Eschrichtius gibbosus*) — if this species did survive in the Atlantic until mediaeval times — may have been the main species taken; both live near the coast and are relatively easy to catch. A decline in their abundance in the late Middle Ages seems likely and may have been caused in part by hunting. Evidence of this early whaling includes references to the availability of whale meat in mediaeval markets and anecdotes about the intervention of saints in whale hunts. Harbour porpoises (*Phocoena phocoena*), now the only cetacean commonly found in the North Sea, were hunted along its southern coast and in the English Channel before the Dutch fishery for them began in the 16th century. Other small cetaceans were probably also taken.

Résumé

Bien que les cétacés ne soient pas actuellement abondants en mer du Nord, les preuves historiques montrent qu'ils y étaient communs, ainsi que dans la Manche, au Moyen Age et durant les périodes antérieures. Les baleines ont probablement été chassées avec régularité dans cette région depuis au moins le 9ème siècle, surtout par les Flamands et les Normands. Les baleines de Biscaye (*Eubalaena glacialis glacialis*) et peut-être aussi les baleines grises (*Eschrichtius gibbosus*) — si cette espèce réussit à survivre dans l'Atlantique jusqu'à l'époque médiévale — peuvent avoir été les principales espèces capturées. Toutes deux vivent près de la côte et sont relativement faciles à capturer; un déclin de leur abondance semble probable à la fin du Moyen Age; il peut avoir été partiellement provoqué par la chasse. Les preuves de cette chasse comprennent des références à la présence de viande de baleine sur les marchés médiévaux et à des anecdotes sur l'intervention des saints pendant la chasse. Le marsouin commun (*Phocoena phocoena*), le seul cétacé actuellement répandu dans la mer du Nord, était chassé le long de sa côte sud et dans la Manche avant le début de la pêcherie hollandaise qui les exploita à partir du 16ème siècle; d'autres petits cétacés étaient aussi probablement capturés.

Extracto

Aunque hoy día los cetáceos no abundan en el Mar del Norte, hay pruebas históricas de que durante la Edad Media y épocas anteriores eran abundantes tanto en ese mar como en el

Canal de la Mancha. Probablemente los flamencos y los normandos se han dedicado a la caza de ballenas en esa zona desde al menos el siglo IX. Tal vez las principales especies capturadas hayan sido la ballena de Vizcaya (*Eubalaena glacialis glacialis*) y quizás también *Eschrichtius gibbosus*, si esta especie sobrevivió en el Atlántico hasta la época medieval; ambas viven cerca de la costa y son relativamente fáciles de capturar. Parece probable que en la tarda Edad Media su abundancia haya disminuido, quizás debido, en parte, a la caza. Entre las pruebas de que ya entonces se cazaba la ballena, pueden citarse las referencias a la venta de carne de ballena en los mercados medievales y algunas anécdotas sobre la intervención de los santos en la caza de las ballenas. La marsopa común (*Phocoena phocoena*), que hoy es el único cetáceo que se encuentra frecuentemente en el Mar del Norte, se cazaba también en la costa meridional del Mar del Norte y en el Canal de la Mancha antes de que los holandeses iniciaran su explotación en el siglo XVI. Probablemente se capturaban también otros pequeños cetáceos.

W.M.A. De Smet
Rijksuniversitair Centrum Antwerpen, Antwerp, Belgium

Introduction

In most books, whaling is given as a classic example of uncontrolled predation by man on a marine resource and the period of Arctic whaling around Spitzbergen for the Greenland whale (1611-1719) is generally cited as the first of the unwise steps in the whaling industry. Several texts, however, also draw attention to the fact that the Biscayans had an active whaling industry in earlier centuries (from the 11th century onward) and that they were probably responsible for the drastic reductions in the numbers of Biscayan right whales in the Northern Hemisphere. Only a few authors are aware of the fact that whaling existed in still earlier days in other European seas, and that it was practised in the North Sea and English Channel during the Middle Ages, certainly from the 9th century onward. Evidence of this early whaling is quite scarce and is distributed throughout a number of texts. Study of the assembled references, however, leads one to the opinion that the North Sea probably had a dense whale population that disappeared because of uncontrolled whaling. These mediaeval whale hunters were mostly Flemish and Norman, and hunted long before

the Biscayans. Slijper (1958) has already referred to this hunting, even suggesting that the Biscayans may have learned whaling from the Flemish and the Normans, who may have learned it from the Norsemen.

Present state of cetaceans in the North Sea

At present, cetaceans are not at all abundant in the North Sea. Only the harbour porpoise, *Phocoena phocoena*, is commonly found, although its numbers have decreased considerably during the last decades. Three other species, *Tursiops truncatus*, *Delphinus delphis*, *Lagenorhynchus albirostris*, are of quite regular appearance, and the pilot whale, *Globicephala melaena*, is normally found in the northern part, even in large schools, but very rare in the southern North Sea. Twenty-two other species have been observed (Schultz, 1970) but always exceptionally or in limited numbers. On the southernmost border of the North Sea, the Flemish coast, 18 species have been found (De Smet, 1974).

Among the whale species, the minke

whale, *Balaenoptera acutorostrata*, is the least rare: Schultz (1970) lists some 80 cases of strandings or catches between 1824 and 1970; this means more or less once every 2 years. The common fin whale, *Balaenoptera physalus*, has been reported some 75 times from 1595 onward (once every 6 years), the blue whale, *Balaenoptera musculus*, 16 times from 1594 onward (once every 25 years), the sei whale, *Balaenoptera borealis*, 7 times from 1590 onward (once every 50 years) and the humpback whale, *Megaptera novaeangeliae*, 5 times from 1545 onward (once every 80 years). The sperm whale, *Physeter catodon*, and odontocete, is of more regular appearance, having been reported some 60 times from 1531 onward, but only old bulls are known to visit the North Sea, sometimes in small herds. The presence of the Biscayan whale, *Eubalaena glacialis glacialis*, has been noted in 1658, 1682, 1751, 1784 and 1872, and there are also cases from the Baltic Sea in 1365 and 1489 (Schultz, 1970). Fossil remains of the grey whale, *Eschrichtius gibbosus*, point to this species' presence in earlier times.

It is clear, therefore, that such a paucity of whales during the last centuries could not sustain any profitable whaling industry. Nevertheless, the North Sea with an area of 575 000 km² and its high productivity offers a proper habitat for several species of whales. During the Miocene and Pliocene periods, whales and other cetaceans were abundant in the sea that corresponds to the North Sea of the Holocene. Huge quantities of fossil bone have been discovered, especially in the region of Antwerp, from which many fossil species have been described.

Evidence of whale abundance in Roman and mediaeval times

Although during recent centuries whales were rare in the North Sea, this was certainly not the case in earlier times. This conclusion is supported by the following facts:

(i) Whale bones, often engraved, have been found in many human settlements around the North Sea and are very numerous in several places (Clark, 1947; van Beneden, 1886).

(ii) The Roman poet Juvenal speaks in his Satire X, verse 14, of a "Britannic whale" (*Ballaena brittanica*), which indicates that:

    (a) a species of whale was found along the coast of Britain, presumably the North Sea coast or the Channel coast;

    (b) this species was well known to the Roman residents of the British Isles, which would hardly be the case if it were a cetacean that was only seen at Sea;

    (c) this species was recognized as being different from the whales seen in the Mediterranean Sea.

(iii) A fishery for large sea animals, which often used harpoons and was in an organized condition certainly as early as 875 (see examples below), existed in several places along the coasts of the North Sea and the Channel.

(iv) Whale meat was normally found in many mediaeval fish markets (see examples below).

(v) In 1004, several ships sank in the English Channel after encounters with whales, a fact cited by Fischer (1881), van Beneden (1886) and Thomazi (1947).

(vi) The mediaeval chronicler Vilhelmus Brito or Guillaume le Breton (12th century) recounts that the Duke of Boulogne had baleen plates on his helmet, taken from whales found in the English Channel.

303

(vii) The making of ornaments from baleen plates was a business activity in the city of Rouen, which was controlled by ordinance of King Charles II in 1403 (Cochin, 1935). This argues that whales were being landed regularly.

(viii) In the village of Saint-Vaast-la-Hougue (in Normandy), whale bones were so numerous that they had many different uses (Thomazi, 1947).

Evidence of whale hunting during the Middle Ages

Several authors (Vaucaire, 1941) have doubted if true whaling existed in the Middle Ages, because a later authority, Olaus Magnus, when speaking of whales in the northern countries in 1555 gave as an example the flensing of a stranded animal. But evidence of whaling in the mediaeval centuries has come to us in several texts. A few of them are just anecdotes about miraculous huntings that involved the assistance of some saint, but they reveal that hunting of whales was a normal undertaking.

In 875, some fishermen, probably form Arras, had a dispute with other fishermen along the Flemish coast as how to divide the catch when a whale was taken. They then went to sea with only 2 ships and, having invoked Saint-Vaast (*Sanctus Vedastus*), they captured an "enormous fish", while the other fishermen with their many ships did not capture any. Other citations in this book on the miracles of the saint lead one to believe that whaling (the text often speaks of the whale) was a common activity during that period along the Flemish coast. The date of 875 has been cited by several authors, but in greatest detail by Vaucaire (1941).

At the end of the 10th century, another story on miraculous whale hunting with the intervention of Saint-Bavo of Ghent speaks of harpooning an animal that came up from the depths and made several appearances at the surface (Lestocquoy, 1948).

A similar miraculous happening in 1116 gives more details: the Flemish fishermen had wounded a whale with their arrows and lances and were encircling the animal with their ships. The angry animal blew rays of water into the air, sprang out of the water, disappeared and reappeared and attacked the hunters so intensely that the latter made an oath to Saint-Arnulf, promising him a piece of meat from this whale (Fischer, 1881; Degryse, 1940; Cochin, 1935; Lestocquoy, 1948).

Other proofs of whaling during the Middle Ages include:

(i) A mediaeval text cited by Beddard (1900), probably of English origin and dating from before 1000, makes it clear that whale hunting was considered to be a dangerous business in which many ships were needed.

(ii) Whaling also went on in other parts of Europe, especially along the Norwegian coast. Many Scandinavian sagas speak of whaling, even the old Edda (Vaucaire, 1941) and especially the King's Mirror, a book on Iceland written in Norway in the 12th century. Best known is a report written around 890, probably in 887 (Beddard, 1900) to the English King Alfred the Great by the Norwegian explorer Octher from Hälogoland, who sailed to the region of present-day Tromsö "as far as the whalers go" and who talks of whales 50 ft long.

(iii) In 1098, a corporation of "wallmanni" (whalers) was founded in Normandy. This corporation has been cited by several authors, including Curvier, but it has often been doubted if it was really whales they hunted or only porpoises. However, since it is known that they presented baleen plates to the Abbey of Montebourg (Lestocquoy, 1948), there can be no doubt that they whaled.

(iv) The fishermen of Boulogne (in a part of the mediaeval Flemish city of Boonen) were obliged to give a portion of the meat from every whale they caught to the Abbey of Saint-Wulmar, and this custom was already practised before 1121 (Degryse, 1944). Many other abbeys had similar rights and priorities (Vaucaire, 1941), but it is not clear if in many cases it concerns caught or stranded animals and if many of the "whales" were not simply porpoises or sharks.

(v) In 1178, when the Flemish Count Philip of Alsace returned from a long voyage, he received from the population of Bruges a monstrous beast (probably a Biscayan whale, see below) that had been taken a few days before (Degryse, 1940).

(vi) According to the mediaeval chronicler Vincent de Beauvais (1272), the inhabitants of the "German" coasts (location indeterminate) hunted whales by making noises with kettledrums and other instruments. Another writer of that period, Albertus Magnus, one of the most illuminated spirits of the Middle Ages, tells that these people hunted with harpoons, thrown by hand or by balistas (Fischer, 1881; van Beneden, 1886; Thomazi, 1947).

(vii) In the 9th century (832), the Abbey of Saint-Denis in Paris owned a place in Normandy on the Cotentin Peninsula where "whales" were hunted and flensed (Degryse, 1940; Lestocquoy, 1948).

(viii) According to Vaucaire (1941), "fishes" 50 ft long were taken in the south of the Seine by order of the Abbey of Jumièges. Other sources, however, mention fishes only 5 ft long and this suggests that the harbour porpoise was the species being taken.

(ix) In 1456, the Duke of Burgundy, Count of Flanders, had a whaling ship on the North Sea (Filliaert, 1944).

(x) Even in 1606 (i.e., 5 years before the start of extensive hunting for the Greenland whale), whaling ships were still leaving the harbour of Bruges.

(xi) Van Deinse (1966) has proved that there are signs of whaling in Holland before the year 1611, the first whaling year in Spitzbergen.

The role of whale meat in the Middle Ages

The role of whale meat in the mediaeval markets may be judged from many details. It is clear from the regularity with which whale meat occurred in these markets that it cannot have come from stranded animals alone and that there must have been regular landings. In many cases, these animals might have been dolphins and porpoises. However, the use of the words "balaena" and "Walfisch" is sufficiently common in the texts, and the quantities of meat such to make it clear that large animals must have been present. Examples of such citations include the following:

(i) In a reference from 1024 on Arras, whale meat is cited as a product with taxation for every 100 portions (Degryse, 1944).

(ii) The city of Boulogne was a centre of whale meat trading from the 11th century onward (Degryse, 1944).

(iii) A report on the market of the Flemish town Nieuwpoort in 1163 includes whale

305

meat ("partem ceti") with taxation on every portion (Degryse, 1944).

(iv) Whale meat was sold in the Flemish city of Damme in 1252 (Slijper, 1958).

(v) The market of Calais also sold whale meat and in 1300, the Count of Artois bought 33 pieces there, weighing a total of 380 lb (Degryse, 1940).

(vi) One hundred "whales" were transported to Paris on the Seine in 1315 (Cochin, 1935).

(vii) At the wedding of the Burgundian Duke Charles the Bold to the Flemish Countess Margareta at Bruges, the wedding meal included a "whale" (perhaps only a porpoise).

(viii) The Flemish Count Louis of Male often sent whale meat to his daughter Margareta at the Burgundian Court, e.g., in 1371 and 1381 (Lestocquoy, 1948; Slijper, 1958).

Species taken

The question remains, however, as to the identity of the whales which were the victims of this mediaeval hunting in the North Sea and the English Channel. Detailed accounts are too few to provide any conclusive evidence but the species involved can be postulated by a process of elimination.

Of the large whales, several can easily be excluded from consideration. First of all, the Greenland whale, *Balaena mysticetus* could hardly have been involved, because this animal normally does not leave the ice fields, although a specimen was found in the Sea of Japan (Nishiwaki and Kasuya, 1970). Fossil remains of this species in the North Sea basin date back to the Ice Ages (Jux and Rosen-

bauer, 1959). The balaenopterids are ocean dwellers and fast swimmers, who until 1864 when the harpoon gun was invented, remained beyond the reach of whalers. Only stranded specimens may have attracted people's interest. The sperm whale, *Physeter catodon*, is a species of tropical waters and only an irregular visitor to the North Sea. It is quite possible that in a few cases, especially when the fishermen had to invoke the assistance of the saints, it was this species that they were hunting, but it could not have been the object of a regular business.

This leaves only 2 species of great whale, the Biscayan or black whale, *Eubalaena glacialis glacialis*, and the grey whale, *Eschrichtius gibbosus*.

The Biscayan whale is a slow-swimming species, approximately 15 m long with a thick layer of fat. It is very widely distributed throughout the world — 3 subspecies are recognized at present. This species has always been an easy prey for whalers in several areas of the world. The quite good documentation of the last century shows that the southern sub-species (or species?), *Eubalaena glacialis australis*, which was considered as being unbelievably abundant in the beginning of the century (Harmer, 1928) was reduced to the verge of extinction some 50 years later. In the latter part of the Middle Ages, each of at least 20 fishing towns on the Bay of Biscay captured some 3 whales per year, until the species became more and more rare, and from the 1600s onward only appeared irregularly. The diminishing numbers of whales in their waters induced the Biscayans to undertake longer voyages into the Atlantic Ocean and perhaps to other seas (van Beneden, 1878, even thinks that they hunted in the English Channel and the North Sea, but no proof of this has yet been provided). It is possible that during their yearly migration the Biscayan whales not only visited the Bay of Biscay but also the English Channel and the North Sea, where they fell an easy prey to the specialized fishermen. The location of Boulogne and Calais near the isthmus between both seas put these cities in a very favourable position for predation on this migrating species.

A few facts support the supposition that *Eubalaena* was involved in this early whaling.

The short description given by a chronicler of a whale 42 ft long, caught near Ostend in 1178 and presented to the Flemish Count Philip of Alsace, says that the snout resembled the beak of an eagle. This probably refers to the typical curvature of the lips of this species (De Smet, 1974).

Many ribs, vertebrae and jaw bones found near the Flemish coast in Belgium and France (Ostend, Furnes, Mardyck, Calais) and even far inshore (Guemps in northern France) and in Zealand, Middelburg (De Smet, 1974) have been identified by van Beneden as belonging to *Eubalaena*. These bones are present in human settlements of different ages and it is possible that they come from stranded animals; it is likely however, that this species was far more abundant than at present.

The other possible species, the grey whale, *Eschrichtius gibbosus*, would seem to be easily ruled out because it is at present extant only in the North Pacific Ocean. But as early as 1936, from evidence derived from the discovery of several remains of this species in Europe, van Deinse and Junge states that the extinction of this species in the Atlantic Ocean may have occurred much later than generally believed. These authors propose that the "scrag-whale" described by Dudley in 1725 as one of the whales hunted near the northeast coast of America, and repeatedly cited by several authors in the 1700, s, is none other than this species. Recently, Fraser (1970) has supported this opinion by finding that a whale described in a 17th century Icelandic book must be the grey whale. Thus, it would not be out of the question to assume that in the Middle Ages the grey whale was still present in the eastern part of the North Atlantic Ocean and that it annually visited the North Sea and even the English Channel for calving. Several places along the coast of the North Sea, e.g., the Rhine-Scheldt-Meuse delta and the Wadden Sea, would provide excellent calving grounds, comparable to those frequented by this species in Baja California. The story that in the mouth of the Seine "fishes" 50 ft long were repeatedly caught would fit quite well with such an assumption. Re-examination of the subfossil whale bones of the European coasts would probably provide more concrete information on the presence or absence of this species in European waters during the Middle Ages.

Hunting of smaller species

The catching of smaller cetaceans has also been a regular business in several places on the coasts of the North Sea. There is no doubt that several of the references given above could refer not to large whales but to porpoises and dolphins.

Before the well-known and large-scale take of harbour porpoises began between the Danish islands in the 16th century, active porpoise hunting had occurred elsewhere. In the Flemish village of Wenduine, hunting started in 1340, after an official order had been obtained, because of the destruction that the harbour porpoise caused to the fishermen's nets (De Smet, 1974). In the markets of the Flemish towns of Ypres, Damme and Nieuwpoort, porpoise meat was among the most common products (Degryse, 1944). In Normandy (Thomazi, 1947) this species was very often caught in several places, especially near Fécamp and at the mouth of the Seine near Jumièges, from where it was exported to Paris, London or other large towns. King Henry VI of England liked the meat of this porpoise very much, according to a charter of 1426 (Slijper, 1958), as did his son, Henry VII.

Mediaeval texts also often refer to an animal called "*crassus piscis*" (fat fish), "craspois" or "graspois". It is not clear if this refers to dolphins, but the animal in question was found regularly in mediaeval markets (Degryse, 1944; Lestocquoy, 1948; Fischer, 1881; Cochin, 1935).

307

Conclusions

If true whales are rare at present in the North Sea and English Channel, despite the fact that Roman and mediaeval texts give the impression of their regular occurrence, it is clear that the number of animals must have decreased considerably during the late Middle Ages. One of the causes of this phenomenon may have been over-exploitation by man. If grey whales still existed in these waters during the Middle Ages (a fact still unproven), they must have been easily killed, just as they were during the 1800s on the east and west coasts of the North Pacific Ocean, even by primitive hunting methods. The numbers of Biscayan whales, whose past occurrence in the North Sea is proved by irregular cases in later centuries and by many subfossil findings, must have declined seriously. As this species apparently recovers very poorly, even with good protection, it is not surprising if hunting over a long time period has caused its numbers to drop consistently. In this area, the pelagic balaenopterids have not taken over the ecological niche left vacant by the demise of the slow-swimming coastal species. Thus, the North Sea is at present uninhabited by large cetaceans.

Acknowledgements

The author feels much obliged to those who have helped him to find several references to mediaeval whaling: his colleague, S. Lefevère, Koninklijk Belgisch Instituut voor Natuurweten-schappen, Brussels, and R. Degryse, Antwerp. He is convinced that a thorough consultation of many mediaeval texts by historians might yield more good examples of whaling activities in these early centuries.

References

BEDDARD, F.E., A book of whales. London, John Murray, 320 p.
1900

BENEDEN, P.J. VAN, Un mot sur la pêche de la baleine et les premières expéditions arctiques. *Bull. Acad. R. Sci. Belg. (2° Sér.)*, 46:966-86.
1878

—, Histoire naturelle de la baleine des Basques (*Balaena biscayensis*). *Mém. Couronnés Acad. R. Sci. Belg.*, 36(5):44 p.
1886

CLARK, J.C.D., Whales as an economic factor in prehistoric Europe. *Antiquity*, 21:84-104.
1947

COCHIN, J., Introduction à l'histoire des pêches maritimes en France. *In* Manuel des pêches françaises. *Mém. Off. Sci. Tech. Pêch. Marit.*, (9):17-83.
1935

DEGRYSE, R., De Vlaamache walvischvangst in de Middeleeuwen. Biekorf, pp. 20-3.
1940

—, Vlaanderens Haringbedrijf in de Middeleeuwen. Antwerpen, *Seizoenen*, (49):108 p.
1944

DEINSE, A.B. VAN, Sedert waneer is de walvisvaart vanuit Nederland bedreven? *Lutra*, 8:17-22.
1966

DEINSE, A.B. VAN and G.C.A. JUNGE, Recent and older finds of the California grey whale in the Atlantic. *Temminckia*, 2:161-88.
1936

FILLIAERT, J., De laatste Vlaamsche Ijslandvaarders. Tielt, Lannoo, 228 p., (2nd ed.).
1944

FISCHER, P., Cétacés du sud-ouest de la France. *Actes Soc. Linn. Bordeaux*, 25:5-219.
1881

Fraser, F.C., An early 17th century record of the California grey whale in Icelandic waters. *In* Investigations on Cetacea, edited by G. Pilleri. Berne, Brain Anatomy Institute, vol. 2:13-20.
1970

HARMER, S.F., The history of whaling. *Proc. Linn. Soc. Lond.*, 140:51-95.
1928

JUX, U. and K.A. ROSENBAUER, Zum Vorkommen von Cetacea-Resten in Jungpleistozänen Flussa-
1959

blagerungen der Niederrheinischen Bucht. *Neues Jahrb. Geol. Palaeontol.*, 108:81-126.

LESTOCQUOY, J., Baleine et ravitaillement au moyen âge.
1948   *Rev. Nord*, 117:39-43.

NISHIWAKI, M. and T. KASUYA, A Greenland right whale
1970   caught at Osaka Bay. *Sci. Rep. Whales Res. Inst.*, Tokyo, (22):45-62.

SCHULTZ, W., Über das Vorkommen von Walen in der
1970   Nord- und Ostsee (Ordn. Cetacea). *Zool. Anz.*, 185:172-264.

SLIJPER, E.J., Walvissen. Amsterdam, Centen, 524 p.
1958

SMET, W.M.A. DE, Inventaris van de walvisachtigen
1974   (Cetacea) van de Vlaamse kust en de Schelde. *Bull. Inst. R. Sci. Nat. Belg. (Biol.)*, 50(1): 156 p.

THOMAZI, A., Histoire de la pêche. Paris, Payot, 645 p.
1947

VAUCAIRE, M., Histoire de la pêche à la baleine. Paris,
1941   Payot, 262 p.

ACTIVIDAD BALLENERA EN EL PACIFICO SUR-ORIENTAL

L.G. ARRIAGA

Abstract

Within the sphere of the Permanent Commission for the South Pacific (CPPS) — Chile, Ecuador and Peru — whaling activity is based mainly on cachalot (*Physeter macrocephalus*) hunting, and in recent years there was an increase in the size of catches of the sei whale (*Balaenoptera borealis*). Of the three previously mentioned countries, Ecuador does not have whaling activity.

In 1950 catches in Chile and Peru came to 769, a figure which increased progressively until it levelled off at about 5 500 per year during the 1959-62 period (Saetersdal *et al.*, 1963; Mejía, 1964; 1965), consisting of an average of 97 percent of cachalots and 3 percent of other species (sei, fin, and humpback whales). In 1963 the volume of catches began to decline and between 1965 and 1970 the yearly average was about 2 000. During this last period and up to the present time, two land-based stations have been operating in the region, one in Chile (Talcahuano, 36°45'S) and the other in Peru (Paita, 05°05'S), having two and three whaling vessels, respectively.

In the 1968-74 period the annual average of cetacean catches in Chile was 283.8. Total catches for the entire period contained 88.3 percent of cachalots and 11.7 percent of other species.

According to the monthly reports received at the CPPS, the catch in Peru (Paita) during the period from December 1974 to May was of the following order:

Species	Size (m)	Males	Females	Total
Fin	more than 15.0	8	3	11
Sei	10.0-11.9	53	58	111
	12.0-14.9	368	318	686
	more than 15.0	2	22	24
Cachalot	9.0- 9.9	240	408	648
	10.0-11.9	662	127	789
	more than 12.0	189	1	190
Total				2 459

For the purpose of comparison, following the methodology of Saetersdal *et al.* (1963) and Mejis (1964, 1965) catch-per-unit effort (day/whaler at sea) was established, resulting in levels of 3.04 cachalots p.u.e. for January-May 1976. Compared to the data of the authors cited for the Paita area, these figures are higher than those of 1964 (2.70) and 1965 (2.55); similar to

311

those of 1962 (3.09) and 1963 (3.03); and lower than the figures for the 1958-61 period (between 3.56 and 4.09).

In the last few years we have noted a paralysis in the investigation of cetacean populations in the region, so that there is a need to recommence these studies.

Résumé

Dans le cadre de la Commission permanente du Pacifique Sud (CPPS), constituée par le Chili, l'Equateur et le Pérou, l'activité des baleiniers repose essentiellement sur la chasse aux cachalots (*Physeter macrocephalus*), et ces dernières années les captures de rorquals de Rudolf (*Balaenoptera borealis*) ont pris de l'importance. Parmi les trois pays susmentionnés, l'Equateur ne participe pas à la chasse à la baleine.

Les prises du Chili et du Pérou, qui atteignaient 769 unités en 1950, ont augmenté progressivement pour se stabiliser aux environs de 5 500 captures par an durant la période 1959-62 (Saetersdal *et al.*, 1963; Mejía, 1964, 1965); sur ce total on comptait en moyenne 97 pour cent de cachalots et 3 pour cent de cétacés d'autres espèces (rorquals de Rudolf, rorquals communs, jubartes). En 1963, le volume des prises a commencé à décliner, et la moyenne annuelle s'est située aux environs de 2 000 unités entre 1965 et 1970. Durant cette période et jusqu'à présent deux stations terrestres ont opéré dans la région, l'une au Chili (Talcahuano, 36°45' de latitude sud) et l'autre au Pérou (Paita, 05°05' de latitude sud), chacune dotée de deux à trois chasseurs de baleines.

Durant la période 1968-74, la moyenne annuelle des captures de cétacés a atteint 283,8 unités au Chili. Pendant toute cette période, les captures totales comportaient 88,3 pour cent de cachalots et 11,7 pour cent d'autres espèces de cétacés.

On trouvera ci-après un tableau des captures faites au Pérou (Paita) pendant la période de décembre 1974 à mai 1976, selon les données mensuelles communiquées à la CPPS.

Espèce	Taille (m)	Mâles	Femelles	Total
Rorquals communs	plus de 15,0	8	3	11
Rorquals de Rudolf	10,0-11,9	53	58	111
	12,0-14,9	368	318	686
	plus de 15,0	2	22	24
Cachalots	9,0- 9,9	240	408	648
	10,0-11,9	662	127	789
	plus de 12,0	189	1	190
Total				2 459

En déterminant, à des fins comparatives, les captures par unité d'effort (jour/chasseur à la mer) en suivant la méthode de Saetersdal *et al.* (1963) et de Mejía (1964, 1965), on a obtenu les valeurs suivantes: 3,04 cachalots par unité d'effort pour 1975 et 2,72 cachalots par unité d'effort pour janvier-mai 1976. Ces chiffres, comparés aux données des auteurs citées pour la zone de Paita, excèdent ceux de 1964 (2,70) et de 1965 (2,55); ils sont analogues à ceux de 1962 (3,09) et de 1963 (3,03) et inférieurs aux chiffres relatifs à la période 1958-61 (entre 3,56 et 4,09).

Ces dernières années, on a constaté que les recherches sur les populations de cétacés dans la région sont au point mort, ce qui rend nécessaire une reprise des travaux.

Extracto

En el ámbito de la Comisión Permanente del Pacífico Sur (CPPS) − Chile, Ecuador y Perú − la actividad ballenera está sustentada principalmente por la caza de cachalotes (*Physeter macrocephalus*), y en los últimos años aumentó la significación de las capturas de ballena boba o sei (*Balaenoptera borealis*). De los tres países, Ecuador no cuenta con actividad ballenera.

Las capturas en Chile y Perú alcanzaron en 1950 a 769 ejemplares, cifra que aumentó progresivamente hasta estacionarse en alrededor de 5 500 por año en el período 1959-62 (Saetersdal *et al.*, 1963; Mejía, 1964 y 1965), comprendiendo en promedio un 97 por ciento de cachalotes y un 3 por ciento de otras especies (ballena sei, ballena fin, ballena de joroba). En 1963 se inició una declinación del volumen de capturas y entre 1965-70 el promedio anual fue de alrededor de 2 000 ejemplares. En este último período y hasta la fecha actual operan en la región dos estaciones terrestres: una en Chile (Talcahuano, 36°45' lat. S) y otra en Perú (Paita, 05°05' lat. S), contando con dos y tres buques cazadores, respectivamente.

En el período 1968-74, el promedio anual de capturas de cetáceos en Chile fue de 283,8 ejemplares. En todo este período, las capturas totales comprendieron un 88,3 por ciento de cachalotes y un 11,7 por ciento de otras especies.

La captura en Perú (Paita) en el período de diciembre de 1974 a mayo de 1976, de acuerdo con los informes mensuales recibidos en la CPPS, fue del orden siguiente:

Especie	Talla (m)	Machos	Hembras	Total
B. fin	más de 15,0	8	3	11
B. sei	10,0-11,9	53	58	111
	12,0-14,9	368	318	686
	más de 15,0	2	22	24
Cachalote	9,0- 9,9	240	408	648
	10,0-11,9	662	127	789
	más de 12,0	189	1	190
Total				2 459

Con fines comparativos y siguiendo la metodología de Saetersdal *et al.* (1963) y de Mejía (1964 y 1965) se determinó la captura por unidad de esfuerzo (día/buque cazador, en el mar), obteniéndose valores de 3,04 cachalotes p.u.e. para 1975 y de 2,72 cachalotes p.u.e. para enero-mayo de 1976. Estas cifras, comparadas con datos de los autores citados para la zona de Paita, son superiores a las de 1964 (2,70) y 1965 (2,55); similares a las de 1962 (3,09) y 1963 (3,03); e inferiores a las del período 1958-61 (entre 3,56 y 4,09).

En los últimos años se ha observado una paralización en la investigación de las poblaciones de cetáceos en la región, siendo necesaria la reiniciación de los estudios.

L. Arriaga
Comisión Permanente del Pacífico Sur, Casilla 16199, Santiago 9, Chile

Introducción

En el ámbito de los tres países que forman la Comisión Permanente del Pacífico Sur (CPPS), la caza de ballenas es realizada por Chile y Perú. Ecuador no ha efectuado capturas hasta la fecha actual. La actividad está sustentada principalmente por el faenamiento de cachalotes (*Physeter catodon*) y, en los últimos años, tiene importante significación la caza de ballena boba o sei (*Balaenoptera borealis*), contándose también una reducida proporción de ballena de aleta o fimbaque (*B. physalus*).

Las más amplias informaciones sobre capturas y condiciones de población de ballenas en el Pacífico SE (correspondientes a estudios efectuados en la región) pueden encontrarse en diferentes trabajos que se citan en la bibliografía, la mayoría de ellos realizados entre 1961-65. Lamentablemente, con la declinación de la actividad ballenera a partir de 1964, también decreció el esfuerzo científico dedicado al estudio de las ballenas. Los datos principales que incluimos en el presente informe corresponden a las actividades recientes y de acuerdo a las informaciones que está recibiendo la CPPS, principalmente de Perú.

Revisión de informaciones (1951-65)

Capturas

Las series estadísticas para el período indicado han sido publicadas por diferentes autores (Saetersdal *et al.*, 1963; Mejía, 1964, 1965), en las que se puede observar que la caza de ballenas en el Pacífico SE creció progresivamente desde 1950 (769 ejemplares) para estacionarse en alrededor de 5 500 ballenas por año entre 1959-62, cuya composición promedio corresponde a un 97 % de cachalotes y el 3 % restante a otras especies (ballena azul, ballena de aleta, ballena jorobada y sei). Se exceptúa 1951, año en que se alcanzó la cifra más alta conocida para el Pacífico SE (7 113 cachalotes + 410 de otras especies), producción originada en la actividad de 4 factorías flotantes con 51 buques cazadores.

Exceptuando 1951 y 1954 (operó una factoría flotante), la actividad ballenera de la región fue realizada con participación de 6 estaciones terrestres (en algunos años 5), 3 localizadas en Chile y 3 en Perú.

A partir de 1963 se inició la declinación de las capturas (1963 = 4 763; 1964 = 3 273 cachalotes) y entre 1965-70 el promedio anual de capturas totales para Chile y Perú fue de alrededor de 2 000 ejemplares. La declinación de las capturas estuvo originada en la reducción del stock de cachalotes (Mejía, 1964, 1965) a un nivel tal que las bajas capturas obligaron al cierre de dos factorías en Perú (1974). Esta situación también fue evidente en Chile, y actualmente operan en la región sólo dos estaciones terrestres, una en cada país.

Población y datos biológicos

La población de cachalotes del Pacífico SE parece mantenerse bastante estacionaria dentro del área comprendida entre las Islas Galápagos y los 40° lat. S, lo cual es particularmente evidente en el caso de las hembras y, probablemente, de los machos jóvenes. Una cierta parte de la población, consistente especialmente de machos de gran talla, se desplaza hacia la Antártida y puede retornar al Pacífico SE (Heinsen, 1957; Saetersdal *et al.*, 1963). Al respecto, por lo menos un individuo de cachalote procedente de la Antártida fue registrado con certeza en Talcahuano, Chile (Clarke *et al.*, 1964).

La distribución y abundancia de la población de cachalotes del Pacífico SE estaría vinculada a la alta productividad del sistema de Corrientes de Humboldt, ocurriendo un desplazamiento hacia el sur a partir de septiembre, lo cual es concordante con la mayor captura observada durante diciembre en la

314

Antártida y, al respecto, Saetersdal *et al.*, (1963) estiman como más probable la existencia de « una pérdida continua de machos hacia las latitudes más altas en lugar de una migración estacional regular ».

La temporada de apareamiento del cachalote en el Pacífico SE se prolonga por 11 meses (abril-febrero), con una ocurrencia preferente entre junio y diciembre (invierno-primavera del H.S.) y un máximo en septiembre. La gestación se prolonga por 518 días (17 meses) y los partos ocurren entre septiembre-julio, especialmente entre noviembre-mayo, con un máximo en febrero (Clarke *et al.*, 1974). Otros datos biológicos basados en el estudio de 2 162 ejemplares (1 235 machos y 927 hembras) de Chile y Perú por Clarke, Aguayo y Paliza (1964) y de 1 200 hembras del Perú (Mejía, 1963), establecen los siguientes resultados:

Machos:
Longitud media a la
madurez sexual (50 %)   9,8 m

Hembras:
Longitud media a la
madurez sexual (50 %) 8,6 m (Clarke *et al.*, Chile-Perú)
8,7 m (Mejía, Perú)

Rango a la primera
madurez   7,6-9,8 m
(Mejía, Perú)

Las informaciones sobre aspectos de poblaciones y biológicos para otras especies de ballenas presentes en el Pacífico SE, que estén basadas en estudios efectuados en la región, son muy escasas. Algunas conclusiones fueron establecidas por Clarke (1962), tales como: la ballena de aleta migra en primavera (H.S.) hacia la Antártida y la misma población que es cazada en aguas de Chile en primavera es capturada durante el verano en la Antártida. Cita también el primer registro probado de una ballena Bryde en el Pacífico SE (Iquique, Chile).

Revisión de información actual disponible en la CPPS

CAZA DE BALLENAS EN CHILE (1968-74)

Las informaciones disponibles (Tabla 1) comprenden datos globales por especie de las capturas efectuadas por la única estación terrestre que opera actualmente en Talcahuano (36°45' lat. S) con la intervención de 2 buques cazadores. Sólo para 1974 se informó de la composición por sexos de las capturas totales: 30,7 % machos y 69,3 % hembras.

Tabla 1. Capturas totales por especies en Chile (1968-74)

	1968	1969	1970	1971	1972	1973	1974
Azul	—	—	1	1	—	—	—
Aleta	25	—	3	3	—	—	2
Jorobada	1	1	—	—	—	—	—
Sei	83	31	17	1	15	14	32
Cachalote	319	221	270	246	337	232	130
Otras	—	—	—	2	—	—	—
Total	428	253	291	253	352	246	164

Fuente: M. Vargas, División de Pesca y Caza, SAG, Chile (com. personal).

CAZA DE BALLENAS EN PERU (1975-76)

A partir de diciembre de 1974 la CPPS está recibiendo regularmente los datos sobre caza de ballenas en el Perú por intermedio de la Sección Peruana de la CPPS. Los datos corresponden a las capturas de la única estación terrestre que opera actualmente en Paita (05°05' lat. S) con el apoyo de tres buques cazadores.

Composición de las capturas por especies

En la Tabla 2 se puede apreciar una mayor participación de la ballena boba (sei) en las capturas totales y la disminución en porcentaje de las capturas de cachalotes en rela-

315

Tabla 2. Composición de las capturas, por especies (Paita)

Año	Especie	Capturas Nº de ejemplares	%
1975	Cachalote	793	59,1
	B. boba (sei)	545	40,6
	B. de aleta (fin)	5	0,3
		1 342	100,0
1976 (enero-mayo)	Cachalote	699	76,1
	B. boba (sei)	213	23,2
	B. de aleta (fin)	6	0,7
		918	100,0

ción al período 1959-62 (97 %). El desplazamiento del esfuerzo de caza hacia especies diferentes al cachalote se inició desde 1963, conforme fue observado por Mejía (1965) como una consecuencia de la declinación del stock de cachalote en el Pacífico SE.

Talla y sexo de las capturas de ballena boba y de ballena de aleta (diciembre 1974-mayo 1976)

Los datos recibidos presentan la talla de cada ejemplar cazado. La Tabla 3 muestra

Tabla 3. Composición por talla (m) y sexo en las capturas de ballena boba y ballena de aleta (Paita)

Mes	Ballena boba (Sei)							Ballena de aleta (Fin)		
	10,0-11,9 m		12,0-14,9 m		más de 15,0 m		Total	más de 15,0 m		Total
	M	H	M	H	M	H		M	H	
diciembre 1974	2	3	34	24	—	—	63	—	1	1
enero 1975	3	—	47	29	—	—	79	—	—	—
febrero	—	—	10	11	—	—	21	—	—	—
marzo	3	—	24	27	—	—	54	—	—	—
abril	4	2	24	22	—	—	52	—	—	—
mayo-julio [1]	—	—	—	—	—	—	—	—	—	—
agosto	—	—	17	8	2	1	28	1	—	1
septiembre	—	—	45	32	—	7	84	1	—	1
octubre	7	8	36	32	—	3	86	—	—	—
noviembre	8	13	25	19	—	—	65	1	—	1
diciembre	2	3	26	39	—	6	76	—	1	1
Total 1975	27	26	254	219	2	17	545	3	1	4
enero 1976	6	9	49	37	—	4	105	2	—	2
febrero	6	4	21	18	—	1	50	3	—	3
marzo	11	13	—	—	—	—	24	—	1	1
abril	—	—	—	1	—	—	1	—	—	—
mayo	1	3	10	19	—	—	33	—	—	—
enero-mayo 1976	24	29	80	75	—	5	213	5	1	6
Todo el período	53	58	368	318	2	22	821	8	3	11
Participación (%)	13,5		83,6		2,9		100			
Composición por sexo (%) machos hembras	47,7 52,3		53,6 46,4		8,3 91,7		51,5 48,5	72,7 27,3		

M = macho; H = hembra.
[1] Los buques permanecieron en puerto.

Tabla 4.   Composición por talla (m) y sexo en las capturas de cachalote (Paita)

Mes	9,0-9,9 m		10,0-11,9 m		Más de 12,0 m		Total
	M	H	M	H	M	H	
diciembre 1974	45	34	22	13	20	1	135
enero 1975	17	14	45	5	16	—	97
febrero	39	47	52	14	31	—	183
marzo	6	13	40	3	20	—	82
abril	7	13	8	1	14	—	43
mayo-julio [1]	—	—	—	—	—	—	—
agosto	—	—	1	—	2	—	3
septiembre	10	12	12	3	2	—	39
octubre	30	47	43	5	4	—	129
noviembre	16	30	48	2	1	—	97
diciembre	8	21	69	8	14	—	120
Total 1975	133	197	318	41	104	—	793
enero 1976	5	32	66	7	16	—	126
febrero	9	36	43	16	10	—	114
marzo	15	53	107	15	7	—	197
abril	11	32	56	19	25	—	143
mayo	22	24	50	16	7	—	119
enero-mayo 1976	62	177	322	73	65	—	699
Todo el período	240 (14,7 %)	408 (25,1 %)	662 (40,7 %)	127 (7,8 %)	189 (11,6 %)	1 (0,06 %)	1 627 (100,0 %)
Participación (%)	39,8		48,5		11,7		100,0
Composición por sexo (%)							
machos	37,0		83,9		99,5		67.1
hembras	63,7		16,1		0,7		32,9

M = macho;   H = hembra.
[1] Los buques permanecieron en puerto.

Tabla 5.   Comparación de capturas de cachalotes por unidad de esfuerzo entre diferentes períodos (Paita)

Fuente	Período	Esfuerzo (día-cazador)	Captura por día-cazador
Saetersdal, Mejía y Ramírez, 1963	1958	327,5	4,09
	1959	543,9	3,56
	1960	476,0	4,00
	1961	547,0	4,05
J. Mejía, 1964	1962	511,2	3,09
	1963	669,2	3,03
J. Mejía, 1965 (de gráfico)	1964	—	2,70
	1965 (7 meses)	—	2,55
Datos este informe	1975	261	3,04
	1976 (enero-mayo)	257	2,72

la composición por talla proporcionada por la estación de Paita, por cuanto sólo en ella figura la separación de frecuencias de sexos. Se observa que en las capturas de ballena boba hay una participación dominante de ejemplares entre 12,0 y 15,0 m (83,6 %) y que todas las capturas de ballena de aleta correspondieron a una talla superior a 15,0 m.

En la Tabla 3 consta la proporción de las capturas por sexo, en la cual observamos una evidente mayor captura de hembras de ballena boba (91,7 %) en las tallas de más de 15,0 m.

Datos sobre capturas de cachalote (diciembre 1974-mayo 1976)

Como en el caso de las dos especies anteriores, con la finalidad de establecer la composición de las capturas por sexo, se mantienen los rangos de talla proporcionados por la estación de Paita.

La mayor participación en las capturas de todo el período corresponde a cachalotes entre 10,0 y 12,0 m (48,5 %). La participación por sexo en las capturas totales, para los rangos indicados en la Tabla 4, presenta una mayor captura de machos entre 10,0 y 12,0 m (40,7 %), seguido de las hembras pertenecientes a la categoría de 9,0 a 10,0 m (25,1 %). Nótese la bajísima participación de hembras mayores a 12,0 m (0,06 %), representada por la captura de un ejemplar en diciembre de 1974.

La observación de la composición por sexos en cada categoría, presentada en la Tabla 4, nos demuestra una alta participación de hembras (63,0 %) en el rango de 9,0 a 10,0 m y la casi ausencia de hembras con más de 12,0 m ya citada. Las mayores capturas fueron efectuadas sobre la población de machos (67,1 %) si se consideran globalmente todas las categorías.

Capturas por unidad de esfuerzo (cachalote)

Los datos sobre el esfuerzo de caza corresponden al total de días en el mar dedicados a la caza de cachalotes por los tres buques cazadores que operan en el período que se analiza para la estación de Paita. Los tres cazadores tienen condiciones similares y por tanto se estima que el esfuerzo de ellos es comparable entre sí, no habiéndose efectuado corrección alguna por tonelaje u otra variable. Así, la unidad de esfuerzo corresponde a día-cazador (en viaje), manteniendo con fines de comparación la metodología usada por Saetersdal et al. (1963) y por Mejía (1964, 1965). Debemos anotar que en algunos casos no fue posible una clara separación del esfuerzo dedicado a la caza de cachalote y el correspondiente a otras especies; sin embargo, siendo muy pocas estas dificultades, estimamos que el error inducido no es apreciable. En la Tabla 5 presentamos los resultados obtenidos, comparándolos con datos anteriores para el área de Paita.

Teniendo en consideración que la relación de captura por unidad de esfuerzo « representa muy bien sobre la abundancia del stock » (Mejía, 1965), no obstante las limitaciones del presente informe, podríamos encontrar en las cifras de captura por día-cazador de 1975-76 una ligera indicación de la recuperación del stock de cachalotes en el Pacífico SE y que los niveles del mismo podrían ser comparables a los de los años 1962-64, al menos para el área de Paita.

Finalmente, debería restablecerse el esfuerzo de investigación sobre las poblaciones de cetáceos del Pacífico SE, aspecto fundamental para que el manejo de su explotación contenga criterios adecuados para la conservación de estos recursos.

Referencias

CLARKE, R., Observación y marcación de ballenas frente
1963   a las costas de Chile en 1958 y desde Ecuador hacia las Islas Galápagos y más allá en 1959. *FAO Fish. Biol. Tech. Pap.*, (32):27 p.

CLARKE, R., A. AGUAYO, L. y O. PALIZA, G., Progress
1964   report on sperm whale research in the southeast Pacific Ocean. *Norsk Hvalfangsttid.*,
53(11):297-302.

HEINSEN, M., H., Stock de ballena esperma en el Pací-
1957   fico Sur, su racionalización y explotación. *Inf.*
Arch. Com. Perm. Pac. Sur (mimeo).

LOESCH, H., Observación de ballenas en aguas ecuato-
1966   rianas. Informe de datos mayo de 1963-febre-
ro de 1964. *Bol. Cient. Téc. Inst. Nac. Pesca*
Guayaquil, 1(4).

MEJÍA, G., J.M., Sobre la madurez sexual de las hembras
1963   de ballena esperma (*Physester catodon* L.)
frente a la costa del Perú. *Inf. Intern. Inst.*

Mar. Perú, (112/1963) (Mimeo Arch. Com.
Perm. Pac. Sur).

–, El estado del stock de cachalotes. *Inf. Intern. Inst.*
1964   *Mar Perú*, (17/64) (Mimeo Arch. Com
Pac. Sur).

–, Efecto de la explotación en el stock de cachalotes
1965   (Mimeo Arch. Com. Perm. Pac. Sur).

SAETERSDAL, G., J. MEJÍA y P. RAMÍREZ, La caza de ca-
1963   chalotes en el Perú. Estadísticas de captura
para los años 1947-61 y un intento de analizar
las condiciones de la población en el período
1954-61. *Bol. Inst. Invest. Recurs. Mar.*,
1(3):45-84.

HISTORICAL RECORDS OF AMERICAN SPERM WHALING

R.C. KUGLER

Abstract

The means and incentive to hunt sperm whales regularly occurred in America in 1750 with the introduction of tryworks aboard the vessel and of the process of manufacturing candles from spermaceti. The number of whales taken before the American Revolution, mostly from the North Atlantic, was modest; the fishery shut down during the war and recovery after it was slow until, following the reopening of the British oil market after the War of 1812 and the growth in population and economic activity in America, the fishery entered a period of prosperity, with intense exploitation of sperm whale stocks around the world, lasting from 1820 to 1860. Most of the sperm oil imported into the United States during this period came from whaling vessels out of Nantucket and New Bedford, though after 1843, when the American fishery began to decline, New Bedford merchants began increasingly to dominate the fishery and those from Nantucket to withdraw from it. Imports of sperm whale oil into America were more than 40 000 barrels in 1821, almost 167 000 barrels in 1843, less than 100 000 barrels in 1850 (for the first time in 18 years) and, after the Civil War, only once exceeded 50 000 barrels. The decline of the fishery was related more to market factors, especially competition from other means of lighting and also a shift of interest to the products of right and bowhead whales, rather than to a belief that sperm whales had become too scarce to hunt — thus, it seems that stocks of sperm whales began their recovery from American whaling as early as 1844. While records of American sperm whaling can be useful in certain ways to those concerned with management of these whales now, this use is probably limited. One source which might yield much of the information hoped to be obtained by abstracting logbooks today, is the chart of world sperm and right whale sightings produced by M.F. Murray in 1852, based on data from about 1 000 logbooks, mostly from the decade of greatest whaling activity, 1835-45.

Résumé

Les moyens et la motivation de la pêche régulière du cachalot apparurent en 1750 en Amérique avec l'installation des fondoirs de graisse à bord des bateaux et la fabrication de chandelles de spermaceti. Le nombre de baleines capturées avant la révolution américaine, surtout dans l'Atlantique Nord, était modeste; la pêcherie ferma pendant la guerre et ne reprit que lentement jusqu'au moment où, à la suite de la réouverture du marché de l'huile britannique après la guerre de 1812, ainsi que de la croissance de la population et de l'activité économique en Amérique, la pêche entra dans une période de prospérité qui dura de 1820 à 1860 avec l'exploitation intense des stocks mondiaux de cachalots. La majeure partie de l'huile de cachalot importée aux Etats-Unis à cette époque provenait de baleiniers basés à Nantucket et New Bedford même si après 1843, quand la pêche américaine commença à décliner, les marchands de New Bedford commencèrent à dominer de plus en plus la pêcherie et ceux de

Nantucket s'en retirèrent. Les importations d'huile de cachalot en Amérique dépassaient 40 000 barils en 1821, atteignaient presque 167 000 barils en 1843, étaient inférieures à 100 000 barils en 1850 (pour la première fois en 18 ans) et, après la guerre de Sécession, ne dépassèrent qu'une fois 50 000 barils. Le déclin de la pêche fut lié plus à des facteurs de marché, spécialement la concurrence d'autres procédés d'éclairage (et aussi un déplacement de l'intérêt vers les produits des baleines franches et des baleines du Groenland) qu'à l'impression que les cachalots étaient devenus trop rares pour qu'on les pêche. Il semble ainsi que les stocks de cachalots commencèrent à reprendre dès 1844, après le ralentissement de l'exploitation américaine. Si les archives de la chasse américaine au cachalot peuvent avoir une certaine utilité pour ceux qui s'intéressent actuellement à l'aménagement de cette ressource, cette utilité est probablement limitée. Une source susceptible de produire une bonne partie des informations que l'on cherche à obtenir aujourd'hui des livres de bord est la carte des repérages visuels de cachalots et de baleines franches établie par M.F. Murray en 1852, fondée sur les données de quelque 1 000 livres de bord provenant en grande partie de la décennie de majeure activité, de 1835 à 1845.

Extracto

La introducción en América, en 1750, de los hornos para fusión de grasa a bordo y del proceso de manufactura de velas de esperma ofreció por primera vez los medios e incentivos necesarios para explotar regularmente las poblaciones de cachalote. El número de cachalotes capturados antes de la Revolución Americana, en su mayoría en el norte del Atlántico, era modesto; la caza se interrumpió durante la guerra, y la recuperación después de ella fue lenta, hasta que, al abrirse de nuevo el mercado británico de grasas, después de la guerra de 1812, y con el aumento de la población y de la actividad económica en Estados Unidos, la caza del cachalote conoció un período de prosperidad que duró de 1820 a 1860, durante el cual se explotaron intensamente las poblaciones de cachalote de todo el mundo. La mayor parte del aceite de esperma que entró en los Estados Unidos durante ese período procedía de los balleneros de Nantucket y New Bedford, aunque después de 1843, cuando la pesquería americana empezó a disminuir, los comerciantes de New Bedford fueron progresivamente dominando la pesquería y los de Nantucket fueron retirándose de ella. Las descargas de aceite de esperma en Estados Unidos fueron de más de 40 000 barriles en 1821, casi 161 000 barriles en 1843 y menos de 100 000 en 1850 (por primera vez en 18 años), y después de la guerra civil sólo una vez superaron los 50 000 barriles. La disminución de la explotación del cachalote se debió más a factores de mercado, especialmente a la competencia de otros medios de iluminación y a un desplazamiento del interés hacia los productos derivados de las ballenas franca y de cabeza arqueada, que a la convicción de que los cachalotes eran ya demasiado escasos para que valiera la pena cazarlos. Según parece, pues, las poblaciones de cachalote comenzaron a recuperarse de la explotación ya en 1844. Si bien los datos existentes sobre la caza de cachalote en los Estados Unidos pueden ser útiles en cierta forma para quienes se ocupan hoy de la ordenación de la explotación de esos cetáceos, su uso es probablemente limitado. Una fuente que podría facilitar buena parte de la información que hoy día se espera conseguir sacando datos de los cuadernos de bitácora es el mapa de los avistamientos de cachalote y ballena franca preparado por M.F. Murray en 1852, basado en datos procedentes de unos mil diarios de navegación, la mayoría de ellos de la década en que la explotación de los cachalotes fue más activa: 1835-45.

R.C. Kugler
Whaling Museum, Old Dartmouth Historical Society, 18 Johnny Cake Hill, New Bedford, MA 02740, USA

American sperm whaling

The beginnings

Intensive exploitation of worldwide sperm whale stocks by Americans occurred during the 40 years from 1820 and 1860, with the peak decade falling between 1835 and 1845. Although the origins of the fishery began as early as 1712, neither the means nor the incentives to pursue sperm whales came into being until 1750.

In that year, two innovations provided both technology and markets (Kugler, 1971). The first consisted of the introduction of the tryworks aboard the vessel, thus permitting the processing of the blubber immediately after the whale was captured and enabling the cruise to be extended into the warmer climates inhabited by sperm whales.

The incentive to use the tryworks also occurred about 1750, when the process of manufacturing candles from spermaceti was introduced into the American colonies. Earlier, the prices offered by merchants for sperm and right whale oil differed little, and few distinctions were made between the two in the market-place. Later, as candleworks proliferated, the two oils were sold as separate products, with the higher prices offered for spermaceti and sperm oil providing the mainspring for the expansion of that fishery.

Before the American Revolution, the number of sperm whales taken was modest, limited by the small size of the vessels then employed and by the restricted amounts of capital available to underwrite a business regarded as more risky than other branches of trade or commerce. The hunting of these whales was confined for the most part to the North Atlantic, although successful voyages had been made to such South Atlantic grounds as Walvis Bay, Brazil Banks and Falkland Islands (Islas Malvinas).

The war itself shut down the fishery. When peace came, the American whalemen found the English market, the only one of consequence, barred by a prohibitive tariff. The industry's recovery was slow and fitful, dependent on a limited domestic consumption or on the ability to penetrate the markets of France and the north European states. For control of the latter markets, Great Britain began to compete vigorously, entering the sperm whale fishery for the first time and offering generous bounties to encourage her merchants and investors. The adeptness of the Americans as sperm whalers could not be matched so rapidly and England's effort to break the former colony's monopoly of the fishery did not survive the war of 1812.

With the peace of 1815, the American whaling industry entered a period of prolonged prosperity, based partly on the growth in population and economic activity at home, and partly on the reopening of the British market to American oils.

Intensive exploitation of sperm whale stocks

In 1821, the annual imports of sperm whale oil by the American fleet exceeded 40 000 barrels for the first time. Table 1, based on Starbuck (1878), Hegarty (1959), and Lyman's Appendix 6.1 to *Nineteenth Report of the International Whaling Commission* (1969), summarizes annual production, from 1800 to 1909:

Sources of production

The bulk of all sperm oil imported into the United States in the period 1820-1860 was produced by vessels employed solely in that fishery by the whaling merchants of Nantucket and New Bedford. The sources of sperm whale production in 1843, a year of near record imports, are analysed below (Table 2).

The sources of sperm oil production changed to some extent between 1843 and 1860, the principal difference being New Bedford's increasing dominance of the fishery and

323

Table 1. US sperm oil production, 1800-1909

Decade	Average yearly US catch, in barrels [1]	Average yearly world catch, in barrels	Average number whales per year 40 and 45 bbls [2]	Add 10 % for whales and oil lost [3]
1800/09	11 910	19 850	496/ 794	546/ 873
1810/19	14 726	22 655	566/ 906	623/ 997
1820/29	64 951	92 787	2 320/3 711	2 552/4 082
1830/39	126 334	147 414	3 685/5 897	4 054/6 487
1840/49	130 628	145 790	3 645/5 832	4 009/6 415
1850/59	85 333	94 814	2 370/3 793	2 607/4 172
1860/69	53 616	59 573	1 489/2 383	1 638/2 621
1870/79	42 454	47 171	1 179/1 887	1 297/2 076
1880/89	24 617	27 352	684/1 094	752/1 203
1890/99	14 321	15 912	398/ 636	437/ 700
1900/09	16 382	18 202	455/ 728	500/ 801

[1] 1 barrel = 31.5 gals = 170 kg.
[2] The computation of whales taken uses 2 averages frequently cited in earlier whaling studies, i.e. 40 and 25 barrels per whale. Logbook research might permit a single, more reliable estimate.
[3] The 10 % allowance for whales killed but not recovered and for oil taken but lost by shipwreck or leakage is also traditional. Lyman (1969), however, uses 15 %.

Table 2. Major sources of sperm oil production, 1843

	Barrels	Average voyage/ months
1. by 28 New Bedford sperm whalers from Pacific or Indian Oceans	49 873	41.3
2. by 17 Nantucket Pacific sperm whalers	29 350	43.3
3. by 7 New Bedford Atlantic sperm whalers	1 470	17.0
4. by 7 Fairhaven Pacific/Indian Ocean sperm whalers	10 046	42.7
5. by 3 Edgartown Pacific sperm whalers	6 400	42.0
6. by 17 Provincetown Atlantic sperm whalers	4 370	9.1
7. by 3 Falmouth Pacific sperm whalers	5 400	44.4
	106 909	
8. Produced by sperm whalers from other ports or by right or mixed whalers	60 076	
Total US sperm oil imported in 1843	166 985	

Nantucket's withdrawal from it. In the latter year, 73 656 barrels of sperm oil were imported, of which New Bedford vessels accounted for about 60 %. If returns from the smaller ports that made up the New Bedford Customs District are included, the figure rises to 78 % of total American production.

The decline

After the near record volume of imports in 1843 — a total exceeded only in 1837 — the American sperm whale fishery began its long decline. In 1850, sperm oil imports fell below 100 000 barrels for the first time in 18 years. By 1860, it had dropped to 73 708 barrels, and in the years following the Civil War only once did it exceed 50 000 barrels. The causes of the decline were largely due to increasing competition from other means of illumination, notably gas but also vegetable oils, animal fats and ultimately petroleum. The whaling interests also contributed to the decline, shifting their efforts to the right whale and bowhead fisheries, to which they were lured by the opening of the North Pacific and western Arctic grounds and by the two marketable commodities — baleen and oil — that these whales afforded. Whatever the importance given to these or other causes, the decline in sperm oil production after 1843 seems not to have been much affected by beliefs among whaling merchants that the species had become too scarce to warrant continued hunting. Their decisions to curtail or to withdraw from sperm whaling were determined by factors that had less to do with attitudes about whale populations than with considerations of the market-place, not least among these being the promise of investment opportunities elsewhere. Further research in the business records of the fishery may confirm or correct this interpretation. In any case it would appear that sperm whale stocks throughout the world began their recovery from American exploitation as early as 1844.

Sperm whaling records

Kenneth Martin (1976) has described the kinds of data on sperm whaling that survive, pointing out some of the opportunities and limitations they offer for those who must assemble estimates about the survival capabilities of these whales today. My own impression is that while these records can tell us certain things, their usefulness should not be overrated. They are largely silent, for example, on matters of natural history and rarely make distinctions on even such simple questions as the sex of captured whales. Whether they can usefully contribute to the determination of the "initial (unexploited) population sizes", used to calculate how many may now be taken, is a matter best explored by those more familiar with determinations of this sort. Compilations from logbooks of the locations where whales

were taken in the past might well assist in identifying the geographical stocks of sperm whales, particularly the nine "divisions" proposed for the Southern Hemisphere in 1972 by the Scientific Committee of the International Whaling Commission.

Maury's Whale Chart

The most carefully plotted record of the American experience of taking sperm whales is perhaps the least consulted of all. Townsend's (1935) work is useful and well-known, but does not match in precise information the whale charts produced under the direction of Lieutenant Matthew Fontaine Maury (1852) at the Navy's Depot of Charts and Instruments. The charts referred to do not include the "Preliminary Whale Chart" of 1851, an early and simplified version of those that followed in 1852. The latter, which are important, consist of 4 sheets, each covering a quarter of the globe with the oceans divided into squares of 5 degrees each. Each square was itself divided by 12 vertical lines representing the months of the year. Within each square and running horizontally across the column of months were three separate graphs: one to show the number of days spent by whaleships cruising within the square, one to show the number of days on which sperm whales were seen, and one to show the same for right whales. One such square – 0°-5°N, 80°-85°W – yields, for example, the following data (Table 3).

A complicated piece of cartography,

Table 3. Monthly sightings of sperm whales in a 5° square (0°-5°N, 80°-85°W)

	Dec	Jan	Feb	Mar	Apr	May	Jun	Jul	Aug	Sep	Oct	Nov
Days of search	125	11	2	7	72	90	155	148	183	138	112	94
No. of days sperm whales seen	18	0	0	1	21	13	20	30	41	37	38	9

these charts offered quantitative data based on abstracts from about 1 000 logbooks, prepared for Maury by a retired New Bedford whaling master (responsible for Massachusetts ports) and a New York agent (who covered those of Connecticut). The data it depicted were derived for the most part from logbooks kept during the decade of greatest sperm whaling activity, 1835-1845, and the sampling was probably large enough to give a reasonably accurate picture of where, when and for how long, whalemen cruised in specific sectors of the oceans and of the number of whales they encountered while there. More than any other single source, these charts contain the accumulated experienced of American whalemen in pursuit of sperm and right whales in the decades of the 1830's and 1840's. If examined with care, I believe they would yield much of the information about sperm whale stocks and densities that might otherwise be obtained, or so it has been suggested, by abstracting logbooks today.

For the historian interested in the whaling industry for its own sake or as one aspect of the American past, the records that survive, whether as logbooks or merchants accounts, are invaluable. For those charged with responsibility for recommending a policy towards the taking of these whales today, these records from an earlier time may have a more restricted use.

References

HEGARTY, R.B., Returns of whaling vessels sailing from
1959   American ports, 1876-1928: a continuation of Alexander Starbuck's "History of the American whale fishery". New Bedford, Old Dartmouth Historical Society.

KUGLER, R.C., The penetration of the Pacific by American
1971   whalemen in the 19th century. *Monogr. Rep. Natl. Marit. Mus. Greenwich*, (2):20-7.

LYMAN, J., Annual catches of sperm whales from 1800 to
1969   1946. *Rep. Int. Comm. Whal.*, (19): Append. 6.1.

MARTIN, K.R., The whaling museum as a source of
1976   historical data for whale population studies, with special reference to initial populations of sperm and right whales: problems and responsibilities. Paper presented to the Scientific Consultation on the Conservation and Management of Marine Mammals and Their Environment, Bergen, Norway, 31 August-9 September, 1976. FAO ACMRR/MM/SC/104:6 p.

MAURY, M.F., Whale chart of the world. Washington,
1852   Bureau of Ordnance and Hydrography, Ser. F, No. (1-4).

STARBUCK, A., History of the American whale fishery
1878   from its earliest inception to the year 1876. Waltham, Mass., by the author.

TOWNSEND, C.H., The distribution of certain whales as
1935   shown by logbook records of American whaleships. *Zoologica, N.Y.*, 19:3-50.

SPERM WHALE SIZE DETERMINATION: OUTLINES OF AN ACOUSTIC APPROACH

B. Møhl, E. Larsen and M. Amundin

Abstract

By measuring the duration of interpulse intervals of the sound clicks of sperm whales and relating them, following the theory of Norris and Harvey (1972), to the length of the spermaceti organ, and then relating this length to total body length, it may be possible to estimate the size of a sperm whale once the pulse interval of its clicks is known. This method and visual measurement have given the same results in the case of a 9-m whale and a 21-m whale — lengths at the lower and the very highest ends of the range of sperm whale lengths, indicating that the acoustic method may be used throughout the entire range of sizes. These 2 estimates also follow an independently derived, empirical mathematical relation between total length and interpulse interval, though no inference should necessarily be drawn from this match. The development of a relation between interpulse interval and age must await clarification of the upper end of the curve relating length to age. It appears that this method is technically feasible and that a station recording sperm whale clicks would have a range of 2 km (12 km²) or more; use of the method would allow the size and perhaps the age of sperm whales to be determined without killing them or even without visual contact. Further research is needed on the acoustic behaviour of sperm whales and on their mechanism of sound generation.

Résumé

En mesurant la durée entre les impulsions des cliquètements des cachalots et en établissant une relation, selon la théorie de Norris et Harvey (1972), avec la longueur de l'organe des animaux, puis en établissant un rapport entre cette longueur et la longueur totale du corps, il pourrait être possible d'estimer la taille d'un cachalot quand on connaît l'intervalle entre les émissions de ces cliquètements. Cette méthode et les mesures visuelles ont donné les mêmes résultats dans le cas d'un animal de 9 m et d'un animal de 21 m, longueurs qui se situent au minimum et au maximum de la gamme de tailles des cachalots, ce qui indique que la méthode acoustique pourrait être utilisée sur toute la gamme de tailles. Ces deux estimations suivent aussi une relation mathématique empirique obtenue de façon séparée, entre la longueur totale et l'intervalle séparant les impulsions, encore qu'on ne doive pas nécessairement tirer des conclusions de cette concordance. L'élaboration d'une relation entre l'intervalle des impulsions et l'âge doit attendre une clarification de la partie supérieure de la courbe liant la longueur et l'âge. Il apparaît que cette méthode est techniquement réalisable et qu'une station enregistrant les cliquètements des cachalots aurait une portée de 2 km (12 km²) ou plus; l'utilisation de la méthode permettrait de déterminer la taille et peut-être l'âge des cachalots sans les tuer et même sans contact visuel. Des recherches plus poussées sont nécessaires pour ce qui est du comportement acoustique des cachalots et du mécanisme de production du son.

Extracto

Midiendo la duración de los intervalos entre impulsos en los sonidos emitidos por los cachalotes y poniéndolos en relación, según la teoría de Norris y Harvey (1972), con la longitud del órgano que segrega el esperma y poniendo luego en relación esa longitud con la longitud total del cuerpo, puede ser posible estimar la talla de un cachalote cuando se conoce el intervalo entre impulsos de los sonidos que emite. Con este método y con mediciones visuales se han obtenido resultados idénticos en el caso de un cachalote de 9 m y otro de 21 m, longitudes que se encuentran en el extremo inferior y superior de la gama de longitudes de los cachalotes, lo que indica que el método acústico puede utilizarse para animales de todas las tallas. En estas dos estimaciones se aplica también una relación matemática empírica, derivada independientemente, entre la longitud total y el intervalo entre impulsos, aunque de esa correspondencia no ha de inferirse necesariamente ninguna conclusión. Para obtener una relación entre el intervalo entre impulsos y la edad es necesario aclarar el extremo superior de la curva que pone en relación la talla con la edad. Parece que este método es técnicamente viable y que una estación receptora de los sonidos emitidos por los cachalotes tendría un radio de acción de 2 km (12 km²) o más; con este método se podría determinar la talla y quizás la edad de los cachalotes sin necesidad de matarlos e incluso sin contacto visual. Son necesarias más investigaciones sobre el comportamiento acústico de los cachalotes y sobre el mecanismo de emisión de sonidos.

B. Møhl
Department of Zoophysiology, University of Aarhus, DK-800 Aarhus, Denmark

E. Larsen
Department of Zoophysiology, University of Aarhus, DK-800 Aarhus, Denmark

M. Amundin
Department of Zoophysiology, University of Aarhus, DK-800 Aarhus, Denmark

A technique that is commonly employed when estimating the effect of exploitation on a given species of marine mammal is the determination of the age distribution within the species. In most cases, this requires that a representative sample of animals be killed to provide teeth, baleen, ovaries and measurement of total length, or whatever parameter or organ that can be processed to indicate the age of the late owner. Clearly, killing a substantial number of specimens for ageing purposes is contra-indicated in most cases, including that of the sperm whale. This paper describes a passive acoustic method that holds promise for the remote size estimation of sperm whales.

The sperm whale owes its peculiar shape to the disproportionate development of the forehead, which has been described as "the biggest nose on record" (Raven and Gregory, 1933). The anatomical structures inside the forehead include the spermaceti organ and a variety of airsacs, respiratory tubes, valves and specially arranged fatty tissues. Widely divergent functions have been attributed to this enigmatic complex. In fact, the very name of the species reflects the idea that the forehead is a formidable sperm cell reservoir. Recently, Norris and Harvey (1972) have proposed an acoustic function for the complex: they suggest that it serves to generate the well-known powerful sound clicks that are presumably used for long range sonar. This theory has many attractions: it attributes homologue functions to homologue anatomical assemblies within the suborder of odontocetes, it provides an explanation for many of the otherwise perplexing anatomical features and it links the fine structure of the sound pulse to the dimensions of the generating mechanisms.

Sperm whale sound clicks consist of trains of brief transients with an exponentially decaying amplitude (Fig. 2). According to Norris and Harvey (1972), this pattern is generated by a single transient, produced in the space between two sound-reflecting mirrors (i.e., air sacs, bounding the anterior and posterior ends of the spermaceti organ as illustrated in Fig. 1). Most of the sound energy

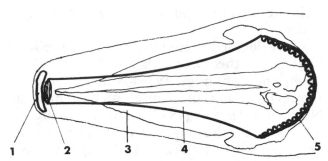

FIG. 1. – Sketch of the relative position of the anatomical structures with a proposed acoustic function in the sperm whale forehead. 1. Distal airsac. 2. Museau. 3. Skull. 4. Spermaceti case. 5. Frontal airsac. The museau is located posterior to the distal airsac and is considered to be the most likely organ for the generation of clicks (Norris and Harvey, 1972).

leaves the system directly, but a fraction of the energy is intercepted by one of the mirrors, returned to the other mirror, and sent back again before it leaves the system. Part of this delayed sound pulse is again intercepted. The repeated interception of a single original sound creates a series of decaying pulses.

The spacing between the pulses in a click will then be solely a function of the distance between the mirrors (i.e., the length of the spermaceti organ) and the speed of sound in spermaceti oil. Assuming the latter to be a constant property and the length of the spermaceti organ to be related to total body length, it follows that the size of a sperm whale may be

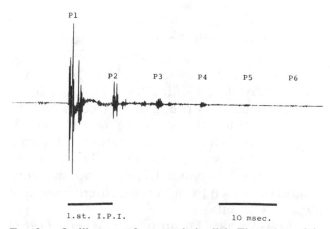

FIG. 2. – Oscillogram of sperm whale click. The pulses of the click are labelled P1 through P6. All interpulse intervals are of equal duration and thought to reflect the size of the whale.

329

estimated once the pulse interval of its clicks is known.

A detailed account of the evidence relevant to this theory (Norris and Harvey, 1972; Møhl and Amundin, 1978) is not given here. However, using the pulse interval method, Norris and Harvey (1972) were able to predict with accuracy the length of a 9 m sperm whale; an independent estimate of its length was obtained using an optical comparison method. Recently Møhl and Amundin (1978) obtained identical photographic and acoustic size estimates of a 21 m specimen. The lengths of these two individuals happen to fall at the lower and the very highest, giving reason to believe that that body lengths throughout the entire range may be established using this method.

The acoustic method is assumed to measure the length of the spermaceti organ, which is roughly equal to the length of the head. Nishiwaki, Ohsumi and Maeda (1963) have established the relationship between the length of the head and the total body length of sperm whales. Using their data, and the double mirror hypothesis of sperm whale sound production, the following empirical relationship has been derived (using a least squares fit to a 2nd order polynomial):

$$TL = 0.76 + 4.64 \times IPI + - 0.259 \times IPI^2 \quad (1)$$

where

TL = total length, m, and
IPI = Interpulse interval, msec.

As Fig. 3 shows, the above mentioned acoustically based estimates follow the independently derived prediction of equation (1). However, since the variance of the data is largely unknown and since equation (1) is used for predicting values considerably above the range of the morphometric data, no inference should be drawn from the coincidence *per se* of the 2 sets of data.

Nishiwaki, Ohsumi and Maeda (1963) also given the relation between total length and number of dentinal growth layers (repre-

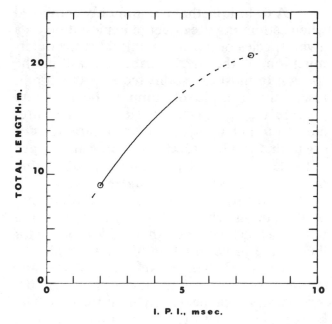

FIG. 3. – Relation between total length and interpulse interval, according to TL = 0.76 + 4.64 × IPI + – 0.259 × IPI²: Solid line shows the range for which the ratio between total length and length of head is known. The two points are acoustic/optical length estimates of two whales.

senting age). The curve they obtain suggests negative growth when a length of 16 m has been reached. This is inconsistent with length records published by other workers (Scammon, 1874; Matthews, 1938; Tomilin, 1957; Berzin, 1972, p. 30-1). Development of a relationship between IPI and age must await clarification of the nature of the upper end of the growth curve.

The acoustic power of sperm whale clicks is fairly high. Levenson (1974) reported it to be of the order of 170 dB re. 1 μ Pascal in the 2 to 8 kHz range. He obtained recordings at a calculated distance of 3.153 km. Assuming an average ambient noise corresponding to sea state 3 (Albers, 1965) and a minimum acceptable signal to noise ratio of 10 dB, a range of 2 km (equal to a coverage of 12 km²) can be expected from a single recording station. This could be increased by an order of magnitude under favourable conditions as also by properly designed recording and analysing equipment.

This paper is intended to describe the principles of the proposed sperm whale sizing technique and its expected range. Its potential for monitoring the size distribution in various sperm whale stocks, both exploited and unexploited, without requiring either visual or physical contact, may justify further research into its validity, accuracy and feasibility. Research is needed into both the acoustic behaviour of sperm whales (including sex, size and herd effects), and the mechanism of sound generation. Further, a number of subtle technical/economical factors have to be considered. Significantly, the technical aspects of the method appears to be well within the state of the art of underwater acoustics (Levenson, 1974; Watkins and Schevill, 1975; Watkins, 1976).

Acknowledgements

This work was supported by grants from the Danish Natural Science Research Council.

References

ALBERS, V.M., Underwater acoustics handbook. 2. Uni-
1965   versity Park, Pennsylvania State University Press.

BERZIN, A.A., The sperm whale. (Kashalot). Jerusalem,
1972   Israel Program for Scientific Translations, IPST Cat. No. 60070 7:393 p. Translated from the Russian published by Izdatel'stvo Pishchevaya Promyshlennost (1971).

LEVENSON, C., Source level and bistatic target strength of
1974   the sperm whale (*Physeter catodon*) measured from an oceanographic aircraft. *J. Acoust. Soc. Am.*, 55:1100-3.

MATTHEWS, L.H., The sperm whale, *Physeter catodon*.
1938   *Discovery Rep.*, (17):93-168.

MØHL, B. and M. AMUNDIN, Sperm whale clicks: pulse
1978   interval in clicks from a 21 meter specimen. *J. Acoust. Soc. Am.*, (in press).

NISHIWAKI, N., S. OHSUMI and Y. MAEDA, Change of
1963   form in the sperm whale accompanied with growth. *Sci. Rep. Whales Res. Inst.*, Tokyo, (17):1-13.

NORRIS, K.S. and G.W. HARVEY, A. theory for the
1972   function of the spermaceti organ of the sperm whale. Washington, D.C., NASA, SP-262.

RAVEN, H.C. and W.K. GREGORY, The spermaceti organ
1933   and nasal passages of the sperm whale (*Physeter catodon*) and other odontocetes. *Am. Mus. Novit.*, (677):1-17.

SCAMMON, C.M., The marine mammals of the north-
1874   western coast of North America together with an account of the American whale fishery. New York, Dover Publications.

TOMILIN, A.G., Zveri SSSR i prilezhashchikh stran. 9.
1957   Kitoobraznye (Animals of the USSR and adjacent countries. 9. Cetacea). Moskva, Izdatel'stvo AN SSSR.

WATKINS, W.A., Biological sound-source locations by
1976   computer analysis of underwater array data. *Deep-Sea Res. Oceanogr. Abstr.*, 23:175-80.

WATKINS, W.A. and W.E. SCHEVILL, Sperm whales
1975   (*Physeter catodon*) react to pingers. *Deep-Sea Res. Oceanogr. Abstr.*, 22:123-9.

THE ADEQUACY OF THE SCIENTIFIC BASIS FOR THE MANAGEMENT OF SPERM WHALES

T.D. SMITH

Abstract

Since 1963, the Scientific Committee of the International Whaling Commission has held several special meetings on sperm whales. Although some progress has been made in understanding the biology of the species, including its life history and division into populations, the results have generally, and particularly in relation to the size and dynamics of populations, been less satisfactory than those for baleen whales, on which much of the work on sperm whales has been based. Estimates of abundance have in recent years been derived from sophisticated analyses of catch-and-effort data. The use of such data is hampered by the difficulty of adequately measuring effort and especially of accounting for factors which cause the search for animals to be non-random. Analogy with the tuna-porpoise fishery in the eastern tropical Pacific, on which much study related to searching and catching effort has been completed, may show how these factors can best be taken into account. Poor knowledge of the sizes of sperm whale populations have in turn made it difficult to correlate changes in vital rates with changes in population size, on which the construction of density-dependent population models depends. A more organized comparison of the population dynamics of different mammals would probably make it easier to understand the dynamics of sperm whale populations and of other mammals difficult to study. Research is needed to resolve the problems of estimating abundance and the relation of changes in it to changes in vital rates. The experimental management of whaling so that different populations are exploited at different rates and the effects on their abundance and vital rates watched, could provide some of the needed information.

Résumé

Depuis 1963, le Comité scientifique de la Commission internationale baleinière a tenu plusieurs sessions spéciales consacrées aux cachalots. Encore que des progrès aient été réalisés dans la compréhension de la biologie de l'espèce, y compris son cycle biologique et la division en populations, les résultats ont été, particulièrement pour la taille et la dynamique des populations, généralement moins satisfaisants que dans le cas des mysticètes sur lesquels s'est fondée une bonne partie des travaux relatifs aux cachalots. Ces dernières années, les estimations de l'abondance ont été tirées d'analyses très élaborées de données sur l'effort et les captures. L'utilisation de ces données est entravée par la difficulté de mesurer l'effort de façon adéquate et, spécialement, d'expliquer les facteurs qui rendent la recherche des animaux non aléatoire. L'analogie avec les pêches thon-marsouin dans le Pacifique tropical-est, sur lesquelles on a effectué de nombreuses études relatives à l'effort de recherche et de capture, peut montrer comment ces facteurs peuvent être pris en considération le plus avantageusement. La médiocre connaissance de la taille des populations de cachalots a rendu, de son côté, difficile

333

de corréler les changements des taux biologiques et les changements de taille des populations, dont dépend la construction de modèles de population dépendant de la densité. Une comparaison plus organisée de la dynamique des populations de différents mammifères faciliterait probablement la compréhension de la dynamique des populations de cachalots et d'autres mammifères dont l'étude est difficile. Il est nécessaire d'effectuer des recherches pour résoudre le problème de l'estimation de l'abondance et de la relation entre les changements qui l'affectent et les changements des taux biologiques. L'exploitation rationnelle expérimentale de la chasse à la baleine, organisée de telle sorte que différentes populations sont exploitées à des taux différents et que les effets sur leur abondance et sur leurs taux biologiques sont observés, pourrait fournir certaines des informations nécessaires.

Extracto

Desde 1963, el Comité Científico de la Comisión Ballenera Internacional ha celebrado varias reuniones especiales sobre el cachalote. Aunque los conocimientos sobre la biología de esta especie, incluido su ciclo vital y su división en poblaciones, han avanzado algo, los resultados, en especial por lo que se refiere a volumen y dinámica de las poblaciones, han sido menos satisfactorios que los conseguidos sobre los mistacocetos, en los que se han basado buena parte de los trabajos sobre el cachalote. En los últimos años se han obtenido estimaciones de la abundancia a partir de complicados análisis de los datos de captura y esfuerzo. El empleo de esos datos se ve dificultado por el problema de medir adecuadamente el esfuerzo y, en especial, de tener en cuenta los factores que hacen que la búsqueda de los animales no proceda al azar. La analogía con la pesquería de atún y delfín del Pacífico oriental tropical, en la que se han realizado muchos estudios sobre búsqueda y esfuerzo de captura, podría indicar cuál es la mejor manera de tener en cuenta esos factores. La falta de conocimientos adecuados sobre el volumen de las poblaciones de cachalote ha hecho difícil, a su vez, establecer una relación entre las variaciones de los índices vitales y los cambios en el volumen de la población, relación de la que depende la construcción de modelos de población dependientes de la densidad. Probablemente una comparación más orgánica de la dinámica de población de distintos mamíferos permitiría entender más fácilmente la dinámica de las poblaciones de cachalote y de otros mamíferos difíciles de estudiar. Es necesario realizar investigaciones para resolver los problemas de la estimación de la abundancia y de la relación entre las variaciones de la abundancia y los cambios en los índices vitales. Regulando experimentalmente la caza de manera que distintas poblaciones se exploten a ritmos diversos y controlando los efectos que eso tiene en la abundancia y en los índices vitales, podrían obtenerse algunas de las informaciones necesarias.

T.D. Smith
NMFS, Southwest Fisheries Center, P.O. Box 271, La Jolla, CA 92037, USA

Introduction

This paper is a report on the special meeting on sperm whales of the Scientific Committee of the International Whaling Commission (IWC), La Jolla, California, 15-25 March 1976. Special meetings to consider the problems of managing the harvest of various whale groups have been held in response to the increasing importance of the role of the Scientific Committee in the IWC, and have allowed world experts to provide their best possible management advice to the Scientific Committee, and to the IWC.

The role of the Scientific Committee began to increase in importance after the Committee of Three (later Four) studied the Antarctic baleen whales from 1961 to 1964, and continued to grow as the Commission moved toward management by species and by individual populations. Special mid-year meetings have occurred frequently and irregularly since that time.

In view of these developments and of the serious questions raised about the adequacy of the Commission's actions in conserving the world whale populations (McVay, 1974), it is important to evaluate the deliberations of the Scientific Committee and to determine the adequacy of its recommendations to the Commission. In this paper, I have indicated my views on these matters, after attending and participating in the 1976 sperm whale meeting.

History of meetings on sperm whales

Since 1963 a series of meetings on sperm whales of part or all of the Scientific Committee has been held every 2 or 3 years. In addition, a subgroup of the Scientific Committee, called the North Pacific Working Group, has from time to time considered North Pacific sperm whales.

The formulation and study of several key lines of research can be traced in the reports of previous meetings, of which the most apparent are:

(i) determination of the population structure of this species;

(ii) determination of size of the various populations;

(iii) determination of the basic life history of sperm whales, including quantification of the vital or demographic rates;

(iv) determination of an accurate ageing technique for use with (iii);

(v) determination of an adequate population dynamics model of sperm whales, along the lines of that developed for baleen whales by the Special Committee of Three (later Four) earlier (IWC 13:32-106, IWC 14:47-60);

(vi) analysis of the current understanding of sperm whale population dynamics, to estimate the impact of different harvesting rates.

Each of these lines or areas of research involves many factors, and they are themselves inter-related. The relationships can be seen in the diagram prepared by Omura and Ohsumi (1974), reproduced in modified form as Fig. 1. The above item numbers are inserted in the appropriate boxes.

In studying the development of these lines of research, it is apparent that the focus of the several meetings on sperm whales has changed as information and techniques have developed.

The current meeting concentrated primarily on estimating population sizes (item 2), exploring population dynamics models (item 5), and analysis of management alternatives (item 6). This latter has been a strong focus for all of the meetings as the Scientific Committee has attempted to provide scientific guidance on management to the Commission.

335

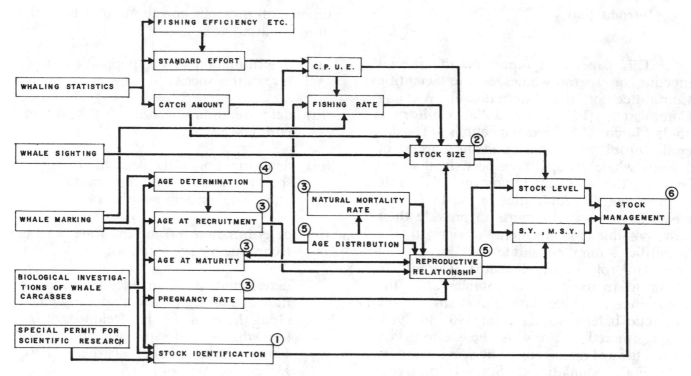

Fig. 1. — Relationship between different lines of research and sources of data in the assessment of populations of whales. The six areas of study noted are indicated by numerals.

It is interesting, incidentally, to note that the Scientific Committee is not an especially rigidly defined group, and the composition of its sub-groups is even less so. Participants are usually aligned according to their nationality, with the addition of the adviser from the Food and Agriculture Organization of the United Nations. Active participation is open to any individuals delegated by participating national governments and to other invited scientists. In Table 1 those individuals who have attended 3 or more meetings are listed. While

Table 1. Major and consistent contributors to the meetings on sperm whales of the Scientific Committee [1]

Participant	Meeting				
	1966 (Honolulu)	1968 (Rome)	1970 (Honolulu)	1972 (Parksville)	1976 (La Jolla)
Bannister (Australia)	X	—	X	X	X
Allen (Canada)	—	—	X	X	X
Ohsumi (Japan)	X	X	X	X	X
Gambell (UK)[2]	X	X	X	X	X
Chapman (USA)	X	X	X	X	X
Ivashin (USSR)	X	—	X	X	X
Best (S. Africa)	—	X	X	X	X
Boerema (FAO)	—	X	X	—	X

[1] From the Reports of the International Whaling Commission.
[2] Now Permanent Secretary to the IWC.

some 23 other individuals have attended one or more meetings, these 8 individuals have had the most active and continuous participation. In several cases these individuals represent other scientists as well, especially in the case of Japan, United Kingdom and USSR.

Problems of abundance estimation

In this series of meetings all of the lines of inquiry listed above have been discussed extensively, and much has been accomplished. It is agreed by most that adequate techniques of age determination have been developed, and that the basic life history is understood. This latter includes the reproductive cycle timing and rates, the mortality rates (at least for mature animals), and spatial and temporal distributions. It is understood that the sexual dimorphism is also reflected in polygamous harem master breeding behaviour and in sex- and age-based migration patterns.

There is reasonably good agreement on the population structure of this species, on a world basis. This is based on tagging and serological work. While this work is still being refined and the conclusions may change markedly, enough is known to allow at least a tentative analysis of impact to proceed at the population level.

There is much less agreement on population sizes, both initial and current. While a variety of techniques have been tried, there is considerable inconsistency in the resulting estimates. Sightings data from Dutch merchant ships were analysed and discarded as unsuitable. Mark-recovery analyses have yielded highly variable results, mainly because of the low number of recoveries. Recently analyses based on catch and effort data have received the greatest emphasis. In the 1968 meeting in Rome, data collection forms for sperm whales were standardized, based principally on those originally developed for baleen whales by the Committee of Four.

These catch and effort data have been subjected to relatively standard techniques for estimating abundance, such as the DeLury method, as used for the Antarctic baleen whales. They have also been analysed with more sophisticated methods developed for North Pacific baleen whales, primarily by Allen (1966, 1969). These latter methods were those most used at the present meeting, but in a considerably refined form. It does not appear that these analyses of the catch and effort data for the sperm whale fisheries are particularly satisfactory. Considerable differences exist in the estimates from different methods and not infrequently a particular method fails to give reasonable estimates at all.

Some of this variability may be seen in the estimates given in Table 1 of the draft report of the La Jolla meeting. For most divisions the estimates obtained from different methods are seen to be rather variable. In Table 13 of that report the corresponding catch per unit effort (c.p.u.e.) based estimates for the North Pacific populations are given. There is considerable difference in the estimates, where the method works, depending on the fleet from which the data being used was obtained.

One new approach was developed during the meeting. This involves comparing the average length of the catch before exploitation to the current average length. The amount of decline in average length can be used to provide an estimate of initial numbers. This method, as it is worked out more completely, will be useful in providing another type of estimate of abundance, and is especially valuable in not being based on the measurement of effort.

Definition of effort

It appears that the primary problem with the estimation of abundance from catch and effort data is the adequacy of the definition of effort. The concept of fishing effort has a long history in fisheries work, and has generally been found useful if sufficient care is taken in defining and measuring the searching process. Thus Russell (1942) in his book *The Overfishing Problem* established his contention by using relatively crude measures of effort such as numbers of days' absence of fishing vessels from port. But as the questions to be answered have become more involved, it has been necessary to refine the measurements of effort. Thus Russell used numbers of hours spent fishing as a more refined measure of fishing intensity in some of these analyses.

The Committee of Four, in considering the Antarctic baleen whale populations, dealt with the definition of effort and noted several problems (IWC, 1962, pp. 45-48). The basic assumption needed in the use of "effort" is that the process of searching for animals is random, and that as such the numbers of animals caught varies directly with the numbers of animals present in the population. There are 3 aspects of this which are important to consider.

Searching time

First, the actual time spent searching and the effectiveness of this time are affected by many factors, such as weather, repairs, non-searching work such as survey, handling of the killed whales, size and speed of the catch or boats, capacity of the factory ships, skill of the gunner, and so forth (IWC, 1962). In their reports the Committee of Three approached the problem of determining the effects of these factors on catching of baleen whales in a quantitative fashion, but concluded this "... would involve a tremendous amount of work". Instead, they settled for the usual number of boat days. They did attempt to correct for changes in efficiency over the years by adjusting for changes in horsepower or tonnage of the catcher boats. Much additional work could be done on this aspect of the definition of effort, although the adequate collection of the data would be costly.

There is the possibility of studying a somewhat similar system on which much data has been collected. The tuna purse seine fishery operating out of California, fishing off Mexico and Central America, has been extensively studied because dolphin schools (*Stenella attenuata* and *S. longirostris*) have been used as indicators of tuna schools (Perrin, Smith and Sakagawa, 1976). There has been extensive data collected on the process of searching for schools of dolphins, on handling times and on the abundances of dolphins and tuna. This data collection scheme was designed to assist in the determination of a useful effort measure to allow estimation of porpoise abundance. In many ways the problem is similar to sperm whale fishing and its detailed study might lend much insight to this problem.

Nature of search

Second, the actual finding of a whale may not involve a completely random search. Allen (1966) justifies the use of effort to measure abundance, primarily of baleen whales in his statement:

"Whaling, although concerned with mammals, more closely resembles fishing in its essential features than it does most forms of terrestrial hunting. The animals sought inhabit an environment which has few permanent small-scale local features, which permits them to move freely in all directions, and which gravely limits the amount of direct observation which is possible. This has the effect of reducing the catching process to one which has many of the features of random sampling ... (p. 1554)".

Allen's randomness is over geographical area. The claim here is that as one searches over the hunting grounds, one is equally likely to encounter a whale in any place. There are 2 levels to this problem of aggregation and randomness. One is that at least female sperm whales tend to occur in small groups called pods. The effect of schooling on catch per unit effort as a measure of abundance was investigated by Paleheimo and Dickie (1964). They conclude that total effort is unsatisfactory in this case.

The second level is the possibility of sperm whales, either singly or in schools, tending to aggregate in localized areas. This has been shown for *Tursiops truncatus* (Evans, 1954). It appears possible that these animals could aggregate on local food abundances, or on more permanent areas of high productivity. Increasing information about cetacean behaviour and sensory systems suggest that this would certainly be possible and evolutionary theory suggests it would be likely.

Schweder (1974) deals with this second aspect of areas of concentrations, as well as the problem of schools discussed by Paleheimo and Dickie (1964). Schweder notes, as the previous authors did, that the relation between abundance and catch per unit of effort is non-linear for schooling animals, and even more so when clusters form.

The fishery for tuna using dolphins described above may also provide an opportunity for studying the existence of and possible cause of local aggregations, and the impact of these on the fishing process (Perrin, Smith and Sakagawa, 1976; Smith, 1975). Extensive information is available on the local distribution of dolphins and tuna, and on oceanographic conditions.

An additional aspect of the randomness of the search process is the development of technology to assist in locating whales. Asdic or sonar has come into wide use, aircraft-spotting planes have been used from land stations, and dedicated fast-sighting vessels to locate concentrations have been used. Also, the structure of the fleet, with catcher boats working for the same factory ships, allows communication between searchers, thus increasing hunting efficiency.

Species preference

The third and last aspect of the utility of effort in estimating abundance is that the sperm whale has generally been only one of several species of large whales caught by the whaling fleet, and not a preferred one at that. This has been recognized in the analysis of the Antarctic data in that only the data from outside the baleen whale season has been used. In the North Pacific data analysis of the data from the whale season is used, as the species preference is not thought to be as marked. Schweder (1974) discusses this problem and notes "one cannot expect to have good (abundance) estimators for species that are partially ignored". Again, the tuna fishing provides an interesting and potentially informative parallel where considerable information has been obtained. Of the 2 species of dolphin principally involved, 1 is thought by the fishermen to be more likely associated with tuna that the other. Thus, when a dolphin school is sighted, if it is the less favoured type, it may be passed by. This is complicated to some degreee because the 2 species frequently occur together. Extensive detailed operational information has been collected on this hunting process, and might provide useful information relative to the sperm whale harvesting.

It is clear that the hunting process involved is not simple, and also that the measurement of a simple aspect of it such as catcher days work is unlikely to provide a consistent and accurate index of abundance. In order to apply this approach to estimating abundance serious work will be necessary, both in collecting data and in developing suitable estimation techniques. Schweder's (1974) work is the first comprehensive approach to the process of searching and hunting, and would be worth following up. His results are quite theoretical, however, and the application of this work would itself involve considerable study.

With those problems with a good measure of abundance, the evaluation of the accuracy of estimates of initial and current population sizes has lagged. This has a direct and devasting impact on the development of an adequate model of the dynamics of sperm whale populations.

Problems of our understanding of population dynamics

The general theory of population dynamics which is assumed to apply is that the abundance of a population is directly dependent on its own density. While the development of this theory of population dynamics has involved extensive controversy, it now has the status in many quarters as described by Haldane, of a "blinding glimpse of the obvious" (quoted in Gilbert, 1973). Some attempts have been made to establish or support this theory on the basis of (i) logical necessity (Gilbert, 1973 and others); (ii) observed correlations of abundance with the magnitude of changes in abundance (Tanner, 1966), and (iii) demonstrated changes in vital or demographic processes in individuals which are directly associated with population size change. The information on baleen whales in the Antarctic provides some of the most clear support of this latter type.

It has been observed that the interval between calves, as estimated by the proportion of the catch which is pregnant, has declined as the Antarctic baleen whale populations have been reduced. Correspondingly, it has been observed that the age of first reproduction has decreased. Both of these changes tend to increase the rate of reproduction and hence, if the mortality rate is constant or decreasing, to cause the population to increase.

In attempting to apply this theory, it is necessary to obtain information on how the reproductive parameters change with abundance and, of course, on abundance itself. If has been observed in the data on sperm whale carcasses that the pregnancy rate is different for different populations, implying different intervals between calving. The problems with estimating abundance to associate with these differences in reproductive rates have been discussed above. This has prohibited determination of how the calving interval has changed with abundance. Additionally, there appear to have been few observations on sperm whale populations where the female population size has been greatly reduced.

While the magnitudes of change in reproductive rates are now known, the absence of accurate abundance estimates prevents the determination of the nature of the relationship between calving interval and abundance. They could be related in a linear manner, a unit change in abundance causing a constant proportional change in calving interval. Alternately the calving interval might decrease at a decreasing rate with abundance. In the absence of information on this point, a range of relationships has been considered.

Lacking convincing information on the dynamics of the vital processes in sperm whale populations, some insight can hopefully be obtained by analogy to populations of other better understood species. The models used for sperm whales to date have been heavily based on analogy to baleen whales, seals, porpoise and elephants.

Generally what is known about the dynamics of different species has not been considered in a comparative sense; rather, individual cases are considered by separate workers. They frequently do not approach the problems in the same manner, and use different mathematical descriptions. It would be valuable to synthesize these various observations in one framework. This would allow easy and reliable reference to comparative data, and determination of the relative reliableness of the various observations. A more important advantage is that it would provide a possible basis for a comparative study of the dynamics of mammalian populations. As in all comparative work, one hope would be to determine general patterns which might be useful in predicting the dynamics of other species.

While the baleen whale studies provide some of the strongest support of the density dependent theory of population dynamics, these same studies are now providing insights into the limitation of this theory. Recent evidence suggests changes in age of sexual maturity of the sei whales prior to their being harvested (Gambell, 1974; Lockyer, 1974).

This suggests that the sei whale populations responded reproductively to declines in other populations of baleen whales, implying a competitive linkage. The changes estimated to have occurred would have been sufficient to cause a substantial increase in sei whale abundance prior to their initial harvesting in 1960 (Smith, 1975a).

Another limitation of the density-dependent theory of population dynamics is the possibility of lags in response to changes in abundance. These lags could be related to a delayed response of food supply as numbers decline, or to a delayed physiological response of the animal to an increased food supply. Additionally, with long-lived animals like whales, if reproductive rates are determined by nutrition at young ages, the average time to maturity and the calving interval will change only gradually as the older animals die. The possible impacts of such lags have not been evaluated.

It is difficult to determine the details of the dynamics of sperm whale populations. The major proximal problem is inadequate estimates of abundance. More distal problems are inadequacies of existing theories, and inadequate knowledge of the marine ecosystem. There are no easy answers to these problems, and for some perhaps no answers at all.

Diversified management strategies

The above discussion of research areas needing additional work has mainly been concerned with the possibilities of retrospective studies of existing data. The aim has been to reanalyse what is known using new and more appropriate tools. Population dynamics of exploited animal species has not traditionally been an experimental science.

In contrast the possibility exists (ACMRR/FAO, 1977) of managing separate populations of sperm whales toward different ends. This might, if properly planned and monitored, provide comparative data from populations at different levels of abundance and allow the determination of the dynamics of sperm whale populations rather directly. One could manage one population to reduce its abundance to low levels, another to reduce its abundance only slightly if at all, and others to intermediate states. This would be a way of capitalizing on the population structure of this species, instead of just allowing this to confound the analyses.

Such experimental management has not been used extensively in fisheries work, but it is not unprecedented. The Inter-American Tropical Tuna Commission engaged in its Experimental Overfishing Programme in the mid-sixties to determine the accuracy of the predicted optimum harvesting strategy, based on a surplus production model. The North Pacific Fur Seal Commission similarly has stopped sealing on St. George Island in order to contrast the response of exploited and unexploited populations. An example of the utility of this approach can be seen in the amount of information on exploited marine populations which has been gained because of the fishing hiatus during the Second World War.

This approach requires, of course, that careful monitoring of the populations is possible. Both abundance and vital rates must be observed. If the population is harvested at all, the vital rates can be fairly well observed. The problems of abundance estimation were discussed above. It would be necessary to overcome these problems if the full potential of this approach were to be realized. This would also be important just to ensure the safety of populations, especially those identified for large abundance reduction.

Summary and conclusions

Were a report of this type being written in 1966 instead of 1976, its tenor would neces-

341

sarily have been much different. While the same 6 areas of investigation could have been identified, the unknowns would have been more numerous. The approaches to be explored in trying to gain understanding would have been more obvious and easier to pattern after the earlier work on baleen whales. Much of this "similar" work has been done or will be in the near future, and the results do not seem as convincing and consistent as those of the baleen whales. In particular, the population abundances and dynamics are poorly understood. I see a great need for additional studies of abundance on a population basis, and a corresponding attempt to correlate possible changes in vital rates with abundance changes. In this fashion, the dynamics of the populations can be understood on a firmer basis so as to allow better management advice to the Commission.

Even in the best situations, however, such studies are unlikely to resolve all of our uncertainties. The refinement of the inferences on population dynamics made by analogy is another important research consideration. Extensive use of this approach has been made to date, and it is important to determine its strengths and weaknesses so as to evaluate this source of information properly.

Finally, in designing management strategies for the future, it is important to consider diversified management for different populations. This would increase the stability of the species as a whole by ensuring that some populations at least would not be damaged, and it could potentially provide detailed information on the dynamic changes in sperm whale populations as abundance changes.

In conclusion, the Scientific Committee's recommendations on the management of sperm whales appear to be based on somewhat weaker scientific ground than those made on the management of baleen whales. The Scientific Committee is quite aware of the limitations of its information, and in fact carried out during the meeting an extensive sensitivity analysis of the mathematical models being used. Some new insights were gained in this process. We must assume that the harvesting of sperm whales is having an impact, even though we cannot determine the magnitude of impact with certainty. Management has too frequently been lacking because of gaps in our understanding. The one thing we do know is that harvesting of large whales can cause rapid and extensive reduction in abundance.

It is important that available information be reanalysed using more adequate models, and that future management strategy allow for obtaining information which increases our understanding.

Acknowledgements

My thanks are due to Dr. Gambell, Convenor of the meeting and Secretary of the IWC, and to the Marine Mammal Commission for the travel support.

References

ACMRR (FAO) Working Party of Marine Mammals, 1977   La Jolla, 21-25 January 1977, Report of the Advisory Committee on Marine Resources Research, Working Party on Marine Mammals. *FAO Fish. Rep.*, (194):43 p.

ALLEN, K.R., Some methods for estimating exploited 1966   populations. *J. Fish. Res. Board Can.*, 23(10):1533-74.

—, An application of computers to the estimation of 1969   exploited populations. *J. Fish. Res. Board Can.*, 26(1):179-89.

EVANS, W.M., Radio-telemetric studies on two species of 1974   small odontocete cetaceans. *In* The whale problem, edited by W.E. Schevill. Cambridge. Mass., Harvard University Press, pp. 385-94.

GAMBELL, R., A review of reproduction parameters and
1974   their density dependent relationship in South-
        ern Hemisphere sei whales. Contributed
        paper, International Whaling Commission
        Scientific Committee Meeting, La Jolla.

HOLT, S.J., Personal communication.
1976

IWC, Report of the International Whaling Commis-
1950-   sion. (1949/50) - to date.
to date

LOCKYER, C., Investigation of the ear plug of the south-
1974   ern sei whale, *Balaenoptera borealis*, as a
        valid means of determining age. *J. Cons.
        CIEM*, 36 (1):71-7.

McVAY, S., Reflections on the management of whaling.
1974   *In* The whale problem, edited by W.E. Sche-
        vill. Cambridge, Mass., Harvard University
        Press, pp. 369-84.

OMURA, H. and S. OHSUMI, Research on whale biology
1974   of Japan with special reference to the North
        Pacific stocks. *In* The whale problem, edited
        by W.E. Schevill. Cambridge, Mass., Harvard
        University Press, pp. 196-208.

PERRIN, W., T.D. SMITH and G. SAKAGAWA, Status of
1976   populations of spotted dolphin, *Stenella atte-*
        nuata, and spinner dolphin, *Stenella longiro-*
        stris, in the eastern tropical Pacific. Paper
        presented to the Scientific Consultation on
        the Conservation and Management of Marine
        Mammals and their Environment, Bergen,
        Norway, 31 August-9 September 1976. Rome,
        FAO, ACMRR/MM/27:19 p.

RUSSELL, E.S., The overfishing problem. Cambridge,
1942   Cambridge University Press, 128 p.

SCHWEDER, T., Transformation of point processes: ap-
1974   plications to animal sighting and catch prob-
        lems, with special emphasis on whales. Ph.D.
        Dissertation, University of California, Berke-
        ley, California.

SMITH, T.D., Estimates of sizes of two populations of
1975   porpoise (*Stenella*) in the eastern tropical Pa-
        cific Ocean. La Jolla, California, National
        Marine Fisheries Service.

–, Calculation of apparent increases in the Antarctic sei
1975a   whale population between 1930 and 1960.
        Working paper, presented to the Scientific
        Committee, International Whaling Commis-
        sion, London.

TANNER, J., Effects of population density on growth
1966   rates of animal populations. *Ecology*,
        47(5):733-45.

LESLIE MATRIX MODELS AND WHALE POPULATIONS

M.B. USHER

Abstract

A Leslie matrix model using constant age-specific fecundity and survival terms is given for blue whales, *Balaenoptera musculus*; it indicated an annual intrinsic rate of natural increase of 0.0036, is simple and inexpensive to use but only describes a population well at low density. An improved model more useful in simulation studies and giving more accurate results can be constructed by making the constant terms functions of the population size; however, there are not enough data on reproduction and natural mortality to determine the exact forms of these functions and obtaining this data would be expensive. The models are not overly sensitive to errors in the values of parameters and the resulting maximum error in predictions can be calculated; they are weakly cyclic with a period of about 10 years.

Résumé

Un modèle de matrice de Leslie, utilisant des termes constants de fécondité et de survie spécifiques de l'âge est décrit pour la baleine bleue *Balaenoptera musculus*. Il indique un taux intrinsèque annuel d'accroissement naturel de 0,0036, est d'une utilisation simple et peu coûteuse mais ne décrit qu'une population à une faible densité. Un modèle amélioré, plus utile dans les études de simulation et donnant des résultats plus précis, peut être construit en établissant les fonctions à termes constants de la taille de la population; cependant, on ne dispose pas d'assez de données sur la reproduction et la mortalité naturelle pour déterminer les formes exactes de ces fonctions et l'obtention de ces données serait coûteuse. Les modèles ne sont pas excessivement sensibles aux erreurs des valeurs des paramètres et on peut calculer l'erreur maximale de prévision qui en résulte; ils sont faiblement cycliques avec une période d'environ 10 ans.

Extracto

Se presenta un modelo matriz de Leslie para la ballena azul (*Balaenoptera musculus*), en el que se utilizan términos constantes para la fecundidad en función de la edad y la supervivencia. El modelo revela un índice intrínseco anual de aumento natural de 0,0036 y es sencillo y poco costoso de utilizar, pero sólo describe bien una población cuando su densidad es pequeña. Es posible preparar un modelo mejor, más útil para estudios de simulación y que permita obtener resultados más exactos haciendo que los términos constantes sean función del volumen de la población; sin embargo, no se dispone de datos suficientes sobre reproducción y mortalidad natural para determinar las formas exactas de esas funciones, y obtener esos datos resultaría costoso. Los modelos no son muy sensibles a eventuales errores en los valores de los parámetros, y es posible calcular el error máximo consiguiente en las predicciones. Son cíclicos, con un período de unos 10 años.

M.B. Usher
Department of Biology, University of York, Heslington, York Y01 5DD, England

This paper gives a corrected form of a Leslie matrix model for a blue whale population. The versatility of this type of model is described.

The basic model

A Leslie matrix model for a blue whale population (*Balaenoptera musculus*) was described by Usher (1972). Basically, the matrix, with some corrections is:

$$A = \begin{matrix} 0 & 0 & 0.19 & 0.44 & 0.50 & 0.50 & 0.45 \\ 0.77 & 0 & 0 & 0 & 0 & 0 & 0 \\ 0 & 0.77 & 0 & 0 & 0 & 0 & 0 \\ 0 & 0 & 0.77 & 0 & 0 & 0 & 0 \\ 0 & 0 & 0 & 0.77 & 0 & 0 & 0 \\ 0 & 0 & 0 & 0 & 0.77 & 0 & 0 \\ 0 & 0 & 0 & 0 & 0 & 0.77 & 0.78 \end{matrix}$$

The derivation of the terms in this matrix is discussed by Usher (1972). The elements in the first row of the matrix are age-specific fecundities, denoting the number of live female blue whales born to a female in the age class concerned. The elements in the sub-diagonal, all 0.77 are age-specific survival terms over a 2-year period. This value was estimated from data given by Laws (1962) for the fin whale (*B. Physalus*). The element in the lower right-hand corner, 0.78, is included since it indicates that a whale entering the final class will have a mean life expectancy of 8 years, which corresponds with the 20 year age estimated for such animals.

If the age-specific population structure at time t is denoted by:

$$a_t = \{a_{0,t}, a_{1,t}, a_{2,t}, a_{3,t}, a_{4,t}, a_{5,t}, a_{6,t}\}$$

where $a_{0,t}$ is the number of whales in their 1st and 2nd years, $a_{1,t}$ the number in their 3rd and 4th years, etc., and $a_{6,t}$ is the number in their 13th and subsequent years, and if a_{t+1} is the number one interval of time (two years) later, then the Leslie matrix model can be written as

$$a_{t+1} = A a_t$$

which can be extended k periods of time into the future

$$a_{t+k} = A a_{t+k-1} = A^2 a_{t+k-2} = \ldots = A^k a_t$$

where $k \geq 1$ is an integer. If the population structure is stable, i.e. if there are no proportional changes in the population vector a from one period of time to another, then

$$Aa = \lambda a$$

where λ is the eigenvalue and a is an eigenvector of A. λ is a scalar, and is greater than 1 if the population size is increasing, equal to 1 if the population size is stable, and is less than 1 if the population size is decreasing. Solving this matrix equation gives

$$\lambda = 1.0072$$

An eigenvector is determined except for a constant multiplier (i.e. the proportions be-

346

tween elements are precisely determined, though their magnitude is not determined). Choosing the first element as unity, the eigenvector a is

$$a = \{1, .764, .584, .447, .341, .261, .885\}$$

Refining the basic model for greater realism

It will be seen above that it is predicted that the population will increase by a factor of 1.0072 every 2 years, and thus that the basic model only predicts an exponential growth or decay of a population. In fact, there is a close correspondence between the exponential growth equation (where r is the intrinsic rate of natural increase) and the basic Leslie matrix model, where

$$r = \ln (\lambda)$$

and hence the value of λ found above would indicate an intrinsic rate of natural increase per year for blue whales of r = 0.0036. The Leslie matrix model suffers from the same disadvantages as the exponential growth model. It describes a population reasonably well if that population is at a low density, but it is not realistic to use the model to predict far into the future.

The model can be improved by making the survival and/or fecundity terms functions of the population size. Such a model was first described by Pennycuick, Compton and Beckingham (1968), and has been discussed further by Usher (1972). Thus, in the matrix the fecundity elements in the first row, f_i, are replaced by functions

$$f_{i,t} = g (a_t)$$

and the survival elements in the subdiagonal, p_i, by functions

$$p_{i,t} = h (a_t)$$

Unfortunately in the case of whales there are insufficient data on natural mortalities to be able to investigate the form of the function h (a). It is known that reproduction in whales tends to occur earlier when they are at very low population density (e.g. Watt, 1968), but there are probably insufficient data to determine the exact form of the function g (a). Also, there are insufficient data to decide whether $f_{i,t}$ and $p_{i,t}$ are functions of a_t or of only one or several of the elements of this vector. Research on these aspects of natural mortality and survival would have considerable use in the construction of a more realistic model of a whale population.

Error, stability and oscillations

Three questions relating to ecological theory are of concern if one wishes to consider the properties of these Leslie matrix models.

What is the effect of error in parameter estimation on the predictions of the model? This has been considered by Usher (1976), who shows that the model is reasonably robust to the sort of parameter estimation errors that are likely to arise. If the error of estimation is known, or can be assessed, then the maximum error in the eigenvalue can be calculated from equations (11) in Usher's paper.

Secondly, how stable are the models? The stability properties of models have been investigated by Beddington (1974), who gives equations for estimating the stability of such models. It is, however, pertinent to ask if the stability properties of an over-simplified model reflect anything about the stability of the population being modelled.

Thirdly, are there any cyclic features of the population or the model? There are no indications that the size of a natural blue whale population is cyclic. The seven roots of the basic model for blue whales are, in decreasing order of modulus, 1.007, 0.162 ± 0.579i, − 0.467, − 0.232 ± 0.359i, and 0.379. Fol-

347

lowing the arguments of Usher (1976) these indicate that the model has a weak cyclic behaviour with a period of about 4.8 steps (approximately 10 years), though oscillatons are strongly damped (since $|\lambda_1/\lambda_2| = 1.68$).

Conclusions

Leslie matrix models are simple to formulate, simple to compute, and require no great algebraic skill on the part of the modeller. However, their role should not be over-estimated when the simplest form of the models is being used.

By changing the matrix elements from constants to functions of the population size, it is possible to build more realistic models. However, the modeller is faced with a dilemma. On the one hand, a simple model can be built that uses only the data that are available at present and that gives an answer that might be approximately correct. On the other hand a more complex model would be more useful in simulation studies, and would give results that are more accurate, but considerable research effort would be required to collect data on which to construct the appropriate functions. Both models are cheap to operate; the more complex one would be very expensive in research time in its formulation. Using Leslie matrix models it should be feasible to adopt a flexible modelling policy that allows for greater or lesser precision depending upon the nature of the specific problem and the availability of financial and research resources.

References

BEDDINGTON, J.R., Age distribution and the stability of
1974   simple discrete time population models. *J. Theoret. Biol.*, 47:65-74.

LAWS, R.M., Some effects of whaling on the southern
1962   stocks of baleen whales. *In* The exploitation of natural animal populations, edited by E.D. Le Cren and M.W. Holdgate. *Symp. Br. Ecol. Soc.*, 2:137-58.

PENNYCUICK, C.J., R.M. COMPTON and L. BECKINGHAM,
1968   A computer model for simulating the growth of a population, or of two interacting populations. *J. Theoret. Biol.*, 18:316-29.

USHER, M.B., Developments in the Leslie matrix model.
1972   *In* Mathematical models in ecology, edited by J.N.R. Jeffers. *Symp. Br. Ecol. Soc.*, 12:29-60.

USHER, M.B., Extensions to models, used in renewable
1976   resource management, which incorporate an arbitrary structure. *J. Environ. Manage*, 4:123-40.

WATT, K.E.F., Ecology and resource management. New
1968   York, McGraw-Hill.

OBJECTIVES OF MANAGEMENT, WITH PARTICULAR REFERENCE TO WHALES

S.J. HOLT

Abstract

The author considers the objectives of management of humans by humans, through regulation of their impact on sea life and makes proposals for amended management policies within the existing structure. He argues for the identification of "conservation" limits of parameter values in keeping with known data, and that these should be the basis for scientific assessments, even if leading sometimes to recommendations for catches to be reduced even more than they have been so far. He explains the difference between MSY (number) and MSY (weight), noting that a 33 percent higher fishing mortality must be caused to obtain MSY (number) than MSY (weight). The overall nominal catch per unit effort is about 35 percent higher.

Résumé

L'auteur examine les objectifs de l'aménagement de l'homme par l'homme, qui réglemente son action sur la vie marine, et formule des propositions visant à modifier les politiques d'aménagement au sein des structures actuelles. Il prône l'établissement de paramètres pour fixer des limites de conservation conformes aux données connues et pouvant servir de base aux évaluations scientifiques, même s'il doit en résulter des recommandations préconisant une réduction des captures plus fortes que celles décidées jusqu'alors. Expliquant la différence entre rendement maximal équilibré (valeur unitaire) et rendement maximal équilibré (valeur pondérale), il note que la mortalité due à la pêche doit être de 33 pour cent plus élevée pour obtenir le rendement maximal équilibré en valeur unitaire plutôt qu'en valeur pondérale. Le total des captures nominales par unité d'effort est d'environ 35 pour cent plus élevé.

Extracto

El autor examina los objetivos del control del hombre por el hombre, mediante la regulación de sus repercusiones en la vida marina, y formula propuestas para la modificación de políticas de ordenación en el marco de la estructura existente. Propugna la determinación de los límites de « conservación » de los valores del parámetro de acuerdo con los datos conocidos y sostiene que deberían constituir la base de las evaluaciones científicas, aun cuando a veces den lugar a que se recomiende una reducción todavía mayor de las capturas. Explica la diferencia entre el RSM (número) y el RSM (peso), observando que para conseguir un RSM (número) mayor que el RSM (peso) haría falta una mortalidad por captura un 33 por ciento más elevada. La captura global nominal por unidad es aproximadamente un 35 por ciento mayor.

S.J. Holt
Threshold Foundation, 7 Regency Terrace, Elm Place, London SW7, England

Introduction

Before humans presume to "manage" marine mammals, we need to learn to manage ourselves. In fact, no wild marine living resources are managed. At best — and that not very widely — there is a degree of management of *some* of the human activities which affect those resources. The distinctions between managing "environments", managing "resources" and managing human activities are not, I believe, merely terminological as some scientists have maintained, nor trivial. This paper concerns the objectives of management of humans by humans through regulation of their impact on sea life, and makes proposals for amended management policies within the existing structure.

Practically all attempts to bring some order into modern fishing and human activity have had as their aim the reduction to acceptable levels of the deleterious effects of competition among present exploiters, and of temporal "competition" between them now and them later, or their descendents or successors. The need for such action has been particularly evident in, but not limited to, the exploitation of "common property" resources. The attempts to "manage" have always involved a *partial* restraint on *some* of the relevant activities by *some* of the participants. The approach has been to try to identify which actions have the most deleterious effect — potentially or in fact — and which participants are most powerful in their impact on the resource, and then to regulate one or a few of the factors of impact. This usually means regulating the factors of exploitation itself, which are considered to have a direct and incontestable effect on the resource.

Experience shows that such regulation can sometimes be effective, by reducing some of the deleterious effects of high intensity competitive exploitation. However, experience also shows that every partial regulation leads to "reactions" which are inevitably opposite in sense. If one kind of fishing is regulated there will be a tendency to shift toward one that is not; if sizes of meshes are regulated in trawling, there will be predictable changes in the amount and distribution of fishing effort; if fishing on some stocks of a species is regulated, there will be a shift toward other stocks of the same or similar species; if fishing or whaling under the flags of some nations is regulated, there will be a shift toward the use of flags of *other* nations, and so on. Furthermore, if some of the interrelated but qualitatively different uses of the sea are regulated and others not, shifts will occur in the pattern of use. These changes are predictable in a general way because, despite the present weakness of economic and social theory, we *do* in most cases know the motives — the search for economic advantage by some human groups over others, and the securing of profit, that is, of yield with respect to input costs. This is so obvious as to make one doubt the value of repeating it — repetition is needed, however, to dispel the haze of myth that surrounds most international discussion of marine resource management, which still obscures the main issues and confuses both the scientific discussion and the decision making.

The "Airlie" proposals

A recent formulation of the objectives for managing activities with respect to wild living resources, eventually summarized in a Resolution of the 1975 General Assembly of the International Union for the Conservation of Nature (IUCN) referred specifically to whales and whaling. It suggests that a number of related objectives should be sought as a package, that not all of these are fully compatible, and not all can now be quantified — some perhaps never. It is proposed that due weight be given to the possible interests of future generations of humans, which may or may not be the same as ours, quantitatively or qualitatively, and that regard should be paid to harmonizing conflict between different present interests, including qualitatively different "uses" of the resource, which can include not regarding natural systems as "resources".

In addition, according to this formulation, to the extent that economic yields are important it is the net yield rather than the gross yield that matters; account is to be taken of inputs, as well as outputs, especially to the extent that these inputs are derived from exploitation of other natural resources, which are not infinite.[1]

---

[1] It appears that in some "commercial" exploitation that continues there is a net loss. According to A.W. Reitze, Jr. and Glenn L. Reitze (1975 "Law, Blood and Seals in the Pribilofs", *Environment*, 17(7):3-4, the Pribilof research and management programme of the US costs the taxpayers US$ 1 000 000 annually more than the value of the seal skins obtained. These costs in fact are indirect contributions to welfare of the inhabitants of St. Paul Island.

Discussion

This and similar formulations of objectives have been criticized, because all the factors cannot be quantified. Such criticism·is not sound, for 2 main reasons. First, "objectives" do not emerge from science as such; science might indicate whether particular objectives are feasible in terms of natural law, and might suggest certain objectives as possible and rational, which had not previously been recognized. But in no area of human governance are decisions taken on the basis *only* of factors which can be measured. If society decides that "protection of an ecosystem" is an objective of management, then an argument that we do not yet know enough about the dynamics of ecosystems seems no impediment to that objective. On the contrary, declaration of the objective is the stimulus to the scientific research needed to implement it.

Second, the distinction between what can and cannot be measured is more blurred than is usually supposed. "Values" of *live* whales have not been measured, even as tourist attractions — the latter because no serious attempt has been made to do so — but, equally, "maximum sustainable yield" of whales has not been measured, to my satisfaction, for any stock. At most, the distinction is a matter of degree.

It is, therefore, entirely appropriate now to consider a multiple, interacting "package" of management objectives such as that adopted by IUCN. It is, in fact, still insufficiently comprehensive; it still does not recognize that management of human activities with respect to wildlife cannot be separated from management of *all* human activities affecting the ecosystem in which that wildlife resides, and cannot be considered without proper account being taken of the natural properties of those ecosystems. The most important property of ecosystems which the package ignores, is their continuous evolution and change, on all time scales. A provision should therefore be added to the IUCN formulation that human impact on wildlife be regulated such that its ability to respond successfully to natural environmental change — to survive adverse changes and to thrive in response to favourable changes — should not be impaired.

The objectives of whaling management

Unfortunately, with respect to whales, we seem in practice to be lightyears away from international acceptance of social and scientific views of the kind embodied in the IUCN resolution, which are themselves already far from what some lawyers, philosophers and social scientists would now find acceptable, being, for them, still too anthropocentric. The International Whaling Commission cannot take — or has not taken — account of *any* of the values of live whales, except to a small degree as the parents of a future catch in the medium term — and that with low priority. The IWC furthermore, is not in a position to consider whales in relation to other predators on krill and fish, for example. It does not yet regulate all whaling — in fact, as total whale yields dwindle, the *proportion* taken by non-members of IWC is gradually increasing; recent accessions to membership have been by non-whaling countries.

What then can be done about the objectives of whaling management through the present institutional arrangements? On a more detailed level, some suggestions for improvement are made below.

CURRENT WHALE MANAGEMENT POLICY

The positive features of recent changes in IWC policy are: (i) the recognition that biologically distinct populations or populations which mix only slowly and/or partially (so-called "species stocks") must be treated separately; (ii) immediate cessation of whaling on a population reduced by whaling to less than a particular fraction of its abundance before whaling started; (iii) the need to regulate rate of reduction of a population by whaling; (iv) the aim of regulation of whaling is stability. However, there are a number of other factors which could very easily have been taken into account but have not, and concerning which there is great resistance from whaling countries, such as taking into account input as well as yield (i.e., whaling effort), and attempting to maximize the amount of meat, oil and other products obtained, by their weight or value rather than the numbers of whales caught.

A positive aspect of the "new management policy" (NMP) of the IWC is that it has built into it cautionary rules for margins of error. Thus, the quota for a "sustained management stock" is set at 90 % of the estimated MSY rather than at MSY itself and there is a graduated reduction in quotas from stocks considered to be near to, but if anything below, "MSY level". On the other side, however, if a stock is thought to be an "Initial management stock", somewhat *above* MSY level, it is permitted to take catches greater than the sustainable yield that corresponds to the estimated stock level, despite the fact that it is in precisely those cases that least is known of the population parameters and dynamics. Furthermore, the application of the NMP is in most cases so far such as to pay minimal attention to the *structure* of a stock (e.g., composition by sex, age, size, maturity), concentrating almost entirely on numbers of individuals.

The rule of "opposite, if not equal, reaction" has also come into play. Before the NMP was promulgated, the Scientific Committee of the IWC always insisted on making so-called "conservative" estimates and recommendations, which have usually been overruled, overlooked or delayed in their application by the Commission itself. Now that the NMP rules contain *some* automatic, cautionary provisions, the Scientific Committee appears to be reverting to making "best" estimates rather than "conservative" ones. The danger of this is that "best" is not likely — because of our broad scientific ignorance — to be much better than something very different, and we have little idea how near "best" is to "correct".

POSSIBLE CHANGES

In these circumstances, those who wish to see the balance of interest in today and in tomorrow redressed even to a small degree in favour of the latter might concentrate on what can be done now within the terms of the NMP. I see 3 possibilities, at least.

Best estimates and safety factors

It may be argued that it is not satisfactory to replace the customary caution in the Scientific Committee recommendations by the "90 % of MSY" rule of the NMP. Thus, if the sustainable yield from a stock in a certain state is, say, one half or less of the Committee's estimate, application of the 90 % rule in determining quotas gives an entirely false sense of "conservation". That errors of that magnitude and in that direction are possible, even likely, is evident from inspection of the procedures of assessment on which regulations are currently based. The quotas for the baleen whales still open to exploitation in the Southern Hemisphere (minke and sei) are based on extremely insecure evidence of the changes in abundance of these stocks since intensive exploitation of them began, and does not consider the structure of those stocks; assumptions as to, not assessments of, the stock "levels" which should give MSY; and values of the net rate of population increase which have a very flimsy data base, and which most likely overestimate the net rate of increase at the "MSY level".

The quotas for sperm whales are now based on the best, and most complex, theoretical models devised for whale stock assessment, but estimates of the values of parameters to be used in those models leave much to be desired. In particular, there are no estimates of the most significant parameter of all, the value of which practically determines the quota for catches of males, and hence most of the total yield. This is the so-called "harem reserve". The estimates of sustainable catches of males are overwhelmingly conditioned by the assumption that a certain number of surplus males, not needed for reproduction, can be taken without affecting the rate of reproduction. The proportion of such males is a pure guess. According to the Scientific Committee the number of socially mature bulls not actively engaged in breeding activity may, in an unexploited population, be up to 4 times the number of harem masters. The Committee wrote, "*There is no evidence* to indicate to what level this ratio could decline under exploitation. A *reasonable ratio might* be 1:1, ... or *alternatively no reserve might* be necessary" (my emphasis). And why may not a ratio of 2:1 be "reasonable"? In such circumstances it cannot be maintained on any scientific basis that, by assuming, alternatively, harem reserves as 0 or 1, one has taken "optimistic" and "pessimistic" limits and thereby straddled the "true" value (the "best" value, used by the Committee now, is 0.3).

In such circumstances, because of the great asymmetry of the consequences of being in error in 1 direction or the other, I suggest that it is essential that one seek to identify the "conservation" limits of parameter values in keeping with whatever real data exist, and make these the basis of scientific assessments, even if this sometimes leads to recommendations for catches to be reduced even more than they have been so far.

Experimental management

I have elsewhere (Holt, 1977) suggested that the only scientific way to better estimation of sustainable yields is through experiment, not by watching the dismal history of overexploitation of whales and afterwards trying to explain how and why each collapse occurred. Further, there are some situations in which such experiments could be done without danger to the species — where, for example, there exist several otherwise similar populations, which are separated from each other, but not completely. Such seems to be the case of the Southern Hemisphere sperm whales, the whales which have the greatest current economic importance, about which least is known and which just might not yet all be overexploited. Such an experiment would involve an immediate moratorium on exploitation of some of the stocks and a controlled intensive exploitation of others. To produce useful results, the experiment would certainly have to last many years, because of the apparent magnitude of the vital rates of sperm whales, the duration of their natural lives and the time pattern of sexual and social maturity.

MSY weight and MSY number

In adopting the NMP, the Commission left open the possibility of adopting as its objective the attainment of "optimum sustainable yield" (OSY) rather than MSY. OSY has been taken so far to mean maximum sustainable yield in weight. However, in 1975 and in 1976 the Commission has failed to adopt such an objective. Indeed, its Scientific Committee has, with respect to sperm whales, gone backward a step in this matter. In its original proposals concerning this species, the Committee referred to MSY by number, but with a weighting factor used in combining the numbers of males and females to take account of the fact that the average weight of a male is 2-3 times that of a female. In its most recent submissions, however, the Committee has given, with little comment, recommendations based alternatively on MSY by weight, and MSY by number, but now with no allowance in the latter for the sex difference. The Commission has, of course, adopted the latter criterion. One has the impression that although individual scientists attempt to reach objective conclusions and are often quite outspoken between themselves about the weaknesses in the pres-

ent assessments, *collectively* the Scientific Committee fears to "go too far" in drawing conclusions from its own studies; there is perhaps a desire not to rock the boat so far that those nations who wish to continue relatively unrestrained whaling for short-term advantage will object to the whole NMP procedure and not abide by majority decisions.

The degree of difference between the MSY number and weight criteria depends on what fraction of the life-long growth in size occurs during the "exploitable phase" — that is, after recruitment. It happens that this fraction is much higher among the exploited toothed whales — sperm and bottlenose[2] — than among the rorquals, although even in the latter, it is not negligible; much depends on the sizes at recruitment, which in turn are partly determined by the legal minimum size limits in force.

It has been argued in the IWC, that the weight criterion is in some way more "theoretical" than the numerical one, because whales cannot be weighed, and because an algebraic expression is used to calculate total weights from length compositions of catches. In fact, the conversion to weight is one of the least "theoretical" of all the arithmetic operations in whale stock assessment, being based on relatively few, but careful observations. This debate, however, obscures the real question, which is how, as a practical matter, to maximize the quantities of useful products from whales. In no case is the amount of product proportional to the number of whales, irrespective of their size. As between species, bet-

---

[2] The report of the special meeting on sperm whales of the IWC Scientific Committee in La Jolla, March 1976, is ambiguous in this respect. Thus while it is demonstrated from the dynamic model that there is a significant difference between the MSY (No.) and MSY (Wt.) situations, in discussing minimum size limits the report says that "the growth curve of both males and females is very flat in the main exploited age groups". Yet the "best" growth curves used in the model indicate that 19 % of the growth in length of both males and females, and 46 % of the growth in weight occurs *after* recruitment.

ween sexes and between ages it is closely dependent on the size of the animal.

The relation of product to catch becomes closer if the number caught of each species is multiplied by the mean length of animals in the catch, the cube of the mean length, the mean of the cubes of length of animals killed, and the best estimate of weight as calculated from the best fitting power functions of length (Table 1).

The calculations of the Scientific Committee with respect to Southern Hemisphere sperm whales, using the Committee's "best"

parameter values, can be summarized as follows:

Total exploitable stock in all 9 Divisions in 1946 (assumed relatively unaffected at that time by any previous whaling) – 229 900 males, 360 700 females and in 1975 106 200 and 298 300 respectively.

Adoption of MSY number criterion gives a total catch of 158 000 tons of which 88.8 % is from males, while the MSY weight criterion gives a total of 163 500 tons of which 98.4 % is from males. Adoption of the number instead of the weight criterion therefore sacri-

Table 1. World catches of sperm whales and production of sperm oil

| Year | Whale catch | | Oil production | | | Estimated efficiency of oil extraction % | Potential oil production Barrels × 10⁻³ |
	No. × 10⁻³	Wt. tons × 10⁻³	Barrels × 10⁻³	Barrels per whale	Barrels per ton		
1948-49	9.02	271.5	344.7	38.2	1.27	74.7	461.4
50	8.22	221.3	285.6	34.6	1.29	75.9	376.3
51	18.27	454.2	630.2	34.5	1.39	81.8	770.4
52	11.56	338.9	447.2	38.7	1.32	77.6	576.3
53	9.58	210.6	304.0	31.7	1.44	84.7	358.9
54	13.56	316.6	429.5	31.7	1.36	80.0	536.9
55	15.59	403.7	539.5	34.6	1.34	78.8	684.6
56	18.59	462.1	651.1	35.0	1.41	82.9	785.4
57	19.16	423.8	589.7	30.8	1.39	81.8	720.9
58	21.85	492.6	725.2	33.2	1.47	86.5	838.4
59	21.30	465.0	691.6	32.5	1.49	87.6	789.5
1959-60	20.34	466.3	650.7	32.0	1.40	82.4	789.7
61	21.13	485.9	643.3	30.5	1.32	77.6	829.0
62	23.32	481.0	716.1	30.7	1.49	87.6	817.5
63	27.86	565.8	848.9	30.5	1.50	88.2	962.5
64	29.26	611.9	898.3	30.7	1.47	86.5	1 038.5
65	25.55	562.0	832.6	32.6	1.48	87.1	955.9
66	27.38	588.2	872.0	31.9	1.48	87.1	1 001.2
67	26.43	593.2	883.7	33.4	1.49	87.6	1 008.8
68	24.08	502.7	712.4	29.6	1.42	83.5	853.2
69	24.22	513.7	771.7	31.9	1.50	88.2	874.9
70	25.83	546.2	822.7	31.9	1.51	88.8	926.5
71	22.41	452.4	723.5	32.3	1.60	94.1	768.9
72	15.35	312.8	577.2	37.6	1.85	(100.0)	(577.2)
73	21.68	372.8	619.5	28.6	1.66	97.6	634.7
74			586.7				(586.7)
75			461.6				(461.6)

Note: Over most of the period for which calculations have been made (1948/49-1972/73) world sperm oil production per whale tended to decline but with large fluctuations in the last years. The decline is associated with a gradual decline in the mean length of whales in the catches. The very low value of 28.6 for 1972/73 is associated with a big increase in the relative numbers of females in the catch. On the other hand the barrels of oil per ton of whale increased throughout the period; this can be explained in terms of improvements in the efficiency of extraction of oil from the carcasses. The coefficient of variation of the oil per whale ratios is 0.07, that of the oil per ton rations is 0.05. Five-year averages of the relevant factors are plotted in Fig. 1.

If the recent values of barrels per ton – about 1.70 – are taken as representing full extraction and index of efficiency of extraction can be calculated from the series of barrels per ton and applied to the barrels per whale series; this can be applied, in turn, to the catches in number, to give estimates of the oil which could have been obtained each year with the present efficiency of extraction; these are shown in the last column of the table.

fices 3.4 % of the potential sperm oil and meal production. However, to obtain the lower yield, 5.7 % *more* whales have to be caught each year, needing correspondingly more whaling effort and therefore at greater cost. *How much* more effort is revealed by calculation of the required fishing mortality rates (F). According to the model used, MSY (number) is obtained when the males are sustained at 32 % of their "initial" level, and females at 79 %; the corresponding fractions for MSY (weight) are 39 % and 97 %. The fishing mortalities, corresponding to these levels are shown in Table 2.

Thus, to obtain MSY (number) a 33 % higher fishing mortality must be caused, requiring correspondingly more whaling effort, than to obtain MSY (weight). The overall nominal catch per unit effort is about 35 % higher at MSY (weight) than at MSY (number).

The small advantage in production to be gained by adopting the weight criterion is seen to be perhaps the least of its advantages. Because the aim is to maintain the numbers and biomass of the stocks at relatively higher levels than required for the number criterion, the consequences of errors in assessments can more easily and rapidly be corrected.

The IWC has taken its decision in favour of numbers "for the time being". However, it is obvious that whereas a shift from the higher to the lower stock level can be made very quickly, without even increasing the number of whaling expeditions, the transition from the lower to the higher level by growth of the whale population would take many years, even if in the interim there were a complete moratorium. If the recovery rate at each stock level is equal to the sustainable F at that level, then it is 0.5 % per year at 0.79 of initial size and 0.1 % at 0.97 of initial size. Taking an average recovery rate of 0.3 % we find it would take 60 years for the females to recover from the MSY (number) level to the MSY (weight) level if there were no whaling, and proportionately longer if some whaling were to continue. The males would recover from 32 % toward 39 % of initial level at a faster rate, but ultimately the increased recruitment of males is based on the recovery of the *female*. It must be remembered furthermore that to give maximum sustained product yield the stocks must have the appropriate steady-state age and size composition as well as the appropriate abundance. Since the age of social maturity of males has been taken as 25 years, the ages at recruitment as 20 and 13 years for males and females, respectively, and the age of maturity of females as between 8.5 and 10 years, it may be necessary to add several decades to the above estimate of partial recovery times.

Thus, from the point of view of ultimately having a sustained and efficient industry for sperm whale products in the reasonably near future, based on thriving Southern Hemisphere stocks, the decision in 1976 to permit nearly 1 000 females to be taken in the coming season could prove disastrous on the basis of the Scientific Committee's own calculations.

If the objective really is to secure as large a sustained industry for sperm whale products as the resources can support, why does the IWC reject the weight criterion? The economic answer is, I think, clear from Clarke (1976) and Price (1976), as well as from other economic studies: the industry and the whaling nations discount the future too heavily in whaling plans and accounting. If one's overwhelming interest is in securing agreement on as big a catch as possible next season, and this has to be done with a regime that specifies that the catch must be sustainable, then one will *always* opt for the number criterion, whatever the particular levels of present stocks, since one will be permitted thereby to catch more whales; their average weight will, that year, be the same as if

Table 2. Fishing mortality

	Males	Females	Both sexes
MSY (No.)	.078	.005	.020
MSY (Wt.)	.073	.001	.015

Table 3. Classified number from 18 "stocks" (9 divisions × 2 sexes)

	MSY (Number) criterion	MSY (Weight) criterion
Initial	6	3
Sustained	8	7
Protected	4	8

the weight criterion – in the very short term only, of course – a greater weight of catch and hence of product. This would be true if whaling were to be based on quotas related to sustainable yields at all existing stock levels.

The discrepancy is aggravated by the various arbitrary thresholds between the different stock categories, as Table 3 demonstrates.

Conclusion

Even with respect to industrial production patterns and even within a sustainable yield rule, the objectives of the present whaling industry are not for the world community in the long run. This, together with the question of appropriate discount rates, illustrates the dilemma the IWC now finds itself in while attempting to implement a long-term policy.

References

CLARKE, C., Economic aspects of renewable resource
1976   exploitation. Paper presented to the Scientific Consultation on the Conservation and Management of Marine Mammals and their Environment, Bergen, Norway, 31 August-9 September 1976. Rome, FAO, ACMRR/MM/SC/65:13 p.

HOLT, S.J., Whale management policy. *Rep. IWC*,
1977   (27):133-7.

PRICE, C., Some economic aspects of marine mammal
1976   management policies. Paper presented to the Scientific Consultation on the Conservation and Management of Marine Mammals and their Environment, Bergen, Norway, 31 August-9 September, 1976. Rome, FAO, ACMRR/MM/SC/85:6 p.

ANTARCTIC ARGUMENTS AND OPTIONS

D. J. TRANTER

Abstract

There are two different interpretations of the possibilities of exploitation of living resources in the Antarctic, both based on the assumption that there exists there a biocenosis depending on krill abundance and already changed through the reduction of baleen whale stocks by man. One proposition states that an equilibrium can be reached allowing the recovery of the whale stocks and the sustained harvesting of them and krill. The other proposition states that krill harvesting and competition from other species may prevent the recovery of some stocks of whales. A comparison of management options based on this second argument indicates that a harvest of whales and krill, the option most likely to occur, is the one least likely to maximize renewable yield.

Résumé

Il existe deux différentes interprétations des possibilités d'exploitation des ressources vivantes de l'Antarctique, toutes deux fondées sur l'hypothèse d'une biocenose dépendant de l'abondance de krills, déjà modifiée par la réduction des stocks de mysticètes provoquée par l'homme. Une thèse avance qu'un équilibre peut être atteint, permettant la reprise des stocks de baleines et leur exploitation soutenue ainsi que celle des krills. D'après l'autre thèse, l'exploitation des krills et la concurrence d'autres espèces peut interdire la reprise de certains stocks de baleines. Une comparaison des options d'aménagement basée sur ce second argument indique qu'une exploitation des baleines et des krills, l'option la plus probable, est l'une de celles qui offre le plus de chances de maximiser le rendement renouvelable.

Extracto

Hay dos interpretaciones diferentes de las posibilidades de explotación de los recursos vivos del Antártico, basadas ambas en el supuesto de que existe en esa zona una biocenosis que depende de la abundancia de krill y que ya ha sido modificada a causa de la reducción de las poblaciones de mistacocetos debida a la acción del hombre. Una de las interpretaciones afirma que puede alcanzarse un equilibrio que permita la recuperación de las poblaciones de ballena y la explotación sostenida de las ballenas y del krill. La otra interpretación afirma que la explotación del krill y la competencia de otras especies puede impedir la recuperación de algunas poblaciones de ballenas. Comparando las distintas opciones de regulación basadas en este segundo argumento se concluye que la que más difícilmente permitirá aumentar al máximo el rendimiento renovable es la explotación contemporánea de las ballenas y del krill, que constituye la opción más probable.

D.J. Tranter
Marine Ecosystems Group, Division of Fisheries and Oceanography, CSIRO, P.O. Box 21, Cronulla, NSW 2230, Australia

359

From the mass of information presented at this meeting, two divergent propositions can be identified, whose acceptance leads to quite different management strategies for marine mammal resources in the Antarctic. The purpose of this paper is to examine briefly the scientific logic on which they are based, identify the point at which they diverge, and assess the management options that follow from one of those propositions.

The common sequence of arguments on which both propositions are based would appear to be the following:

— The total biomass of Antarctic baleen whales is now much less than it was 50 years ago (ACMRR, 1976; Gulland, 1976; SCO-SCAR, 1976).

— During this period several of these species and others which share the same food resource (krill) (notably crabeater seals), underwent important changes in reproductive biology such as a marked decrease in age at first maturity (Gambell, 1976; ACMRR, 1976a; Lockyer, 1972).

— This suggests that krill abundance may have been an important limiting factor in the carrying capacity of the environment, and that the withdrawal of large numbers of baleen whales from the krill biocenosis has increased the potential for increase in those populations which feed on krill (ACMRR, 1976, a; Gulland, 1976; Gambell, 1976).

At this point, the arguments diverge:

a) The populations of depleted stocks of baleen whales will probably increase if whaling pressure is reduced (ACMRR, 1976; Allen, 1976).
True. But so will the populations of all other species in the krill biocenosis (squid, fish, birds, seals, etc.) (ACMRR, 1976b).

b) This increase will probably lead eventually to a form of recovery i.e., numbers will increase until an equilibrium is established at the carrying capacity of the environment (Gulland, 1976; Allen, 1976).
Recovery may begin but this will be shortlived. The smaller species in the krill biocenosis will have faster specific rates of growth (May, 1976) than the large baleen whales. Their total biomass will increase at the expense of that of baleen whales.

c) Some harvesting of krill may take place without prejudice to the recovery of the stocks of baleen whales.
Krill harvesting will lower the carrying capacity of the krill biocenosis. Those elements of the biocenosis with the fewest numbers will be at a disadvantage.

d) An equilibrium can be reached at which the yield of krill and baleen whales will be near the optimum (Horwood, 1976).
At equilibrium, the numbers of baleen whales in some stocks may be too low for their permanent survival (May, 1976), e.g., one or more of the humpback stocks.

If the second proposition should prove to be the stronger, it would follow that management decisions should be based upon the krill biocenosis. In particular, since the present biomass of crabeater seals in the Antarctic is of the same order as that of the present biomass of baleen whales (SCOR-SCAR, 1976; ACMRR, 1976a), there is a case for considering not only the joint exploitation of krill and whales (Horwood, 1976) but a joint krill/baleen whale/crabeater seal system. The following management options are available, arranged in order of their likelihood of maximizing the yield on a renewable basis from all components of this system.

(i) Harvest seals

(ii) No harvest of seals, krill or whales

(iii) Harvest seals + krill

(iv) Harvest krill

(v) Harvest whales + seals

(vi) Harvest whales

(vii) Harvest whales + krill + seals

(viii) Harvest whales + krill

It is therefore distressing to contemplate that *what could, perhaps, be the least desirable option of all (viii) is the very one which, on present indications (FAO, 1976), is most likely to eventuate in the near future*. It would seem to be a useful function of this meeting to draw attention to this possibility and to recommend what could be better options from a biological point of view.

References

ACMRR (FAO), Draft report of the Ad Hoc Group I on
1976   Large Cetaceans. Paper presented to the Scientific Consultation on the Conservation and Management of Marine Mammals and their Environment, Bergen, Norway, 31 August-9 September 1976. Rome, FAO, ACMRR/MM/SC/2:76 p.

—, Draft Report of the Ad Hoc Group III on Seals
1976a   and Marine Otters. Paper presented to the Scientific Consultation on the Conservation and Management of Marine Mammals and their Environment, Bergen, Norway, 31 August-9 September 1976. Rome, FAO, ACMRR/MM/SC/4:182 p.

ACMRR (FAO), Draft report of the Ad Hoc Group IV
1976b   on Ecological and General Problems. Supplement 1. Paper presented to the Scientific Consultation on the Conservation and Management of Marine Mammals and their Environment, Bergen, Norway, 31 August-9 September 1976. Rome, FAO, ACMRR/MM/SC/5. Suppl. 1:12 p.

ALLEN, K.R., The optimization of management strategy
1976   for marine mammals. Paper presented to the Scientific Consultation on the Conservation and Management of Marine Mammals and their Environment, Bergen, Norway, 31 August-9 September 1976. Rome, FAO, ACMRR/MM/SC/57:12 p.

FAO, The problems of harvesting and utilization of
1976   Antarctic krill. An outline prepared for the SCOR + SCAR Conference on the Living Resources of the Southern Ocean. Woods Hole, U.S.A. August 1976.

GAMBELL, R., A note of the changes observed in the
1976   pregnancy rate and age at sexual maturity of some baleen whales in the Antarctic. Population biology and Antarctic. Population biology and the management of whales. *Appl. Biol.*, 1:247-343.

GULLAND, J.A., Antarctic baleen whales: history and
1976   prospects. *Polar Rec.*, 18(112):5-13.

HORWOOD, J.W., On the joint exploitation of krill and
1976   whales. Paper presented to the Scientific Consultation on the Conservation and Management of Marine Mammals and their Environment, Bergen, Norway, 31 August-9 September 1976. Rome, FAO, ACMRR/MM/SC/116:6 p.

LOCKYER, C., The age at sexual maturity of the southern
1972   fin whale (*Balaenoptera physalus*) using annual layer counts in the earplug. *J. Cons. CIEM*, 34(2): 276-94.

MAY, R., Factors controlling the stability and break-
1976   down of ecosystems. *Nature, Lond.*, 263:91 p.

SCOR-SCAR Working Group of Specialists on the
1976   Southern Ocean. Draft BIOMASS proposal. Prepared at Woods Hole, U.S.A., August 1976.

ON THE JOINT EXPLOITATION OF KRILL AND WHALES

J.W. HORWOOD

Abstract

A simple model of the joint exploitation of krill, *Euphausia superba*, and baleen whales in the Antarctic where the whales feed on the krill, is constructed, using different equations of logistic growth functions and constant predation and reproductive rates. From the resulting equilibria values under various constant rates of fishing, it is shown that whales decrease in abundance with exploitation of krill but not so much as when the whales are harvested directly and that, to a far lesser extent, krill yields are reduced by the presence of whales. With a maximum fishing mortality rate of krill of 0.3, the total yield in weight is greatest when whales have been exterminated and only krill is harvested; if whales are ten times more valuable than krill, the yield is greatest when mostly krill are harvested but whales are allowed to persist and, if whales are worth 100 times more than krill, when a substantial harvest of whales is taken. Thus, there may be a conflict of interest between separate fisheries for krill and whales and it would be best if they were managed together.

Résumé

Un modèle simple d'exploitation conjointe des krills, *Euphausia superba*, et des mysticètes dans l'Antarctique où les baleines se nourrissent de krill est construit en utilisant différentes équations de fonctions logistiques de croissance, de prédation constante et de taux de reproduction. D'après les valeurs d'équilibre qui en résultent avec divers taux constants de pêche, on montre que l'abondance des baleines diminue avec l'exploitation du krill mais pas autant que lorsque les baleines sont exploitées directement et que, dans une mesure nettement inférieure, les rendements de krill sont réduits par la présence des baleines. Avec un taux maximal de mortalité par capture des krills de 0,3, le rendement total en poids est le plus élevé quand les baleines ont été exterminées et quand on n'exploite que les krills; si les baleines sont dix fois plus précieuses que les krills, le rendement est le plus grand quand on exploite surtout les krills mais quand on laisse subsister les baleines et, si les baleines valent 100 fois plus que les krills, quand une quantité substantielle de baleines est capturée. Il peut donc exister un conflit d'intérêts entre les pêcheries distinctes de krills et de baleines et il serait avantageux de les exploiter rationnellement ensemble.

Extracto

Se presenta un modelo sencillo para la explotación conjunta de krill (*Euphausia superba*) y mistacocetos en el Antártico, donde las ballenas se alimentan de krill, utilizando diferentes ecuaciones de funciones logísticas de crecimiento e índices constantes de depredación y reproducción. Los valores de equilibrio resultantes con varios índices constantes de explotación muestran que las ballenas disminuyen en abundancia si se explota el krill, pero no tanto como cuando se explotan directamente las ballenas, y muestran además que, en medida mucho menor, el rendimiento del krill se ve reducido por la presencia de ballenas. Con una mortalidad máxima del krill (por pesca) del 0,3, el rendimiento total en peso es mayor si se han exterminado las ballenas y sólo se explota el krill. Si las ballenas son diez veces más valiosas que el krill, el rendimiento es mayor cuando se explota sobre todo el krill pero se permite que subsistan las ballenas, y si las ballenas son 100 veces más valiosas que el krill, cuando se explotan éstas en forma sustancial. Así pues, puede haber un conflicto de intereses entre pesquerías separadas de krill y ballena, y sería mejor regular su explotación conjuntamente.

J.W. Horwood
Fisheries Laboratory, Ministry of Agriculture, Fisheries and Food, Lowestoft, Suffolk NR33OHT, England

Introduction

The history of large scale commercial fishing shows a pattern of heavy exploitation of a species, followed by its decline, and the switching of fishing effort on to new species. More recently this redirection of effort has been forced onto the industries by management policies designed both to protect stocks and to achieve maximum sustained yields. The new fisheries being investigated include that for Antarctic krill (*Euphausia superba*).

Estimates of total world potential marine fish catches are of the order of 100 million tons (FAO, 1974) and the estimated yield from krill is considered to be between 100 and 200 million tons, from a standing stock of between 800 and 5 000 million tons. It is clear that krill provide an important potential fishery. Already Japan and the USSR have fleets experimentally exploiting this resource and many other countries have been involved in commercial studies.

However, krill are also utilised by the Antarctic baleen whales which, in turn, are subject to exploitation by the whaling fleets. This is another example of mixed fishery (Pope, 1976), with, in this case, exploitation of both predator and prey. This study considers the joint exploitation of a predator and its prey and the relative yields that can be obtained from different fishing strategies.

Method

If no predation or exploitation occurs on the krill then their dynamics may be represented by the familiar logistic equation such that

$$dz/dy = rz(1 - z/K),$$

where

z   is the amount of krill in millions of tons,

r   is the instantaneous rate of increase observed at low z, and

K   is the carrying capacity of the environment.

Predation occurs by the exploited whales and also by seals and other whales and this must be incorporated into the model along with fishing, giving

$$dz/dt = rz(1 - z/K) - awz - bz - F_1 z,$$

where

w   is the amount of whales in millions of tons,

a   is the instantaneous predation rate of the whales,

b   is the instantaneous predation rate of the other predators multiplied by their average biomass, and

F_1 is the instantaneous fishing mortality rate on the krill.

Values for some of these parameters may be estimated from the FAO report (FAO, 1974). If the standing stock is considered to be between 800 and 5 000 million tons then the carrying capacity must be somewhat above this and so a value of $K = 10^4$ million tons is probably reasonable. If the consumption of seals is about 250 million tons with a standing stock of krill 5×10^9 tons then $b \simeq 0.05$. To give reasonable yields to the whale stock a must be 10^{-4} approximately, but this is only justifiable in light of the model for the whale population discussed below. F_1 will be varied and r is arbitrarily set at 1.0.

The whale population dynamics can be similarly expressed with a logistic growth function and a constant fishing rate, such that

$$dw/dt = r^1 W (1 - w/K^1) - F_2 w,$$

where

r^1 is the instantaneous rate of increase of the whales at low w,

K^1 is the carrying capacity of the environment, and

F_2 is the instantaneous fishing mortality rate.

Assuming that the ratio of the standing stocks is about equal to that of the loss rates

from the two trophic levels, the value for K^1 may be guessed by considering the ecological efficiency (α) from one trophic level to the next. Although Cushing (1975) has shown that this varies seasonally, a value of 10 per cent is accepted here; hence $K = \alpha z$ where $\alpha = 0.1$. The rate of increase (r^1) is smaller than that for krill and a figure of $r^1 = 0.05$ has been chosen, similar to that used in recent sei whale assessments. Incorporating the formulation for K^1 leads to the new equation

$$dw/dt = r^1 w (1 - w/\alpha z) - F_2 w$$

Thus the predator-prey and exploitation model has been formulated but, because of the seasonal nature of both predation and exploitation, they are better expressed as difference equations and hence they become

$$Z_{t+1} = Z_t + r Z_t (1 - Z_t/K) - a w_t Z_t - b Z_t - F_1 Z_t \ldots (1)$$

and

$$w_{t+1} = w_t + r^1 w_t (1 - w_t/\alpha Z_t) - F_2 w_t \ldots (2)$$

Results

The equilibrium states for the two equations are given by

$$z = (r - b - F_1)/((r/K) + a \alpha (1 - F_2/r^1)$$

and

$$w = \alpha Z (1 - F_2/r^1)$$

It can be seen that for whale stocks to exist F_2 must be less than r^1, that is, the instantaneous fishing mortality rate must be smaller than the maximum rate of increase. Further, for the equations to be considered valid, the equilibrium solutions should be stable to small perturbations. May (1973) describes tech-

365

niques for testing the local stability of difference equations. If we use these techniques, the solutions prove stable for the range of F_1 and F_2 considered in this study and the other parameter values previously defined.

Figs. 1a and 1b show the equilibrium values of z and w for different fishing rates. In Fig. 1a it is shown that krill increases with increased exploitation of whales and the consequent reduction in predation. Krill itself decreases as exploitation on it increases. Fig. 1b shows the converse situation with the whale stocks decreasing with exploitation of krill, but the effect is less noticeable since exploitation on the whales greatly decreases their standing stock.

The total yields from both krill and whales, calculated as $F_1z + F_2w$, are shown in Fig. 2a. A maximum of 30 per cent exploitation of krill has been assumed because of the difficulties inherent in the fishery, and consequently only a range of F_1 of between 0 and 0.3 has been considered. Within this range yields increase rapidly as exploitation of krill increases, and reach a maximum when whaling has exterminated whales. This is consistent with the knowledge that greater yields can be obtained the lower down the trophic level harvesting occurs.

While this may be true for yield by weight, it is not necessarily so economically; markets tend to pay proportionately more for luxury foods from the higher trophic levels. Whales may not be classified as a luxury food but weight for weight they are better value than krill, at least at present. One advantage of the whale is that it has already searched out the krill and concentrated it in growth, and so yield per unit of effort may be an order of magnitude greater in the whaling industry. Fig. 2b illustrates the relative financial returns against different levels of fishing for whales where their value is 10 times that of krill ton for ton. Fig. 2a can be regarded as the same illustration with whales and krill having a similar economic value. From Fig. 2b it can be seen that the harvest from krill still dominates, although better economic yields are obtained

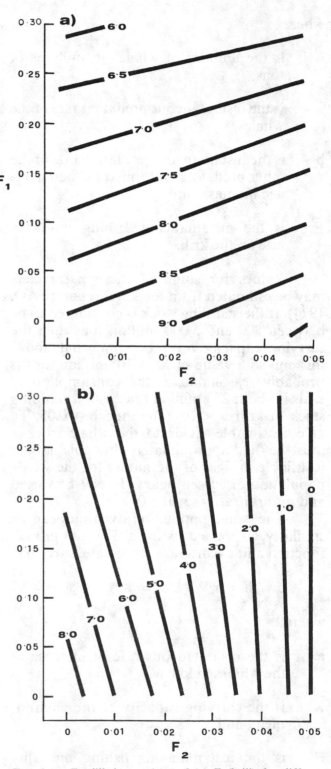

FIG. 1. — Equilibrium values of (a) Z (krill) for different instantaneous fishing mortality rates of krill (F_1) and whales (F_2), values in tons $\times 10^{-9}$; and (b) W (whales) for different instantaneous fishing mortality rate of krill (F_1) and whales (F_2), values in tons $\times 10^{-8}$.

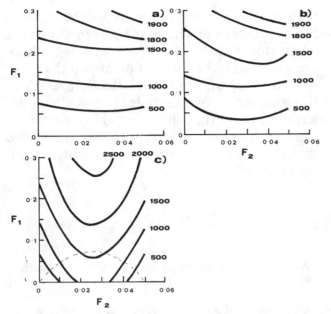

FIG. 2. — (a) Total yield from krill and whales calculated as $F_1Z + F_2W$ in tons $\times 10^{-6}$.
    (b) Relative total economic yield calculated as $F_1Z + 10 F_2W$.
    (c) Relative total economic yield calculated as $F_1Z + 100 F_2W$.
    The dotted line is where whales provide 50 % of the total economic catch.

of both predator and prey. If yield in weight is considered as the sole criterion for exploitation of the system, then obviously maximum yields may be taken from harvesting the prey species and minimizing predator numbers. However, the separate industries are based upon maximizing their independent economic yield and, if the predator is substantially more valuable than the prey, then a balance between the exploitation of both species has to be achieved in order to maximize total economic yield. The model presented here illustrates this feature. It is a simple model and undoubtedly can be improved; but I suspect that the qualitative results will not change. The quantitative results are based on best guesses, some of which may be bad, and the quantitative nature of the results should not be relied upon. What the model demonstrates is that the economic criteria of the fishery have important consequences for its management.

It may be argued that with the great reduction of the Antarctic whales there is much surplus krill in the ocean and that this may be harvested without detriment to the whale stocks. Two counter arguments to this point of view are given here. First, the amount of food available to the protected whale species would be reduced by harvesting the krill and the already slow rate of recovery of the whale species might be further slowed down. Second, if the whale populations showed no functional responses to exploitation then harvesting would not be possible. As it is, the whales do respond with a decrease in the age at maturity and an increase in the pregnancy rate of mature females. It is these responses that allow exploitation without driving the stocks to extinction and one explanation of the increased reproductive rate may be the increased per capita food supply. Consequently, it may be argued that at least some of the excess krill is already being harvested.

These points emphasise the purpose of this study which is to demonstrate that the management of the krill and whale fisheries will be difficult, needing to take into account the economic aspects of harvesting and the

if whales are allowed to persist. However, a ratio of 10:1 is likely to be a gross underestimate and a factor of 100:1 may be more reasonable. The effect of such a ratio is shown in Fig 2c. The situation has now been drastically altered, it being necessary to exploit the whale stocks in such a way as to achieve substantial returns. From all the results it is clear that a potential conflict of interests might occur if the two industries operated independently: both would certainly profit by the other's not existing.

Discussion

The model presented in this study is a simple predator-prey system with exploitation biological interactions between the species. Since, as recently pointed out by FAO's Advisory Committee on Marine Resources Research (ACMRR (FAO), 1976), some of the baleen whales are almost completely dependent on Antarctic krill, the whale stocks could not but be adversely affected by exploitation of krill, especially as there is no alternative food source of sufficient extent and concentration. Conversely, but to a far lesser extent, the yields from the krill fishery will be reduced by the presence of whales. At present Antarctic whale stocks are low but, if protection methods are successful, then it is to be hoped that they will increase substantially. Because the two fisheries are interdependent and their interests may conflict, they should be considered together for management purposes. It is hoped that this study will further the interest in this problem.

References

ACMRR (FAO), Mammals in the sea. *Ad hoc* Group I
1976   on Large Cetaceans. Report. Supplement 2. Paper presented to the Conservation and Management of Marine Mammals and their Environment, Bergen, Norway, 31 August-9 September, 1976. Rome FAO, ACMRR/MM/SC/2 Suppl. 2:7 p.

CUSHING, D.H., Marine ecology and fisheries. Cam-
1975   bridge, England, Cambridge University Press, 278 p.

FAO, Report on informal consultation on Antarctic
1974   krill. *FAO Fish. Rep.*, (153):14 p.

MAY, R.M., Stability and complexity in model ecosys-
1973   tems. Princeton, N.J., Princeton University Press, 235 p.

POPE, J.G., The effect of biological interactions on the
1976   theory of mixed fisheries. *Sel. Pap. ICNAF*, (1):157-62.

THE WHALING MUSEUM AS A DATA SOURCE FOR WHALE POPULATION STUDIES: PROBLEMS AND RESPONSIBILITIES

K.R. Martin

Abstract

Whaling museums and libraries contain records of at least 25 percent of the voyages made by Yankee whaling vessels, mostly in logbooks and journals. Entries on, for example, species encountered and taken, whale behaviour, weather and, occasionally, the oil yield per whale, can, with more recent data, provide information on the history of the fisheries, including the initial size of the population exploited: studies of this sort have been or are being conducted on fisheries for the California gray whale, the bowhead whale and the sperm whale. Useful information is also available in business records and accounts, so far little used. It is important for biologists to work closely with museum or library staff in order to avoid wrong assumptions about the nature of the fishery and to choose researchers best qualified to work with the manuscripts. One limitation of the use of logbooks and journals is the assumption that whale movements were the same in the 19th century as they are today; another difficulty, the whereabouts of ships for which no logbook or journal is available, can be partly solved by using references in other sources. Museums and libraries, and private collectors, are mainly interested in collecting and conserving records rather than in research using them. The resources for such studies must, therefore, come from outside the collections; adequate financing and coordination of all studies are important to obtaining useful results.

Résumé

Les musées de la chasse à la baleine et les bibliothèques contiennent les traces écrites d'au moins 25 pour cent des voyages effectués par les baleiniers Yankee surtout sous forme de livres de bord et de journaux. Les notes sur, par exemple, les espèces rencontrées et capturées, le comportement de la baleine, le temps et, occasionnellement, le rendement d'huile par baleine peuvent, avec les données plus récentes, fournir des informations sur l'histoire des pêches, y compris la taille initiale de la population exploitée: des études de ce type ont été ou sont effectuées sur la pêche de la baleine grise de Californie, de la baleine du Groenland et du cachalot. On trouve aussi des informations utiles dans les archives et la comptabilité commerciales qu'on a si peu utilisées jusqu'ici. Il est important pour les biologistes de travailler en contact étroit avec le personnel du musée ou de la bibliothèque pour éviter les hypothèses erronées sur la nature de la pêche; il importe aussi de choisir les chercheurs les mieux qualifiés pour le travail sur manuscrits. Une des limitations de l'emploi des livres de bord et des journaux réside dans l'hypothèse que les déplacements des baleines étaient les mêmes au 19e siècle qu'aujourd'hui. Une autre difficulté, l'emplacement des bateaux pour lesquels il n'existe ni livres de bord ni journaux, peut être résolue en partie par des références à d'autres sources. Les musées, les bibliothèques et les collectionneurs privés s'intéressent avant tout au rassemblement et à la conservation des pièces plutôt qu'à la recherche qui les utilise. Les ressources pour ces études doivent par conséquent venir du dehors; il est important d'assurer le financement et la coordination de toutes les études si l'on veut obtenir des résultats utiles.

Extracto

Los museos y bibliotecas balleneras contienen datos principalmente en cuadernos de bitácora y diarios de bordo, sobre al menos un 25 por ciento de los viajes hechos por balleneros yanquis. Combinando los datos así obtenidos sobre las especies encontradas y capturadas, el comportamiento de las ballenas, el tiempo y, en algunas ocasiones, el rendimiento en aceite por ballena con otros datos más recientes, es posible conseguir información sobre la historia de la explotación e incluso sobre el volumen inicial de la población explotada. Se han hecho o están haciendo estudios de este tipo sobre la caza de ballena gris de California, ballena de cabeza arqueada y cachalote. También se encuentra información útil en archivos y contabilidades comerciales, poco aprovechados hasta la fecha. Es importante que los biólogos trabajen en estrecha colaboración con el personal de los museos o bibliotecas, para evitar partir de hipótesis falsas sobre la naturaleza de la explotación, y que los investigadores estén debidamente calificados para trabajar con manuscritos. Una de las limitaciones del uso de los cuadernos de bitácora y los diarios de bordo es la suposición de que los movimientos de las ballenas eran iguales en el siglo XIX y hoy; el problema de los movimientos de los barcos sobre los que no existen cuadernos de bitácora o diarios de bordo, puede resolverse en parte utilizando referencias contenidas en otras fuentes. Los museos y bibliotecas, y los coleccionistas privados, se interesan principalmente por recoger y conservar documentos, más que por hacer investigaciones sobre ellos. Los recursos para estas investigaciones, pues, habrán de buscarse en otra parte. Para obtener resultados útiles es importante disponer de financiación adecuada y coordinar todos los estudios.

K.R. Martin
The Kendal Whaling Museum, Sharon, MA 02067, USA

For decades, whaling museums and libraries in the USA have been accumulating hundreds of important primary manuscripts[1]. Most notable of these are period whaling logbooks and whalemen's journals. Although Lieutenant M.F. Maury's celebrated study (Maury, 1855) made creative use of whaling accounts and pointed the way for subsequent research, these documents have attracted little serious attention until lately (Hohman, 1928). This paper discusses the value of North American manuscript collections as indicators of whale populations.

About 14 000 whaling voyages were made by Yankee vessels, and records of at least 25 % of these remain in museum or library collections, enough to give a representative sampling of hunted stocks on every nineteenth century whaling ground (Sherman, 1965).

The plenitude of statistically reliable whaling manuscripts is a by-product of the trade's peculiarities. Most voyages were long and characterized by idleness. Boredom apparently promoted painstaking journalism. Experienced crew members were often working their way toward a command. They kept reference accounts, similar in format to official logbooks, as did established officers, since much whaling expertise was a matter of business secrecy and experience (Sherman, 1965). The format of these documents is standardized, convenient for researchers, and includes daily weather, navigational and whale behaviour data. Many are illustrated with whale symbols that enable a scholar to read at a glance the species involved, the outcome of an encounter, and occasionally the oil yield per whale. A systematic study of such material, integrated with more recent data, can shed light on a number of population questions. For example, a noticeable decline in whale size (based on oil yield) over a measured period might suggest overfishing of the grounds concerned – provided the migratory routes of the animals are known, and no behavioural causes of scarcity are involved. (These are matters not covered by period sources.) In view of current discussions about "initial stocks" of sperm whales, primary sources of the subject would seem to be vital (S.J. Holt, pers. comm., 1976).

The limitations of logbooks and journals should be noted. Documents should be studied in the light of specific, regional, whale stocks, a process which presumes that in 1830 whale movements were much the same as in 1860, 1890, or today. Researchers working on a specific whale stock may trace the movement of animals from ground to ground, plotting changes in biomass (based on oil tally), and numbers against hunting activity. But while a fairly accurate annual whaleship and cargo roster is available, it is often difficult to learn the seasonal whereabouts of a whaler for which no logbook or journal has survived[2]. On which grounds did such a vessel get her whales?

Some help can be obtained from other ship's logs or journals. When one whaler crossed another's path, it was customary to exchange reports of cargo, and log the encounters. A single logbook or journal will, therefore, often reveal the names and fortunes of nearby vessels. In addition, seasonal whaling reports in such newspapers as *The Friend* (Honolulu) or *The Whalemen's Shipping List and Merchants' Transcript* (New Bedford) carried information on vessels on the grounds.

---

[1] Major collections may be found at the Whaling Museum and New Bedford Free Public Library, both in New Bedford, Massachusetts; the Nicholson Whaling Collection, Providence Public Library, Providence, Rhode Island; the Kendal Whaling Museum, Sharon, Massachusetts; Mystic Seaport, Mystic, Connecticut; the Nantucket Whaling Museum, Nantucket, Massachusetts; the Dukes County Historical Society, Edgartown, Massachusetts. A major microfilm repository is located at the International Marine Archives, Inc., Nantucket. Another, the Pacific Manuscripts Bureau, Australian National University, Canberra, maintains a growing microfilm collection of New England logbook journals which includes many of the museum collections listed above.

[2] A remarkable manuscript which gives approximate abstracts of American whalets' seasonal whereabouts is Dennis Wood's (Wood, MS, 1831-73).

Judicious integration of manuscript references to ship spoken, abstracts of voyages, and shipping news can go a long way toward determining relative whale population dynamics.

The foregoing limitations have shaped the course of manuscript research to date. D.A. Henderson's resourceful use of statistics on California gray whaling enabled him to make a persuasive case that initial stocks of grays and their predation by Yankees have been historically overestimated and that, under protection, the gray whale population now approaches initial stock numbers (Henderson, 1972). It was, however, facilitated by the predictability of gray whale movements, the confined locale of gray whaling grounds, and the plenitude of data. J.R. Bockstoce is currently engaged in important primary research on bowhead whaling in the Alaskan-Siberian Arctic. He estimates that 20 % of all commercial bowhead whaling voyages are covered in logbook or journal collections. Because that fishery was routinized, well-reported in trade newspapers and confined to a knowable region, he believes that available collections provide a representative sample from which total bowhead population figures can be extrapolated (J.R. Bockstoce, pers. comm.). He is less confident that accurate population estimates can be extrapolated for sperm and right whales.

G.W. Shuster is more optimistic. Using published statistics of the American whale fishery, he has concluded that perhaps the most important factor in its decline was its overfishing of sperm whales (Shuster, 1972). Shuster's research, still in its early stages, suggests that yield per individual sperm whale declined over the 19th century. Therefore, a declining scale of yield per whale might be necessary to estimate numbers taken and from this, initial populations (Shuster, 1975). With sperm whales it is important to know the sex of the groups described in logbooks or journals, a fact often omitted in such documents. As for other quarry, whalemen did not always differentiate between right whales and bowheads. Thus, educated guessing by a cautious researcher is inevitable.

Logbooks or journals, their shortcomings notwithstanding, have an added value: they often record aberrant whale behaviour and/or physical characteristics. This should be of value to individual research projects. Much insight into the dynamics of whaling, and therefore whale populations, can be gleaned from business records and accounts. These documents, also available in quantity, are the least used by scholars. They have little human interest and no anecdotal appeal. But knowledge of the cost and price structure of whaling, its margin of profit, investment in technology and the like, is of course essential to any general understanding of the fishery.

Researchers using available collections would need to maintain close contact with maritime museum staff. Proper use of sources involves knowing the conditions under which they were kept, archaic nautical and geographical nomenclature, and the natures of the journalists. Assumptions about the American whale fishery are often based upon popularized notions with little basis in fact. Contact with biological researchers has revealed a widespread mystique about Yankee whalemen: they are presumed to have been canny experts whose cetological identifications and judgements can be taken as unquestionable. In fact, New England whalers were manned to a large extent by very young, inexperienced landlubbers, who seldom made more than a single voyage. Mere sightings by such look-outs should not necessarily be construed as very accurate. Researchers may have to be critical of their sources. Such work is slow-going and time is of the essence.

Productive research should begin quickly. How much research time might it take? Shuster estimates 2-3 h per logbook or journal (G.W. Shuster, pers. comm.). Museums should be consulted before outside researchers are assigned the tedium of such research. Extracting whale population data is not simply a matter of mechanical recording of statistics. Errors could be so large that researchers must have a well-developed sensitivity to the character of manuscripts, to say nothing of a strong academic bent. Museums may be able to help locate research candidates with the appropriate blend of skills.

What are the overall responsibilities of maritime museums or libraries in supplying whale population statistics? Their fundamental obligation is to collect and conserve records and to date, collection has been impressive. But manuscripts have recently become prohibitively expensive. Private collectors who hold hundreds, if not thousands, of logbooks and journals must be persuaded of the importance of donating or depositing their collections in museums, or otherwise cooperating with scholars. In the less glamorous area of conserving manuscripts, museums have not performed so well, although most major collections are on microfilm. In the USA logbook or journal indexing by the several whaling museums ranges from complete to none at all. Almost every whaling museum is a repository of local history. Few, then, have any great commitment either to cataloguing general data or to devoting resources to general questions such as current whale stock studies. This is more than a question of appreciating the continuity of history — it is a matter of finance.

I believe that, individually, these institutions are responsible agencies. My experience suggests to me that they have not made the promotion of general research a high priority, which is understandable, given their parochial character and limited funds. It is more puzzling to me why, in view of the recent superheated speculation on whale populations, no general, *unified* research has been undertaken by the International Whaling Commission or any other agency. The results would seem to be worth the effort. Adequate financing and unity of method are the two keys. Both lie beyond the reasonable range of museum or library responsibility. Despite the inevitable uncertainties, logbooks and journals could be of colossal importance in reaching accurate estimates of initial or relative whale populations. But the nature of the research is such that the time, money and sensibility for such work must come from outside. I hope that whale population research using historic sources will begin soon. As a museum director and historian, I believe there could be no better justification for years of collecting than this.

References

HENDERSON, D.A., Men and whales at Scammon's Lagoon. Los Angeles, Dawson's Book Shop, Baja California Travel Series, vol. 29.
1972

HOHMAN, E.P., The American whaleman: a study of life and labour in the whaling industry. New York, Longmans, Green & Co.
1928

MAURY, M.F., The physical geography of the sea. New York, Harper.
1855

SHERMAN, S.C., The voice of the whaleman, with an account of the Nicholson whaling collection. Providence, R.I., Providence Public Library.
1965

SHUSTER, G.W., Environmental affairs. Boston, Boston College Environmental Law Center, vol. 2:345-57.
1972

—, Population effects of 19th century American sperm whaling. Paper presented at the National Whale Symposium, Indiana University, Bloomington, Indiana, 10 November 1975.
1975

WOOD, D., Abstracts of whaling voyages from the United States 1831-73. New Bedford, New Bedford Free Public Library (Melville Whaling Room), 7 vols. (MS).
1831-73

ENERGY INPUTS AND OUTPUTS IN AN AUSTRALIAN COASTAL WHALING OPERATION

K.R. ALLEN

Abstract

Comparison of the calorific values of the fuel and power consumption to that of the main products (oil, meals and solubles) of the Cheyne's Beach Whaling Station, Western Australia, for the 1973, 1974 and 1975 seasons gives an approximate total energy output-input ratio of 0.929, considerably higher than that typical of a number of other sectors of the fishing industry. Catching and towing account for approximately 2/3 of the energy input and processing on shore for the other 1/3. Oil accounts for about 80 percent of the total energy output; however, it is mostly used for lubrication where its energy content is not directly utilized. This produces a lower value of 0.184 for the perhaps more relevant ratio of consumable energy output to total energy input.

Résumé

On a déterminé un rapport approximatif production/consommation totale d'énergie égal à 0,929 par comparaison de la valeur calorique du combustible et de la consommation d'énergie avec les principaux produits (huiles, farines et produits solubles) à la station de chasse à la baleine de Cheyne's Beach (Australie Occidentale) pendant les campagnes de 1973, 1974 et 1975. Ce rapport est sensiblement supérieur à celui qui caractérise un certain nombre d'autres secteurs de l'industrie de la pêche. Capture et remorquage représentent environ les deux tiers de la consommation d'énergie tandis que la transformation à terre consomme le reste. L'huile représente environ 80 pour cent de la production totale d'énergie; cependant, elle est surtout employée pour le graissage où sa teneur énergétique n'est pas directement utilisée. Par suite, on obtient une valeur, faible, de 0,184 pour un rapport production d'énergie consommable/consommation totale d'énergie, sans doute plus pertinent.

Extracto

Comparando el valor calorífico del consumo de combustible y energía eléctrica con el de los principales productos obtenidos (aceite, harina y sustancias solubles) en la Estación de Caza de Ballenas de Cheyne, en Australia occidental, durante las temporadas 1973, 1974 y 1975, se obtiene una razón entre la producción y el consumo total de energía del orden de 0,929, que es considerablemente superior a la de otros muchos sectores de la industria pesquera. La captura y el remolque representan aproximadamente dos tercios del consumo de energía, y la elaboración en tierra el tercio restante. El aceite representa cerca del 80 por ciento de la producción total de energía. Sin embargo, se utiliza principalmente para lubricación, por lo que su contenido de energía no se aprovecha directamente. Teniendo eso en cuenta, la

375

razón producción de energía consumible/consumo total de energía, que quizás sea más reveladora, da una cifra menor: 0,184.

K.R. Allen
Division of Fisheries and Oceanography, CSIRO, P.O. Box 21, Cronulla, NSW 2230, Australia

The Cheyne's Beach Whaling Company in Western Australia has kindly made available to the author data on the fuel and power consumption of its catching and factory operations for the 1973, 1974 and 1975 seasons. It has also provided data on the calorific value of its fuel and of the principal products, oil, meal and solubles.

From these data it is possible to calculate approximately the energy input-output balance for these operations, following the advice of Leach (pers. comm.) that a useful first approach can be made by simply looking at direct fuel consumption in comparison with the catch.

The results, averaged over the 3 years, are summarised in Table 1. The conversion factors supplied by the company for fuel oil have been multiplied by an efficiency factor of 1.134 as suggested by Leach (1975) to allow for production and transportation losses. Similarly an efficiency multiplier of 4.0 has been applied to the conversion for electric power.

The table shows that the total energy content of the products at the factory gate is slightly less than the energy input in fuel and electric power to the catching and processing operations combined. Much the largest part of the energy output (about 80 %) is contained in the calorific value of the oil. Since however this oil is largely used for lubrication in which its energy content is not directly utilised, it may be more relevant to consider the energy output ratio for the other products, meal and solubles, which are probably used mainly in foodstuffs. The ratio is then rather under 20 %.

The table also shows that approximately 2/3 of the energy input is utilised in catching and towing in the whales and the other third in processing on shore. The amount used in aerial searching for whales is negligible, about 2 % of that used in catching.

The total output: input ratio for energy appears to be considerably higher than that typical of a number of other sectors of the fishing industry. Leach (1975) quotes values for this ratio ranging between 0.0061 and 0.058. The higher ratio for this coastal whaling operation seems to arise from 2 factors. Firstly the gross weight of the catch per tonne of oil required for its capture is high. Taking the approximate total annual weight of the catch at 20 000 tonnes, the catch per tonne of catcher fuel is about 4.6 tonnes, or, for all fuel used, about 3.2 tonnes. By comparison, in the fisheries examined by Leach (1975) the catch per tonne of fuel ranged from 0.14 to 2.1 tonnes, the overall figure for the UK fishing industry being 1.5 tonnes per tonne of fuel. Secondly, the energy content of the ultimate products seems for whales to be unusually high in relation to the gross weight of catch. Again taking the gross catch at 20 000 tonnes the energy yield of the products is 14.4 GJ[1] per tonne, compared with a figure for UK fisheries on Leach's data of 1.75 GJ per tonne of landed catch.

The high catching efficiency is substantially due to the fact that this operation is usually able to take most of its whales within a fairly small radius (often less than 30 miles) of its base and regularly uses aircraft to keep searching to a minimum.

---

[1] GJ = Giga-joule.

Table 1. Energy inputs and outputs of the Cheyne's Beach Whaling Station, Western Australia, averaged on 1973, 1974 and 1975 seasons

Energy input		
Aerial searching		
Fuel oil	82.5 t at 49.4 GJ/t	0.004 MGJ
Catching		
Fuel oil	4339 t at 47.7 GJ/t	0.207 MGJ
Processing		
Fuel oil	1841 t at 48.5 GJ/t	0.089 MGJ
Electric power	672710 kwh at 14.4 MJ/kwh	0.010 MGJ
Total processing input		0.099 MGJ
Total input		0.310 MGJ
Output		
Weight of catch (approx.)		20 000 MGJ
Energy output		
Oil	5792 t at 39.9 GJ/t	0.231 MGJ
Meal	1573 t at 11.7 GJ/t	0.018 MGJ
Solubles	2673 t at 14.6 GJ/t	0.039 MGJ
Energy content of consumable products		0.057 MGJ
Total energy output		0.288 MGJ
Total weight of products		10 038 t
Ratios		
Energy - total output: total input		0.929
consumable output: total input		0.184
input per tonne of products		30.9 GJ
input per tonne caught		15.5 GJ
Weight of products per tonne of fuel (all purposes)		1.60 t

The high energy yield per gross tonne caught seems however to require further examination. If significant comparisons are to be made between the energy conversion efficiency of whaling and of other fishing of food producing industries, it is essential to be sure that the outputs being compared are actually of the same nature.

References

LEACH, G., Energy and food production. London,
1975   International Institute for Environment and
        Development, 151 p.

GROWTH AND ENERGY BUDGETS OF LARGE BALEEN WHALES FROM THE SOUTHERN HEMISPHERE

C. Lockyer

Abstract

The growth, migration, seasonal feeding and fattening, metabolism and associated morphological, physiological and behavioural parameters and energy budgets for blue and fin whales are discussed and comparisons with sei, minke and humpback whales are made whenever data are available.

Published material on growth in length with time for balaenopterid foetuses is reviewed and growth in weight with time is discussed by reference to weight-to-length relationships. The growth formulation,

$$W^{1/3} = a (t - t_0),$$

(where W = body weight in grammes, a = growth velocity constant, t_o = time constant in days since conception prior to the linear growth phase, t = time in days since conception) of Huggett and Widdas (1951) is found to describe well the very rapid foetal growth phase which occurs after the fifth month of the 11-12 month gestation period. The values found for blue, fin and sei whales (0.52, 0.47 and 0.35, respectively) agree closely with estimates given by Fraser and Huggett (1974).

Published data on body dimensions relative to body length throughout postnatal growth for blue, fin, sei and humpback whales are compared. In all species, the anterior end of the body appears to increase disproportionately in size relative to the rest of the body while a negative trend is apparent in the posterior region, the net effect being isometric with increase in body length. Growth curves of length at age and weight at age are constructed for blue, fin and sei whales, and the von Bertalanffy (1938) growth equation,

$$L_t = L_\infty (1 - e^{-k(t + t_e)}),$$

(where L_∞ = mean length at physical maturity in metres, t = age in years, L_t = length in metres at age t, t = time constant, k = growth velocity constant) is fitted to curves for blue, fin and sei whales as follows:

blue male:	$L_t = 25.0 (1 - e^{-0.216(t + 4.92)})$
blue female:	$L_t = 26.2 (1 - e^{-0.240(t + 4.50)})$
fin male:	$L_t = 21.0 (1 = e^{-0.221(t + 5.30)})$
fin female:	$L_t = 22.25 (1 = e^{-0.220(t + 4.80)})$

sei male: $\qquad$ $L_t = 14.8 \,(1 - e^{-0.1454(t+9.36)})$

sei female: $\qquad$ $L_t = 15.3 \,(1 - e^{-0.1337(t+10.00)})$

The modification of the von Bertalanffy formula by Laws and Parker (1968) to weight at age is found to describe the estimated curves of growth in weight well. The formulae,

$$W_t = W_\infty \,(1 - e^{-k(t+{}^te)})^3,$$

(where W_∞ = body weight in tonnes at physical maturity, W_t = body weight in tonnes at age t) are as follows:

blue male: $\qquad$ $W_t = 102 \,(1 - e^{-0.216(t+4.92)})^3$

blue female: $\qquad$ $W_t = 117 \,(1 - e^{-0.240(t+4.50)})^3$

fin male: $\qquad$ $W_t = 55 \,(1 - e^{-0.221(t+5.30)})^3$

fin female: $\qquad$ $W_t = 64.5 \,(1 - e^{-0.220(t+4.80)})^3$

sei male: $\qquad$ $W_t = 18 \,(1 - e^{-0.1454(t+9.36)})^3$

sei female: $\qquad$ $W_t = 19.5 \,(1 - e^{-0.1337(t+10.00)})^3$

These formulae illustrate that the pattern of growth in the balaenopterid whales is probably not complex. Estimated curves of growth in weight show that the largest net weight gain during life takes place during the first year following birth, with weight increasing more than tenfold; it is also apparent that both sexes in the three species have attained at least 70 percent of their mature weight at sexual maturity, assumed to be 5, 6 and 8 years of age in blue, fin and sei whales, respectively. Comparison of data on length and apparent growth parameters for fin whales in six different Antarctic areas show a variance with locality in the mean maximum sizes attained, size at sexual maturity and growth rates.

Analyses of published data on whale mark returns, seasonal variations in abundances of whales on the whaling grounds and patterns of diatom fouling by *Cocconeis ceticola* of the skin of whales in the Antarctic, show that balaenopterid whales of both sexes undertake extensive annual migrations, moving southward in the austral summer to polar waters and northward to lower latitudes for the winter, and tending to remain within the same longitudinal sector throughout the year. Peak densities in the Antarctic are reached in December to February, and in low latitudes around June to August. Immature whales appear to reach peak abundances later than mature animals, suggesting a hierarchal order in migration, probably also involving differences in sexual condition; the average stay in the Antarctic is about 120 days. An ordered succession of arrival of species is also apparent, blue preceding fin which are followed by sei, blue whales moving further south to the pack ice along with minke whales. In recent years, there has been good evidence that the sei whale is arriving earlier in the Antarctic than it was observed to do in the thirties, perhaps because of long-term oceanographic changes toward warmer sea temperatures and reduced interspecific competition for food due to diminution of other balaenopterid species. This observation is perhaps indicative of the interrelation between the distribution, abundance and migrations of balaenopterids and its dependence on production and oceanographic conditions.

Published information on seasonal variation in blubber thickness is reviewed and new data are analysed, both in relation to migration patterns. Amongst blue, fin, sei and humpback whales, it is apparent that pregnant females are exceptionally fat relative to resting females and adult males, lactating females are very lean and juvenile whales are usually fatter than adult males and nonpregnant females. Variations in weights of blue, fin and humpback whales between November and April are compared. Weight data are standardized through the use of the body weight/body length ratios for humpbacks, and the body weight/bone weight ratio for blue and fin whales. Mean ratios with time suggest a doubling of body weight in humpbacks throughout the Antarctic feeding season, and fattening in blue and fin whales of about 50 and 30 percent of body weight, respectively, although the increase in the latter species is probably similar to that in blue whales. These estimates of fattening are in harmony with those demonstrated for North Pacific gray whales and appear reasonable from comparisons with the extent of fattening in hibernators and weight losses anticipated during starvation in various mammals. The distribution of weight increases in the various body tissues for fin and blue whales is calculated in the same way as for total body weight increases in these species: the greatest overall increase occurs in the musculature, then the blubber, followed by the viscera, chiefly because of these tissues' relatively greater presence by weight, respectively, while the greatest percentage occur in the opposite order.

Food taken in different localities is reviewed, the conclusion being that feeding is generally opportunistic and that swarming planktonic crustacea are preferred. In true polar waters, krill, *Euphausia superba*, is the main food species. Review of data on occurrence and quantities of stomach contents of balaenopterid whales indicates that in the Antarctic, feeding reaches an incidence level of about 85 percent and food intake may average 30-40 g/kg body weight/day; in low latitudes, the incidence level generally falls to less than 50 percent and probably less than 20 percent in adults, and food is consumed at about 10 percent or less of the summer rate. In a whale feeding at these rates, 3.5-5 times the body weight is consumed in a year and an average of 12 g/kg body weight/day/year is eaten. The feeding pattern of sei whales present in sub-Antarctic latitudes appears to be less seasonal with a more varied diet. Evidence for a diurnal rhythm of feeding controlled by vertical migration of the prey is discussed, together with information on digestion rates.

The dimensions of the mouth in balaenopterids are compared and the mechanics of feeding are discussed. The baleen plates seem to be shorter and narrower in the largest species when compared with the equivalent sizes of animals of the smaller species. The jaws of blue, fin and minke are very similar in shape, whereas those of the humpback and sei are the widest and narrowest, respectively, of all the Balaenopteridae. Calculations of mouth volume in blue and fin whales based on that of minke whales show that when feeding in waters where production is 2 kg/m^3 in a band 1.22 m thick, the mouth volume filtered repeatedly for a few hours is sufficient to provide adequate nourishment. However, it is clear that annual migration to the productive waters of the Antarctic where the necessary higher densities of near-surface euphausiid swarms are present, is essential for survival.

The heart and kidney weights of cetaceans appear to follow the usual relationship to body weight found for other mammals. These organs are closely associated with metabolism, so the basal metabolic rate, Q, might also be expected to follow the relationship observed for other mammals,

$$Q = 70.5 W^{0.7325} \text{ kcal/day,}$$

(where W = body weight in kilogrammes). Three methods of calculating surface area, A, are discussed. The two most promising methods are by Parry (1949),

$$A = 11.1 \times W^{0.67},$$

(where W = body weight in grammes, A is given in square centimetres), and Brody (1968),

$$A = 1\,000 \times {}^{0.685}$$

(where W and A are in kilogrammes and square centimetres, respectively), both of which give predictions close to actual observations. A third method by Guldberg (1907) assumes that the body of the whale roughly conforms to two cones placed end to end and gives a result about 2/3 of the areas calculated by the other two methods. The basal phase rate/body weight relationship in whales is probably similar to that of other mammals, and there is a possibility that bradycardia may occur during diving as observed in small marine mammals. The blood weight, nearly equivalent to blood volume, in adult baleen whales is assumed to be about 10 percent of body weight. From published data, the oxygen capacity of the blood appears variable, but generally falls in the range of 20-30 percent. The vital capacity of cetaceans is estimated to be a very high proportion of total capacity.

Data on blowing rates and diving times for different whale species are considered. In rorquals, blowing rates may fall to one per minute in resting animals and increase with activity; smaller whales generally blow more frequently than larger ones. Rorquals do not appear to dive deeper than 300 m according to published information. Patterns of sleeping and activity in cetaceans appear unclear, though apparent sleep is characterized by a marked fall in the respiratory rate and the adoption of certain postures in the water. Well-fed whales appear to be slow and relaxed in their environment suggesting that sleep could possibly occur after feeding and that the sleeping period may therefore be determined by the possible diurnal rhythm of the prey.

The resting metabolic rate in blue, fin, sperm and bottlenose whales is calculated with the assumptions that the average Respiratory Quotient (RQ) is 0.82, that utilization of inspired air is 10 percent and that estimates of lung capacity, averaging 2.5-2.8 percent of body weight, and respiratory rhythms, are as observed. The predicted values for large cetaceans are in agreement with extrapolations of data for small cetaceans and pinnipeds. Calculations of energy expenditures needed to overcome resistance at different swimming speeds suggest that 0.01 hp/1b (1.64×10^8 ergs/kg) operative musculature is a valid basis on which to estimate the active metabolic rate, representing energy expenditure at top cruising speeds, using a conversion factor, 1 hp = 1.54×10^4 kcal/1b/day. The basal metabolic rate is calculated (1) using the formula,

$$Q = 70.5W^{0.7325},$$

(W = body weight in kilogrammes), (2) as about 85 percent of the resting rate, and (3) as 10 percent of the active rate. These three methods give remarkably similar results, though the first method generally gives values lower than the latter two.

Energy budgets are estimated for blue and fin whales at different stages of development. Growth efficiencies (gross and net) are found to fall with increase in age, from levels of about 30 percent in the unweaned calf to 1 percent in near adults.

High estimates of assimilation efficiency, around 80 percent, may be inaccurate due to indirect calculation. The ratio of energy expenditure in metabolism to energy capture in terms of developmental growth is found to increase from about 2:1 to nearly 100:1 between birth and adulthood, indicating that relative production efficiency falls off dramatically with age. Calculation of the balanced energy budgets shows the necessity of the seasonal body weight increases (at least 50 percent) demonstrated earlier. Pregnant females appear to require an increase of about 60-70 percent in order to survive the large energy drain imposed by lactation and therefore require a longer Antarctic feeding period; the energy budget for pregnant females is necessarily only meaningful over a two-year cycle.

Résumé

L'auteur étudie la croissance, les migrations, les cycles saisonniers d'alimentation et d'engraissement, le métabolisme et les paramètres morphologiques, physiologiques et de comportement connexes, ainsi que les bilans énergétiques des baleines bleues et des rorquals communs et procède à des comparaisons, quand les données disponibles le permettent, avec les rorquals de Rudolf, les petits rorquals et les mégaptères.

Il passe en revue la documentation publiée sur la croissance linéaire en fonction du temps pour les fœtus des baleinoptères et examine la croissance pondérale dans le temps en tenant compte des rapports poids-longueur. La formule de croissance,

$$W^{1/3} = a\,(t - t_0),$$

(où W = poids corporel en grammes, a = constante de vitesse de croissance, t_0 = constante de temps, en jours, entre la conception et la phase de croissance linéaire, t = temps, en jours, depuis la conception), établie par Hugget et Widdas (1951) décrit correctement la phase de croissance fœtale très rapide qui se produit après le cinquième des 11-12 mois de la période de gestation. Les valeurs observées pour les baleines bleues, les rorquals communs et les rorquals de Rudolf (respectivement 0,52, 0,47 et 0,35) concordent étroitement avec les estimations fournies par Fraser et Hugget (1974).

L'auteur compare les données publiées sur les dimensions corporelles par rapport à la longueur du corps tout au long de la croissance post-natale chez les baleines bleues, les rorquals communs, les rorquals de Rudolf et les mégaptères. Dans toutes les espèces, la partie antérieure du corps semble s'accroître en taille de façon disproportionnée par rapport au reste du corps tandis qu'une tendance négative caractérise la région postérieure, la résultante nette étant isométrique avec l'augmentation de la longueur du corps. Les courbes de croissance de la longueur et du poids par rapport à l'âge sont établies pour les baleines bleues, les rorquals communs et les rorquals de Rudolf et l'équation de croissance de von Bertalanffy (1938),

$$L_t = L_\infty (1 - e^{-k(t + t_e)}),$$

(où L_∞ = longueur moyenne à la maturité physique, en mètres, t = âge en années, L_t = longueur en mètres, à l'âge t, t = constante de temps, k = constante de vitesse de croissance) est appliquée aux courbes pour les baleines bleues, les rorquals communs et les rorquals de Rudolf, comme suit:

baleine bleue mâle: $L_t = 25.0\,(1 - e^{-0.216(t+4.92)})$

baleine bleue femelle: $L_t = 26.2\,(1 - e^{-0.240(t+4.50)})$

rorqual commun mâle: $L_t = 21.0\,(1 - e^{-0.221(t+5.30)})$

rorqual commun femelle: $L_t = 22.25\,(1 - e^{-0.220(t+4.80)})$

rorqual de Rudolf mâle: $L_t = 14.8(1 - e^{-0.1454(t+9.36)})$

rorqual de Rudolf femelle: $L_t = 15.3(1 - e^{-0.1337(t+10.00)})$

La modification de la formule de von Bertalanffy par Law et Parker (1968) pour le poids en fonction de l'âge, convient à la description des courbes estimatives de la croissance pondérale. Les formules,

$$W_t = W_\infty (1 - e^{-k(t + t_e)})^3,$$

(où W_∞ = poids corporel, en tonnes, à la maturité physique, W_t = poids corporel, en tonnes, à l'âge t) sont les suivantes:

baleine bleue mâle:

$$W_t = 102 (1 - e^{-0.216(t+4.92)})^3$$

baleine bleue femelle:

$$W_t = 117 (1 - e^{-0.240(t+4.50)})^3$$

rorqual commun mâle:

$$W_t = 55 (1 - e^{-0.221(t+5.30)})^3$$

rorqual commun femelle:

$$W_t = 64.5 (1 - e^{-0.220(t+4.80)})^3$$

rorqual de Rudolf mâle:

$$W_t = 18 (1 - e^{-0.1454(t+9.36)})^3$$

rorqual de Rudolf femelle:

$$W_t = 19.5 (1 - e^{-0.1337(t+10.00)})^3$$

Il ressort des formules ci-dessus que le schéma de croissance des baleinoptères n'est probablement pas complexe. Les courbes estimatives de la croissance pondérale montrent que le maximum de gain pondéral net survient la première année qui suit la naissance, le poids augmentant alors plus de dix fois; il apparaît aussi que, chez les trois espèces considérées, les animaux des deux sexes atteignent au moins 70 % de leur poids mature à l'âge de la maturité sexuelle, qu'on pense être de 5, 6 et 8 ans, respectivement, pour les baleines bleues, les rorquals communs et les rorquals de Rudolf. La comparaison des données sur les paramètres de la longueur et de la croissance apparente des rorquals communs dans 6 zones antarctiques différentes révèlent des écarts selon le lieu dans les dimensions maximales moyennes, la taille au moment de la maturité sexuelle et les taux de croissance.

Les analyses des données sur les marques renvoyées, les variations saisonnières de l'abondance sur les lieux de chasse à la baleine et les schémas des dépôts laissés par les diatomées *Cocconeis ceticola* sur la peau des baleines dans l'Antarctique, montrent que les baleinoptères des deux sexes entreprennent d'importantes migrations annuelles, se dirigeant vers le sud durant l'été austral en direction des eaux polaires et vers le nord en direction de latitudes plus basses en hiver et qu'ils tendent à demeurer dans le même secteur longitudinal tout au long de l'année. Dans l'Antarctique, les plus fortes densités se rencontrent de décembre à janvier et, aux latitudes plus basses, de juin à août. Les baleines immatures semblent atteindre leur abondance maximale plus tardivement que les animaux matures — il se pourrait donc que les migrations se fassent selon un ordre hiérarchique et qu'interviennent aussi des différences en rapport avec l'état sexuel; la durée moyenne du séjour dans les eaux antarctiques est d'environ 120 jours. L'arrivée des diverses espèces semble également suivre un certain ordre: les baleines bleues précèdent les rorquals communs, que suivent les rorquals de Rudolf, les premières se déplaçant davantage vers le sud en direction de la banquise, en même temps que les petits rorquals. Il est nettement apparu ces dernières années que les rorquals de Rudolf arrivent dans l'Antarctique plus tôt qu'on ne l'avait observé au cours des années 30, du fait peut-être de modifications océanographiques à long terme entraînant un réchauffement des températures de l'eau et d'une moindre concurrence interspécifique pour la nourriture, consécutive à la diminution des effectifs d'autres baleinoptères. Cette observation dénote peut-être la relation réciproque entre la distribution, l'abondance et les migrations des baleinoptères et sa dépendance à l'égard de la production et des conditions océanographiques.

L'auteur passe en revue les renseignements publiés sur les variations saisonnières de l'épaisseur du lard en fonction des schémas de migration et analyse de nouvelles données sur ce dernier point. Chez la plupart des baleines bleues, des rorquals communs, des rorquals de Rudolf et des mégaptères, les femelles pleines sont exceptionnellement grasses par comparaison avec les femelles au repos et les mâles adultes; les femelles en lactation sont très maigres et les juvéniles sont habituellement plus gras que les mâles adultes et les femelles non fécondées. L'auteur étudie les variations de poids des baleines bleues, des rorquals communs et des mégaptères entre novembre et avril. Les données pondérales sont normalisées sur la base du rapport poids corporel/longueur corporelle pour les mégaptères et du rapport poids corporel/poids du squelette pour les baleines bleues et les rorquals communs. Les rapports moyens en fonction du temps donnent à penser que le poids corporel des mégaptères double pendant la saison d'alimentation dans l'Antarctique et que celui des baleines bleues et des rorquals communs augmente d'environ 50 et 30 pour cent, respectivement, encore que l'accroissement soit probablement semblable chez ces deux espèces. Ces estimations de l'engraissement concordent avec les chiffres établis pour les baleines grises du Pacifique nord et semblent raisonnables d'après les comparaisons faites sur l'ampleur de l'engraissement des animaux hibernants et des déperditions de poids prévues durant les périodes de jeûne chez divers mammifères. La distribution des augmentations pondérales des divers tissus corporels des rorquals communs et des baleines bleues est calculée comme dans le cas de l'accroissement du poids corporel total chez ces espèces: l'augmentation globale maximale affecte la musculature, puis le lard et ensuite les viscères, essentiellement en proportion de la prépondérance pondérale relative de ces tissus, tandis que cette progression est inversée pour les augmentations maximales en pourcentage.

L'étude de la nourriture présente en différents sites révèle que les animaux consomment en général ce qu'ils trouvent à leur disposition et qu'ils préfèrent les crustacés planctoniques vivant en bancs. Dans les eaux polaires proprement dites, *Euphausia superba,* est la principale espèce alimentaire. L'examen du contenu stomacal — nature et quantités — des baleinoptères montre que dans l'Antarctique le taux de fréquence de l'alimentation atteint environ 85 % et que la quantité de nourriture absorbée est en moyenne de 30 à 40 grammes par kilo de poids corporel et par jour; sous les latitudes basses, le taux de fréquence tombe généralement au-dessous de 50 % et probablement au-dessous de 20 % chez les adultes; la consommation de nourriture est inférieure d'environ 10 % — ou moins — à ce qu'elle est en été. Dans le cas d'une baleine qui s'alimente de cette façon, la quantité de nourriture absorbée en un an représente 3,5 à 5 fois le poids corporel et la consommation s'établit en moyenne à 12 g/kg de poids corporel/jour/an. Le schéma d'alimentation des rorquals de Rudolf sous les latitudes subantarctiques semble avoir un caractère moins saisonnier et leur régime est plus varié. L'auteur examine les preuves d'un rythme diurne de l'alimentation régi par les migrations verticales des proies ainsi que les données relatives aux taux de digestion.

Les dimensions de l'appareil buccal des baleinoptères sont comparées et le mécanisme de l'alimentation est examiné. Les fanons des mysticètes semblent plus courts et plus étroits chez les grandes espèces que chez les animaux de dimensions équivalentes des espèces plus petites. Les mâchoires des baleines bleues, des rorquals communs et des petits rorquals sont morphologiquement très semblables alors que celles des mégaptères et des rorquals de Rudolf sont, respectivement, les plus larges et les plus étroites de toutes celles des baleinoptères. Le volume buccal des baleines bleues et des rorquals communs, calculé sur la base de celui des petits rorquals, permet d'avancer que dans les eaux où la production est de 2 kg/m³ dans une bande de 1,22 m d'épaisseur, la quantité d'eau filtrée de façon répétée pendant quelques heures suffit à assurer une alimentation adéquate. Toutefois, il est évident que la migration annuelle vers les eaux productives de l'Antarctique, où se trouvent près de la surface les communautés très denses nécessaires d'euphausiidés, est indispensable à la survie des animaux.

Le rapport entre le poids du cœur et des reins et celui du corps semble du même ordre de grandeur chez les cétacés et les autres mammifères. Ces organes sont étroitement associés au métabolisme, de sorte qu'on peut s'attendre que le métabolisme de base Q suive le rapport observé pour les autres mammifères:

$$Q = 70,5W^{0.7325}\,\text{kcal/jour},$$

(où W = poids corporel, en kilogrammes). Trois méthodes de calcul de la surface développée, 385

A, sont examinées. Les deux qui semblent offrir le plus de possibilités sont la méthode Parry (1949)

$$A = 11.1 \text{ x } W^{0.67},$$

(où W = poids corporel, en kilogrammes, A étant exprimé en centimètres carrés) et celle de Brody («968)

$$A = 1\,000 \text{ x } W^{0.685},$$

(où W et A sont exprimés, respectivement, en kilogrammes et en centimètres carrés); ces deux méthodes donnent des valeurs très proches des observations réelles. Une troisième méthode, celle de Guldberg (1907) suppose que le corps de la baleine ressemble assez à deux cônes placés base contre base; la superficie ainsi calculée représente les deux tiers du résultat obtenu par les deux autres méthodes. Le rapport pouls de base/poids corporel chez les baleines est sans doute semblable à celui des autres mammifères et il se peut qu'une bradycardie survienne pendant la plongée, comme on l'a observé chez les petits mammifères marins. Le poids du sang, approximativement équivalent au volume sanguin, correspondrait à environ 10 % du poids corporel chez les mysticètes adultes. Selon les données publiées, le taux de l'oxygène sanguin est variable mais il est généralement de l'ordre de 20-30 pour cent. On estime que le taux vital des cétacés est très important par rapport au taux total.

L'auteur examine les données sur le rythme de soufflement et la durée de plongée de diverses espèces de baleines. Chez les rorquals, le rythme de soufflement peut descendre à 1/minute pour l'animal au repos et augmenter selon l'activité; en général, les petites baleines soufflent plus fréquemment que les grandes. D'après les données publiées, les rorquals ne semblent pas plonger au-dessous de 300 m. La répartition des périodes de sommeil et d'activité chez les cétacés est mal connue; toutefois, le sommeil apparent est caractérisé par une baisse marquée du rythme respiratoire et par l'adoption de certaines attitudes dans l'eau. Les baleines bien nourries ont des mouvements lents et sont détendues dans leur environnement, ce qui laisse penser que le sommeil pourrait éventuellement apparaître après l'ingestion de nourriture et que la période de sommeil pourrait donc être déterminée par le rythme diurne des proies.

Le taux de métabolisme, au repos, des baleines bleues, des rorquals communs, des cachalots et des hyperoodons est calculé en fonction de l'hypothèse que le quotient respiratoire (QR) moyen est de 0,82, que l'utilisation de l'air inspiré est égale à 10 pour cent et que les valeurs estimatoires de la capacité pulmonaire — en moyenne 2,5-2,8 pour cent du poids corporel — et du rythme respiratoire correspondent aux valeurs observées. Les valeurs prévues pour les grands cétacés concordent avec l'extrapolation des données disponibles pour les petits cétacés et les pinnipèdes. Les valeurs calculées des dépenses énergétiques requises pour surmonter la résistance à différentes vitesses de nage donnent à penser que le chiffre de 0,01 ch par livre (453 g) de musculature fonctionnelle (1,64 x 10^8 ergs/kg) est une base valable pour estimer le taux de métabolisme actif, représentant une dépense énergétique aux vitesses de croisière maximales avec un coefficient de conversion de 1 ch = 1,54 x 10^4 kcal/livre/jour. Le taux de métabolisme de base est calculé de trois façons: (1) en utilisant la ormule:

$$Q = 70,5W^{0.7325},$$

(où W = poids corporel, en kilogrammes), (2) comme représentant environ 85 pour cent du taux au repos et (3) comme 10 pour cent du taux d'activité. Ces trois méthodes donnent des résultats remarquablement semblables, encore que les valeurs fournies par la première soient généralement inférieures à celles des deux autres.

Le bilan énergétique des baleines bleues et des rorquals communs est estimé à différents stades de développement. Les efficacités de croissance (brutes et nettes) diminuent à mesure que l'âge augmente, passant d'environ 30 pour cent chez le jeune non sevré à 1 pour cent chez l'animal presqu'adulte. Les fortes estimations de l'efficacité d'assimilation — environ 80 pour cent — sont peut-être inexactes du fait que leur calcul est indirect. Le rapport entre la dépense énergétique du métabolisme et la capture d'énergie en termes de croissance augmente d'environ 2:1 à près de 100:1 de la naissance à l'âge adulte, ce qui indique une chute spectaculaire de l'efficacité relative de production avec l'âge. Le calcul des bilans énergétiques équilibrés révèle la nécessité d'accroissements saisonniers du poids corporel (au moins 50 pour cent), phénomène démontré précédemment. Les femelles pleines ont apparemment besoin d'une augmentation pondérale d'environ 60-70 pour cent pour résister à la forte déperdition énergétique imposée par la lactation; elles ont donc besoin d'une plus longue période d'alimentation dans l'Antarctique; le bilan énergétique des femelles pleines ne peut avoir de sens que s'il est établi sur un cycle de deux ans.

Extracto

En este trabajo se examinan el crecimiento, los movimientos migratorios, la alimentación y engorde estacionales, el metabolismo y los parámetros morfológicos, fisiológicos y de comportamiento correspondientes, y el balance de energía de la ballena azul y de aleta, haciendo comparaciones, siempre que se dispone de datos, con la ballena boba, la ballena enana y la ballena jorobada.

Se examinan los datos publicados sobre el crecimiento longitudinal de los fetos de balenoptéridos y se estudia el aumento de peso atendiendo a la relación entre éste y la talla. Los resultados indican que la fórmula de crecimiento de Huggett y Widdas (1951),

$$W^{1/3} = a\,(t - t_0),$$

(donde W = peso corporal en gramos, a = constante de velocidad de crecimiento, t_0 = constante de tiempo en días desde la concepción hasta que comienza la fase de crecimiento linear, t = tiempo en días desde la concepción) describe bien la fase rapidísima de crecimiento del feto que se produce al cabo del quinto mes de los 11/12 que dura la gestación. Los resultados obtenidos con la ballena azul, de aleta y boba (0,52, 0,47 y 0,35, respectivamente) coinciden muy de cerca con las estimaciones hechas por Fraser y Huggett (1974).

Se comparan, para la ballena azul, de aleta, boba y jorobada, los datos publicados sobre las dimensiones del cuerpo en relación con la talla durante la fase de crecimiento posnatal. En todas las especies, la parte anterior del cuerpo aumenta desproporcionadamente de tamaño con respecto al resto del mismo, mientras en la región posterior se manifiesta una tendencia negativa, lo que produce un claro efecto de isometría al aumentar la talla. Se trazan curvas de crecimiento que describen la longitud según la edad y el peso según la edad de la ballena azul, de aleta y boba y se ajusta la ecuación de crecimiento de von Bertalanffy (1938),

$$L_t = L_\infty\,(1 - e^{-k(t + t_c)}),$$

(donde L_∞ = talla media en el momento de la madurez física, en metros, t = edad en años, L_t = longitud en metros a la edad t, t = constante de tiempo, k = constante de velocidad de crecimiento) a las curvas para la ballena azul, de aleta y boba, en la forma siguiente:

Macho de ballena azul: $L_t = 25,0\,(1 - e^{-0,216(t + 4,92)})$

Hembra de ballena azul: $L_t = 26,2\,(1 - e^{-0,240(t + 4,50)})$

Macho de ballena de aleta: $L_t = 21,0\,(1 - e^{-0,221(t + 5,30)})$

387

Hembra de ballena de aleta: $\qquad L_t = 22,25\,(1 - e^{-0,220(t+4,80)})$

Macho de ballena boba: $\qquad L_t = 14,8\,(1 - e^{-0,1454(t+9,36)})$

Hembra de ballena boba: $\qquad L_t = 15,3\,(1 - e^{-0,1337(t+10,00)})$

Los resultados obtenidos indican que la modificación de la fórmula de von Bertalanffy hecha por Laws y Parker (1968) para determinar el peso según la edad describe bien las curvas estimadas de crecimiento en peso. Las fórmulas,

$$W_t = W_\infty\,(1 - e^{-k(t+t_e)})^3$$

(donde W_∞ = peso corporal en toneladas en el momento de la madurez física, W_t = peso del cuerpo en toneladas a la edad t) son las siguientes:

Macho de ballena azul: $\qquad W_t = 102\,(1 - e^{-0,216(t+4,92)})^3$

Hembra de ballena azul: $\qquad W_t = 117\,(1 - e^{-0,240(t+4,50)})^3$

Macho de ballena de aleta: $\qquad W_t = 55\,(1 - e^{-0,221(t+5,30)})^3$

Hembra de ballena de aleta: $\qquad W_t = 64,5\,(1 - e^{-0,220(t+4,80)})^3$

Macho de ballena boba: $\qquad W_t = 18\,(1 - e^{-0,1454(t+9,36)})^3$

Hembra de ballena boba: $\qquad W_t = 19,5\,(1 - e^{-0,1337(t+10,00)})^3$

Estas fórmulas indican que la estructura de crecimiento de los balenoptéridos no es probablemente compleja. Las curvas estimadas de aumento de peso muestran que el mayor incremento neto de peso durante la vida de estos animales se produce durante el primer año después del nacimiento, en el que el peso aumenta más de 10 veces; se observa también que los animales de ambos sexos de las tres especies examinadas alcanzan al menos el 70 por ciento de su peso de animales maduros en el momento de la madurez sexual, que se supone se produce a los 5, 6 y 8 años de edad para la ballena azul, de aleta y boba, respectivamente. Comparando los datos sobre longitud y crecimiento aparente de ballenas de aleta de seis zonas distintas del Antártico se observa una variación, según la zona, de la talla media máxima, la talla en el momento de la madurez sexual y el índice de crecimiento.

Analizando los datos publicados sobre devoluciones de marcas de ballenas, variación estacional de la abundancia de ballenas en las zonas de caza y acumulación de diatomeas en la piel de las ballenas del Atlántico debido a *Cocconeis ceticola*, resulta que los balenoptéridos de ambos sexos realizan amplias migraciones anuales, desplazándose durante el verano austral hacia el sur a aguas polares, y durante el invierno, hacia el norte, a latitudes menores y tienden a permanecer dentro del mismo sector longitudinal durante todo el año. Las máximas densidades en el Antártico se alcanzan entre diciembre y febrero y en latitudes menores en torno a junio-agosto. Parece que el máximo de abundancia de ballenas inmaduras se produce más tarde que el de animales maduros, lo que parece indicar un orden jerárquico en los movimientos migratorios, que probablemente entraña también diferencias según el estado de evolución sexual; la permanencia media en el Antártico es de unos 120 días. Se observa también un orden de sucesión en la llegada de las distintas especies: la ballena azul precede a la de aleta, y a ésta sigue la ballena boba, que se desplaza más hacia el sur, hasta los bancos de hielo, junto con la ballena enana. En los últimos años ha habido buenos indicios de que la ballena boba llega al Antártico antes de lo que se había observado en los años treinta, debido quizás a cambios oceanográficos a largo plazo, con un aumento de la temperatura del agua, y a la reducción de la competencia interespecífica por los alimentos debido a la disminución de otras especies de balenoptéridos. Esta observación puede indicar quizás la existencia de una interrelación entre la distribución, abundancia y movimientos migratorios de los balenoptéridos y su dependencia de la producción y de las condiciones oceanográficas.

Se examinan los datos publicados sobre las variaciones estacionales del espesor de la capa de grasa y se analizan algunos datos nuevos, teniendo siempre presentes los movimientos migratorios. En la ballena azul, de aleta, boba y jorobada se observa que las hembras preñadas son excepcionalmente gordas respecto a las hembras maduras no preñadas y los machos adultos, las hembras en lactación son muy delgadas y las formas juveniles están de ordinario más gordas que los machos adultos y las hembras no preñadas. Se comparan las variaciones de peso de la ballena azul, de aleta y jorobada entre noviembre y abril. Para uniformar los datos sobre peso se utiliza la razón peso corporal/longitud corporal en el caso de la ballena jorobada y la razón peso corporal/peso de los huesos en el caso de las ballenas azul y de aleta. La razón media en función del tiempo indica que la ballena jorobada duplica su peso corporal durante la temporada de alimentación en el Antártico, mientras la ballena azul y de aleta registran un engorde del 50 y el 30 por ciento aproximadamente de su peso corporal, aunque probablemente en esta última especie el aumento es análogo al de la ballena azul. Estas estimaciones del proceso de engorde coinciden con los datos correspondientes a la ballena gris del norte del Pacífico y, comparadas con el engorde de los animales en hibernación y las pérdidas de peso de varios animales durante las épocas de hambre, parecen razonables. La distribución del aumento de peso entre los distintos tejidos del cuerpo de la ballena de aleta y azul se calculan de la misma forma que el aumento total del peso corporal de esas especies: el mayor aumento se produce en la musculatura, luego en la capa de grasa, y después en las vísceras, debido principalmente a la mayor proporción relativa de esos tejidos en peso, mientras los mayores aumentos porcentuales se producen en el orden contrario.

Se examinan los alimentos que consumen en diferentes zonas, llegándose a la conclusión de que en general son oportunistas y prefieren los crustáceos planctónicos que viven en grandes concentraciones. En aguas polares, la principal especie de que se alimentan es *Euphausia superba*. Examinando los datos sobre la presencia y cantidad de alimentos en el contenido de los estómagos de balenoptéridos se observa que en el Antártico la alimentación alcanza un nivel de incidencia del 85 por ciento aproximadamente y la ingesta de alimentos puede llegar por término medio a 30-40 g/kg de peso corporal/día; en latitudes menores, el nivel de incidencia disminuye en general a menos del 50 por ciento y probablemente a menos del 20 por ciento en los adultos, y el consumo de alimentos es del orden del 10 por ciento o menos del promedio estivo. Una ballena que se alimente según esos índices consumirá en un año de 3,5 a 5 veces su peso corporal y comerá un promedio de 12 g/kg peso corporal/día/año. La alimentación de las ballenas bobas que se encuentran en latitudes subantárticas resulta menos estacional, con una dieta más variada. Se examinan los datos que parecen indicar la existencia de un ritmo diurno de alimentación, dependiente de las migraciones verticales de sus presas, y la información disponible sobre índices de digestión.

Se comparan las dimensiones de la boca de los balenoptéridos y se examinan los aspectos mécanicos de su alimentación. Las barbas de los animales de las especies de mayor talla parecen más cortas y estrechas si se comparan con las de animales de talla equivalente de las especies menores. Las mandíbulas de la ballena azul, de aleta y enana tienen forma muy similar, mientras las de la ballena jorobada y boba son, respectivamente, las más anchas y más estrechas de los balenoptéridos. Los cálculos del volumen de la cavidad bucal de la ballena azul y de aleta, basados en el de la ballena enana, muestran que cuando se alimentan en aguas donde la producción es del orden de 2 kg/m^3 en una banda de 1,22 m de espesor, el volumen de la cavidad bucal, filtrado repetidamente durante algunas horas, es suficiente para ofrecer una alimentación adecuada. Es evidente, sin embargo, que las migraciones anuales hacia las aguas productivas del Antártico, donde se encuentran en densidades mayores y cerca de la superficie los eufasiáceos que necesitan para alimentarse, son esenciales para su supervivencia.

La proporción del peso del corazón y los riñones de los cetáceos respecto al peso corporal parece ajustarse a la relación que se encuentra de ordinario en otros mamíferos. Como estos órganos están estrechamente relacionados con el metabolismo, es de esperar que también el índice de metabolismo basal, Q, se ajuste a la relación observada en otros mamíferos:

$$Q = 70,5W^{0,7325} \text{ kcal/día}$$

(donde W = peso corporal en kilogramos). Se examinan tres métodos de calcular la superficie, A. Los dos métodos más prometedores son los de Parry (1949),

$$A = 11,1 \times W^{0,67}$$

(donde W = peso corporal en gramos, y A resulta en cm^2), y Brody (1968),

$$A = 1\,000 \times W^{0,685}$$

(donde W y A se expresan en kilogramos y cm^2, respectivamente). Ambas fórmulas dan resultados muy próximos a los datos observados de hecho. En un tercer método (Guldberg, 1907), se supone que el cuerpo de las ballenas semeja aproximadamente a dos conos unidos por sus extremos y se obtiene un resultado de unos 2/3 de la superficie calculada por los otros dos métodos. La relación entre el ritmo de pulsación basal y el peso corporal de las ballenas es probablemente análoga a la de otros mamíferos y es posible que se produzca bradicardia durante la inmersión, como se ha observado en los pequeños mamíferos marinos. El peso de la sangre de los misticetos adultos, equivalente aproximadamente al volumen de la sangre, se estima en un 10 por ciento, aproximadamente, de su peso corporal. En los datos publicados, la capacidad de oxigenación de la sangre resulta variable, pero en general es del orden del 20-30 por ciento. Se calcula que la capacidad vital de los cetáceos representa una altísima proporción de la capacidad total.

Se examinan los datos disponibles sobre índices de resoplido y tiempo de inmersión de diversas especies de ballenas. En los rorcuales, el índice de resoplido puede disminuir a 1/minuto en los animales en reposo, aumentando con la actividad; en general, las ballenas de menor talla resoplan con mayor frecuencia que las mayores. Los rorcuales no parecen inmergirse a profundidades de más de 300 m, a juzgar por los datos publicados. La estructura de sueño y actividad de los cetáceos parece poco clara, aunque el sueño aparente se caracteriza por una notable disminución del ritmo respiratorio y por la adopción de determinadas posturas en el agua. Las ballenas bien alimentadas son lentas y dan la impresión de estar relajadas en su medio ambiente, lo que indica que quizás el sueño de produzca después de la alimentación y, por tanto, es posible que el período de sueño esté determinado por el posible ritmo diurno de las especies que les sirven de presa.

El índice metabólico en reposo de la ballena azul y de aleta, del cachalote y de la ballena hocico de botella se calcula suponiendo que el cociente respiratorio medio es 0,82 y la utilización del aire inspirado del 10 por ciento y que la estimación de la capacidad pulmonar, con un promedio del 2,5-2,8 por ciento del peso corporal, y el ritmo respiratorio corresponden a lo observado. Los valores previstos para los grandes cetáceos coinciden con lo que habría de esperarse a partir de extrapolaciones de los datos correspondientes a los pequeños cetáceos y los pinnípedos. Un cálculo de la energía necesaria para superar la resistencia nadando a diversas velocidades indica que una musculatura operativa de 0,01 hp/libra (1,64 x 10^8 ergios/kg) constituye una base válida para estimar el índice metabólico activo, que representa el gasto de energía a las velocidades máximas de crucero, utilizando un coeficiente de conversiòn de 1 hp = 1,54 x 10⁴ kcal/libra/día. El índice de metabolismo basal se calcula (1) como el resultado de la fórmula,

$$Q = 70,5W^{0,7325}$$

(W = peso corporal en kilogramos), (2) en un 85 por ciento del índice en reposo y (3) en un 10 por ciento del índice en actividad. Los tres métodos dan resultados notablemente análogos, aunque con el primero se obtienen en general cifras menores que con los otros dos.

Se calcula el balance energético de la ballena azul y de aleta en diversas fases de su desarrollo. La eficiencia de crecimiento (bruta y neta) disminuye al aumentar la edad, pasando desde un 30 por ciento aproximadamente en los cachorros sin destetar hasta un 1 por ciento cuando se acercan a la edad adulta. Las estimaciones más elevadas de la eficiencia de asimilación, del orden del 80 por ciento, podrían ser inexactas debido a que se han calculado por métodos indirectos. La razón entre el consumo de energía en el metabolismo y la absorción de energía en forma de crecimiento aumenta de 2:1, aproximadamente, a casi 100:1 entre el nacimiento y la fase adulta, lo que indica que la eficiencia relativa de producción disminuye drásticamente con la edad. El cálculo del balance de energía muestra la necesidad de los aumentos estacionales de peso corporal (al menos 50 por ciento) indicados más arriba. Las hembras preñadas necesitan un aumento del orden del 60-70 por ciento para poder soportar el enorme consumo de energía derivado de la lactación, y necesitan, por tanto, un período más largo de alimentación en el Antártico; el balance energético de las hembras preñadas sólo puede determinarse con sentido a lo largo de un ciclo de dos años.

C. Lockyer
British Antarctic Survey, Madangely Rd., Cambridge CB3 OET, England

Introduction

The understanding of an ecosystem with its complex of plant and animal interactions and the energy flow into, within, and out of the system, begins with an intimate knowledge of the biology of its components.

In the Antarctic sea regions, defined as within the limits of the Antarctic Convergence, there are reactions between residents and migrants, semi-terrestrial, marine and airborne. Apart from the territorial aspects, all these creatures are involved with each other directly or indirectly via the food web. In order to gain an ultimate overview of the whole ecosystem, each creature can first be considered in isolation as a unit part.

There are five balaenopterid species which frequent the Antarctic Ocean for a few months each year. These are blue, fin, sei, minke and humpback whales. Each of these species has its preference for timing of migration, proximity to ice, food and other oceanographic features, although all species overlap considerably in these aspects.

The growth, development, migration, seasonal feeding and fattening, metabolism and energy budgets for blue and fin whales will be discussed in detail, and comparisons with the sei, minke and humpback whales will be made whenever data are available. It is hoped that this study will be a useful lead into more complex analyses of the Antarctic marine ecosystem which are needed for evaluating fishery resources, particularly krill, in the southern ocean. The energy budgets of the large whales form only a small section of the total Antarctic energy cycle, but nevertheless, the whales are substantial consumers of the energy resources, and take energy out of the Antarctic system for many months in each year.

Growth

FOETAL GROWTH

Growth in Length

Curves of increase in length from time of conception have already been constructed by several authors for blue, fin, sei and humpback whale foetuses, although the information given by Laws (1959) for blue and fin whales, and by Gambell (1968) for sei whales has been drawn upon mainly here. Laws (1959) observed that the foetal growth pattern in blue, fin, sei and humpback whales is exponential after about the initial 5 months from conception. The entire gestation period in balaenopterid whales takes about 11 months according to Mackintosh and Wheeler (1929) and Laws (1959). Laws suggested that the foetal growth pattern may be influenced by the seasonal migratory and feeding habits of the pregnant females, because the exponential growth phase of the foetus coincides with the arrival of the adult whales on the feeding grounds.

Weight/Length Relationship

There are data on both weight and length for foetuses of blue, fin, sei and hump-back whales. Nishiwaki and Hayashi (1950) calculated a formula for estimating the body weight of a foetus from its length, in blue and fin whales; their formula, converted to metric units, is

$$W = 20.52 \times 10^{-6} L^{2.83},$$

where W = body weight in kilogrammes, and L = body length in centimetres. A formula for

sei whale foetuses, based on a sample of 134 weighings (115 from the Antarctic, 19 from Durban) recorded in biologists' logbooks at the Institute of Oceanographic Sciences (IOS), and calculated using the method of least-squares regression analysis, is

$$W = 14 \times 10^{-6} L^{2.9198}.$$

In many instances, large baleen whale foetuses have had to be cut up into two or more pieces before weighing and a proportional allowance of 1-2 % was made at the weighing for blood loss. A provisional formula describing the weight/length relationship in foetal humpbacks,

$$W = 74 \times 10^{-6} L^{2.6704},$$

is based on only 9 weighings.

Growth in Weight

In Fig. 1, curves for mean body weight with time from conception have been constructed for blue, fin and sei whales. These curves result from a comparison of length at age, and weight at length curves from both published and unpublished sources. Laws (1959) discussed foetal growth in detail for different species with reference to the theory of Huggett and Widdas (1951), which was that a relationship existed between gestation time and body weight. Laws, however, because of inadequate foetal weight data, modified the Huggett and Widdas method to enable him to test for relationship between gestation time and body length. The original formula was $W^{1/3} = a (t - t_0)$, where W = body weight in g, a = growth velocity constant, t_0 = time constant in days since conception prior to the linear growth phase, t = time in days since conception. The modification of Laws was $L = a (t - t_0)$, where L = body length in centimetres. Laws assumed a cubic relationship between weight and body length, recognizing that the values of t_0 calculated using length

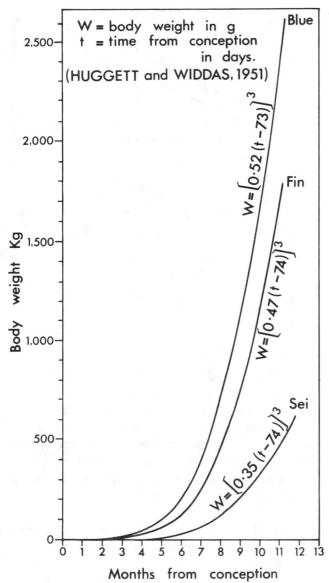

FIG. 1. — Predicted growth in weight with time for blue, fin and sei whale foetuses.

uous between about a month after conception and birth. This phase included an initial short linear phase followed by a phase from the fifth month, best described as linear only when logarithms of length were used. Gambell (1968), however, found that both the initial phase and the latter phase were both linear for the sei whale. In attempting to establish which growth phase, if any, was described by the Huggett and Widdas formula, Laws concluded that balaenopterid foetal growth was complex, and an exception to the usual mammal pattern, and that the growth phases appeared not to be adequately described by the Huggett and Widdas theory.

Rice and Wolman (1971) pointed out that the estimates of t_0 calculated by Laws for the initial growth phase of balaenopterid foetuses were too low when compared with predicted values, and were probably incorrect because Laws mistakenly assumed an inverse rather than a direct relationship between t_0 and t_g (the duration of gestation in days).

The evaluated formula, $W^{1/3} = a(t - t_0)$, for each species shown in Fig. 1, was calculated by plotting the cube root of body weight against time since conception. Unlike the plot of body length against time since conception, the latter growth phase of the foetus from the fourth or fifth month onwards was reasonably described by a straight line. The gradient of this line gave the value of a, and the point where the line, produced backward, cut the time axis at $W^{1/3} = 0$, gave the value of t_0. This method of determining a and t_0 is open to question, because doubt still exists as to which growth phase should be used to determine the growth constants. However, it is clear that this latter growth phase of balaenopterid foetuses follows the Huggett and Widdas theory. These authors stated that an expected value of t_0 for gestation periods of 100-400 days might be $0.2 \times t_g$, so that for a whale with a gestation period of 335-360 days, t_0 could be about 67-72 days. The estimated values shown in Fig. 1 are all close to the higher predicted value of t_0. Huggett and Widdas predicted values of about 0.5 for a in large cetacea, and in fact, this

might differ from those using weight, because of an imperfect cubic weight/length relationship. In fact the weight/length relationships discussed in the previous section indicate that the exponent of length is more probably between 2.5 and 3.0.

Laws (1959) found that the modified Huggett and Widdas formula did not describe balaenopterid foetal growth well, chiefly because of the two-phase growth pattern conspic-

393

estimate seems reasonable with reference to the values of a for blue and fin whales in Fig. 1. The values of a given in Fig. 1 are also close to estimates given by Frazer and Huggett (1974) who listed values of 0.58, 0.46 and 0.36 for blue, fin and sei whales respectively. This means that balaenopterid whale foetuses have the highest growth rate of all mammals (Huggett and Widdas, 1951; Frazer and Huggett, 1974). Fig. 1 shows the phase of very rapid growth after the fifth month, already observed by Laws. It should be noted that the actual near-term and birth weights are not as accurate as the intermediate values, mainly because of lack of data.

POST-NATAL GROWTH

Length and Age

There are two main types of growth occurring in an animal, that concerned with development and differentiation, and that involved in maintenance and tissue repair. In the young animal growth is conspicuous. The creature gains in weight and volume, and its dimensions change.

In Fig. 2, the possible variations in body proportions during growth and development are shown. The length of a particular part of the body, l, is plotted as a proportion of total body length, L. A positive trend indicates that the region under consideration is of greater relative size in larger than in smaller animals. Conversely, a negative trend indicates the opposite growth relationship. Because size is generally dependent on age, growth trends and changes can be correlated with development from juvenile to adult. Complex growth patterns may be observed where a combination of trends occurs. Sometimes changes in growth patterns in mammals can be correlated with puberty (Brody, 1968), and the overall trend may be a resultant of several values of K, the coefficient of proportionality, having both positive and negative values at different phases of growth.

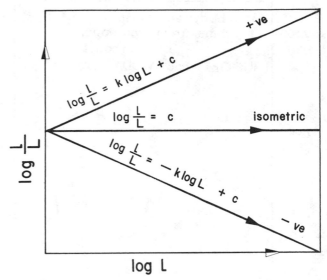

FIG. 2. – Diagram showing the possible variations in body proportions during growth and development; L = total body length, and l = length of the body part being compared.

Mackintosh and Wheeler (1929) and Matthews (1937, 1938) used the same system and notation for taking external measurements in blue, fin, humpback and sei whales. Five of their measurements have been used and compared here as follows. (They are the most relevant to overall growth in body dimensions):

(1)   total body length from the tip of the snout to the notch between the tail flukes;

(2)   severed head length from the tip of the snout to the occipital condyles;

(3)   skull width at the widest part;

(4)   post-anal length (anus to the notch between the tail flukes);

(5)   depth of the tail below the dorsal fin (measure of girth).

In Figs. 3, 4, 5 and 6, these body dimensions have been plotted logarithmically as percentages of total body length for whales of various sizes. Data from South Africa and South Georgia have been combined, because they are very similar, but the sexes have been kept separate in each species.

Fig. 3 shows that linear growth relative to body length in blue, fin, sei and humpback whales of both sexes has a negative trend in the posterior region of the body. Fig. 4 shows that there is a positive trend in the anterior region, although this is not clear in sei whales. Overall, the relative growth in the anterior and posterior regions tends to balance out, the net effect being isometric with increase in body length. No direct measurements of the trunk region are available, but presumably the relative growth of the trunk is isometric with increase in body length.

Skull width in blue and fin whales appears to increase relatively more than body length in Fig. 5, although in sei whales there is no clear trend, and in humpbacks the trend is negative. The relative growth in girth posteriorly is also variable, being different for the sexes as well as the species, with reference to blue and fin in Fig. 6.

The overall growth pattern emerging in all four species and in both sexes, is that shape alters with increase in body length, and that the region of greatest growth activity and expansion is the head. This may be a reflection of the great food requirements of the older and larger whales whose jaws must accommodate the continuously growing plankton sifting apparatus, the baleen plates. Mackintosh and

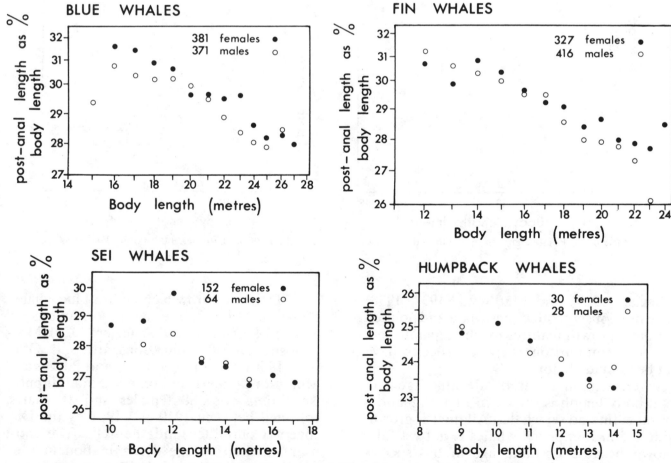

FIG. 3. — Relative growth in length in the posterior region of the body of blue, fin, sei and humpback whales.

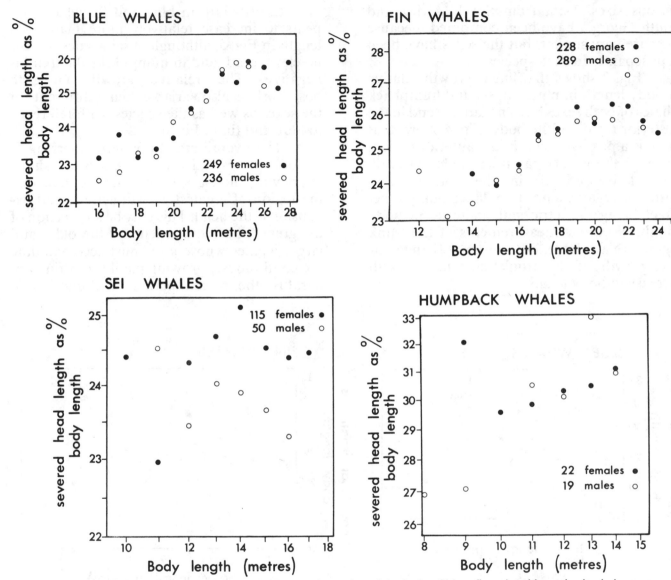

FIG. 4. — Relative growth in length in the anterior region of the body of blue, fin, sei and humpback whales.

Wheeler (1929) and Matthews (1937, 1938) have previously plotted these data and pointed out these growth features for the separate species, but comparisons between species have not been made before.

Growth in terms of absolute age and total body length is shown in Figs. 7, 8 and 9 for blue, fin and sei whales. Allometric growth with age in humpback whales is not yet fully known because absolute age in this species cannot yet be determined with certainty.

Estimated growth curves for blue whales, shown in Fig. 7, are based on published data in Mackintosh and Wheeler (1929), Mackintosh (1942), Ruud, Jonsgård and Ottestad (1950), Laws (1959, 1961) and Purves and Mountford (1959), and on additional unpublished data for 388 females and 16 males, collected between 1930 and 1961 by the Discovery Committee and the IOS. The most usual method of age determination in blue whales was from annual ridge counts in the

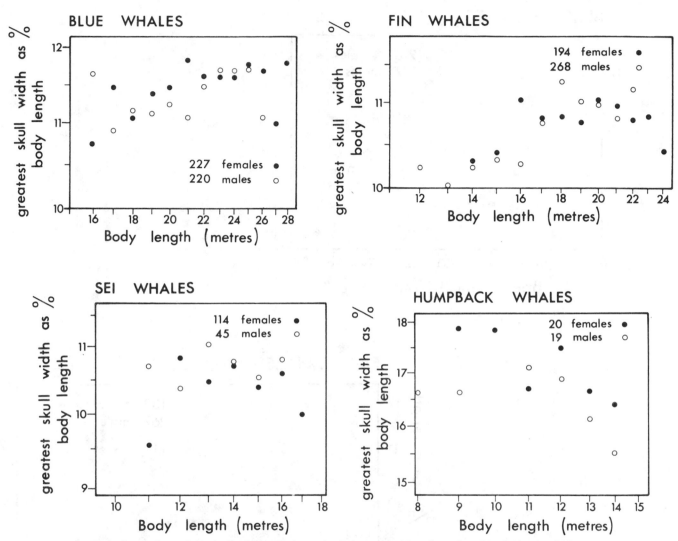

FIG. 5. — Growth in skull width relative to body length in blue, fin, sei and humpback whales.

baleen plates (Ruud, Jonsgård and Ottestad, 1950), which are reliable only for juvenile whales in which the baleen has not worn appreciably. Recent data for 20 females and 16 males include layer counts in the ear plug, a keratinized tissue in the external auditory meatus. The core of the ear plug contains alternating light and dark laminae which continue to form from the basal epithelium throughout life. These laminae are thought to form seasonally in a similar manner to the ear plug tissue of fin whales, where a light and dark lamina together are known to constitute an annual growth layer (Roe, 1967, 1967a).

Furthermore, the study by Nishiwaki (1952) on the baleen plates from blue and fin whales showed that in both species ridges form annually. Because baleen plate ridges and ear plug growth layers correspond exactly in juvenile fin whales, by analogy it would seem logical to assume that the same would be true for blue whales. The age readings from these 36 blue whales, using ear plugs, support previous estimates of age at sexual maturity by Ruud, Jonsgård and Ottestad (1950) and Nishiwaki (1952) as 5 years.

Ovary observations are available for a further 368 females for which the corpora

397

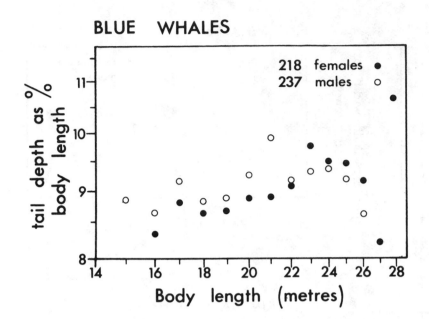

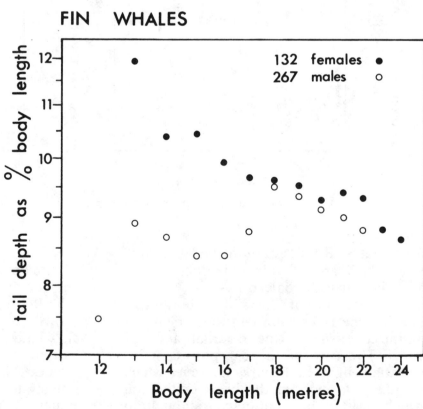

FIG. 6. — Growth in girth relative to body length in blue and fin whales.

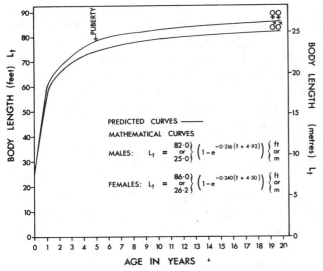

FIG. 7. – Mean body length with age in Southern Hemisphere blue whales.

number gives an estimate of relative age. Peters (1939), Laurie (1937), Mackintosh (1942) and Ruud, Jonsgård and Ottestad (1950) gave estimates of ovulation rates of about one corpus per year, which seems rather high. This implies that all females experience an ovulation in the second year of their breeding cycle during which gestation lasts about 11 months and lactation about 6-7 months.

Gambell (1968, 1973) found that the

incidence of post-partum ovulations are 17.9 % and 11.1 %, and incidences of postlactation ovulations are 39 % and 12.5 % in fin and sei whales respectively. Chittleborough (1954) found 18 % of postlactation phase humpback whales were ovulating. Gambell (1968) suggested that the higher incidence of ovulation in fin and humpback whales might be explained by the fact that they were subjected to the pressures of over-fishing before the sei whales. Blue whales have been grossly over-exploited and so a high incidence of post-partum and post-lactation ovulations might also be expected. However, Laurie (Purves and Mountford, 1959) amended his earlier estimate of one corpus per 0.88 year and claimed that this is the ovulation rate per breeding cycle and that the true rate, allowing for pregnancy and lactation, is one corpus per 1.44 years. Lockyer (1972) found that the ovulation rate in fin whales was 1 per 1.43 years. Because blue and fin whales are so similar in physiology and ecology, and are closely related, Laurie's revised ovulation rate has been accepted as being more likely than earlier estimates.

In Fig. 7 the length at weaning in calves of both sexes has been determined from an analysis of 63 blue whales. All the data were collected between 1930 and 1960 chiefly at South Georgia or off South Africa. The presence or absence of milk and other foods in the stomach, parasites in the gut, and type of faeces were noted in estimating length at wean-

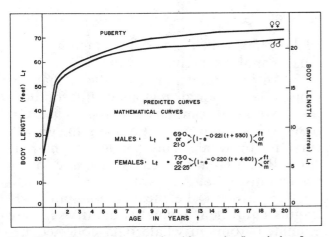

FIG. 8. – Mean body length with age in fin whales from Antarctic Areas II (0°-60°W) and III (0°-70°E), and Durban, South Africa.

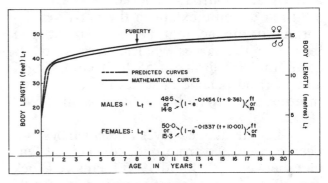

FIG. 9. – Mean body length with age in sei whales from Durban, South Africa.

ing. The length at weaning is in fair agreement with that implied by Mackintosh and Wheeler (1929).

The growth up to the end of the second year of life was estimated from an analysis of monthly length data for 211 juveniles. In this way it was possible to distinguish first and second-year animals, although the very young animals of less than 14 m length were rather scarce.

The length at age curves for fin whales in Antarctic Areas II (0°-60°W) and III (0°-70°E) have been recently revised on the basis of a reassessment of the age at sexual maturity (Lockyer, 1972). The fin whales in these areas seem to attain sexual maturity at an earlier age than that previously estimated. Laws (1962) suggested, and Lockyer (1972) found that fin whales are now also growing faster so that the growth curve age scale has been shortened. The length at age curves for fin whales shown in Fig. 8 are based on the recent age at sexual maturity of 6 years for catches in the sixties.

The methods used to determine age include counting the layers in the ear plugs. Each annual growth layer, as already mentioned, consists of a pale fatty and dark keratinized lamina (Roe, 1967, 1967a). Ovarian corpora counts have also been used for females, on the basis of an ovulation rate of one every 1.43 years in Areas II and III (Lockyer, 1972; Gambell, 1973). The curve shown here involves only length at age data from 1960 onwards, for 828 females and 548 males collected from South Georgia, Antarctic Areas II and III and Durban. The curves have been adjusted to some extent in the early years to correct for the fact that the mean size at age is biased to the larger whales, because of the minimum size limits of 55 ft at land stations and 57 ft on pelagic factories imposed by the International Whaling Commission (IWC) (1952, 1955).

Length at age curves for sei whales have been constructed from data collected at Durban, between 1960 and 1965 inclusive (Lockyer, 1974). The original sample sizes for the length at age analysis for males and females are small, 152 and 92 respectively, so that these curves may be less reliable than for other species. There are more data available from the Antarctic, but the Durban data have been chosen for analysis because they contain a wider spectrum of ages. However, only the very large animals are found further south (Gambell, 1968). Therefore, the mean maximum sizes attained are likely to be significantly greater than those given here.

The von Bertalanffy (1938) growth equation, describing expected growth increments and rates between birth and physical maturity,

$$L_t = L_\infty (1 - e^{-k(t + t_0)})$$

fits the blue, fin and sei whale length at age data fairly well although not within the early period from birth to year 2. Here, L_∞ = mean length at physical maturity, t = age in years, L_t = length at age t, t_0 = time constant, k = growth velocity constant. This treatment of length at age data was developed for describing growth in fish rather than mammals; whales are exceptional in their atypical mammalian appearance. The non-agreement between observed and mathematical growth curves in early years is accentuated by the bias to larger sizes in the whale catches, as already mentioned. However, logarithmic plots of $\dfrac{(L_\infty - L_t)}{L_\infty}$ with t give good linear relationships which can be evaluated by least-squares analysis to give the value of k.

An important point to note is that, although data from all regions of the southern seas have been combined for blue whales in order to boost the amount of data, the growth curve for fin whales only applies to Antarctic Areas II and III and South Africa, and for sei whales only to South Africa.

In Table 1 the mean maximum body lengths, equivalent to von Bertalanffy's L_∞, are estimated from fin whale data collected between 1955 and 1969 and include measurements of length and age by the IOS and the

Table 1. Maximum mean length (L_∞), mean length at sexual maturity (L_m), and immediately pre-pubertal growth rates (dL/dt) for southern fin whales in different Antarctic Areas

Sex	ANTARCTIC AREAS								
	Area I (120°W-60°W)			Area II (60°W-0°)			Area III (0°-70°E)		
	L_∞	dL/dt	L_m	L_∞	dL/dt	L_m	L_∞	dL/dt	L_m
Females	67.3 ft or 20.5 m	0.83 ft/year or 0.25 m/year	61.0 ft or 18.6 m	68.5 ft or 20.9 m	0.86 ft/year or 0.26 m/year	63.5 ft or 19.4 m	67.8 ft or 20.7 m	0.86 ft/year or 0.26 m/year	62.6 ft or 19.1 m
Females	72.9 ft or 22.2 m	0.94 ft/year or 0.29 m/year	64.7 ft or 19.7 m	73.3 ft or 22.3 m	1.14 ft/year or 0.35 m/year	65.5 ft or 20.0 m	73.0 ft or 22.3 m	1.13 ft/year or 0.34 m/year	65.6 ft or 20.0 m

Sex	Area IV (70°E-130°E)			Area V (130°E-170°W)			Area VI (170°W-120°W)		
	L_∞	dL/dt	L_m	L_∞	dL/dt	L_m	L_∞	dL/dt	L_m
Males	67.3 ft or 20.5 m	0.64 ft/year or 0.19 m/year	61.7 ft or 18.8 m	66.6 ft or 20.3 m	0.71 ft/year or 0.22 m/year	62.8 ft or 19.1 m	65.4 ft or 19.9 m	0.52 ft/year or 0.16 m/year	61.5 ft or 18.7 m
Females	72.5 ft or 22.1 m	0.92 ft/year or 0.28 m/year	65.6 ft or 20.0 m	71.5 ft or 21.8 m	0.97 ft/year or 0.30 m/year	65.6 ft or 20.0 m	71.0 ft or 21.6 m	0.64 ft/year or· 0.19 m/year	64.5 ft or 19.7 m

Whales Research Institute in Tokyo. The mean lengths at sexual maturity for fin whales according to Ichihara (1966) are also given. The pre-pubertal growth rates appear to be higher in Areas II and III compared with other Antarctic areas, and also the values of L_∞.

Weight at Length and Age

Published data on body weights for blue, fin, sei and humpback whales were reviewed by Lockyer (1976). Weight/length relationships for these species are exponential, and previously published formulae describing them were derived by the method of least-squares regression analysis of the logarithmic values of weight and length. The logarithmic form of the weight/length equation is

$$\log_{10} W = b \log_{10} L + \log_{10} a$$

although it is nearly always expressed in the form

$$W = a L^b,$$

where W = body weight, L = body length, b = exponential constant, a = constant factor. Lockyer (1976) gave numerical formulae for these four species, based on selected actual weights, and adjusted the formulae to allow for about 6 % observed blood and fluid loss during flensing for the piece-meal weighings.

401

These formulae are given in Fig. 9, where they accompany a graphical illustration of the weight/length relationship for blue, fin, sei and humpback whales.

Growth in Weight

No ages were recorded for the animals weighed, and in any case, these might not be very useful because of the variations likely in such a small number of weighed specimens. However, a weight at age guide can be constructed from a cross-reference between weight/length and length/age curves.

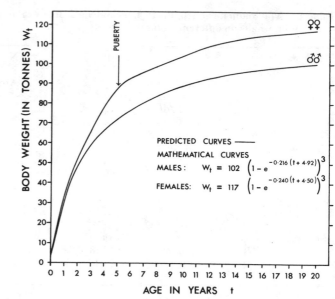

FIG. 11. — Predicted growth in body weight with age for southern hemisphere blue whales.

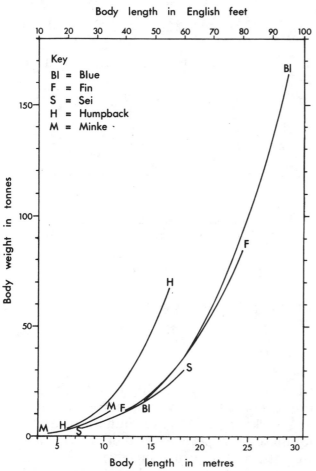

FIG. 10. — Body weight at length in blue, fin, sei, minke and humpback whales [after Lockyer, 1976]

Figs. 11, 12 and 13 show the estimated body growth in weight with time for blue, fin and sei whales respectively. These curves are based on the length/age curves shown in Figs. 7, 8 and 10, substituting the weight at each age using the formulae given in Fig. 9. Because of the lack of precise information on age in humpbacks, the construction of curves of

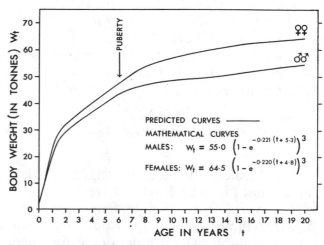

FIG. 12. — Predicted growth in body weight with age for fin whales from Antarctic Areas II (0°-60°W) and III (0°-70E), and Durban, South Africa.

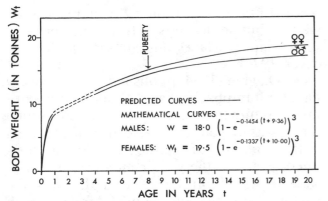

FIG. 13. — Predicted growth in body weight with age for sei whales from Durban, South Africa.

yet fit well later. Both the length and cubic weight Bertalanffy formulae serve to illustrate that the pattern of growth in balaenopterid whales is probably simple, and not a complex one such as is observed in some other mammals; for example, male elephant seals (Laws, 1953), and male sperm whales (Best, 1970), where the male commences a new phase of growth after puberty. This growth spurt in these two species is associated with social maturity or attainment of harem status. The bulls grow larger than the cows, the opposite of the situation in balaenopterids for whom monogamy is more common than polygamy.

growth in weight with time is not possible for this species except on a relative scale.

The chief features of the curves in Figs. 11 to 13, are firstly, that in the initial year of post-natal life the body weight increases more than tenfold, which represents the largest net gain in weight during the entire life phase; secondly, that at puberty the animal has reached between 70 % and 75 % of the mature body weight in blue and fin whales, and over 75 % in sei whales. These values are high in comparison with those for terrestrial mammals where between 30 % and 75 % of mature body weight is attained at puberty, for example, cattle, sheep, chimpanzee and man (Brody, 1968), and African elephant (Laws and Parker, 1968).

Laws and Parker (1968) used a multiple of the von Bertalanffy formula,

$$W_t = W_\infty (1 - e^{-k(t+t_0)})^3,$$

where W_∞ = body weight at physical maturity, W_t = body weight at time t, to describe growth in body weight with time in elephants. The body weight is related to the cube of body length, and in the formula the exponential is cubed. This type of formula is shown in Figs. 11 to 13. However, as in the length at age analysis, the mathematical curves of weight at age in whales do not fit the estimated curves of body weight with age in the very early years,

Correlation of seasonal feeding migration and variations in blubber thickness and body weight

Migration

Evidence from Whale Marking

The seasonal migration of balaenopterid whales in the Southern Hemisphere has been demonstrated by several methods. The most direct and reliable method is that employing whale marking and recapture. The Discovery Committee initiated the marking schemes and since the thirties, marking programmes have continued, the mark placement and recapture data being published from time to time.

Rayner (1940) gave mark return data for blue and fin whales, which showed that from South Georgia these species tend to migrate several hundred miles south and southwest during the whaling season in the summer months. Mark returns from other Antarctic regions showed that movement south up to 900 miles is not unusual during the southern summer, and the whales tend to follow the retreating pack-ice edge. Rayner pointed out that longer-term mark recoveries show that whales returned to virtually the same longitudinal sector of the Antarctic in which the animals were marked, perhaps several years previously. Often the whales appear to be in the same company.

Mark recoveries also show that movements occur northward. For fin and humpback whales Rayner presented evidence that some individuals had migrated northward between latitudes 65°S and 22°S, distances of well over a thousand miles.

Brown (1954), in a review of whale mark recoveries in the Southern Hemisphere, confirmed all that Rayner had stated for blue and fin whales. Furthermore, he was able to demonstrate that marked whales were recovered from the same area of marking, and at the same time of year, 17 years after marking. Generally, Brown concluded that dispersion on the feeding grounds does not exceed 50° of longitude east or west from the original location of marking. Brown also demonstrated that fin whales marked off Brazil in winter were later killed off South Georgia in the summer. Thus, the movement between latitudes, first described by Rayner, was found to be two-way, suggesting migratory behaviour in these whales.

Brown (1957) and Dawbin (1959) both gave the same details of the mark recovery in a humpback whale which was caught in the Antarctic in 68°S, but marked in 21°S off New Zealand. Chittleborough (1959) also gave information showing that not only did humpback whales move between Antarctic and sub-tropical latitudes, but that this migration can occur within a few months. In January 1954, a tin containing a message was thrown overboard in 64°5'S, and by chance was swallowed by a humpback which was later killed off west Australia in June 1954, when the object was recovered from the intestine.

Brown (1959, 1960, 1961, 1962, 1962a, 1962b, 1966) gave mark recovery data for balaenopterid whales, chiefly fin and humpback, which showed that these animals migrated between 20° and 30°S, and 50° and 65°S within a few months, most probably seasonally. The mark returns showed that the same migration pattern was likely for both sexes. Little information is available for sei whales, but some mark return data tends to show that sei whales migrate seasonally (Lockyer, 1974).

Dawbin (1964) published mark recovery data for humpback whales. From a number of marks returned within a year, he concluded that it was evident that these animals, marked in the Antarctic during the summer, had migrated north for the winter months when they were captured off New Zealand. Thus the whaling grounds off South Africa, South America, Australia and New Zealand could be linked with the relevant Antarctic whaling grounds. Also, the humpbacks observed swimming in schools along the coasts of Australia and New Zealand regularly each year (Dawbin, 1956, 1966), could now be considered to be passing through on their migration routes. The general pattern of migration was therefore seen as a northerly autumn movement and a southerly spring movement.

Evidence from Seasonal Variations in Whale Density

Records kept of monthly abundance of whales from sightings and catches on the whaling grounds show that the numbers of balaenopterid whales present in the Antarctic increase during the summer (Mackintosh and Wheeler, 1929; Harmer, 1931; Matthews, 1937, 1938; Mackintosh and Brown, 1956; Mackintosh, 1965; Gambell, 1968). These authors and Mackintosh (1942) also gave evidence that the percentage of immature whales increases as the summer season progresses and that the relative abundance of adult males, resting, pregnant and lactating females also varies with the month. Dawbin (1966) showed that humpback whales migrate south in an orderly procession, according to maturity and sexual status. Pregnant females are usually seen to pass through first when swimming south, and also seem to be the last in the northerly migration at the end of the Antarctic summer.

This order of migration appears to be

reflected in the pattern of arrival in the Antarctic, and would fit in with the information found by Mackintosh and Wheeler (1929), Laurie (1937) and Mackintosh (1942) for blue and fin whales, and by Matthews (1937) for humpback whales, and Matthews (1938) and Gambell (1968) for sei whales. Similarly, Kawamura (1974) showed that variations in seasonal abundance of sei whales in different latitudes reflects their tendency to move south during the southern summer to feed in the Antarctic waters. Sei whale abundance, however, tends to reach a peak rather later than blue and fin whales (Harmer, 1931; Matthews, 1938; Nemoto, 1962).

The blue whales generally reach peak abundance earlier than fin in the Antarctic, and Fig. 14 gives an analysis of the cumulative percentage of the catch of each of these species, both adult and juvenile. The data are taken from the logbooks kept by the Discovery Committee, now held by the IOS, for the seasons between 1925 and 1931 inclusive. During this period no restrictions were enforced as to the length of the whaling season, or the size or sex of the whales caught. Adults and juveniles in Fig. 14 were broadly classified according to size: for blue whales, females > 23.5 m and males > 22.5 m were considered adult; for fin whales, females > 20.0 m and males > 19.4 m were considered adult. These sizes are based on the lengths at puberty shown in Figs. 7 and 8. Figure 14 also shows that the juveniles reach peak abundance in the catches about half to a full month later than the adults.

There is good evidence from Nemoto (1962), Bannister and Gambell (1965) and Gambell (1968) that sei whales have altered their timing of migration in recent years. Since the late fifties, sei whales have been observed to reach peak abundance in January rather than March as originally observed in the thirties by authors such as Harmer (1931) and Matthews (1938). Speculatively, this may be connected with the rapid diminution of blue, fin and humpback whale species (Nemoto, 1962; Lockyer, 1974). These species are all known to feed on *Euphausia superba* which the

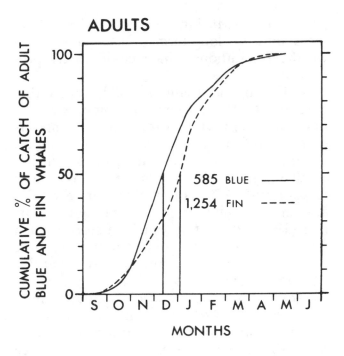

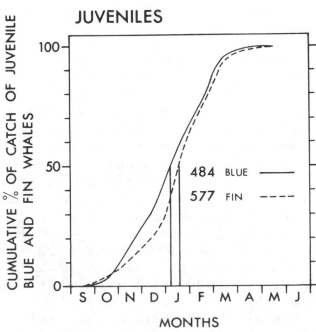

FIG. 14. — Cumulative catch frequencies by month from the period 1925-1930 in the Antarctic, showing the timing of peak catches for adult and juvenile blue and fin whales.

sei whale also consumes (Brown, 1968) when south of the Antarctic Convergence (Kawamura, 1974). Water temperature is closely linked with food availability, particularly in polar waters (Mackintosh, 1972, 1973), where there may be long-term cyclical changes in weather conditions which could affect production.

Kemp and Bennett (1932) found that blue and fin whales could to some extent be mutually exclusive according to their preference for different environmental conditions. Certainly blue whales seem to travel further south than fin whales. Ohsumi, Masaki and Kawamura (1970) found that minke whales also tend to migrate right to the ice edge. Harmer (1931) found that during a season when fin whales were in good abundance, blue whales were in poor abundance, and that this situation could be reversed in other seasons. He found that this inverse abundance relationship was correlated with the September (spring) temperature in the Antarctic. He also noted that during seasons of good abundance of fin whales, the animals were often relatively lean.

Nemoto (1959, 1962) and Kawamura (1970, 1974) showed that water temperature is important in the pattern of distribution and migration of sei whales. Significantly, these authors found that the food type of the sei whale, which is more sub-Antarctic in habit, varied according to the temperature and latitude, and that moving southward toward the Antarctic Convergence, copepods of the genus *Calanus* were displaced by *Euphausia vallentini* and *Parathemisto gaudichaudi,* and by *Euphausia superba* in the true polar waters. The whales appeared to feed opportunistically on whichever food species was most prevalent in the area. Nemoto (1962) reviewed data on food habits of other balaenopterid whales and mentioned that blue, pigmy blue and fin whales have been observed to take *Euphausia vallentini* and *Parathemisto gaudichaudi* off Kerguelen and Crozet Islands, and in areas south of New Zealand, outside the Antarctic Convergence. Ohsumi, Masaki and Kawamura (1970) found that minke whales, like sei whales, take calanoid copepods in the sub-Antarctic. The overall picture, therefore, is that the feeding habits of rorquals tend to overlap when the different whale species are all present in the same area. This raises the important consideration that, during relatively warm summer seasons, more sei whales than usual will penetrate high latitudes frequented by the blue, fin and minke. On the other hand, in relatively cold summers when blue, fin and minke whales are unable to migrate far south because of sea-ice, some animals may be forced to take the food type present in lower Antarctic latitudes along with the sei whales. Nemoto (1959) discussed the former possibility, and noted that in 1957 and 1958 the early January peak abundance of sei whales in the Antarctic seemed to be correlated with a marked seasonal rise in sea temperature, the sei generally being most abundant at the peak temperatures. Nemoto (1962) postulated that this shift in the time of peak abundance of sei whales in recent seasons could be partly the result of real long-term oceanographical changes in the Antarctic and also the result of reduced interspecific competition for food with other balaenopterids.

Bannister and Gambell (1965) and Gambell (1968) presented data on aerial observations of sei whale movements during the whaling season at Durban. Overall, it appeared that sei whales swam northward in May and June and southward in August and September. Gambell (1968) pointed out that there was much random local movement confusing the picture, but that these findings agreed with the trend of movements found by Best (1967) for sei whales off Cape Province.

Bannister and Gambell (1965) also compared the abundance of balaenopterid whales, especially fin and sei, at South Gerogia and Durban. From actual aerial sightings and catch per unit of effort, they showed that peak abundance in the Antarctic was approximately 6 months out of phase with that of Durban. Presenting other supportive information on the condition of the whales taken, Bannister and Gambell suggested that these whaling grounds were in latitudes entered by the whales towards each end of their north-south annual migrations.

Evidence from Patterns of Infection by Ectoparasites

Further indirect information on whale movements is gained by an examination of their ectoparasites. Among these are species which are quite specific as to environment. Pennellae and Cirripedia generally attack the skin and blubber in warmer waters, whilst diatom fouling of the skin occurs in polar waters (Bennett, 1920; Barrett-Hamilton, 1915 *in* Hinton, 1925). The latter is generally ascribed to an infection by *Cocconeis ceticola* Nelson, although at least a further dozen species have been found on whale skin (Hart, 1935).

The infective microspores of *Cocconeis ceticola* are more saprophytic than parasitic, and have never been found free-living (Hart, 1935). The spores require about one month for germination and the diatom films are most prolific from February onward. All whale species and all parts of the anatomy are liable to fouling by diatoms if the whales penetrate Antarctic waters for long enough. Harmer (1931) mentioned that juveniles are often quite diatom-free late in the summer season suggesting that many immature whales arrive after January.

Hart was of the opinion that diatom presence is not a clear indication of total time spent in the Antarctic as the extent of the fouling can be very variable. However, diatom-covered whales are relatively fatter than clean whales, according to Bennett (1920). Omura (1950a) mentioned the increase in percentage diatom growth in fin whales in the Ross Sea as the season progressed. He noted that the infection pattern is more consistent in adults than juveniles.

In Figs. 15 and 16 the percentage monthly incidence of diatom infection is shown for blue and fin whales, by sex and by area. Data have been collected from several Antarctic areas and South Africa for seasons between 1925 and 1965. In Figs. 15 and 16, South Georgia falls into Area II, and Area I covers 60°-120°W. The separate seasons have been compared, and although the overall incidence of infection is higher in some seasons than others, probably due to climatic variations from year to year, the monthly trends of infection are similar. The data from all seasons have, therefore, been combined. In some instances, Antarctic regions have been kept separate because consistent variations in trends have been observed, probably due to local oceanographic conditions. Omura's data (1950a) on diatom infection in fin whales in Antarctic Areas IV and V (70°-130°E and 130°-170°W, respectively) are shown graphically in Fig. 16.

The general trend in Fig. 15 for blue whales is that, after January, the incidence of clean, diatom-free whales in the Antarctic falls sharply, suggesting that there are few new arrivals after this time, particularly of adults. The fall-off in immature whales is not as sharp as in adults and a high incidence of clean female juveniles occurs well into March. The pattern at South Georgia shows a general fall-off in clean animals as the season progresses. Off South Africa, diatom infected whales start appearing around August, and these are all immature animals. There is little information on adults, but it is clear that these whales have migrated to warmer waters from the Antarctic, and that arrivals reach a peak during winter.

Fin whales also show a fall-off in the abundance of diatom-free individuals everywhere in the Antarctic as the summer season proceeds. The peak of abundance of clean immature whales appears to occur after that in mature ones, at least in females, suggesting that the former arrive after the latter. The appearance of diatom-covered juveniles off South Africa during winter shows that these animals have migrated from far south.

Judging from the time taken for the peak number of clean animals to be reduced to a very low level, it seems likely that most blue and fin whales remain on the Antarctic grounds for upward of three months. Probably four months or more is likely, considering that one month normally elapses before microspore germination occurs, although Nemoto (1958)

407

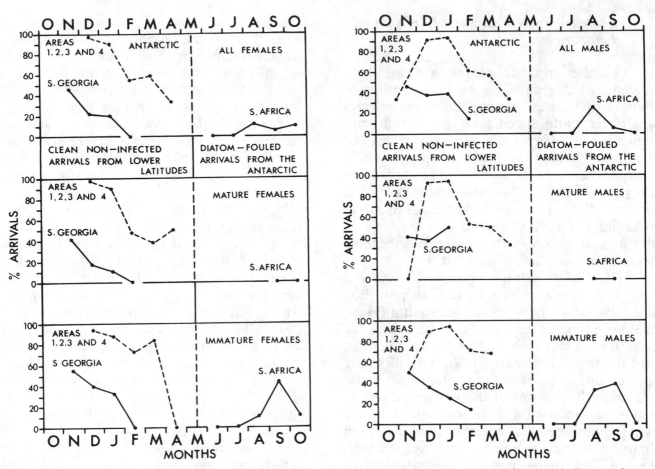

FIG. 15. – Monthly arrivals of blue whales using incidence of Antarctic diatom-fouling of the skin as an indicator of recent whereabouts.

suggested that a shorter germination period is possible. The lack of data for mature whales off South Africa does not imply that adults overwinter in the Antarctic, but that they are elsewhere than off the South African coast.

In conclusion, it can be stated that balaenopterid whales undergo extensive annual migrations, generally wintering in low latitudes whilst entering polar waters for the summer season where they seek the shoals of krill.

SEASONAL VARIATIONS IN BLUBBER THICKNESS

Measurement of blubber thickness have been recorded from whales taken in the Ant-

arctic and also at land stations in lower latitudes. Seasonal increases in blubber thickness occur, correlated to the annual migration to feeding grounds. Among authors writing on this subject are Mackintosh and Wheeler (1929), Slijper (1948), Nishiwaki and Hayashi (1950) and Ash (1956, 1957) on blue and fin whales, Risting (1912, 1928), Olsen (1915), Matthews (1937) and Ash (1957) on humpback whales, Nishiwaki and Oye (1951), Mizue and Murata (1951), Ohno and Fujino (1952) and Kakura, Kawakami and Iguchi (1953) on baleen whales (chiefly blue and fin), and Matthews (1938) and Gambell (1968) on sei whales.

The evidence in this paper for seasonal fattening is confined to the baleen whales such as blue, fin, humpback and sei whales. A

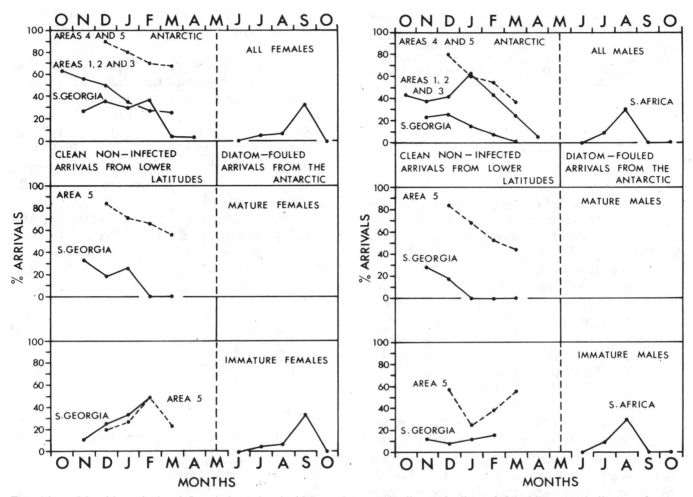

FIG. 16. — Monthly arrivals of fin whales using incidence of Antarctic diatom-fouling of the skin as an indicator of recent whereabouts.

common feature of all these, noted by most of the above authors, is the unusual relative fatness of pregnant females and juveniles, and a tendency to leanness in lactating females. This feature is most marked in humpbacks (Matthews, 1937). Some authors believe that the blubber fat increase slows after February (Slijper, 1948), and that there is a temporary decrease in thickness shortly after arrival in the Antarctic (Mizue and Murata, 1951). All authors are agreed that whales appear to become measurably fatter on the feeding grounds of the Antarctic than in lower latitudes.

Blubber measurements from different geographical areas are not always directly comparable owing to differences in the sites of measurement on the carcasses, and non-random sampling of age groups in a catch. Also localities can vary enormously in food availability, ice-cover and so forth even within close distances (Mackintosh, 1972, 1973).

An analysis of post-war data from 58 humpback whales caught in the Antarctic Areas II, III and V reveals inconclusive results for seasonal blubber thickness variations, owing to the small sample size. However, general results are in agreement with the findings of Matthews (1937). Individuals have been grouped according to maturity and sexual condition, and the mean ratio between blubber thickness and body length between December and March is shown in Table 2. A clear

409

Table 2. Percentage ratio of blubber thickness and body length in humpback whales

Immature males and females	Mature males and females (>12.0m) (>12.5m)	Pregnant females
%	%	%
1.0882	0.7661	0.8543

seasonal pattern of fattening might be expected to emerge from larger samples, since humpbacks show a steadier wave of migration than do other baleen whales, even though it is protracted over two months (Matthews, 1937).

All age groups of the humpback pass close to the coasts of the southern continents during their migrations between low and high latitudes. As in other rorquals, the migration is an orderly procession of whales grouped by maturity and sexual condition (Dawbin, 1966), the pregnant females being the first to arrive in, and also the last to leave, the Antarctic.

Matthews (1938) found that for sei whales grouped by sex and maturity, the maximum and minimum values of blubber thickness attained are as shown in Table 3. The blubber thins appreciably once the Antarctic feeding season is ended. Gambell (1968) had more data on which to work but owing to different points of blubber measure-

Table 3. Percentage ratio of blubber thickness and body length in sei whales (according to Matthews, 1938)

Season and Region	Immature Males	Immature Females	Immature Males >14.0m	Mature Females >14.5m Resting	Mature Females >14.5m Pregnant
	%	%	%	%	%
Antarctic late summer	0.510	0.495	0.380	0.360	0.490
Durban winter	0.315	0.320	0.300		0.275

ment, he was unable to equate South African and Antarctic data as directly as could Matthews.

As we have seen in the previous section on migration, blue and fin whales are present on the Antarctic feeding grounds any time between October and April, but adults tend to reach peak abundance around January, and juveniles rather later. Ingebrigtsen (1929) and Kellogg (1929) have written that the migrations of the immature whales are often erratic compared with adults.

An analysis of data collected during whaling operations between 1946 and 1965 on blubber thickness in blue and fin whales is shown in Figs. 17, 18, 19 and 20. For blue whales, some pre-war data, mainly from the

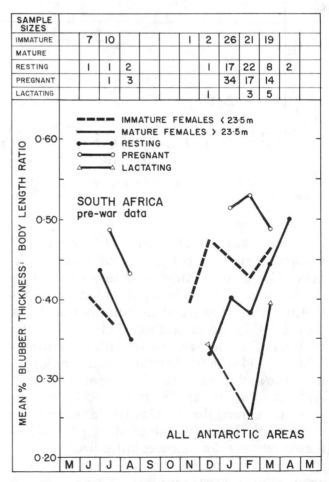

FIG. 17. — Mean monthly blubber thickness: body length ratios in female blue whales.

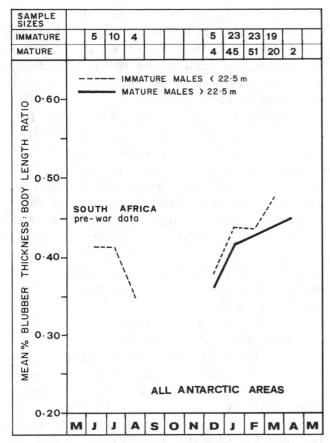

SAMPLE SIZES											
IMMATURE	5	10	4			5	23	23	19		
MATURE						4	45	51	20	2	

FIG. 18. — Mean monthly blubber thickness: body length ratios in male blue whales.

thickness to increase as the Antarctic summer progresses. The most marked increases seem to occur after December or January, and even after February in some blue whales. For nearly every class shown, the increases are not straightforward but consistently show temporary decreases during December or January in fin whales suggesting that this is the result of a bulk influx of lean whales at this time, perhaps representing the peak migration wave. The data for blue whales does not extend over as long a season as for fin whales, so that detailed inferences could be unreliable. The reason for the sudden apparent decrease in blubber thickness in March in female fin whales could either be the absence, and therefore obvious departure, of the fattened large mature fe-

1930's, is included for South Africa. The data have been analysed separately according to sex, sexual maturity and sexual condition. Unfortunately, data from South Africa and the Antarctic are not directly comparable because blubber measurements were made at different points on the body. Blubber thickness was found to increase fairly constantly with body length (if state of maturity and sex are allowed for) by Mackintosh and Wheeler (1929), and expressing blubber thickness as a percentage of body length is a convenient way of combining and comparing data from different sizes of whales.

The most noticeable feature at South Africa and in the Antarctic is the relative leanness of lactating females and fatness of pregnant females of both species. In virtually all sexual classes there is a trend for blubber

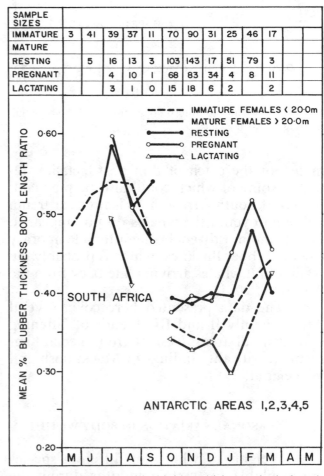

SAMPLE SIZES												
IMMATURE	3	41	39	37	11	70	90	31	25	46	17	
MATURE												
RESTING		5	16	13	3	103	143	17	51	79	3	
PREGNANT			4	10	1	68	83	34	4	8	11	
LACTATING			3	1	0	15	18	6	2		2	

FIG. 19. — Mean monthly blubber thickness: body length ratios in female fin whales.

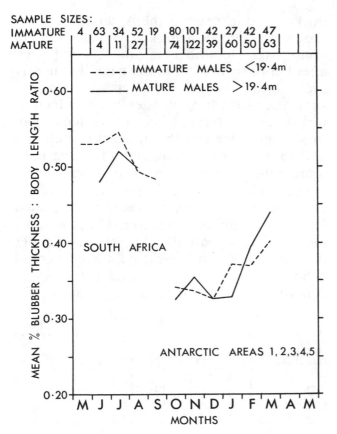

SAMPLE SIZES:

| IMMATURE | 4 | 63 | 34 | 52 | 19 | | 80 | 101 | 42 | 27 | 42 | 47 | |
| MATURE | | 4 | 11 | 27 | | | 74 | 122 | 39 | 60 | 50 | 63 | |

----- IMMATURE MALES <19·4m

—— MATURE MALES >19·4m

SOUTH AFRICA

ANTARCTIC AREAS 1, 2, 3, 4, 5

MEAN % BLUBBER THICKNESS : BODY LENGTH RATIO

MONTHS

M J J A S O N D J F M A M

FIG. 20. – Mean monthly blubber thickness: body length ratios in male fin whales.

males, or the catch of thin post-lactation females, some of which could also be pregnant.

Off South Africa there is no clear trend, although overall the whales do not appear to fatten, and mostly get leaner after a temporary rise in blubber thickness which is probably the result of fat whales arriving late back from the Antarctic.

The data presented here cover several Areas, mainly II and III. Trends of fattening appear in all areas. Results are in general agreement with the findings of Mackintosh and Wheeler (1929).

SEASONAL VARIATION IN BODY WEIGHT

Weights of whales of different species are available for certain months during the Antarctic summer, but not for the winter months. Weights at the beginning of the summer have therefore been compared with those at the end. The difference represents a seasonal increase. As seen in the previous section, there are valid reasons for expecting that weight might vary seasonally.

Adult Southern Hemisphere blue, fin, sei and humpback whales undergo extensive annual migrations to the Antarctic seas to feed during the summer for periods of at least three to four months. During winter, whales taken in lower latitudes have been found to be feeding at greatly reduced rates (Mackintosh and Wheeler, 1929; Matthews, 1937; Gambell, 1968). In order to sustain the body during the winter, the whale must store energy during the summer in a manner similar to that of hibernators accumulating fat.

There is evidence both qualitative and quantitative for the storage of energy in the body during the feeding season. Blubber (fat) thickness increases as the summer season progresses. Oil yields from whale carcasses taken in the Antarctic are usually greater per foot length of whale than in lower latitudes and increase with the length of the Antarctic season (Ash, 1953, 1955, 1955a, 1956, 1957; Risting, 1912, 1928; Olsen 1915 and Harmer, 1931). The increased oil yield is likely to be reflected in noticeable weight increases in the body, which logically must reach a peak immediately prior to the whale leaving the feeding grounds.

Slijper (1962) mentioned that the fat storage occurs initially in the blubber and later in other tissues such as muscle, viscera and bone. This is substantiated to a certain extent in a later section, where oil yields exceed those attributable to the blubber alone. There is probably a critical maximum to fat deposition in the subcutaneous region otherwise overheating would be incurred. Hibernators rapidly accumulate subcutaneous and then visceral fat prior to winter dormancy, which they subsequently lose during the fast. This process is common to all hibernating mammals (Friant, 1942; Kayser, 1953; Grizzell, 1955).

All pregnant whales regardless of the time of season are exceptionally heavy and fat compared with other adults. For this reason any data on body fattening in pregnant females have been kept separate from other data used in analyses of variation in body weight. Lactating whales, which are frequently very lean according to analyses of blubber thickness, are not represented in the data. Juvenile whales are usually relatively fat compared with adults, but not as fat as the pregnant cows.

Data are taken from Appendix Table 1 where the weights of blue, fin and humpback whales are shown together with body lengths and dates of capture, and information on sexual status where available. In comparing body weights and tissue weights of different individuals, a method using ratios has been used. This method reduces all data to a weight per known standard, such as tonne of bone or foot body length. Bone weight must be reflected in the relative weights of other tissues since bone is the supporting skeleton, and body length is dependent on skeletal growth. Of these two possible standards, the former is fully three-dimensional and is preferable, allowing for variation in body build, whereas the latter is only one-dimensional. The bone tissue is unlikely to vary much seasonally in weight, because any oil storage which occurs must take place internally where it will replace water. The bone weight will, therefore, remain fairly stable. The matter of oil storage in bone will be discussed later in this section. By calculating the tissue weight to bone weight ratio or to foot body length ratio in different whales, animals of various sizes may thus be directly compared.

Humpback Whales

According to Matthews (1938) and Dawbin (1966) the humpbacks tend to migrate to and arrive in the Antarctic in a more or less continuous stream, so that fattening appears fairly steady in the population. Hinton (1925) mentioned that nearly all humpbacks arrive at South Georgia between October and January, reaching peak numbers from November onwards. Ash (1953) produced a formula for predicting increased product yields with time from January:

$$W = 1.70L - 46.4 + 0.9t,$$

where W = body weight in long tons, L = body length in feet, and t = number of weeks elapsed since 1 January. The humpback weights obtained by Ash are all approximate in that they were obtained from cooker fillings. Although any analysis investigating seasonal weight variation has pitfalls because different animals of different sexes and different years are inevitably involved, all the whales considered by Ash were of similar body length and taken in only two seasons, so that the variability was reduced to a minimum.

The data in Appendix Table 1 given by Ash have been used here to determine the average weight increase in time for a 40 ft (12 m) whale. In order to do this, average body weight per foot body length of whale for animals, all between 38 and 43 feet (11.5 and 13 m) in size, has been calculated for different dates, and simply multiplied by 40. Data for bone weight are not available. The mean body weights of a 40 ft (12 m) whale calculated for several dates in a period of about two months commencing 28 December, are shown in Fig. 21, which indicates that the humpback whale of 40 ft (12 m) increases in body weight by about 10 tonnes in under two months.

The initial lean body weight of the humpback is probably heavier for length than that of other baleen whales. Data on relative blubber thickness in the previous section indicated that humpback blubber is generally at least twice as thick as that in other balaenopterids. The earlier comparison between species of relative skull width in Fig. 5, showed that humpbacks have the greatest girth. Lockyer (1976) found that in a weight/length comparison with other rorquals, the humpback appeared heaviest, and this is apparent in Fig. 10. The lean body weight of the humpback is thus

413

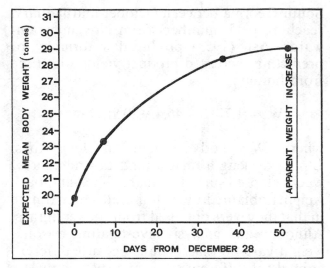

FIG. 21. – Expected body weight increase with time for a 40 ft humpback whale.

unlikely to be less than 10 tonnes which is slightly less than the expected mean weight for other species at this length (Lockyer, 1976). Matthews (1937) showed that peak catches of humpbacks at South Georgia occurred between November and January, coinciding with the information on arrival times given by Hinton (1925), so that most humpbacks probably arrive well before the end of December. The lean body weight prior to summer feeding is, therefore, more probably about 15 tonnes if the whale arrives around mid-December. Increase in weight appears to slacken off during February so that overall increases probably amount to about 15 tonnes. The whale, therefore, appears to double its body weight on the feeding grounds, where according to Dawbin (1966) it may remain for up to 5 1/2 months. This order of weight increase is not unusual in hibernating animals, as will be discussed later.

Blue and Fin Whales

The half-monthly mean body and tissue weight/bone weight ratios in blue and fin whales calculated from data in Appendix Table 1 are shown in Figs. 22 and 23. The

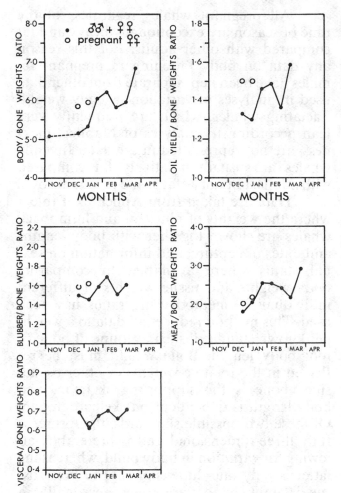

FIG. 22. – Half-monthly mean ratios of weights of body organs: bone weight for male and female blue whales.

variability shown is very great, although generally where the time scale is long, there is a definite upward trend in the mean ratio indicating a weight increase. There is a consistent drop during February for both blue and fin whales, and from the information on the pattern of migration of blue and fin whales discussed in an earlier section, this could well be the result of a secondary influx of lean whales, late in the season.

The order of increase in body weight, ignoring pregnant animals, could be about 35 % in blue whales (see Fig. 22) between November and March inclusive, and at least 12 % between December and February inclusive in fin whales in Fig. 23. However, the degree of

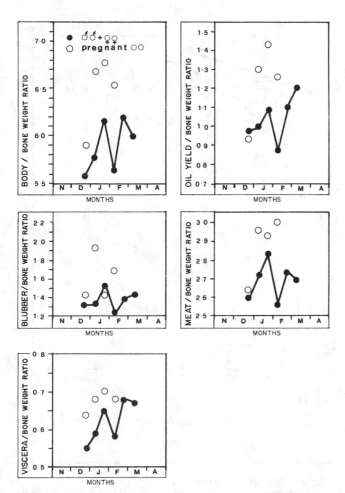

FIG. 23. – Half-monthly mean ratios of weights of body organs: bone weight for male and female fin whales.

fattening shown is incomplete in both blue and fin whales, because there is only a small amount of data available, and also because of the successive waves of immigration. Fattening definitely occurs, however, as is also shown by the increase in oil yield (data given in Appendix Table 1) in Figs. 22 and 23.

In order to assess the real fattening of these whales over the time periods shown in Figs. 22 and 23, with the small body of data available, data from all months have been combined and are assumed to refer to the same whale population and to be a random sample of this. The data then comprise a heterogeneous sample of fattened and lean whales, and whales undergoing fattening. These various categories of whales can be detected when cu-

mulative frequencies of weight ratios are plotted on arithmetical probability paper. Figs. 24 and 25 show this treatment of body weight data for blue and fin whales respectively. The data do not lie along a single straight line as would be expected for a normal distribution, but appear to form a complex of lines. The fattest and thinnest whales are, therefore, not covered by the variation within a single normal distribution, but constitute the lean and fattened conditions for which there are separate means and standard deviations. The central mass of data in Figs. 24 and 25 represent whales in varying stages of fattening. The lean and fat body weight/bone weight ratios can then be identified according to the method of Harding (1949) and be used to predict the body weights before and after fattening for different sizes of whales. These estimates of lean and fat conditions are shown in Fig. 26 and 27 for blue and fin whales.

The standard errors can be calculated from the standard deviations given in Figs. 26 and 27, and then applied to the body weight/bone weight ratios to determine whether fattening is significant. For blue whales lean and fat ratios are 4.66 ± 0.10 and 6.94 ± 0.43 respectively, and for fin whales, they are 5.07 ± 0.06 and 6.51 ± 1.08 respectively. The fattening is thus significant and averages 49 % of lean body weight in blue whales and 28 % in fin whales. The blue whale data may give a better indication of the order of body weight increase because of the longer time span of their data, and the consequent inclusion of individuals at the ends of the range of increase. Because blue and fin whales appear similar in their biology, it would seem likely that body weight increases in both these species are of the order of at least 50 % initial lean weight. This value is low by comparison with the value just calculated for the humpback whale. Rice and Wolman (1971) estimated a weight loss of 11-29 % of the body in the North Pacific gray whales between the southward and northward migrations. This estimate of weight loss was calculated indirectly from girth measurements which could

415

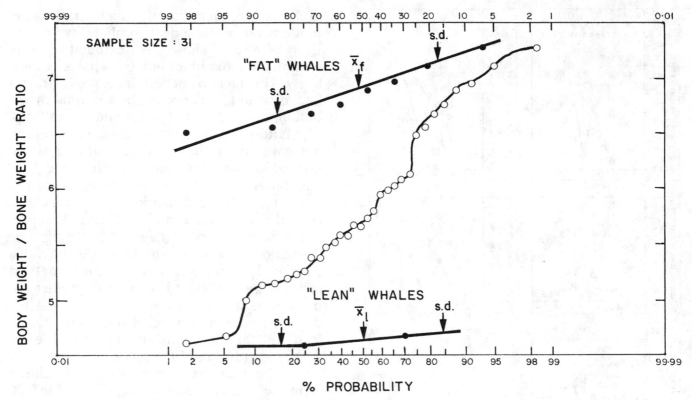

FIG. 24. — Cumulative frequency distribution of lean and fattening blue whales during the Antarctic feeding season.

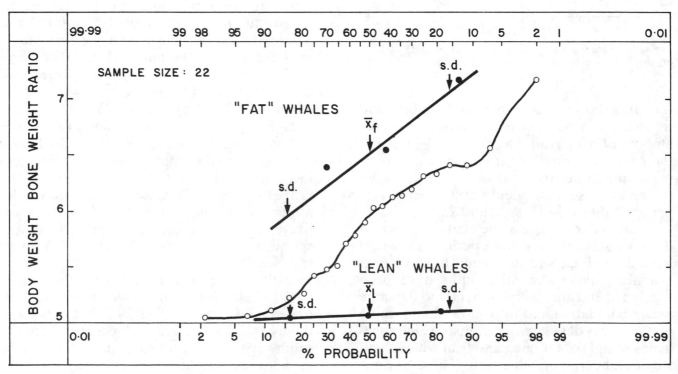

FIG. 25. — Cumulative frequency distribution of lean and fattening fin whales during the Antarctic feeding season.

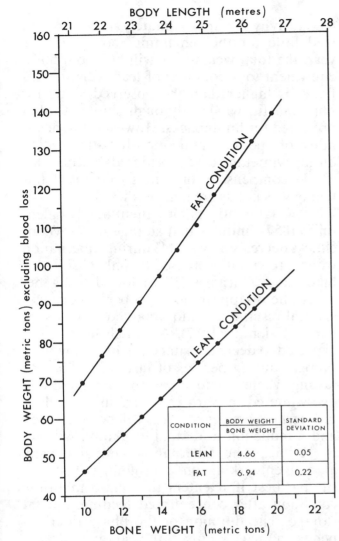

FIG. 26. – Estimated body weight in lean and fattened Southern Hemisphere blue whales.

occurrence was near the peak period of the southerly migration, and indicated that food might be taken during migration if it was available.

In hibernators, the maximum weight is observed immediately before hibernation. This is nearly twice the minimum post-hibernation weight in ground squirrels (Kayser, 1952). Kayser found a weight loss of 28.3 % in ten squirrels during 132 days of hibernation, and Kalabukhov (1956), Valentin (1857), Camus and Gley (1901) and Johnson (1928) mentioned that weight losses in bat, woodchuck, hedgehog and ground squirrel are 30-49 % during 127-163 days of hibernation.

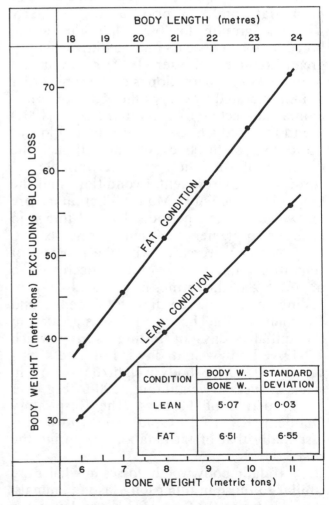

FIG. 27. – Estimated body weight in lean and fattened Southern Hemisphere fin whales.

be translated to weight values by using a formula relating girth and body weight, based on actual weight and girth data. This weight loss in the gray whale is equivalent to a weight increase on lean weight of 12-41 %. The gray whale, although in a different family to that of blue and fin whales, is known to undertake extended south-north migrations in the Northern Hemisphere, while feeding little, if at all during the migrations and winter months. Sund (1975) observed gray whales consuming small bait fish from a school off Monterey, California. He pointed out that the time of this

417

These losses are equivalent to 43-96 % of post-hibernation weight. The daily weight losses amount to about 0.22 % body weight: Valentin gives 0.21 % for the woodchuck, Johnson gives 0.28 % for the ground squirrel, Camus and Gley give 0.25 % for the hedgehog and Kalabukhov gives 0.21 % for the bat. Rice and Wolman estimated that the daily weight loss during migrations in the North Pacific gray whale represents 0.21-0.37 % body weight.

Herter (1933) and Mayer (1954) found that in the hedgehog and ground squirrel body weight increased noticeably at the breeding period and very markedly prior to hibernation. This pattern of fattening resembles that in Southern Hemisphere rorquals which are in prime fat condition prior to the northerly winter migration to lower latitudes where breeding mostly occurs and feeding activity is greatly reduced. Kayser (1961) believed that weight loss in hibernation is of the same order as that tolerated by every other homeothermic animal subjected to starvation. Noé (1901) found that hedgehogs can withstand a loss of up to 36 % of the body weight, although the duration of starvation may be variable according to environmental conditions and the amount of activity. Many hibernators are known to awaken periodically and forage in order to supplement their energy reserve.

Activity decreases with the progress of starvation, and heat evolution decreases to about 75 % of the former level after 14 days of fasting in steers according to Benedict and Ritzman (1927). The weight loss is greatest in the initial 3 days of fasting, during which 1.2-1.4 % body weight/day is lost, and subsequently, 0.5-0.7 % body weight/day is lost in steers of average weight range 590-670 kg, over a fast period of 14 days. The Respiratory Quotient (R.Q.) falls rapidly to 0.70 which is essentially that of fat katabolism, during the first 3 days.

If daily body weight losses in whales are similar to those in other long fasting animals, that is between 0.2 % and 0.3 % per day over a prolonged period, and blue and fin whales feed heavily for approximately 4 months and feed little for the remaining months of the year, the total weight loss will be about 50 %, equivalent to a doubling of lean weight. This figure is higher than the observed value for blue and fin whales although similar to that predicted for humpbacks. However, the possibility of sporadic and opportunistic feeding during winter must be considered, because this would compensate any energy deficit if fattening was not as great as predicted.

Kalabukhov (1956) mentioned Valentin's (1857) findings that storage fat in woodchucks decreased by 99.3 % during hibernation whilst brown fat decreased by only 68.7 %. Bibikov and Zhirnova (1956) found a lag between the disappearance of subcutaneous and visceral reserve fat, and obtained results similar to Valentin (1857). Although heat loss is expected to decrease, subcutaneous fat will disappear during periods of fasting, so that the fasting whale would be at an advantage if it overwintered in warmer waters in lower latitudes where heat energy loss will be minimized and a reduction in dermal insulation will not be serious. The hibernating animal or any animal embarking on a prolonged fast will survive best if it enters the phase in prime condition after concentrated feeding and fat storage. The duration of the phase then depends on the economy of the energy budget.

Distribution of Weight Increases in the Body

The lean and fattened conditions of the various body tissues can be assessed in the same way as for total body weight. Data on tissue weight/bone weight ratios from different months can be combined and the cumulative frequencies plotted on arithmetical probability paper in the same way as for total body weight. The mean lean and fattened conditions with standard deviations have been determined in this manner, and are shown in Figs. 28 and 29 for blue and fin whales. In all tissues the fattening is significant.

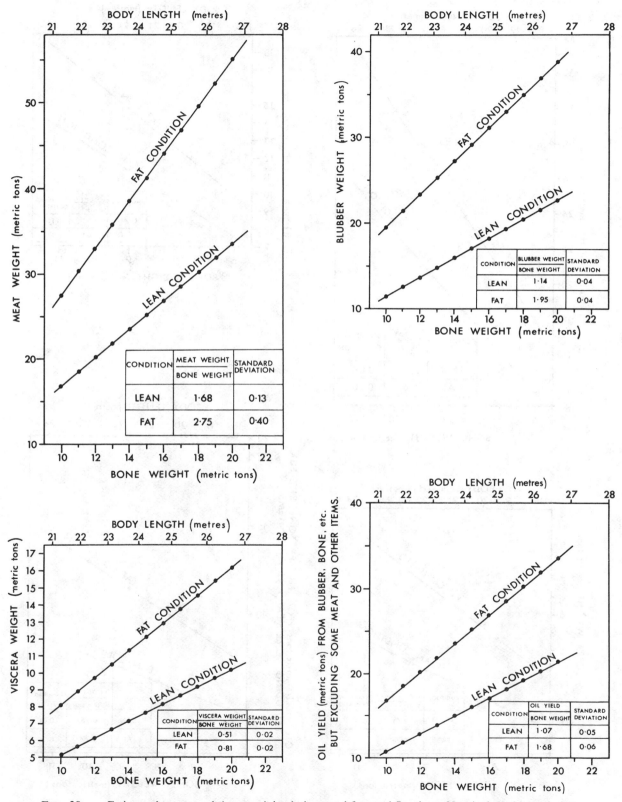

FIG. 28. – Estimated organ and tissue weights in lean and fattened Southern Hemisphere blue whales.

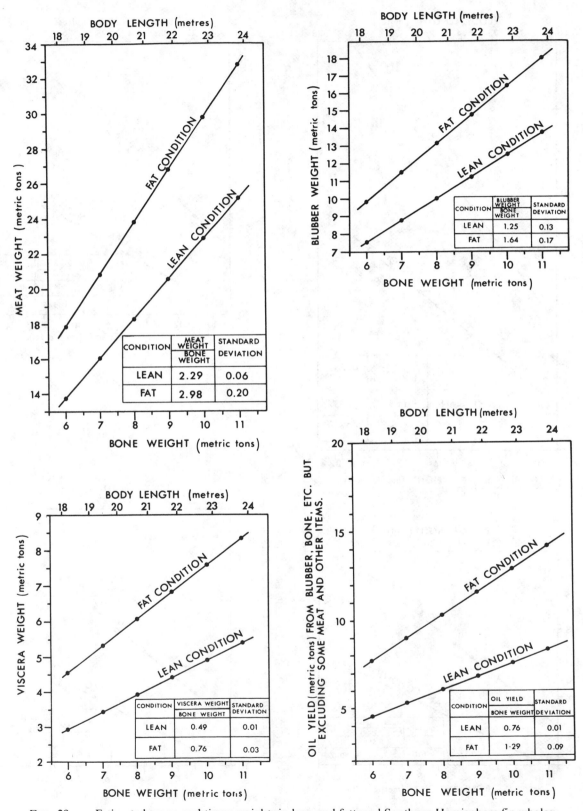

Fig. 29. – Estimated organ and tissue weights in lean and fattened Southern Hemisphere fin whales.

In Table 4 the overall distributions of weight increases in the body are summarized for blue and fin whales, and the oil yield increase is also given. It is clear from this that the greatest observed increases in weight occur in the internal musculature during the sampling period.

The data for fin whales cover even less of the summer season than the blue whales, as already stated, so that relatively lower estimates of tissue weight increases must be expected. These estimates are not necessarily precise measures of the increases, but serve to show that substantial weight increases do occur.

The weight increases in the individual tissues in Figs. 28 and 29 show that blubber content increases 71 % and 31 %, muscle 64 % and 30 %, and viscera 82 % and 59 % in blue and fin whales respectively. These increases are very substantial.

Data of the IOS on oil content of bone tissue from different regions of the body of a blue whale show that oil can constitute 56-69 % and water 8-13 % of the bone. As oil must presumably replace water in the bone, it is unlikely ever to exceed 80 % of bone tissue, because of the finite internal structure of bone. From

Table 4. Average increase in body weight during the Antarctic feeding season

	Blue			Fin		
	x^1	%	$\frac{x}{A}$	x	%	$\frac{x}{A}$
Lean body weight, A	4.66	100		5.07	100	
Fat body weight, B	6.94	149		6.51	128	
Net body weight increase (B-A)	2.28	49		1.44	28	
Subcutaneous weight increase:						
blubber	0.81	17		0.39	8	
Internal weight increase:						
muscle	1.07	23		0.69	14	
viscera	0.30	6		0.27	5	
Oil product increase	0.61	13		0.53	10	

[1] x: body, tissue or oil weight/bone weight.

Table 5. Mean proportions by weight of different body tissues in blue and fin whales

Species	Blubber	Muscle	Viscera	Bone
		(% of total body weight)		
Blue	27	39	12	17
Fin	24	45	11	17

Table 5 where the average tissue constitution of the body weight regardless of season is shown using data from the Appendix Table 1, it is clear that bone generally forms 17 % of total body weight. Therefore, oil content in the bone normally constitutes about 10 % of total body weight. Even if it were to increase, the net effect would probably be to lighten the bone density, because of the low specific gravity of oil. Bone weight, therefore, probably remains fairly stable although much energy can be stored within the bones where it cannot be accounted for in terms of weight increase. The important point is to know how much extra energy has been stored in the form of fat and oils during the summer. This is difficult to assess without routine chemical analyses of body tissues throughout the whaling season. The increases in weight in blubber, meat and viscera are thus assumed to be mainly due to fat deposition, especially since the unit oil product yield increase.

Food and feeding patterns

FOOD OF WHALES

The food source of whales is an important factor in the present study because the seasonal migration, growth and energy budget hinge on this. According to Marr (1962) the chief food of southern baleen whales is krill, originally a Norwegian term for the euphausiid *Euphausia*

421

superba which occurs in the polar waters south of the Antarctic Convergence (Mackintosh and Wheeler, 1929; Deacon, 1937; Marr, 1956, 1962 and Mackintosh, 1972).

This crustacean is not their sole food. Mackintosh and Wheeler (1929), Mackintosh (1942), Nemoto (1959, 1962) and Kawamura (1974) mentioned that *Parathemisto gaudichaudi,* an amphipod, is not infrequently taken by blue and fin whales when it is locally abundant, and Nemoto and Kawamura both showed that *Parathemisto* is an important food for sei whales. Mackintosh and Wheeler (1929) also mentioned that off South Africa *Euphausia recurva, Euphausia lucens* and *Nyctiphanes africanus* are taken by blue and fin whales, and Matthews (1937) mentioned that off Saldanha Bay, fish are sometimes eaten, particularly by humpbacks. Bannister and Baker (1967) reported that balaenopterid whales off Durban were found to be feeding on *Thysanoessa gregaria* and *Euphausia recurva,* along with various other species of euphausiids. Best (1967) reported that copepods and euphausiids are taken by sei whales off South Africa.

Other organisms retrieved from the stomachs of fin and blue whales in the Antarctic include fish ranging in size from 5 to 46 cm in length, and a jellyfish 38 cm in diameter (Mackintosh, 1942). However, these creatures were present with large quantities of krill and were probably taken by chance rather than choice.

Mackintosh (1942) suggested that *Euphausia crystallorophias* may replace *Euphausia superba* in the Ross Sea area where it occurs abundantly in shoals, and Marr (1956) reported that blue and minke whales feed on this former species. Nemoto and Nasu (1958) found that fin and humpback whales consume *Thysanoessa macrura* in Areas I and VI (120°-170°W) of the Antarctic.

Sei whales which tend to favour the sub-Antarctic do not appear to be very selective in their food. Matthews (1932, 1938), Rayner (1935) and Nemoto (1959) mentioned that sei whales will take *Munida gregaria* and *Grimothea* off Patagonia, Tierra del Fuego and the Falklands. Mackintosh (1942) and Kawamura

(1974) emphasized the importance of *Calanus tonsus* and other copepods in the diet of sei whales, many of which rarely venture far south into the Antarctic. Nemoto (1962), Doi, Ohsumi and Nemoto (1967) and Kawamura (1974) have shown that *Euphausia vallentini* is an important food source in the sub-Antarctic for blue, fin and sei whales. Ohsumi, Masaki and Kawamura (1970) found that the minke whale feeds on *Euphausia superba,* and also on *Euphausia spinifera* and *Calanus tonsus* as far south as 61°27'S.

In conclusion, the rorquals present in the Antarctic, particularly far south, feed mainly on *Euphausia superba,* but feed, perhaps opportunistically, on other species of Crustacea where these replace *Euphausia superba,* in lower latitudes. Generally, rorquals take shoaling Crustacea in any area when they are feeding (Kawamura, 1974). The true krill, *Euphausia superba,* would still appear to be the major food species for blue, fin and humpback whales, the other types being supplementary.

INCIDENCE OF FEEDING

Summer Months

Southern Hemisphere rorquals, as seen above, migrate south to the Antarctic in the spring and remain in polar seas throughout the summer. Mackintosh and Herdman (1940) and Mackintosh (1972, 1973) illustrated the mean pack ice edge in the Antarctic spring and summer when the krill, *Euphausia superba,* and other plankton are largely exposed by the retreating ice. The rorquals are thus able to penetrate the ice-free areas and feed.

Matthews (1937) summarized data on the feeding incidence in humpbacks between 1925 and 1931. At South Georgia 82 % of 33 whales had been feeding and krill were present in the stomachs. An analysis of data collected in the Antarctic between 1951 and 1962 by the IOS from 52 whales of both sexes, indicates that 96 % had food in the stomachs.

Mackintosh and Wheeler (1929) found that the incidence of feeding in blue and fin whales is high at South Georgia. Their findings are summarized in Fig. 30. The incidences of feeding in blue and fin whales respectively in the Antarctic Areas II and III, are shown for the post-war years in Figs. 31 and 32. The level of incidence varies from year to year, but generally by only small amounts. Individual seasons have, therefore, been combined. The feeding incidence for both blue and fin whales is about 85 %, and is similar to the data shown in Fig. 30. These appears to be no real difference in feeding incidence between males and females and very little, if any, between mature and immature whales, in the Antarctic.

Gambell (1968) showed that the majority of sei whales caught and examined in the Antarctic, were feeding. Gambell's data for both sei and fin whales in the Antarctic are shown in Table 6. Matthews (1938) mentioned that the southward migration of sei whales appeared to be correlated with feeding, although Matthews (1932), Rayner (1935), and more recently Kawamura (1974) commented on the ready feeding of sei whales in lower latitudes of the sub-Antarctic, where feeding incidence may average between 42 % and 54 % in a day, according to the area.

Mackintosh and Wheeler (1929) pointed out the difficulties of using data on the presence of food in the stomach as an indication of feeding activity, because whales occasionally vomit when captured. However, Kawamura (1971) found that of 434 sei whales observed during capture, only 1.1 % vomited, and similarly, of 84 fin whales only 0.78 % vomited. Mackintosh and Wheeler also mentioned that it

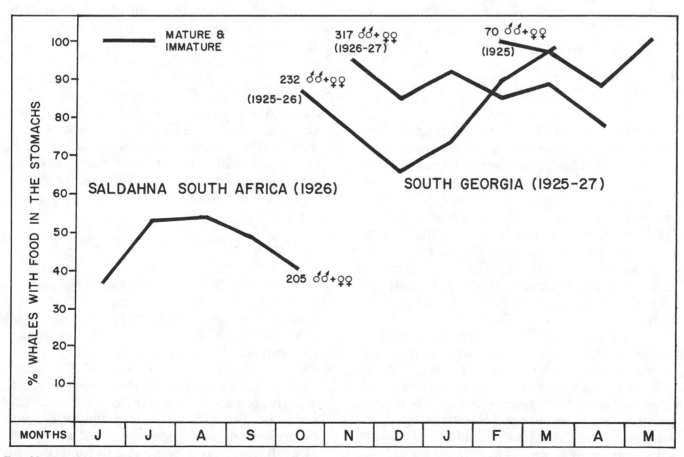

FIG. 30. — Incidence of food in the stomachs of Southern Hemisphere blue and fin whales, according to data from Mackintosh and Wheeler (1929).

Table 6. Feeding incidence in sei and fin whales

Locality		% whales' stomachs containing food		Reference
		Sei	Fin	
Durban Cape Province	S. Africa	4-37	11-42	Bannister & Baker (1967)
		47-56	30-46	Best (1967)
South Georgia	Antarctic	50 in males 65 in females	37	Gambell (1968)
0°-70°W		92	82	Gambell (1968)

is often only the first stomach chamber which is examined whilst the other stomachs and intestine may be full. The presence of food in the stomach is, therefore, only a guide to very recent feeding which appears to be very common in the Antarctic seas.

Winter Months

Matthews (1937) found that, unlike the humpback whales examined at South Georgia, those at Saldanha Bay showed little evidence of recent feeding; only 11.8 % of 17 whales had been feeding, and these contained fish. Dawbin and Falla (1953) and Dall and Dunstan (1957) also gave evidence that the humpback feeds little during winter when in the coastal waters of lower latitudes. Between 1961 and 1963, of 32 humpback whales examined at Durban by biologists of the IOS, only 12.6 % had food in their stomachs.

In Figs. 31 and 32 the feeding incidence of blue and fin whales is shown for areas off South Africa. The immature whales appear to feed more readily than the mature whales, but even they seem to be feeding at only about half the frequency found in high latitudes. The average feeding incidence in adults off South Africa probably amounts to about 20 % or less. In Fig. 30, data of Mackintosh and Wheeler

(1929) are shown, but the breakdown of these by species and maturity in Fig. 31 is perhaps more relevant because of the difference between the mature and immature categories. Information from Bannister and Baker (1967) indicates that the incidence of feeding is comparatively low around Durban, in both fin and sei whales (see Table 6). However, data of Best (1967) indicates that feeding incidence is generally higher off Cape Province, thus indicating, perhaps, that where food is available in lower latitudes, it is readily taken.

Seasonal Feeding Pattern

The pattern of feeding during the year appears to be seasonal for humpback, blue and fin whales, and probably to a certain extent for sei whales. The incidence of feeding during winter in low latitudes is generally considerably less than 50 %, and is probably less than 20 % amongst adult whales. The level of incidence of feeding in low latitudes is clearly very variable according to locality and food availability. On the other hand, feeding during the summer in high latitudes appears very successful with frequencies of 85 % or so in the polar waters. However, sei whales present in sub-Antarctic latitudes during summer may not feed at quite such high frequencies as those in the Antarctic.

Quantities and Rates of Feeding

In the Southern Hemisphere there is much evidence to show that of all feeding whales, those in lower latitudes take little food compared with those in the Antarctic waters. Mackintosh (1942) showed that amongst blue, fin, sei and humpback whales off Durban in 1930, an average of 96 % had empty stomachs and the remainder's stomachs contained little food. In Table 7 it is clear that off South Georgia stomachs containing food are generally full whereas off South Africa food contents even in feeding whales are very small.

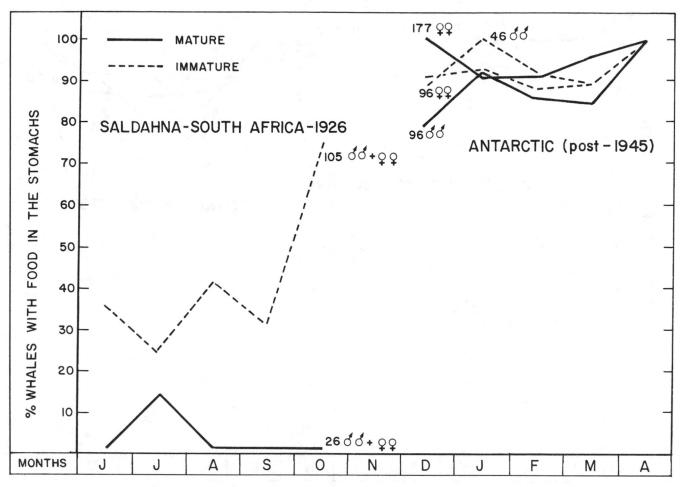

FIG. 31. – Incidence of food in the stomachs of Southern Hemisphere blue whales.

Table 7. A summary of fullness, determined qualitatively, of stomachs of baleen whales according to Mackintosh and Wheeler (1929)

Locality	% stomachs containing food	% food-containing stomachs	
		full/ moderately full	nearly empty
South Georgia (Oct. to May)	87	73	27
South Africa (June to Oct.)	43	24	76

In the Antarctic, some estimates of maximum fullness of the first stomach have been given from actual weighings. In the sei whale for example, Brown (1968) found 305 kg of krill in a 14.7 m female, and 175 kg in all three stomachs of a 15 m male. Zenkovich (1969) mentioned that 300 kg of food is a usual amount in Antarctic sei whales. In the sub-Antarctic, Kawamura (1970, 1974) found a maximum fullness of about 180-230 kg of food in 13-17 m sei whales.

Off South Africa, usual amounts are considerably less. Gambell (1968) found that 16 kg of shrimps represents observed maximum

425

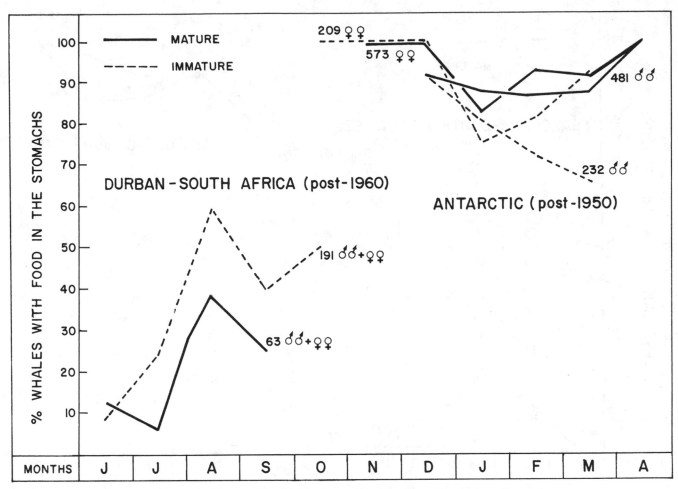

FIG. 32. – Incidence of food in the stomachs of Southern Hemisphere fin whales.

fullness of the stomach of a 14.1 m sei whale at Durban. Best (1967), however, found that 59 kg represents the largest meal for a sei whale off Cape Province. Frequently, considerably less quantities are observed, 3-5 kg being a moderate meal for a 14-15 m sei whale off Durban (Best, 1967 and Gambell, 1968). In any case, the largest meal during winter can only amount to a fraction of that taken during the summer feeding season. Gambell (1968) found that the food quantity consumed in the Antarctic summer amounted to about 5 to 19 times that taken during winter off South Africa. Thus, considering that the incidence of feeding in summer is at least twice that in winter, as already discussed, the average amounts consumed off South Africa probably amount to

about 10 % or less of that in the Antarctic.

The largest quantities held by the stomach of Antarctic fin whales amounted to 130-885 kg food in 18-23 m length animals (Kawamura, 1970, 1974). Sal'nikov (1953) maintained that a full stomach of a 21-22 m fin whale holds 800-900 kg of food. In the North Pacific Betesheva (1954) found that 50-250 kg is a normal meal for fin whales and Ponomareva (1949) mentioned that 450 kg of food is the usual quantity in 18-19 m fin whales. Zenkovich (1969) estimated that a meal in the Antarctic for a fin whale approximates to 700 kg. From observations and calculations Zenkovich also gave estimates of a meal of 1 000 kg for blue whales (an estimate also given by Collett, 1911-12, for Atlantic blue whales) and 500 kg for humpback whales in the Antarctic. A previously unpublished record of a 24.7 m pregnant blue whale carrying 400 kg of krill was made aboard the *Eastern venturer* in March 1946 in the Antarctic, and this quantity probably represents only a moderate amount. Betesheva (1955), Nemoto (1957) and Kawamura (1971) found that the quantity of food discovered decreases with chasing of the whale. This is because of eruction of stomach contents, and also gradual digestion and cessation of feeding during chasing, which all contribute to observing less than the potential maximal food quantities.

Kawamura (1971) gave some evidence that in the North Pacific feeding is likely to occur maximally only twice daily at intervals of 10-15 hours or at any rate the equivalent of two replete meals a day, and more likely only one full meal a day; the numbers of fin, sei and sperm whales caught with full stomachs is reduced to 70 % in the first half of the day and 50 % by mid-afternoon. He believed that 25-30 % of the stomach contents are digested in 5 hours, so that the remaining food would be digested in the following 14-15 hours. He had monitored chased whales closely and had been able to allow for vomiting.

Kawamura (1970, 1974) had evidence that in the Antarctic, the sei whale feeds chiefly in the morning and sporadically throughout the day, so that feeding may be a semi-continuous process until evening, with a clear feeding peak only once a day. Nemoto (1959) suggested that feeding occurs more frequently in the morning and evening in the North Pacific and Antarctic. Off Cape Province, Best (1967) found that the highest incidence of feeding occurred during the day. Ohsumi, Masaki and Kawamura (1970) found that for minke whales in the Antarctic, feeding activity is at a peak in the early morning and gradually decreases towards mid-day, and increases again slightly in the evening. They believed that the feeding activity of the whales is regulated by the diurnal migrations of the food planktons rather than by the whales themselves, so that feeding might be continuous if food was readily available at all times.

The consensus of information, therefore, indicates that whilst feeding activity may take place at all times of the day, feeding generally reaches a peak once or perhaps twice a day, and is correlated with the abundance of food plankton at these times.

It is worth noting the stomach structure in relation to food quantities. As already noted, the stomach comprises three chambers: forestomach deficient in secretory glands, where food is stored and mechanically ground up; main stomach equipped with pepsin and hydrochloric acid glands; pyloric stomach equipped with digestive glands (Slijper, 1962). One school of thought suggests that whales are evolved from ungulate animals which also have multiple chambered stomachs. The whale, however, is a carnivore in who, the development of such a stomach may have allowed it to adapt to feeding continuously and massively when food is abundant and thus to exploit the situation to its fullest.

When whales are examined for stomach contents, frequently only the forestomach is examined. Certainly only in the forestomach can the exact prey species eaten and size distribution of these species ever be identified before digestion attacks the food in the other stomach chambers. However, examination of the forestomach only for evidence of recent feeding and quantities taken may be misleading as already discussed.

The knowledge of the maximum quantity of food a stomach can hold is only a step toward estimating total daily food intake. Availability of food, digestion rate and energy requirements are all controlling factors in daily food consumption.

Daily food consumption rates are even more uncertain than maximal quantities of food contained in the stomach. Klumov (1961) claimed that 1 000 to 1 500 kg of food is consumed daily by an 18-19 m fin whale in the North Pacific. Zenkovich (1969) stated that blue whales take 4 000 kg, fin whales 2 800 kg and humpbacks 2 000 kg of food daily, if they feed about 4 times daily as he supposed in the cold food-rich areas of the Antarctic. He mentioned that sei whales may feed up to 5 times daily so consuming 1 500 kg of food in the Antarctic. These figures apply during the 120 days feeding season for blue, fin and humpback and 100 days for sei, and also assume that digestion of food in the stomach after a meal requires 3-4 hours. Tomilin (1967) estimated that the digestion rate in dolphins is fast and that a full meal would take 14 hours to reach the large intestine from the mouth. There is no reason to suppose that Cetacea feeding on a carnivorous diet would differ much in their digestive capabilities. Kawamura (1974) at least, believed that total digestion in sei whales would not differ greatly from that in dolphins.

There are further data on quantities of food in whale stomachs from the Antarctic and North Pacific, given by Collett (1911-12), Heyerdahl (1932), Nishimoto, Tozawa and Kawakami (1952), Betesheva (1954, 1955), Nemoto (1959, 1962), Sergeant (1969), Shevtsov (1963) and Tomilin (1967). However, these observations do not necessarily indicate the potential maximal feeding capacity of Cetacea, but more probably indicate the variability in feeding capacity. Sergeant (1969) produced an interesting and possibly meaningful correlation between heart weight, body weight and maximal food weight consumed in unit time. He found that the food weight/body weight ratio for adult dolphins and small whales approximates 4 %, also that the heart weight/body weight ratio was about a tenth of the food weight/ratio, 0.4 %. He predicted that food weight/body weight ratios amongst the large rorquals, during the summer feeding season, would be about 4 %.

Gray (1964) made the observation that a full grown captive *Tursiops truncatus* weighing between 225 and 275 kg normally consumes about 7-9 kg daily (about 3 % body weight), and even if fed to capacity five times daily will only consume about 11.5 kg of fresh herring and eels. However, he recorded a captive 180 kg female *Tursiops* being given 20 kg of fish a day (about 11 % body weight), and another

taking a similar amount. Rowe (1968) commented that two adult female dolphins in captivity consumed about 11.5 kg of fish daily. Newman and MacGeer (1966) mentioned that a subadult male killer whale, weighing 1 040 kg, consumed 45-90 kg of salmon and cod daily (about 6 % body weight) whilst in captivity. Wahrenbrock *et al.* (1974) mentioned that a captive 10 months old gray whale calf weighing about 6 150 kg consumed up to 820 kg of squid daily, representing about 13 % of body weight. This seems a very great quantity compared with dolphins, but may be so only in newly weaned whales, and, therefore, inapplicable to older animals. Sergeant (1969) summarized similar data for a variety of dolphins and small whales, which are similar to those just mentioned, and found that 4 % body weight is generally consumed daily. Food requirements of captive Cetacea are likely to be variable, however, according to the amount of activity demanded by the individual dolphinaria. The usual daily food intake in dolphins and small whales is therefore about 4 % of body weight or 40 g of food per kg of body weight.

Klumov (1963) estimated that 30-40 g of food per kg live body weight is consumed daily by baleen whales during the feeding season. This estimate, and that for small Cetacea, are comparable with estimates of approximately 30-35 g/kg/day in man (Brody, 1968) amounting to 55 kcal/kg/day. The captive elephant consumes 13 g/kg/day amounting to about 56 kcal/kg/day according to Benedict (1936), and 10.5 kg/day equivalent to 42 kcal/kg/day in the wild according to Petrides and Swank (1965). Cattle according to Brody (1968) require 20-32 kcal/kg/day representing 5-8 g/kg/day of digestible nutrients (dry weight) for maintenance.

Brody (1968) discussed the annual food consumption in man, cattle, elephants and chickens, which represents between four and nine times the body weight, the food being in a crude state. Brody also showed that relative energy needs decrease with an increase in body size, the daily requirement of digestible nutrients being related to body weight by a power of approximately 0.73 of body weight in most animals. The whale, therefore, might be expected to consume annually a quantity of foodstuff nearer the lower multiple of four times the body weight. If the whale were to consume approximately 30-40 g food/kg/day during the feeding season of 120 days, equivalent to 12 g food/kg/day over a year, about 3.5-5 times the body weight would be eaten in a year.

METHODS OF FEEDING
IN BALEEN WHALES

Nemoto (1959, 1968 and 1970) arbitrarily classified whales into three feeding categories: swallowing, skimming, swallowing and skimming, according to the anatomical detail of the jaws and baleen type. The lattermost group comprises the sei and gray whales. Skimming types are all the right whales, and the swallowing types which are of most interest here, include the blue, fin, humpback, minke and Bryde whales.

According to Nemoto (1959, 1970), the swallowing method of feeding in rorquals is to take food into the mouth by gulping plankton from concentrated swarms. Sal'nikov (1953) and Kawamura (1974) stressed the necessity of concentrated food swarms for the feeding of rorquals. The planktonic foodstuff is retained in the mouth whilst water is strained off through the fringed baleen plates in the gums. Rorquals other than sei, however, rarely if ever swim open-jawed sifting plankton as do the skimming type right whales.

The baleen plates are keratinized epithelial structures with a central cortex of hollow tubules (Ichihara, 1966). Teeth are present only as rudimentary buds in the foetal gums. On average, about 300 plates are present in the upper jaws of balaenopterids, although there is considerable variation in the number, both within and between species. In Figs. 33 and 34 some recent data of maximum lengths and widths of baleen plates collected between 1960

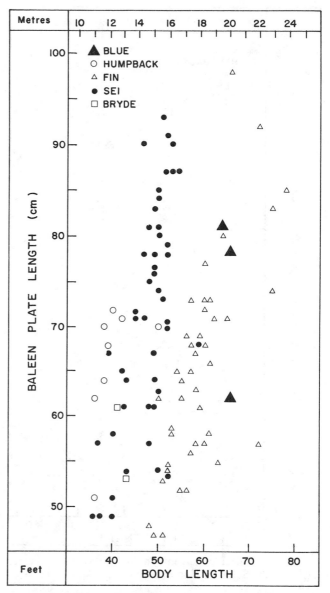

FIG. 33. – Baleen plate length in different species of baleen whales in the Southern Hemisphere.

and 1965 from South Georgia, on board a factory ship in the Antarctic, and at Durban, are shown. The lengths shown are total lengths including the portion normally embedded in the gum, and the fringe, and because the measurements have been made in this way, they are consistently higher at a given body length than the results of Mackintosh and Wheeler (1929) and Matthews (1937, 1938). For any given body length, the largest species,

429

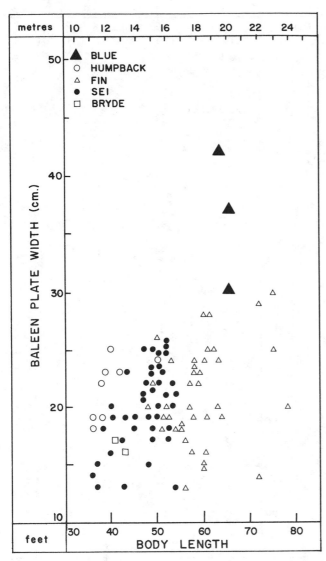

FIG. 34. – Baleen plate width in different species of baleen whales in the Southern Hemisphere.

of flesh decreases with an increase in body weight, W:

$$M = 70.5W^{-0.266}$$

where W is in kilogrammes, M is kcal/kg.

The normal sizes of krill retained by these baleen plates are 20 to 65 mm in length, the smaller sizes normally escaping through the baleen filters (Heyerdahl, 1932; Marr, 1962). Laurie (1933) and Klumov (1961) believed that baleen whales do not dive deeper than 100-130 m in search of krill which is not generally abundant below this depth.

According to Klumov (1961), Slijper (1962) and Nemoto (1970), once the mouth is full the jaws close and the buccal cavity is reduced in dimension in order to expel water through the baleen filtration system. The floor of the mouth is strongly muscular and effects alterations in internal mouth volume with the aid of the tongue; the longitudinal ventral grooves under the throat may also play a part in changing the shape of the floor of the mouth by the ability to expand and contract. The tongue which is fixed to the floor of the mouth rises and in some way pushes food down the oesophagus. The diameter of the latter, which may be up to about 26 cm, can easily accommodate whole items of food much larger in size than krill, such as squid and herring-type fish.

Estimation of Swallowing Volume in Rorquals

Williamson (1972) and Cousteau and Diolé (1972) have some photographs of balaenopterid whales swimming underwater, showing the true head profile in life. In Fig. 35 three photographs of the minke whale, observed by Dr G.R. Williamson in 1970 off Japan, are reproduced and show in A and B the typical appearance of the head whilst swimming. In C the animal is dead and the lower jaw has sagged open displaying the baleen plates in the manner that might be ex-

blue and fin, carry shorter and narrower baleen plates than the corresponding sized animal of the smaller species, sei and humpback. The length and width of the baleen plates must be a function of the filtration capacity, and this may be a reflection of comparatively higher metabolic rates and thus higher food requirements per kilogramme of flesh needed by smaller animals. Brody (1968) formulated the rule that the amount of energy, M, expended in basal metabolism per kilogramme

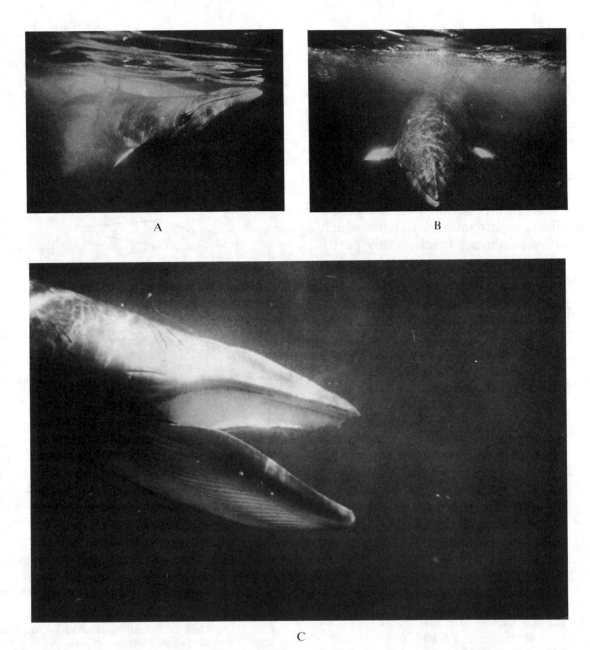

A
B
C

FIG. 35. — Photographs of a minke whale, by courtesy of Dr G.R. Williamson, showing in A, the side view of the head with contracted throat grooves; B, the dorsal view of the rostrum; C, the side view of the head with jaws gaping, displaying the baleen filters.

pected during feeding. The most significant features of the balaenopterid head are as follows:

(1) the expanding muscular sac below the scooplike jaw must effect alterations in the internal mouth volume;

(2) the rigid narrow upper jaw fits into the lower jaw like a lid with the baleen plates housed inside the lower jaw sac;

(3) the long baleen plates at the rear of the upper jaw gradually taper to a very short length at the jaw tip so that water can

431

easily enter the mouth without requiring the jaws to open widely;

(4)   the long baleen plates at the posterior of the mouth must ensure that water flows continuously into the mouth whilst swimming forward.

The skeletal parts of the jaws and the baleen plates represent the constant immutable dimensions of the mouth. The floor of the mouth in life is a soft muscular tissue which can expand and contract (presumably partly by using the ventral grooves and the tongue) so altering the capacity of the lower jaw space. Nemoto (1959) found that the ventral grooves of rorquals usually extend 45-58 % back along the body from the head so permitting much distension.

In Figs. 36 and 37 the profiles and cross sections respectively of the head and lower jaw sac are shown. These are constructed from measurements of the lower jaw bones of a minke whale in the British Museum (Natural History), shown in Fig. 36, and from photographs by Dr G.R. Williamson; also from a photograph of a dead whale initially captured by the *Anglo-norse* in January in a position 61°S, 27°W. The photograph of this dead minke whale which was suspended vertically by the tail stock, was taken by Mr J.E. Hamilton. The original negative is no longer available for reproduction. However, the profile of the whale has been outlined in Fig. 36. The sag in the lower jaw of this whale has been taken as some kind of indication of the potential expansion of the sac, although the actual shape compared with that in life may be distorted because of the weight of internal organs bearing down on the throat region. Until reliable evidence on the true shape of the lower jaw of whales during feeding is available, the lower jaw shape of the dead whale will be used as an indication of the possible shape. The profile C in Fig. 36 has therefore been taken to illustrate the degree of distension of the lower jaw sac when swallowing and filtering. The internal lower jaw space or swallowing/filtering vol-

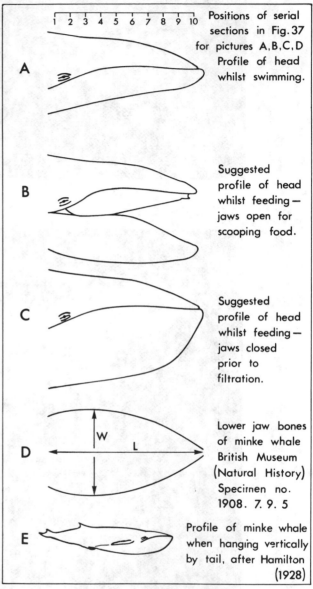

FIG. 36. — Generalized outline diagrams of the shape of the head of a minke whale, showing in A, B and C, the throat in the contracted state, and also expanded as anticipated when feeding. The lower jaw is also outlined in D, to the same scale, to show the length and breadth of the mouth. In E, the profile of a minke whale photographed by Hamilton in 1928 is shown.

ume can then be estimated from the difference between mouth volumes of the whales shown in A and C of Fig. 36.

It is worth noting that long before photographic evidence was available demonstrating the normal contracted shape of the lower

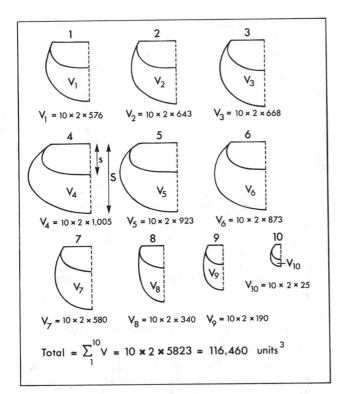

FIG. 37. — Transverse serial half-sections of the lower jaw sac in a minke whale. The positions of the numbered sections are shown in Fig. 36. The terms used are as follows: s = the depth of the jaw sac as in Fig. 36, A; S = the depth of the jaw sac as in Fig. 36, C; v = the estimated volume of the jaw sac between adjacent serial sections; each serial section is 10 units apart from adjacent ones. The text gives fuller explanation.

jaw of rorquals, Mr J.E. Hamilton had already described to the "Discovery" Committee in 1928, in some detail in his log, that the head shape of the minke whale during swimming, was very streamlined and quite unlike the baggy-throated creatures depicted in drawings at that time.

The serial sections of the lower jaw sac in Fig. 37 correspond to the positions indicated in Fig. 36. The difference in mouth volume of the contracted and expanded lower jaw sac, has been calculated assuming that the overall jaw length is 100 units and that the sectional areas are at intervals of 10 units commencing 5 units and ending 95 units back from the jaw tip. The approximate overall volume difference calculated from Fig. 37 is, therefore, $10 \, \Sigma_1^{10} \, v$, where v is the space measured and contained within

the two sac outlines in Fig. 37 at each interval of 10 units. This swallowing/filtering volume, $10 \, \Sigma_1^{10} \, v$, found to be 116 460 units from Fig. 37, is related to maximum jaw length L, maximum jaw width W, and (S-s) the difference in maximum depth of the sac at the point of greatest jaw width as follows:

$$1)^1 \quad 116 \, 460 = K \, (100 \cdot 54.5 \cdot (47\text{-}21)),$$

therefore

$$2) \quad 10 \, \Sigma_1^{10} \, v = 0.82 \, (L \cdot W \cdot (S\text{-}s)).$$

Predicted filtering/swallowing volumes for different jaw lengths are shown in Fig. 38. This assumes that the jaw proportions remain fairly constant as the jaw length increases.

---

[1] Measurements taken from Figures 36 and 37.

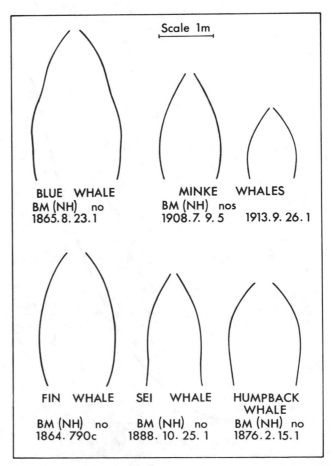

FIG. 38. — Scale outlines of the lower jaw bones of rorqual specimens housed in the British Museum of Natural History.

In Fig. 39 the jaw shapes, based on measurements of specimens in the British Museum (Natural History) of several rorquals have been drawn to scale for comparison. It is clear that whilst the rostral shape may differ, lower jaw shapes are very similar in all rorquals except in the humpback where the jaws are immensely wide.

In Table 8, the relationship between jaw length and width is shown for the jaws drawn in Fig. 39. It is clear that the jaws of minke, blue and fin whales are similar in shape, whereas those of humpback whales are relatively broader and those of sei are narrower. The humpback whale is, of course, in a different genus, and the sei whale is observed to feed in a slightly different manner from the

Table 8. Relationship of jaw dimensions in different baleen whales

Species	Maximum jaw width as % total jaw length	Jaw width at midlength as % half jaw length
Humpback	71.5	130
Blue	58	102.5
Minke	56.5	103.5
Fin	52	110
Sei	52.5	93.5

other rorquals. If the jaw lengths in Fig. 38 are extrapolated the corresponding calculated mouth volumes are a probable indication of jaw capacity in the blue and fin whales.

The average jaw lengths and jaw widths in relation to body lengths in different rorquals are shown in Table 9. These figures are based on the same data used in the body proportion analyses in Figs. 4 and 5 given earlier. From a cross-reference between Table 9 and Fig. 38, swallowing/filtering volumes in different lengths of minke, blue and fin whales can be estimated.

If during a feeding cycle the whale opens the mouth, expands the buccal cavity so drawing water in by vacuum effect, closes the jaws, contracts the buccal cavity and expels the water through the baleen filters, and then swallows, there is no need for the whale to swim very fast. However, this method of feeding requires a very dense concentration of plankton. Klumov (1961, 1963) has mentioned the problems of such feeding in waters where production is low and of the low yield of food per filtration – even though he estimated that the total mouth volume open during feeding is 6 m³ in 18-19 m fin whales. Other estimates of average swallowing volumes for baleen whales are 10-15 m³ (Nemoto, 1970) and 4.5 m³ (Fraser, unpublished, quoted in Marshall and Orr, 1955, verbal comm.).

Klumov (1961) reckoned that in waters where production is low the whale may sift food while swimming along, although there

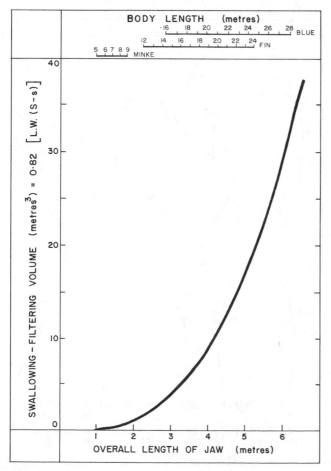

FIG. 39. – Calculated swallowing/filtering volume in different sizes of rorquals.

Table 9. Average length (snout to gape) and width (gape to gape) in metres of lower jaws of baleen whales, males and females combined [1]

Body length (metres)	Blue		Fin		Sei		Humpback	
	jaw length	jaw width	jaw length	jaw width	jaw length	jaw width	jaw length	jaw width
	(in metres)		(in metres)		(in metres)		(in metres)	
8- 9							1.83	1.36
9-10							2.34	1.69
10-11							2.40	1.87
11-12					2.31	1.19	2.80	1.91
12-13			2.19	1.23	2.37	1.29	2.98	2.15
13-14			2.39	1.24	2.62	1.43	3.23	2.19
14-15			2.55	1.39	2.86	1.51	3.49	2.31
15-16			2.84	1.59	2.95	1.57		
16-17	2.79	1.79	3.09	1.71	3.23	1.66		
17-18	3.09	1.90	3.38	1.83	3.42	1.71		
18-19	3.30	1.97	3.68	1.97				
19-20	3.54	2.13	3.95	2.06				
20-21	3.79	2.27	4.13	2.16				
21-22	4.08	2.44	4.26	2.26				
22-23	4.37	2.51	4.55	2.43				
23-24	4.72	2.61	4.87	2.40				
24-25	4.84	2.80	5.13	2.50				
25-26	5.21	2.93						
26-27	5.25	2.98						
27-28	5.39	3.01						
28-29	6.10	3.30						

[1] Data from Mackintosh and Wheeler (1929) and Matthews (1937, 1938).

must be a critical maximum swimming speed to avoid turbulence. Gunther (1949) mentioned instances in the Antarctic where whales have almost certainly been feeding and swimming along slowly on one side rather than upright. The reason for this is not known, but could possibly be a function of the density and depth of a krill swarm and also the time of day which must influence the dispersion of the plankton. The whale swimming on one side is more likely to zig-zag laterally through a swarm rather than up and down, an advantage if the depth of the krill swarm is very reduced as it may be at certain times during the day.

Gill and Hughes (1971) have observed a 13.3 m female sei whale feeding on saury, Co-lolabis saura, off California in the 2-1/2 hours prior to capture. The saury were present in large schools at the sea surface. The sei whale appeared to approach the saury from 10 m below the surface at a shallow angle of ascent whilst maintaining a 3 knot (ca 5.5 km/h) swimming speed. Once beneath the saury, the whale rolled to the surface amidst the dispersing sauries. After capture, the observations of feeding behaviour were confirmed by the presence of 227 kg of saury in the three stomachs, those in the first stomach being very fresh.

Klumov (1961) estimated that feeding fin whales swim forward at 6 km/h or less. Other estimates of speeds of swimming are given in Table 15 in a later section on metabolism. The most likely swimming speed for any feeding rorqual is probably in the range 2-6 km/h, the slower speed possibly occurring in dense shoals. The concentration of food in the water must be critical for efficient feeding by filtration.

Klumov (1961) based his calculations of feeding rates of fin whales on the observation that plankton concentrations in Kurile waters were only 500 mg/m³ in 1956. He came to the conclusion that this density was insufficient for efficient feeding of rorquals. Nemoto (1970) quoted the findings of low euphausiid concentrations in the North Pacific of 100-200 mg/m³, and maximum concentrations in the North Pacific of 2 000 mg/m³ according to Bogorov and Vinogradov (1956). Even in the Antarctic the standing stock only averages 0.75-1.5 g/m³ according to Nemoto (1970). In a personal communication, Dr N.A. Mackintosh provisionally estimated that the density of certain swarms of krill observed and sampled in the Weddell drift in 1931 amounted very roughly to 2.4 kg/m² of swarm surface area, say 2 kg/m³, in swarm 1.22 m (4 ft) thick. He was uncertain whether there is any uniformity in density throughout a swarm. It is not known what factors induce close swarming behaviour in krill, and what factors dispel it.

Supposing Mackintosh's estimate of 2 kg/m³ density was representative of typical swarms, then the potential quantity of food filtered in one filtration volume can be calculated as in Table 10. The small minke whale seems to require more filtrations in order to get an adequate food supply than the large rorquals which can, according to the calculations, amass sufficient food for a day in 70-130 filtrations. Klumov (1961) estimated that one filtration would take about 30 seconds so that if the whale were to feed continuously for about an hour, adequate nourishment could be taken. Presumably slightly lesser densities of plankton, which are more frequently encountered, would also supply sufficient nourishment, whilst still permitting time to search out new feeding areas. However, it would seem essential that the food plankton swarms near the surface for at least a few hours daily so that the whales can feed adequately. In areas where plankton densities are only 500 mg/m³ (Klumov, 1961) then time spent feeding would be 4 000 times that spent amongst dense plankton of 2 kg/m³. The seasonal feeding pattern of many rorquals is thus a necessity, and migration into the Antarctic grounds essential for seeking these swarming types of plankton. A further point to emphasize the importance of swarming plankton is that the baleen filtration system of rorquals is inefficient in retaining very small specimens, and the engulfing of food in bulk must help to reduce this loss. Nevertheless, on these calculations the whale could acquire the food quantities it requires in the Antarctic.

The implication by Klumov (1961) that baleen whales may encounter difficulties in seeking sufficient nourishment in Kurile and North Pacific water through their inability to find dense swarms of euphausiids must be partly discounted because there is much evidence (Zenkovich, 1937) that baleen whales readily accept herring, capelin and other shoal fish when they are available.

Kawamura (1974) calculated that sei whales feeding during summer on copepods in the sub-Antarctic could not achieve maximum stomach fullness, however long they spent filtering, because the swarm densities are insufficient. He observed that although copepods formed the most frequently found foodstuff in stomachs, maximum contents did not exceed 100 kg, whereas euphausiids and amphipods were usually present in quantities of 150-200 kg and 150-250 kg respectively. Kawamura's conclusion was that from calculations based on Sergeant's correlation between heart weight, body weight and maximal food consumption, sei whales must be permanently hungry. This statement may be well founded for certain areas where sei whales have been caught and examined, but the fact remains that many sei do penetrate far south to the Antarctic where they certainly feed to capacity on euphausiids and amphipods.

Considering that all rorquals have been

Table 10. Filtering and feeding capacity assuming a krill density of 2 kg/m³

Species	Body length (m)	Body wt [1] (kg)	Food required daily at 30-40 g/kg/day (kg)	Filter vol. (m³) (Fig. 39)	Filter capacity (kg)	Calculated no. of filtrations required for satiation
Minke	5	2 041	71	.10	.2	355
	7	4 440	155	.35	.7	220
	9	5 602	196	.8	1.6	122
Fin	14	16 852	590	2.85	5.7	104
	18	34 927	1 222	7.2	14.4	85
	22	62 503	2 188	14.9	29.8	73
Blue	16	23 749	831	3.2	6.4	130
	20	49 045	1 717	10.65	21.3	81
	24	88 702	3 105	18.00	36.0	86
	28	146 390	5 124	32.6	65.2	79

[1] Calculated from Lockyer (1976).

found to be feeding very little during winter compared with summer, and that the amount of food ingested per day in the summer is probably of the order of 3-4 % body weight per day, overall daily food consumption averaged throughout the year must amount to considerably less than 4 % body weight. Sergeant (1969) in fact noted that although the large rorquals would seem to consume the equivalent of 4 % of body weight daily, they only feed during the summer months.

Maximal stomach distension with food is perhaps not necessary for ample nutrition, although the advantage afforded by a capacious multiple stomach is there when food is seasonally abundant. It is perhaps important to stress that no animal will store fat unless it has been consuming an inordinately excessive amount of food. Certainly any whale which can increase its body weight by 50-100 % during the feeding season must be considered to be "overeating" at this time with respect to maintenance needs.

Therefore, I conclude that although feeding during the summer months for a period of 120 days only amounts to about 4 % of body weight daily, the average throughout the year, allowing for reduced winter feeding at about a tenth of summer amounts (0.4 % body weight daily) is probably closer to 1.5-2 %.

ENERGY CONTENT AND COMPOSITION OF KRILL

Heyerdahl (1932) estimated that whole krill contains 78-89.3 % water, 1.48 % ash and 1.15-4.61 % crude fat. Il'ichev (1967) gave estimates of 75-82 % water, 3.2-3.5 % ash, 1.2-3.4 % crude fat and 13.7-17.8 % crude protein in wet krill. Hirano *et al.* (1964) gave the composition as 76.6-79.8 % water, 1.48-3.29 % ash, 2.12-2.65 % crude fat and 13.88-19.63 % crude protein for the edible parts of krill. Vinogradova (1967) gave water content of krill as 75-80 %. Water content of *Meganyctiphanes norvegica* was estimated at 78 % by Raymont, Srinivasagan and Raymont (1971), and Mauchline and Fisher (1969) gave similar values for other species of euphausiids. Actual dry weight of krill is thus about 22 % of wet weight. Vinogradova (1960, 1967) estimated that the dry weight of krill comprises 6.74-10.00 % ash, 11-26 % crude fat and 52-67 % protein. Vinogradova (1960) found 9.4-35.5 % fat in *Euphausia crystallorophias*. In *Meganyctiphanes norvegica*, Raymont, Srini-

437

vasagan and Raymont (1971) found that the dry weight consisted of 16.1 % ash, 17.2 % lipid, 56.6 % protein, 2.0 % carbohydrate and 4.2 % chitin. It is likely that there is negligible free carbohydrate in krill, and the polysaccharides in chitin are probably indigestible. Il'ichev (1967) found that krill chitin constituted over 26.4 % of wet weight. This estimate is unreliable as the dissection of the krill seemed crude and incomplete. Values for chitin content in other euphausiids amounted to 5.3-12.8 % of dry weight (Jerde and Lasker, 1966). In *M. norvegica* the indigestible portion of ash and chitin constitutes about 20 % of dry weight. By comparison with other euphausiids, the actual chitin content of krill is probably far less than 10 % dry weight, especially in larger krill of 30 mm length upward where body tissue becomes bulkier. The krill is a far larger euphausiid than the other species mentioned here. If ash constitutes a maximum of 10 % dry weight in krill, the overall indigestible or non-utilizable fraction in terms of energy assimilation is about 20 % of the dry weight or approximately 4.4 % wet weight. Whole krill would thus consist on average of about 78 % water, 5 % ash and chitin, 4 % crude fat and 13 % crude protein. The heats of combustion of fat and protein are approximately 9 400 kcal/kg and 5 700 kcal/kg respectively (Brody, 1968), so that the calorific value of whole krill can be calculated to be about 1 100 kcal/kg. Vinogradova (1967) found that the calorific value of dried krill was 4 000-4 780 kcal/kg, equivalent to about 880-1 052 kcal/kg whole krill. Macquillan (1962) gave a calorie yield of 460 kcal from 1 lb (453 g) of whole krill, equivalent to just over 1 000 kcal/kg. Nemoto (1970) gave a similar calorific value for whole krill. This energy comes chiefly from the digestible soft tissues, so that an extremely high assimilation efficiency might be expected in animals feeding on krill; at least 75 % which is not uncommon in carnivorous animals (Phillipson, 1960). If 30-40 g krill/kg body weight/day is consumed by a whale, this quantity represents 30-40 kcal/kg body weight/day.

Metabolism

There is limited direct information available regarding metabolic rates in Cetacea; that which is known is based chiefly on observations made on dolphins and porpoises. There ae more physiological data for Pinnipedia than Cetacea, chiefly because seals are generally more convenient for laboratory handling than the latter. The two orders of marine mammals can be closely compared however, because in many ways they are physiologically similar: both inhabit the same environment and are adapted to exploit it in many similar morphological and anatomical ways. In this way much of what is known about metabolism in seals and also for dolphins can be extrapolated, and used to predict metabolic rates in whales.

Throughout this section, a variety of information concerned directly or indirectly with metabolism in Cetacea has been considered on a comparative basis. Metabolic rate is measured as the total energy expended due to metabolic processes in the body in a given time period. This rate is governed by such parameters as body form, organ weight, tissue structure and internal chemistry, and influenced by environment, behavioural activity and feeding habits. For this reason, several data relevant to these subjects are considered in deciding the likely ranges and limits of metabolic rate in different whale species.

MORPHOLOGICAL PARAMETERS GOVERNING METABOLIC RATES

Visceral Organ Weights Related to Body Weight

Weight data for heart and kidneys in a variety of Cetacea are shown in Figs. 40 and 41. These data fit quite closely into the relationship formulated by Brody (1968) for all adult mammals, where

heart weight, $H = 0.00588W^{0.984}$,

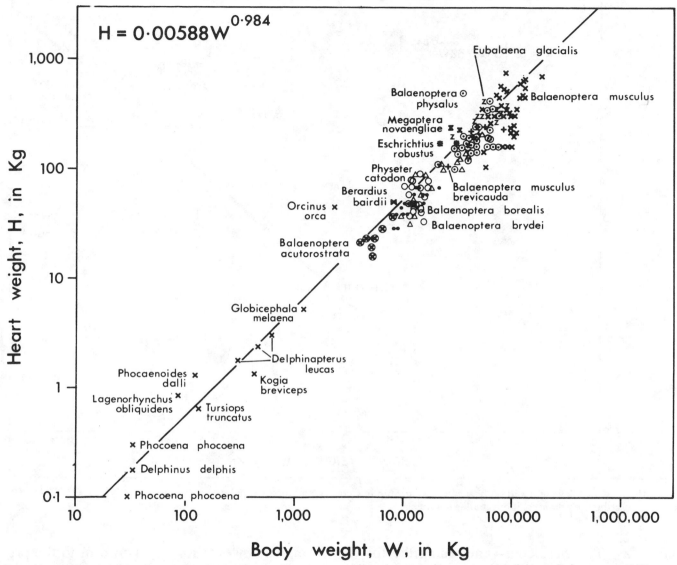

$$H = 0.00588W^{0.984}$$

FIG. 40. — Heart weight and body weight relationship in Cetacea, and the extrapolation of the formula, $H = 0.00588W^{0.984}$, used to describe the relationship in smaller mammals.

and

kidney weight, $K = 0.00732W^{0.846}$,

W being the body weight in kilogrammes. The heart and kidneys are organs of considerable importance metabolically, and it might be expected that metabolic rates of Cetacea also depend on body weight relationships such as those formulated by Adolph (1946) and quoted in Brody (1968) for all mammals ranging in size from mice to elephants. Basal metabolic rates could perhaps be computed for whales from an extrapolation of this equation,

$$Q = 70.5W^{0.7325},$$

where Q is the basal metabolic rate in kcal/day, and W is body weight in kilogrammes. Laurie (1933) calculated from total body surface area of a 122 tonne blue whale, that basal metabolic rate would be 2.75×10^5 439

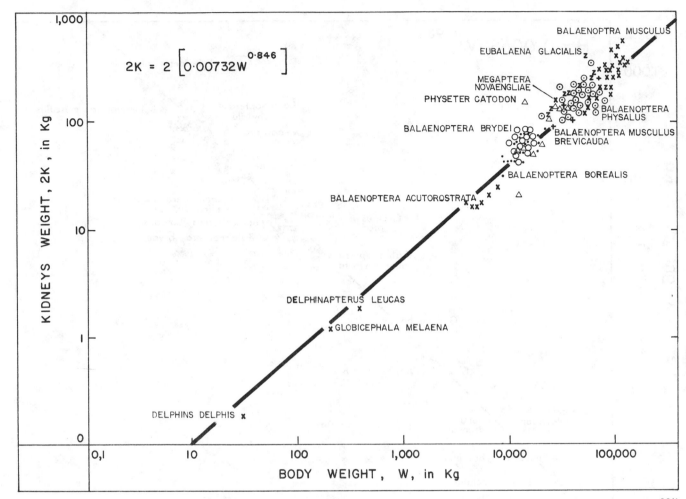

FIG. 41. – Kidneys weight and body weight relationship in Cetacea, and the extrapolation of the formula, $2[K = 0.00732W^{0.846}]$, used to describe the relationship in smaller mammals.

kcal/day. The predicted value using the formula above is 3.75×10^5 kcal/day which is of a similar order of magnitude.

Lung Weight and Capacity

In Table 11 the lung weights and total lung capacities of different whales can be compared with two land mammals, a seal, and with each other. The lung weight/body weight ratios in all whales show considerable agreement and are rather less than the values given for man and elephant. The porpoise however appears to have a high lung weight/body weight ratio unlike the large whales. All lung

weights however roughly conform to the predictions of the formula,

$$Lu = 0.0113W^{0.986},$$

where Lu is lung weight in kilogrammes and W is body weight in kilogrammes (Brody, 1968) so that an average lung weight as a percentage of body weight would be about 1 %.

Most available measurements of lung capacity in whales, which are very few, were not made directly. Laurie (1933) estimated that the vital lung capacity of a 22 m fin whale is 7 000 litres from thoracic cavity dimensions, and for a 27.2 m blue whale weighing 122 tonnes is 3 050 litres from an extrapolation of

Table 11. Lung weight and total lung capacity as percentages of body weight

Species	Average body weight of sample (kg)	Mean lung weight as % body weight	Mean total lung capacity as % body weight
Elephant	4 400	2.55 (Brody, 1968)	
Man	60	1.76 (Spector, 1956)	7.4 (Scholander, 1940)
Porpoise	140	3.69 (Scholander, 1940)	7.1 (Scholander, 1940)
Bottlenose whale	1 400	0.86 (Scholander, 1940)	2.8 (Scholander, 1940)
Sperm whale	35 000	0.91 (Omura, 1950)	1.6 (see Table 14, Sperm 1)
Blue whale	90 000	0.73 (Nishiwaki, 1950)	2.5 (Scholander, 1940)
Fin whale	53 000	0.71 (Nishiwaki, 1950	2.7 (Scholander, 1940)
Sei whale	12 500	0.85 (Omura, 1950)	
Minke whale	5 300	0.85	
Black right whale	60 000	0.65 (Omura, 1958; Omura *et al.*, 1969)	
Gray whale	6 150		7.0 (Wahrenbrock *et al.*, 1974)
Weddell seal	425		4.8 (Kooyman *et al.*, 1971)
Pilot whale	450		10.0 (Olsen *et al.*, 1969)
Bottlenose dolphin	138		5.1 (Ridgway *et al.*, 1969)
Bottlenose dolphin	200		5.5 (Ridgway *et al.*, 1969)

body weight and lung capacity data known for other mammals. Krogh (1934), however, assumed a value of 14 000 litres for this same blue whale. Scholander (1940) found the approximate lung capacity for fin whales by inflating the lungs maximally with air. He obtained values of 800 litres for a 500 ft whale and 2 000 litres for a 72 ft whale. By this reckoning, Laurie's estimate for the blue whale would probably be quite close to reality, the predicted value considering the vital capacities of these two fin whales being about 4 000 litres. The estimate by Laurie for the 22 m fin whale appears too high. Clarke (1970) estimated that the vital capacity of sperm whale lungs varies between 300 and 1 000 litres. The weights of two sperm whale lungs, empty and full of water, both for 10.1 m females gave differences equivalent to volumes of 182 and 96 litres. These weighings were performed at Durban by courtesy of the Union Whaling Company in September 1970 and February 1972. These estimates are almost certainly too low owing to the difficulties of manoeuvring the lungs whilst filling with water under pressure, and leaks from damaged areas of the tissue. Scholander (1940) found that the lung capacity of another toothed whale, the bottle-nose whale, of body length 5.7 m and weight 1 400 kg, was 40 litres by inflation of the lungs with air. This estimate is probably quite realistic because the lungs were inflated to the maximum length of the pleural cavity, and Scholander considered this to be maximal inflation. Scholander also calculated that the lung capacity of *Phocoena* was 1.4 litres in a 19 kg animal, by measuring tidal volumes after various forced dives when only expiration was possible. All these data on lung capacity in Table 11 show that there is generally a direct proportional relationship between lung weight and volume in whales. The lung capacity in whales appears lower than that in man and porpoise, ranging between 2.5 and 2.8 % of body weight, except in sperm whales where the unreliability of the very low value has already been discussed.

Haynes and Laurie (1937) found that there are histological differences between cetacean lungs and lungs of land mammals, the chief of these being the absence of epithelium in the alveoli of the former. Other important features are the extreme thickness and elasticity of the pulmonary pleura, and the profusion of myo-elastic fibres (Wislocki and Belanger, 1940) within the lung tissue of large whale

441

species. Haynes and Laurie and also Scholander (1940) noted how completely the cetacean lungs collapsed on deflation, and considered this to be the result of the intense elasticity of the tissue. It would seem reasonable to assume that considering the force and speed with which Cetacea blow on surfacing, there can be relatively little residual air in the lungs on expiration and vital capacity may be very close to total lung capacity.

According to Brody (1968), the residual lung air in man doing maximal work is about 28 % of total lung volume, so that vital capacity is approximately 72 % of total lung capacity. Scholander (1940) found that in a porpoise, *Phocoena,* the vital capacity amounted to about 1.2 litres and residual air was about 200 cc so that vital capacity was about 86 % of total. Irving, Scholander and Grunnell (1941) found a vital capacity of 80 % total lung capacity in *Tursiops truncatus,* also a high value. Olsen, Hale and Elsner (1969) found that a pilot whale weighing 450 kg had a tidal volume which ranged between 9 and 39.5 litres. Residual air after expiration amounted to about 5.6 litres, and total capacity of the lungs was 45.1 litres. However, the animal could renew up to 88 % of the lung air in one breath. Kooyman *et al.* (1971) found that the Weddell seal had an inspiratory lung volume of 18.8 litres and a total capacity of 20.4 litres, so that the seal was able to renew up to 92 % of the lung air. Wahrenbrock *et al.* (1974) found a lung volume of 428 litres in a live 6 150 kg gray whale calf, but found that at rest, the tidal volume was only about 50 % of the resting lung volume. This seems low, but may increase markedly with activity. Spencer (1970) found that during activity, the tidal volume of a killer whale increased by 30-56 % (50-60 litres increasing to 78 litres).

External Body Surface Area

Metabolic rate is more or less proportional to total body surface area (Brody, 1968) when comparing mice and elephants. However, the body surface of any animal is continually varying with respiration and other movements, and the proportionality with metabolic rate can effectively be altered and increased by changes in internal physiology such as increased peripheral blood flow for example in the flukes, fin and flippers of Cetacea. Therefore, although body surface area is not directly useful for predicting basal metabolic rate, it can be used in a variety of ways to predict other parameters associated with metabolism, such as active metabolic rate.

Guldberg (1907) and Laurie (1933) compared the baleen whale body to the shape of two cones joined base to base at the occipital condyles. For the average blue and fin whale, the widest diameter d of the body is level with the condyles. The anterior cone has a base of diameter d and vertical height h_1 from here to the snout; the posterior cone has a base of diameter d and vertical height h_2 from here to the tail flukes. The overall body length is $(h_1 + h_2)$. Mean ratios of dimensions are as follows:

$$d:h_1 \approx 1:2; \quad d:h_2 \approx 1:4; \quad d:(h_1 + h_2) \approx 1:6.$$

The total surface area, A, of a whale is thus

$$\frac{nd}{4} (\sqrt{d^2 + 4 h_1^2} + \sqrt{d^2 + 4 h_2^2}).$$

This method is clearly too simplified for accuracy, but illustrates a method using dimensions easily measured on any cetacean.

Parry (1949) used the relationship between body weight and body surface area,

$$K = \frac{Area}{(weight)^{2/3}} = \frac{A}{W^{2/3}} \quad \frac{sq \cdot cm}{g}$$

His results for two fin whales using the perimeter method of surface area calculation were values for K of 8.3 in a 160 kg foetus and 11.1 in a 43 000 kg adult. Parry's values of K in adult dogs weighing between 1.07 kg and 27 kg, using methods of skinning, triangulation, moulding, covering with paper and perimetry, ranged between 10.1 and 12.3. Brody (1968) gave a formula relating body weight, height and surface area in humans and compared it to a formula relating these factors given by Du Bois and Du Bois (1916). Brody considered that in a man of normal build, the equation,

$$A = 1\,000W^{.685}$$

was adequate, A in square centimetres, W in kilogrammes, and needed no allowance for variability in height.

Applying these three formulae to a 19.8 m blue or fin whale of weight approximately 47 000 kg, predicted from Fig. 9, we obtain values of A as follows. The Guldberg and Laurie method gives:

$$A = \frac{22 \times 3.3}{7.4} (3.3^2 + 4 \times 6.6^2 + 3.3 + 4 \times 13.2^2)$$
$$= 104.3 \text{ m}^2;$$

The Parry method gives:

$$A = 11.1 \times 47\,000\,000^{.67} = 1\,533\,305 \text{ cm}^2$$
$$= 153.3 \text{ m}^2;$$

The Brody method gives:

$$A = 1\,000 \times 47\,000^{.685} = 1\,586\,306 \text{ cm}^2$$
$$= 158.6 \text{ m}^2.$$

These values and those calculated for other lengths and weights of large rorquals (blue or fin) are shown in Fig. 42. Kanwisher and Sundnes (1966) measured actual surface areas of an 18.3 m fin whale and a 160 kg dolphin, and found values of 137 m² and 0.75 m² respectively. Ridgway (1972) gave average body surface areas for *Tursiops truncatus, Lagenorhynchus obliquidens* and *Phocoenoides dalli,* each of body weight 100 kg, as 1.85 m², 1.70 m² and 1.50 m² respectively. According to predictions by Parry and Brody, the surface area of a 100 kg dolphin would be between about 2.35 and 2.50 m². Recently, Kawamura (1975) cal-

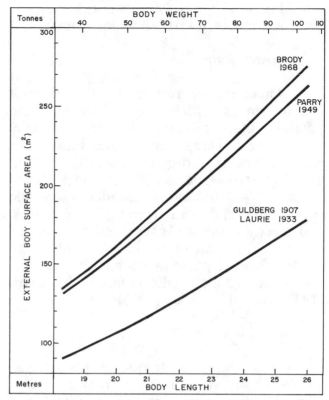

FIG. 42. – Body surface area in various sizes of large rorquals, using three different methods of calculation.

culated from measurements of a fin whale carcass, that the surface area of the body of an 18.1 m fin whale, excluding the head, was 89.9 m². He assumed that such an animal would weigh 34 tonnes. This estimate, allowing for the fact that the head area is ignored, and that of Kanwisher and Sundnes (1966) suggest, with reference to Fig. 42, that the methods of Parry and Brody are applicable at least to large Cetacea, and are more reliable than the "cones" method which gives a value of only two thirds of those predicted by the other two. The formula of Brody is the easiest to use, although body weights of large Cetacea are not always available, whereas linear measurements which can be used in the "cones" method are often at hand. The "cones" method could be used, however, by applying a correction factor of x 1.5 to the surface area formula in order to bring results into line with the other methods.

Physiological Parameters Associated with Metabolic Rate

Heart Rate

There are few recordings of heart pulse rate in whales, which give a guide as to the potential pulse rate, but do not necessarily indicate the average pulse rate. Pulse rates have been measured by electrocardiogram on a resting gray whale calf (Smith and Wahrenbrock, 1974), and on a stranded fin whale (Kanwisher and Senft, 1960). These values, and those for other Cetacea and a variety of other mammals are shown in Fig. 43 where the predicted heart pulse rate for body weight is shown according to the equation by Clark (1927) quoted in Brody (1968):

$$f = 217W^{-0.27},$$

where f is heart beat frequency per minute, and W is body weight in kilogrammes. Considering the likely variations in pulse rate due to emotional excitement during the recordings cited on Cetacea, it is not surprising to find that all the cetacean data lie well above the predicted basal rate in Fig. 43. Smith and Wahrenbrock (1974) observed that the pulse rate for the gray whale calf was higher than anticipated, and explained this by assuming that the calf was excited. The 13.7 m fin whale examined by Kanwisher and Senft (1960) appeared to be under stress while beached, and died shortly afterwards, so that pulse rate was probably not indicative of that in the natural environment. Smith and Wahrenbrock mentioned that variation within species was frequently very great, and affected by such factors as stress and general body fitness. We may conclude therefore, that Cetacea are unlikely to differ very much from other mammals in respect of basal pulse rate/body weight relationships. However, there remains the possibility that like several other diving animals, bradycardia may occur during diving, when a comparatively low pulse rate would be observed.

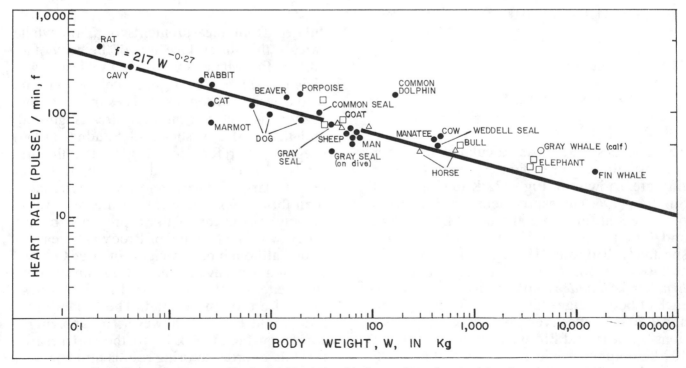

FIG. 43. — Heart rate and body weight relationship in a variety of terrestrial and marine mammals.

In Fig. 43, the average value of the heart beat during rest is shown for marine mammals, chiefly because of the irrelevance of any comparison during bradycardia which might occur during diving. Kooyman and Campbell (1972) showed that immediately pre- and post-dive heart rates in the Weddell seal could average about 84 beats/min whilst heart rate during dives exceeding five minutes could fall as low as 16 beats/min. They also found that there was an inverse relationship between heart beat and dive duration and they suggested that seals may anticipate the length of the dive. Elsner, Kenney and Burgess (1966) observed bradycardia in the bottlenose dolphin, but did not find any indication of anticipation of dive duration.

Respiration Physiology

One aspect of cetacean physiology which might be expected to differ from land mammal physiology is respiration and the associated blood chemistry. This is because of the ability of Cetacea to perform long and often deep diving excursions, when oxygen storage in the body must be at a premium.

Lockyer (1976) commented that blood volume in Cetacea appeared to be high in comparison with terrestrial mammals. Recently, Gilmartin, Pierce and Antonelis (1974) were able to determine the blood volume in a live female gray whale calf, not then one year of age. Although it is likely that relative blood volume may alter as the animal develops, it is nevertheless interesting that at the age of nine months, at a weight of 4 409 kg, the blood volume was 61 ml/kg, and at the age of nearly one year, at a weight of 5 364 kg, the blood volume was 81 ml/kg. These volumes were determined by isotopic dilution techniques by administering radioiodinated human serum albumen into a brachial vessel in the right pectoral fin. Lockyer (1976) came to the conclusion that in adult baleen whales, especially rorquals, blood weight, nearly equivalent to volume, formed about 10 % of body weight.

The chemistry of the blood of Cetacea has been widely investigated. There are several estimates of blood and muscle oxygen-capacity, haemoglobin and myoglobin content, and studies of the oxyhaemoglobin dissociation curves. The general conclusion, referring to Table 12 where some values of oxygen capacity of blood and muscle in a variety of Cetacea and some land mammals are shown, is that the oxygen capacity of the blood is variable, but is generally about 20-30 ml/100 ml in most mammals. The oxygen capacity of the muscle, however, appears to be highest in the deep diving sperm and bottlenose whales. Scholander (1940) estimated that the muscle in seals absorbs up to 47 % of the total oxygen store during diving, and in bottlenose and sperm whales up to 50 %, so that blood oxygen capacity is only a partial indicator of diving potential and endurance of prolonged apnoea. However, there are too few data to say more than this with regard to behavioural activity. Gilmartin, Pierce and Antonelis (1974) maintained that the general fitness of Cetacea, especially those kept in captivity, can affect blood parameters, so that some findings may not be applicable to Cetacea in the wild where many behavioural activities are necessarily different.

From a study of dissociation curves in small Cetacea, Horvath et al. (1968) found that an increased affinity of the blood for oxygen appeared to be related to feeding behavioural habits. There is no evidence, however, that this is especially true for whales, and Gilmartin, Pierce and Antonelis believed that there is no such correlation in large Cetacea.

Diving in marine mammals is frequently associated with bradycardia (Andersen, 1966). Scholander (1940) found that the blood sugar level in infant and adult tamed seals falls during a dive, and increases on recovery. He also noticed that the lactic acid builds up in the muscles chiefly on recovery, and is not released into the blood during the dive, perhaps because of vaso-constriction during the dive. Murdaugh et al. (1968) found that during a dive, bradycardia, decrease in cardiac output,

445

and arterial constriction all occurred in *Phoca vitulina.* By preventing bradycardia during diving by means of an intra-cardiac pacer, they showed that it was the arterial constriction that was important in conserving the oxygen, by restricting perfusion of peripheral tissues and retaining the oxygen for the central nervous system. The peripheral tissues switched to anaerobic glycolysis when the oxygen store was used up. They suggested that bradycardia helped to restrict venous return. Scholander found that the seal's rectal temperature was found to drop from 36.7°C to 35.8°C during a six minute submersion with apnoea in water ranging in temperature from 3°C to 28°C. This reflects a general slowing of all metabolism during prolonged apnoea in order to conserve the oxygen store.

The Respiratory Quotient (R.Q.), that is the ratio of volumes of carbon dioxide evolved to the volumes of oxygen consumed, alters during the dive and recovery period, and also according to the nutritional status. Scholander (1940) found values of R.Q. for seals and *Phocoena communis,* of 0.70-0.74 during dives, yet over 1.00 during recovery depending on anaerobic or aerobic conditions respectively. An average value of R.Q. for *Cystophora cristatus,* including both dive and recovery periods, is about 0.83. Kooyman *et al.* (1973) found that the R.Q. for *Leptonychotes weddelli* averaged 0.69 (range 0.64-0.77), but the R.Q.

immediately after a dive fell to between 0.5 and 0.7; although this rose to 1.0, and even exceeded 1.0 after long dives (over 20 minutes), after the first five minutes from surfacing. Irving *et al.,* (1935), Irving (1939) and Irving and Hart (1957) found an R.Q. of 0.70-0.75 in *Phoca vitulina* and *Phoca groenlandica.* Values less than 0.70 have been recorded during starvation. Kanwisher and Senft (1960) recorded an R.Q. value between 0.8 and 0.9 for a 13.7 m stranded fin whale which seemed rather high for an animal not recently fed and in a chilled condition when the rectal temperature was 33°C and temperature inside the 15 cm thick blubber was 27-31°C, compared with a normal deep body temperature of about 36°C (Irving, 1939).

The actual oxygen consumption as a percentage utilization of the inspired air in *Phocoena communis* was found to be about 8 % by Scholander (1940). In all his calculations on large Cetacea he assumed about 10 % utilization of the inspired air. Spencer (1970) found that oxygen consumption in a 5.18 m killer whale weighing 2 000 kg, was 3-4 litres when resting with a tidal volume of 50-60 litres, and 9 litres after activity with a tidal volume of 78 litres. This represents about 6-8 % after activity. For resting gray whale calves weighing between 3 000 and 5 000 kg, oxygen consumption averages between 9.5 % and 6.5 % respectively of tidal volume, from interpola-

Table 12.   Range of oxygen capacity of muscle and blood, and maximum recorded diving times for various marine mammals

Species	Volume % oxygen capacity		Absolute maximum diving time (min.)	References
	Muscle	Blood		
Balaenoptera physalus	3-5	20-25	30	Scholander (1940); Irving (1939)
Physeter catodon	6-7.6	25-30	75	Green and Redfield (1933); Irving (1939); Scholander (1940)
Hyperoodon rostratus	7.5-8.5	30	120	Irving (1939)
Phocoena phocoena	4	20-24	12.5	Green and Redfield (1933); Scholander (1940)
Halichoerus grypus	3.7-5.5	18-25	15	} Scholander (1940); Irving (1939)
Cystophora cristatus	3.7-5.5	18-29	30	
Dog		21.8	4.5	} Irving (1935, 1939)
Man		20.4-21.5	4.5	

tion of data given by Wahrenbrock *et al.* (1974). Therefore, in all the following calculations an estimate of 10 % utilization of inspired air has been considered reasonable, allowing for the fact that the whale would not be restrained in the natural environment.

An average value of 0.82 has been adopted here as the R.Q. throughout the respiratory cycle in whales. This gives a heat equivalent of 1 litre of oxygen at this R.Q. of 4 825 kcal (Brody, 1968). Wahrenbrock *et al.* (1974) also assumed an R.Q. of about 0.8 in gray whales, because they converted oxygen consumption rates to metabolic rates in a factor of 1 litre of oxygen evolving 4.8 kcal.

BEHAVIOURAL PARAMETERS
AFFECTING METABOLIC RATE

Respiratory Rhythm and Diving

There is an abundance of data on blowing frequency and duration of dive for large Cetacea. Mostly these data are the results of shipboard observations, and even when concerned with the same species, they frequently vary. This is not unexpected because behavioural patterns change with different activities; and also because the whales' normal pattern of behaviour is disrupted by the presence of vessels in their vicinity, especially those equipped with ASDIC — a fact obtained from personal observations of sperm whales from aircraft and ships. The most important information in estimating metabolic rates, apart from the actual respiratory and diving times is therefore the qualifying estimation as to whether the whale is considered to be resting, feeding, active or alarmed, and an estimation of swimming speed.

Gunther (1949) gave a most detailed account of shipboard observations on fin whales, mentioning the timing of the respiratory cycle and dive sequence. Laurie (1933) estimated that rorquals do not (or need not) descend more than 300 m during dives, in sharp contrast to my observations on the toothed sperm

and bottlenose whales which frequently reach depths beyond 1 000 m (see also Heezen (1957) who discussed deep-sea entanglements of whales in cables).

Size and age of a whale appears to affect diving and respiratory performance according to my observations of sperm whales (Lockyer, 1977). Only the large bulls over 13.7 m appear to dive very deeply for periods of an hour or more. Aerial observations of large and small (about 9.5 m) sperm showed that small sperm blow more frequently than large ones. Other data collected by staff of the IOS during whale marking cruises between 1972 and 1975 inclusive are shown in Table 13, where it is apparent that smaller animals in other species tend to blow more frequently than large ones. Gambell, Lockyer and Ross (1973) commented on the ability of a new born sperm calf to dive for a period of seven minutes, so that a certain potential ability is inherent from birth, which clearly improves with development.

In Table 13, it is apparent that after a prolonged chase, as in the case of three sei whales, the blowing rate increases, giving an indication of higher metabolic rate. However, although blowing rate increases, tidal volume has probably increased also, so that a direct correlation between swimming activity and metabolic rate cannot be calculated. Although apparently alarmed when given chase, no rorquals appeared to be greatly fatigued from 40 minutes or more of swimming at about 12 knots (22.2 km/h), because they easily out-ran the pursuing vessel.

From Table 13, one may conclude that in rorquals, blowing rates may fall to one per minute in the relaxed resting animal, but are higher when the whale is actively swimming. The blowing rate in sperm whales does not appear greatly to exceed six per minute in extreme exertion, though a study of observations made between 1972 and 1975 would indicate that the length of the recovery period after a 20 min. repeat dive may be two or three times that of the recovery period after a 10 min. single dive. Perhaps the same pattern is true for rorquals.

447

Table 13. **Observations on respiratory and diving behaviour in whales**

Species	m	Body length ft	Observed dive duration (min.)	Number of blows in recovery period after dive	Average number of blows/min. when not diving	Comments
Blue	17.5	58			2-3; 3-4	
	18	60	4-4½	5-10 in 2 min.		Observations from ship
	18	60	2-3		1.3	
	21	70	3-9		0.7-2 (resting) 3-6 (during chase)	Chased for 1 hr 43 min. with ASDIC on
	22.5	75	3-7	7-11 in 12½ min.	3.5-6	Chased for 40 min. by vessel
Fin	14	47	3-4		1-2	Chased with ASDIC turned on
	19.5	65	4-6 (before chasing)		6-8	
	22.5	75	6; 9	3-5 in 1 min.		Observation from ship
Sei	14.4	48			0.7 (before chase) 1.3 (after chase)	Observation from ship
	12.9	43			1.4 (before chase) 2 (after chase)	Chased for 1hr by vessel
	14.1	47			0.8 (before chase) 1 (after 1 hr chase)	
Bryde	11.7	ca 39			0.5-1	Whale zig-zagging after shoal of fish, and unconcerned with vessel
Humpback	13.8	46	2-3	4-6 in 1-1½ min.	1.7	
Minke	8.4	28			0.5	Observations from ship
	3.6	12			1	
Sperm	9	30	up to 6 up to 45		4-6 6-7	Observation from aircraft Observation from ship
	13.5	45	up to 6 up to 65		3-4 6	Observation from aircraft Observation from ship

Daily Activity Pattern

The whale's daily activities in the summer are probably principally centred around the necessity of feeding. We already know that they vary seasonally with respect to feeding and migration. In order to gain some kind of estimate of daily energy expenditure, however, it is essential to have an idea of the proportion of time in a day involved in different occupations. These would include feeding, resting, sleeping, and a variety of activities such as swimming from one area to another, mating, nursing and playing. All the active occupations could possibly be grouped, so that the whale's day would include periods of feeding, resting or sleeping, and unspecified activities.

In general, Brody (1968) has classified

alternating periods of rest and activity as monophasic, that is one period of rest followed by one period of activity in 24 hours such as for example in adult humans and canaries; or polyphasic, where many periods of rest and activity occur within 24 hours such as for example in rabbits and young dogs. Whales may fit into either category especially in very high latitudes where light may not be an influencing factor. In dolphinaria, short rests are reported after feeds rather than prolonged ones, but this behavioural pattern may be an unnatural régime enforced by captivity.

Malm (1938) observed that *Tursiops truncatus* is unlikely to be a nocturnal animal because in aquaria its peak activity is always in daytime. In the Florida oceanarium, *Tursiops truncatus* always sleeps near the surface for most of the night; during the day sleep generally occurs as an hour-long nap after each feed (Slijper, 1962). The sleeping animals periodically open their eyes for 1-2 seconds and close them again for 15-30 seconds. In these calm conditions, the sleeping animals remain almost parallel to the surface with the drooping caudal peduncle beating slowly and rhythmically every half minute so that the body breaks surface and then sinks smoothly to 30 cm below. Tomilin (1967) reckoned that dolphins must be able to sleep normally in any sea conditions every day, since they possess a natural respiratory reflex of opening the blowhole when the medium changes from water to air on surfacing, and maintain vertical tail movements.

Tomilin (1967) recorded that a pilot whale in captivity always slept at the surface with only the blowhole, back and dorsal fin above water. While it was asleep, which was chiefly at night for 2 hour periods but could also occur at any time of day, the tail rarely moved. Scammon (1869) described an entire school of pilot whales sleeping at the surface of a calm sea, well spaced out at various distances from each other and not visibly spouting.

There are a number of records of large Cetacea observed to be sleeping. Allen (1916) recorded a fin whale sleeping at night at the surface of Placentia Bay (Newfoundland) from the Norwegian whaling steamer *Puma*. Tomilin (1967) quoted Foletarik and Chirkova (1930) on the observation of a school of belugas apparently sleeping at the sea surface. Geptner (1930) also recorded seeing sleeping belugas. Gray (1889) (quoted *in* Tomilin (1967)), Porsild (1922) and Degerbøl and Freuchen (1935) all recorded seeing narwhals apparently solitary and motionless in sleep. Gray (1927) (quoted *in* Tomilin (1967)) twice found Greenland right whales sleeping near ice and oblivious to sounds of approaching danger. When sleeping, the whales do not blow visibly but remain at the surface quietly and maintain equilibrium with the flippers. Scoresby (1825) (quoted *in* Tomilin (1967)) recorded that right whales remain at the surface for 2-30 minutes, immobile. A sleeping Biscayan right whale in the South Atlantic only awoke on being aroused by a passing ship's bow wave (Slijper, 1962).

Slijper (1962) mentioned several observations of deep slumber at the sea surface in sperm whales. These records were usually derived from collisions between ships under way and sperm whales which were presumed to be sleeping because they had not avoided the vessel.

The only record of daily activity in a large cetacean in captivity is that of a minke whale maintained in a seaquarium for just over one month (Kimura and Nemoto, 1956). However, this animal did not appear to sleep, and instead increased respiration and activity occurred between nightfall and midnight, the respiration rate falling off after then. Vincent (1960), however, reported that common dolphins in captivity in Monaco had daily respiratory rhythms of 6 blows/minute and nightly rhythms of 3-4 blows/minute.

Probably, as in most wild carnivorous animals, cetacean sleep follows a large feed. Ingebrigtsen (1929) and Kawamura (1974) mentioned that feeding sei whales appear preoccupied and slow, and do not heed the approach of ships. Gunther (1949) reported similar behaviour by feeding fin whales. Ka-

449

wamura (1974) stated that replete whales are easily approachable whilst hungry ones are agile and shy of approach, suggesting that feeding induces a soporific effect which dulls the senses. Because, as we have seen earlier, daily feeding habits are governed by the distribution of the prey, resulting in a tendency to bulk feeding once a day in the Antarctic (Kawamura, 1974), one might expect the whales feeding thus to doze for a prolonged period after this main feed. This rest period would probably occur in the day if the prey were nocturnal, and at night if diurnal.

Nearly all ship encounters with sleeping fin whales and indeed with most other whale species have been at night. Aquaria observations are unreliable, because the diurnal rhythm of the animals is upset and becomes artificial. However, aquaria observations give information on the extent of activity while resting and on the sleeping posture. Vincent (1960) gave the information that blowing occurs about 1 half to 2 thirds as often during sleep as when awake in common dolphins, representing a substantial drop in metabolic rate, perhaps close to basal rate.

METABOLIC RATE

Resting Metabolic Rate

Estimates of resting metabolic rate in a number of whale species, calculated using the method of lung capacity (Scholander, 1940), are given in Table 14. Items under the headings of R.Q., air utilization, lung capacity and respiratory cycle are based on findings from the previous discussions. The animals listed are those for which the lung capacity and body size and/or weight are known either from published sources or from new data. In Fig. 44, these values have been plotted along with those for directly measured rates in other, mainly small, Cetacea, and Pinnipedia. On the whole, the calculated estimates appear reasonable compared with values one would expect by extrapolation of the direct measure-

ments. Resting metabolic rate is difficult to define, so that limits of what is called resting must be flexible. The calculated rates for the large whales are probably realistic.

Basal Metabolic Rate

These rates are very difficult to assess in nature, because the true basal rate is hardly ever observed in day to day living. Basal rates are probably only applicable to animals in deep slumber, starvation (Benedict and Ritzman, 1927), or in late hibernation. However, estimates of basal metabolic rate can be made by three methods.

Brody (1968) gave a formula,

$$M = 70.5W^{0.7325},$$

relating basal metabolic rate, M, and body weight, W, in land mammals, as noted earlier, which can be extrapolated to include Cetacea. Brody mentioned that basal metabolic rate has little relevance for immature growing animals whose lowest metabolic rate must, because of the growth energy requirements, exceed those in adults who must meet only maintenance costs. Brody found that resting metabolic rates of growing animals can approximate up to twice the basal rates of adults. The true resting metabolic rate in adult animals is the maintenance energy expenditure rate, of which about 85 % is the basal metabolic rate concerned with life processes which operate even during starvation, and the remaining 15 % is the expense in muscular effort necessary in essential moving about and feeding. Basal metabolic rate can thus be calculated as approximately 85 % of resting values.

According to Dill (1936), the oxygen consumed by man running at top speed is 10 to 20 times that at the basal metabolic rate. Brody (1968) quoted several examples substantiating this observation but mentioned that brief peak effort under stress can consume up to 100

Table 14. Some estimates of resting metabolic rates in whales using the method of lung capacity (Scholander, 1940)

Species of whale	m	Body length ft	Body weight kg	Assumed mean RQ	% air utiliza- tion	Lung capacity litres	Average respiratory cycle				Resting metabolic rate QR in kcal/day
							Number of breaths	Breathing duration min.	Dive duration min.	Total cycle min.	
Blue	26.7	89	122 000	0.82	10	3 050	12	4	20	24	1.06×10^6
Fin 1	15	50	30 000	0.82	10	800	5	5	8	13	2.18×10^5
Fin 2	21.6	72	70 000	0.82	10	2 000	10	10	16	26	5.34×10^5
Sperm 1	9.9	33	11 380	0.82	10	182	6	1.5	7	8.5	8.93×10^5
Sperm 2	9.9	33	[11 500]	0.82	10	96 [1]	6	1.5	7	8.5	4.71×10^4
Sperm 3 (hypothetical)	[9.9]	[33]	[11 500]	0.82	10	300	6	1.5	7	8.5	1.47×10^5
Bottlenose	5.4	18	1 400	0.82	10	40	30	7	30	37	2.26×10^4

[1] Lung capacity considered very dubious.

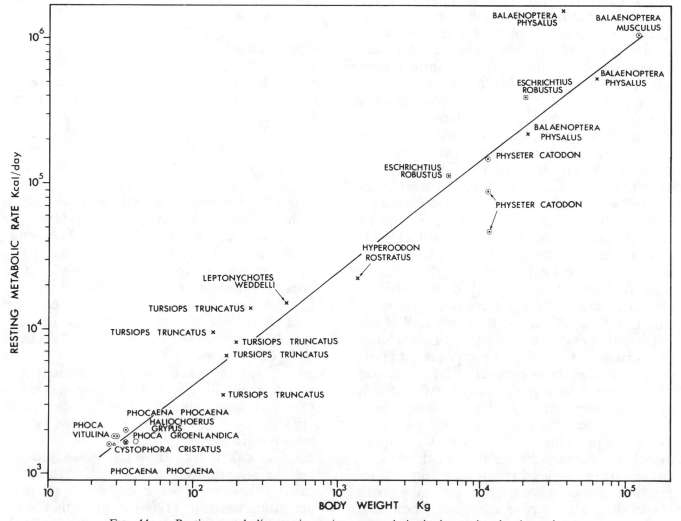

FIG. 44. — Resting metabolic rates in marine mammals, both observed and estimated.

times the basal energy expenditure. However, active metabolic rates for sustained hard work, such as heavy labouring in work animals and man are about 8 times the resting rates. If active metabolic rates are taken to apply at top cruising speed in swimming whales, then basal metabolic rate can be estimated as approximately 10 % of the active rate.

Active Metabolic Rate

Gray (1968) estimated that during activity, the power developed in mammalian muscle was about 0.01 hp/lb muscle equivalent to 1.64×10^8 ergs/kg muscle. He assumed that this probably applied to marine mammals also. Lang (1966) calculated the top swimming speeds for different lengths of time for Cetacea of sizes ranging between dolphins and blue whales, assuming that: (a) full laminar flow occurred during swimming, and (b) turbulence occurred as for a towed rigid body. From a comparison of observed top swimming speeds with predicted ones, he concluded that either dolphin muscle was 4 or 5 times more powerful than suggested, or that nearly full laminar flow occurred. He estimated the power developed for different durations of exertions. Assuming from body weight data for bottlenose and other dolphin species given by Tomilin (1967) that muscle constitutes about 33 % of body weight, the power values given by Lang can be converted to a form directly comparable to Gray's estimate. Thus, for the operative musculature (Gray estimated that only half of the musculature could function at once) the values for horsepower developed per pound of operative musculature in periods of 0.5 sec, 5 sec, 15-120 min. and a day are 0.21, 0.072, 0.018 and 0.0068, respectively. The exceedingly high values are for brief peak exertions only, such as in jumps. The values nearer 0.01 hp/lb muscle are for sustained effort, when top speeds are also considerably less than during brief exertion. Brody (1968) gave data for work horses during 5 sec peak exertions. These animals were able to develop the equivalent of

between 0.1 and 0.2 hp/lb operative muscle, similar to the Cetacea during brief peak exertion.

Kramer (1960) and Aleyev (1965, 1970) found running deformation waves in the skin of swimming dolphins, which decreased the hydrodynamic resistance. Purves (1969) suggested that the structure of the tail flukes might be related to laminar flow in Mysticeti. Purves also found diagonal dermal ridges on the body of Cetacea, and he believed that the skin was sufficiently plastic to permit the formation of the surface ripples whilst swimming. All these findings support the theory that laminar flow must be present in the swimming of the dolphin. Lang (1966) found that between 40 % and 100 % laminar flow occurs in swimming dolphins from experimentation, perhaps on average 70 %, so considerably lowering the drag coefficient and increasing swimming speed.

The likely resistance encountered during different activities can be calculated from the formula,

$$R = \frac{C_d PAV^2}{2_g}$$

where R = resistance in pounds; C_d = coefficient of drag which may be calculated from the Reynold's number, $R_1 = VL/\nu$, and assuming 70 % laminar flow; P = weight of cubic feet of water, 64.1 lb for seawater; A = surface area in square feet; V = swimming velocity in feet per second; g = gravity constant 32.2 ft/sec/sec; L = body length in feet; $\nu = 1.3 \times 10^{-5}$ ft²/sec, the viscosity of seawater at 13°C.

Swimming speeds can be gained from the data of Gilmore (1961), Dawbin (1966), Gray (1968), Tomilin (1967), Zenkovich (1937a), Lang (1966) and Williamson (1972), shown in Table 15.

In calculating the total power output, certain assumptions have been made. The usual efficiency of work in horses and man, according to Brody (1968), is 20 %; that is, for every 100 kcal of muscle energy expended only

Table 15. Comparative ranges of swimming speeds in Cetacea

Species of whale	Swimming speeds km/hr		
	Feeding	Migrating and cruising	Maximum when alarmed
Blue	2-5.5	5-33	48
Fin	2-6.5	5-30	41
Sei	2-6.5	5-30	
Bryde	2-6.5	5-30	
Humpback	2-4	5-15	27
Minke	2-6	5-26	
Right	2-4	5-11	16
Gray	2-4	4-7.5	16
Sperm		5-12	26
Killer		14-22	37
Dolphins		16.5-24	46

20 kcal can be regained in useful work owing to internal body resistances. This efficiency has been assumed for Cetacea, although Kawamura (1975) assumed an efficiency as high as 25 %, so that the actual energy expended in the muscle in order to overcome the calculated resistance is 5 times that calculated from the formula for power developed during swimming,

$$Q_0 = \frac{RV}{550} \text{ hp}$$

(1 hp = 550 ft lb/sec and 1 hp = 7.45×10^9 ergs/sec).

In Table 16, the resistance encountered and power developed in the muscle whilst swimming at various speeds have been calculated for different sizes of blue or fin whales. The energetic heat equivalents, on a daily basis for the whole animal, are also given where relevant, assuming 1 hp = 1.54×10^4 kcal/lb/day. These calorific values are the extra energy expenditure above basal metabolic rate incurred by sustained activity.

The values for cruising energetics in Table 16 approach 0.01 hp/lb operative muscle, so that it would appear reasonable to accept Gray's (1968) estimate in calculating active metabolic rate. The values of effort at maximum levels are comparable with those already mentioned, and would be exceeded for split second activities such as leaping.

Basal metabolic rates would be approximately one tenth of the active rates during cruising speeds; for example, in a 50-59 ft whale, basal rate would be between 1 and 2×10^5 kcal/day. The extra energy required for feeding amounts to rather more than 15 % of

Table 16. Energetics of swimming in blue or fin whales

				SWIMMING SPEEDS											
Body length ft m	Body weight lb kg	Half muscle weight lb kg	Surface area ft²	Feeding: 5 ft/sec (5.5 km/hr)				Cruising: 22 ft/sec (24 km/hr)				Maximum brief effort 44 ft/sec (48 km/hr)		
				Assumed C_d ft²/sec	R lb	$5Q_0$ hp/lb muscle	$5Q_0$ kcal/whale/day	Assumed C_d ft²/sec	R lb	$5Q_0$ hp/lb muscle	$5Q_0$ kcal/whale/day	Assumed C_d ft²/sec	R lb	$5Q_0$ hp/lb muscle
50 −1 277	47 300 .0511	10 000	881	1.2×10^{-3}	26	.0001	1.54×10^4	9.0×10^{-4}	384	.0077	1.19×10^6	7.5×10^{-4}		
15	21 290	4 500												
60 18	81 400 36 630	17 100 7 695	1 473	1.2×10^{-3}	44	.0001	2.63×10^4	9.0×10^{-4}	642	.0075	1.98×10^6	7.5×10^{-4}	2 140	.0501
70 21	129 800 58 410	27 250 10 260	1 945	1.2×10^{-3}	58	.0001	4.20×10^4	9.0×10^{-4}	848	.0062	2.60×10^6	7.5×10^{-4}	2 825	.0415
80 24	195 800 88 110	46 400 20 880	2 580	1.2×10^{-3}	77	.0001	7.15×10^4	9.0×10^{-4}	1 127	.0049	3.50×10^6	7.5×10^{-4}	3 760	.0324

Key; C_d = coefficient of drag; R = resistance; Q_0 = energy developed.

453

Table 17. Estimations of metabolic rates for blue, fin and sperm whales

Species	Body length ft	m	Body weight (including blood) kg	M = 70.5W$^{0.7325}$	Basal rate kcal/day Active rate $\times 10^{-1}$	Resting rate $\times 0.85$	Resting or maintenance rate from vital lung capacity kcal/day	"Active" rate from energy expenditure 0.01 hp/lb muscle kcal/day
Blue	45	13.5	13 460	7.44×10^4	7.55×10^4	7.75×10^5	9.12×10^4	7.55×10^5
	50	15	18 950	9.60×10^4	1.12×10^5	1.16×10^5	1.36×10^5	1.12×10^6
	55	16.5	25 830	1.21×10^5	1.59×10^5	1.63×10^5	1.92×10^5	1.59×10^6
	60	18	34 290	1.49×10^5	2.16×10^5	2.21×10^5	2.60×10^5	2.16×10^6
	65	19.5	44 440	1.80×10^5	2.88×10^5	2.98×10^5	3.50×10^5	2.88×10^6
	70	21	56 560	2.14×10^5	3.75×10^5	3.87×10^5	4.55×10^5	3.75×10^6
	75	22.5	70 790	2.53×10^5	4.80×10^5	5.06×10^5	5.95×10^5	4.80×10^6
	80	24	87 300	2.97×10^5	6.17×10^5	6.29×10^5	7.40×10^5	6.17×10^6
	85	25.5	106 300	3.39×10^5	7.75×10^5	7.82×10^5	9.20×10^5	7.75×10^6
	90	27	122 000	3.92×10^5	9.40×10^5	9.61×10^5	1.13×10^5	9.40×10^6
Fin	45	13.5	14 960	806×10^4	1.03×10^5	1.38×10^5	1.62×10^5	1.03×10^6
	50	15	20 290	1.01×10^5	1.44×10^5	1.79×10^5	2.10×10^5	1.44×10^6
	55	16.5	26 750	1.23×10^5	1.97×10^5	2.21×10^5	2.60×10^5	1.97×10^6
	60	18	34 430	1.48×10^5	2.54×10^5	2.75×10^5	3.23×10^5	2.54×10^6
	65	19.5	43 410	1.76×10^5	3.33×10^5	3.29×10^5	3.87×10^5	3.33×10^6
	70	21	53 830	2.07×10^5	4.22×10^5	3.91×10^5	4.60×10^5	4.22×10^6
	75	22.5	65 770	2.39×10^5	5.21×10^5	4.63×10^5	5.45×10^5	5.21×10^6

Species	length ft	Body m	Body weight (including blood) kg	M = 70.5W$^{0.7325}$	Basic rate kcal/day Active rate $\times 10^{-1}$	Resting rate $\times 0.85$ 182 litres lung capacity	300 litres lung capacity	Extrapolated resting or maintenance rate from vital lung capacity kcal/day 182 litres lung capacity	300 litres lung capacity	"Active" rate from energy expenditure 0.01 hp/lb muscle kcal/day
Sperm	30	9	7 580	4.90×10^4	5.40×10^4	5.64×10^4	9.35×10^4	6.63×10^4	1.10×10^5	5.40×10^5
	35	10.5	12 380	7.05×10^4	7.75×10^4	8.09×10^4	1.37×10^5	9.50×10^4	1.61×10^5	7.75×10^5
	40	12	18 930	9.60×10^4	1.07×10^5	1.11×10^5	1.90×10^5	1.31×10^5	2.24×10^5	1.07×10^6
	45	13.5	27 520	1.26×10^5	1.42×10^5	1.50×10^5	2.53×10^5	1.76×10^5	2.98×10^5	1.42×10^5
	50	15	38 480	1.62×10^5	1.83×10^5	1.93×10^5	3.27×10^5	2.27×10^5	3.85×10^5	1.83×10^6
	55	16.5	52 110	1.97×10^5	2.31×10^5	2.37×10^5	4.16×10^5	2.79×10^5	4.90×10^5	2.31×10^6
	60	18	68 730	2.46×10^5	2.85×10^5	2.87×10^5	5.16×10^5	3.37×10^5	6.07×10^5	2.85×10^6

the basal rate, which is about what would be expected if basal rate is 85 % of resting or maintenance rate. Similar results are obtained for the other sizes of whales.

We may now return to the problem of estimating basal metabolic rate, and using the 3 methods described, values are given in Table 17 for basal metabolic rate in different sizes of whales. Resting metabolic rate is also given, based on interpolation of vital lung capacity data shown in Table 14 and Fig. 44. Active rate is based on Gray's (1968) estimations. Overall, there is considerable agreement and correlation of values gained by various methods. However, the formula prediction of Brody (1968) for basal metabolic rate generally seems low for Cetacea with reference to the other estimates, and does not follow the same gradient as the other estimates in the logarithmic relationship of metabolic rate and body weight shown in Fig. 45. This is not unusual compared with other mammals, since many do not follow

the formulated relationship exactly (Brody, 1968).

Nevertheless, the calculated values of basal, resting and active metabolic rates probably give a fairly realistic guide to energy expenditure, so that they may be used with reasonable confidence in estimating energy budgets for large Cetacea.

Energy Budget

UTILIZATION OF AVAILABLE ENERGY

The energy content of the food ingested is generally only partially utilized. Only a portion of the energy consumed is assimilated, the remainder often being in an indigestible form which is rejected from the body. Of the amount assimilated, the majority is required

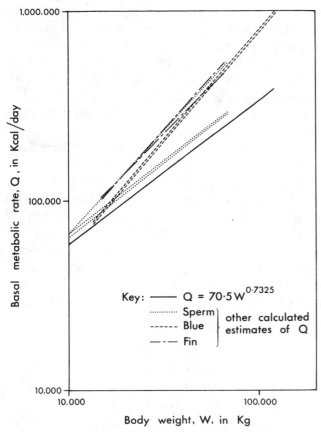

FIG. 45. — Basal metabolic rates calculated and compared for blue, fin and sperm whales.

for metabolic processes and tissue maintenance, whilst any excess is available for growth and development. The proportion of assimilated energy needed for growth in young animals is always relatively high compared with adults, and depends largely on the stage of developmental growth, so that growth efficiency relative to energy consumed is closely linked to, and decreases with, age.

The amount of energy utilized in metabolism varies considerably according to the extent of muscular activity, and external environmental conditions.

Heat Loss from the Body to the Environment

In the central body region, in blue and fin whales, the thickness of the lateral blubber

represents about average overall thickness. However, the dorsal and ventral blubber thicknesses of the areas adjacent to the dorsal fin and anogenital openings, and in the tail, are up to twice that of the lateral blubber (Slijper, 1948). The majority of the body surface area, excluding appendages, is in the central region, so that an estimate of heat loss via the body blubber can be approximated using the lateral blubber thickness as a standard in the formula

$$\Sigma \; \frac{-dQ}{dt} \; = \; \frac{-kA}{\Delta h} \frac{\Delta T}{},$$

where k is the thermal conductivity coefficient of the insulator (blubber), Δh is the thickness of this insulator, ΔT is the temperature gradient between the body and the environment, A is the surface area of the body and dQ/dt is the rate of heat loss per unit area of body surface (Stacy et al., 1955). Parry (1949) and Scholander et al. (1950) estimated that the thermal conductivity of lateral blubber discs, taken from a dead fin whale and harbour seal respectively, was approximately 10 kcal/m²/day/°C/cm in icewater and 6 kcal/m²/day/°C/cm in air at 0°C. The mean body temperature in Cetacea is about 36°C according to the data in Table 18. Ignoring sexual and seasonal variations in blubber thickness, the mean thickness of flank blubber

Table 18. Deep body temperature in Cetacea

Species	Deep body temperature °C	Reference
Balaenoptera (unspecified)	36.6-36.9	Morimoto, Takata & Sudzuki (1921)
Balaenoptera musculus	35.1	Laurie (1933)
Balaenoptera physalus	36.3	Irving (1939)
Megaptera novaeangliae	36.3	Spector (1956)
Eschrichtius robustus	36.5	Spector (1956)
Physeter catodon	35.8	Irving (1939)
Tursiops truncatus	36.5	Kanwisher & Sundnes (1966)

is approximately 0.425 % of body length in blue and fin whales, according to Fig. 17 to 20. Thus blubber thicknesses can be calculated for different sizes of whales. Assuming that average summer water temperatures in the Antarctic are around 0°C and winter off Durban about 25°C, ∆T is 36°C and 11°C respectively. In Table 19, the calculated heat loss through the lateral blubber is shown for different sizes of whales.

Hart and Irving (1959) found that living blubber conducts heat about 50 % more efficiently than dead blubber as investigated by Scholander *et al.* (1950), and ascribed this increase to heat transport via circulation of the blood. Kanwisher and Sundnes (1966) reported that heat flow from *Tursiops truncatus* on the side blubber was 25-60 % more than from dead blubber. Thus the figures for heat loss in Table 19 should be doubled. The heat loss is remarkably low, 10 % or less, so that most heat must be dissipated via the pectoral and dorsal fins, and tail flukes. The fact that the complex vascular tissue in these organs is for the purpose of thermoregulation has now been well established by Ommaney (1932), Parry (1949), Tomilin (1951), Scholander and Schevill (1955) and Belkovich (1961). Kanwisher and Sundnes (1966) found that normally heat flow from these areas was x 3 to x 5 that dissipated through the lateral body blubber of porpoise in water temperatures from 25-27°C. Parry (1949) and Scholander *et al.* (1950) estimated that heat loss via the lateral blubber was only 35-60 % of total heat loss in live animals, so that they assumed that the rest was lost via the fins and tail. Kanwisher and Sundnes, however, found that in common harbour porpoise during periods of near basal activity, in equilibrium with the surrounding water at 8°C, the heat loss via the lateral blubber was greater than that through the extremities.

Heat losses via the blubber are therefore negligible as part of the energy expenditure, and because the blubber is highly efficient as a thermal insulator, hyperthermia during activity is probably more of a problem than hypothermia. The fins and tail flukes must rid the whale of virtually all excess heat, and be more efficient in this process than in small Cetacea and Pinnipedia. Kanwisher and Sundnes (1966) estimated that less than 5 % of total heat losses are through evaporation in the lungs.

GROWTH AND ENERGY REQUIREMENTS OF SUCKLING BLUE AND FIN WHALE CALVES

Growth

Tomilin (1946) estimated that suckling blue whales gain 81.3 kg/day in body weight and consume 90 kg/day of milk during the 7-month nursing period. Kirpichnikov (1949) (quoted *in* Tomilin (1967)) calculated a weight gain of 100 kg/day and a length increment of 4.3 cm/day during the unweaned growth

Table 19. Estimated heat loss through the blubber of the body of the whale, exluding tail flukes, flippers and dorsal fin

Body length ft *m*	Body surface area m²	Mean blubber thickness cm	Total heat loss kcal/day		Heat loss as % of basal metabolic rate	
			Antarctic	Durban	Antarctic	Durban
50 *15.2*	82	6.8	6.55×10^3	2.00×10^3	5.0	1.5
60 *18.3*	137	8.2	9.02×10^3	2.76×10^3	4.0	1.0
70 *21.3*	168	9.6	9.45×10^3	2.89×10^3	2.5	1.0

phase. Referring to Figs. 7 and 11 the average weight and length increments for male and female calves are probably of the order of 81 kg/day and 3.45 cm/day respectively during nursing. This value of weight gain agrees with that of Tomilin (1946), so that his estimate of milk consumption has been accepted.

Ruud (1937) estimated that suckling fin whales grow 60 kg/day and about 3 cm/day if lactation lasts 6 months. Tomilin (1946), however, estimated that sucklings grow 53 kg/day and consume 72.3 kg/day of milk if lactation lasts 7 months as it does in blue whales. Figs. 8 and 12 show that the weight gain is 52.5 kg/day and the length increment is 2.4 cm/day during 7 months of suckling.

Calories of Growth

Analyses of red meat from different parts of the body of Antarctic fin whales are recorded in the logbooks of the IOS. Results give a water content of 60-75 % and fat content of 2.75-16.00 % depending on the region of the body. Arai and Sakai (1952) estimated that

baleen whale meat consists of approximately 70.75 % water, 21.75 % protein, 6.5 % fat and 1 % mineral ash. This would represent a calorific value of about 1 800 kcal/kg, assuming that the heat equivalents of protein and fat are 5 650 kcal/kg and 9 450 kcal/kg respectively (Brody, 1968). The fat content of muscle can vary considerably according to the season and region of the body, from 0.13-34.68 %. Arai and Sakai found that whilst adult animals varied in water content from 40 % to 70 %, and embryos could hold up to 90 %, muscle in all animals contained up to 75 % water and 6-20 % fat. Petrides and Swank (1965) and Brody (1968) gave a calorific value of mammal flesh of approximately 1 500 kcal/kg.

In estimating the calories of growth, it has been assumed that most energy has been channelled into making muscle and visceral tissue. The calorific value of the growth has been estimated in Table 20.

Metabolic Energy Expenditure

Referring to Table 17, the resting metabolic rates of fin and blue whale calves of sizes

Table 20. Energy budget for calves prior to weaning

Item	Blue m + f	Fin m + f
Energy input		
milk consumed	90 kg/day	72.3 kg/day
calorific value of milk consumed, at 4 137 kcal/kg	372 330 kcal/day	299 105 kcal/day
Total energy input	372 330 kcal/day	299 105 kcal/day
Energy utilization and output		
(1) growth increment during 7 months	17 tonnes	11.5 tonnes
daily growth increment	81 kg/day	53 kg/day
calorific value of growth increment, at 1 500 kcal/kg	121 500 kcal/day	79 500 kcal/day
(2) possible average metabolic rate	200 000 kcal/day	200 000 kcal/day
Total energy utilization	321 500 kcal/day	279 500 kcal/day
Energy not utilized		
input − output = faeces	(372 330-321 500) = 50 830 kcal/day	(299 105-279 500) = 19 605 kcal/day
Gross growth efficiency	32.6 %	26.5 %
Assimilation efficiency	86 %	93 %
Net growth efficiency	37.8 %	28.5 %

25-50 ft are unlikely to exceed 2×10^5 kcal/day.

Calories of Milk Consumption

Clowes (1929) found that blue whale milk consisted of 34-37 % fat, 41-51 % water and 14-22 % solid residue comprising protein, a little sugar and mineral salts. Kirpichnikov (1949) (quoted *in* Tomilin (1967)) estimated 50.07 % fat and 49.93 % water and solids in blue whale milk. Clowes found that the fat content of dolphin and pilot whale milk was 43-46 %.

Fin whale milk was found to contain 30.2 % fat, 54.19 % water and 15.61 % solids by Clowes (1929), 31-33 % fat, 53-55t% water and 13-14 % solids by Ohta *et al.* (1955), of which the solids constituted 11.95-13.3 % protein, 0.2-1.79 % sugar and 1.43-2-16 % ash in the milk. An analysis of fin whale milk recorded in the logbooks of the IOS, showed that the milk consisted of approximately 48-52 % fat, 40 % water, 10 % protein and the remainder inorganic salts. Tomilin (1946) estimated that fin whale milk contains 22.24-44.40 % fat.

In both blue and fin whale milk, fat and protein together constitute up to 50 % of the milk by weight. If an average sample of whale milk consists of 36 % fat, 13 % protein, 1 % mineral ash and 49 % water, and the heats of combustion of fat and protein are 9 450 kcal/kg and 5 650 kcal/kg respectively, the calorific content of milk is 4 137 kcal/kg. Tomilin (1946) gave a calorific value of rorqual milk of 3 657-4 305 kcal/kg, which covers the calculated value just given. The calorific values of milk consumption by blue and fin whale calves are given in Table 20.

Energy Budget and Growth Efficiencies

A basic breakdown of energy input and output for blue and fin whale calves prior to weaning is shown in Table 20. Gross growth efficiency, defined as,

$$\frac{\text{Calories of net growth} \times 100}{\text{Calories of food consumed}}$$

and net growth efficiency defined as,

$$\frac{\text{Calories of net growth} \times 100}{\text{Calories of food assimilated}}$$

are given for each species, and fall within the expected ranges observed for other suckling and very young mammals, shown in Table 21. Rubner (1908) gave an estimate of 37 % for net growth efficiency, although the animal type was not mentioned. The calculated high assimilation efficiencies for whales expressed as percentage of calories consumed therefore seem likely, comparing Rubner's estimate to those in Table 20.

Table 21. **Gross growth efficiencies in domestic animals (based on Brody, 1968)**

Species	Body weight kg	Age	% gross growth efficiency	Comments
Rat	0.084		0.6	weaned
	0.089		8.4	and adult
	0.070		7.6	animals
	0.070		11.4	
	0.070		13.6	
Jersey cattle	35.5	1 month	37	suckling
	117.0	6 months	20	
	206.5	12 months	9	↓
	290.5	18 months	7	
		24 months	5	weaning
Holstein cattle	49.15	1 month	35	suckling
	147.7	6 months	17	
	206.0	9 months	10	↓
	260.0	12 months	5.4	
	375.0	18 months	7.0	weaning
Chicken	0.0627	1 week	35	
	0.053		26	
	1.490	24 weeks	2.5	

GROWTH AND ENERGY REQUIREMENTS OF JUVENILE BLUE AND FIN WHALES

The feeding pattern of weaned calves must be considered before the energy requirements of these animals can be estimated. The weaned calf probably commences feeding on krill in the 7th or 8th month of its life, when it accompanies its mother to the southern feeding grounds, and continues to feed on krill for a further 4 or more months. Mackintosh and Wheeler (1929), Harmer (1931) and Bannister and Baker (1967) suggested that after this feeding period in the Antarctic, the calves migrate northwards again to the lower latitudes where feeding continues, although at a greatly reduced rate owing to the decreased availability of suitable food. The young calf, however, probably feeds regularly and continuously throughout the first year of life. After the first year, the apparent growth rate slows appreciably, perhaps reflecting a change in habits from regular to irregular feeding. There does not seem to be any noticeable alteration in growth rate at weaning, even though the diet changes and the whale may need to expend more energy in actively seeking krill shoals. However, from the evidence of full stomachs of whales in the Antarctic (see Tables 6 and 7), food appears to be abundant there, so that feeding rates in the weaned calf are unlikely to differ greatly from those in the suckling whale.

On weaning, the calf probably makes initial contact with several of the parasites which infest cetaceans. In analysis of the gut contents of 4 suckling calves, 16 weaned calves and 112 adults taken in 1930/31, none of the first group, 94 % of the second group and 95 % of adults carried gut parasites such as acanthocephalans, cestodes and nematodes. These parasites very probably only reach the secondary whale host when eaten. Because of the almost total infestation of the weaned whale population by parasites, the general health and rate of growth and development of the animals cannot be seriously impaired. Other external skin parasites and baleen plate commensals probably affect the growth and metabolism of the whale even less.

After the first year of life, the feeding pattern becomes distinctly seasonal. The section on "Correlation of seasonal feeding migration and variations in blubber thickness and body weight" shows that the whale stores much fat during the feeding season in order to survive the poor feeding conditions throughout the remainder of the year.

Growth

The juvenile whale is growing both in terms of development and temporary deposition of fat for the period of poor feeding conditions. The temporary fat depot amounts to between 30 % and 50 % of lean body weight, and the distribution and extent of fattening in various tissues can be estimated from Table 4 and Figs. 26 to 29 inclusive. In Tables 22 and 23 the calorific values of the energy stored during the pre-pubertal year for blue and fin whales respectively are calculated assuming the body sizes indicated in Figs. 7 and 8. The maximum fat tissue content has been assumed at 80 %.

Analyses made during the Antarctic summer of blubber oil content of rorquals, recorded in the logbooks of the IOS, show that the regions of greatest oil yield (76-82 %) are in the dorsal body parts, whilst the head and throat regions yield only 32-39 % oil. The connective tissue forms 6-22 % of the blubber. Clowes (1929a) found that rorqual blubber sampled in summer consists of up to 80 % liquid oil and wax. Recent analyses of the blubber from 2 specimens of pigmy blue whales stranded off Albany in Western Australia during winter, kindly performed by Dr R. Morris of the IOS, show that lateral blubber consists of 73.1-83.3 % lipid, 4.9-8.4 % protein and fibres, and 8.3-22 % water. It would seem likely that a fixed percentage of the fat tissue stored in any of the body organs must be connective tissues similar to those in blubber

459

Table 22. Yearly energy budget for blue whales at puberty

Item	SEX	
	male	female
Energy input:		
(1) food consumed in Antarctic at 35 g/kg body weight/day	$72\,000 \times 0.035 = 2\,500$ kg/day	$87\,000 \times 0.35 = 3\,000$ kg/day
calorific value of food consumed at 1 000 kcal/kg	2.5×10^6 kcal/day	3.0×10^6 kcal/day
total calories consumed during 120 days' feeding in Antarctic	$120 \times 2.5 \times 10^6 = 3.00 \times 10^8$ kcal	$120 \times 3.0 \times 10^6 = 3.60 \times 10^8$ kcal
(2) possible reduced rate of food consumption in low latitudes at 3.5 g/kg body weight/day	$72\,000 \times 0.0035 = 250$ kg/day	$87\,000 \times 0.035 = 300$ kg/day
calorific value of food consumed at 1 000 kcal/kg	2.5×10^5 kcal/day	3.0×10^5 kcal/day
total calories consumed during 245 days' feeding in low latitudes	$245 \times 2.5 \times 10^5 = 0.61 \times 10^8$ kcal	$245 \times 3.0 \times 10^5 = 0.74 \times 10^8$ kcal
Total annual calorie intake	$(3.00 + 0.61) \times 10^8 = 3.61 \times 10^8$ kcal	$(3.60 + 0.74) \times 10^8 = 4.34 \times 10^8$ kcal
Energy utilization and output:		
(1) total growth increment during pre-pubertal year (see Figure 11)	6 000 kg	8 000 kg
blubber weight gain as 27 % total growth increment	1 600 kg	2 100 kg
calorific value of blubber oil and visceral fats at 9 450 kcal/kg	$1\,600 \times 9\,450 = 1.5 \times 10^7$ kcal	$2\,100 \times 9\,450 = 2.0 \times 10^7$ kcal
meat weight gain as 39 % total growth increment	2 300 kg	3 100 kg
calorific value of meat at 1 500 kcal/kg	$2\,300 \times 1\,500 = 0.3 \times 10^7$ kcal	$3\,100 \times 1\,500 = 0.4 \times 10^7$ kcal
Total calories of growth during pre-pubertal year	$(1.5 + 0.3) \times 10^7 = 0.18 \times 10^8$ kcal	$(2.0 + 0.4) \times 10^7 = 0.24 \times 10^8$ kcal
(2) daily metabolic energy expenditure at low activity (resting rate, see Table 17) at puberty	5.7×10^5 kcal/day	6.8×10^5 kcal/day
230 days in low latitudes at this metabolic rate	$5.7 \times 10^5 \times 230 = 1.31 \times 10^8$ kcal	$6.8 \times 10^5 \, 230 = 1.56 \times 10^8$ kcal
120 days in Antarctic at this metabolic rate	$5.7 \times 10^5 \times 120 = 0.68 \times 10^8$ kcal	$6.8 \times 10^5 \times 120 = 0.82 \times 10^8$ kcal
daily metabolic energy expenditure at high activity (active rate, see Table 17) at puberty	4.5×10^6 kcal/day	5.7×10^6 kcal/day
15 days (or time equivalent) of continuous migration at this metabolic rate	$4.5 \times 10^6 \times 15 = 0.68 \times 10^8$ kcal	$5.7 \times 10^6 \times 15 = 0.86 \times 10^8$ kcal
Total minimum expected metabolic energy expenditure	$(1.31 + 0.68 + 0.68) \times 10^8 =$ 2.67×10^8 kcal	$(1.56 + 0.82 + 0.86) \times 10^8 =$ 3.24×10^8 kcal
(3) temporary energy storage in year (see Figure 26)	27 000 kg	33 000 kg
calorific content of fat storage at 7 560 kcal/kg (maximum fat content of tissue assumed 80 %)	$27\,000 \times 7\,560 = 2.04 \times 10^8$ kcal	$33\,000 \times 7\,560 = 2.50 \times 10^8$ kcal
Total energy storage in year	2.04×10^8 kcal	2.50×10^8 kcal
Total energy utilization (neglecting (3))	2.85×10^8 kcal	3.48×10^8 kcal
Energy not utilized:		
input − output = faeces	$(3.61\text{-}2.85) \times 10^8 = 0.76 \times 10^8$ kcal	$(4.34\text{-}3.48) \times 10^8 = 0.86 \times 10^8$ kcal
Gross growth efficiency	5.0 %	5.6 %
Assimilation efficiency	79 %	80 %
Net growth efficiency	6.3 %	6.9 %

Table 23. Yearly energy budget for fin whales at puberty

Item	SEX	
	male	female
Energy input:		
(1) food consumed in Antarctic at 35 g/kg body weight/day	$43\,500 \times 0.035 = 1\,500$ kg/day	$47\,500 \times 0.035 = 1\,660$ kg/day
calorific value of food consumed at 1 000 kcal/kg	1.5×10^6 kcal/day	1.7×10^6 kcal/day
total calories consumed during 120 days' feeding in Antarctic	$120 \times 1.5 \times 10^6 = 1.80 \times 10^8$ kcal	$120 \times 1.7 \times 10^6 = 2.03 \times 10^8$ kcal
(2) possible reduced rate of food consumption in low latitudes at 3.5 g/kg body weight/day	$43\,500 \times 0.0035 = 150$ kg/day	$47\,500 \times 0.0035 = 166$ kg/day
calorific value of food consumed at 1 000 kcal/kg	1.5×10^5 kcal/day	1.7×10^5 kcal/day
total calories consumed during 245 days' feeding in low latitudes	$245 \times 1.5 \times 10^5 = 0.37 \times 10^8$ kcal	$245 \times 1.7 \times 10^5 = 0.42 \times 10^8$ kcal
Total annual calorie intake	$(1.80 + 0.37) \times 10^8 = 2.17 \times 10^8$ kcal	$(2.03 + 0.42) \times 10^8 = 2.45 \times 10^8$ kcal
Energy utilization and output:		
(1) total growth increment during pre-pubertal year (see Figure 12)	3 000 kg	3 500 kg
blubber weight gain as 24 % total growth increment	720 kg	840 kg
calorific value of blubber oil and visceral fats at 9 450 kcal/kg	$720 \times 9\,450 = 6.8 \times 10^6$ kcal	$840 \times 9\,450 = 7.9 \times 10^6$ kcal
meat weight gain as 45 % total growth increment	1 350 kg	1 575 kg
calorific value of meat at 1 500 kcal/kg	$1\,350 \times 1\,500 = 2.0 \times 10^6$ kcal	$1\,575 \times 1\,500 = 2.4 \times 10^6$ kcal
Total calories of growth during pre-pubertal year	$(6.8 + 2.0) \times 10^6 = 0.09 \times 10^8$ kcal	$(7.9 + 2.4) \times 10^6 = 0.10 \times 10^8$ kcal
(2) daily metabolic energy expenditure at low activity (resting rate, see Table 17) at puberty	3.6×10^5 kcal/day	3.9×10^5 kcal/day
230 days in low latitudes at this metabolic rate	$3.6 \times 10^5 \times 230 = 0.83 \times 10^8$ kcal	$3.9 \times 10^5 \times 230 = 0.87 \times 10^8$ kcal
120 days in Antarctic at this metabolic rate	$3.6 \times 10^5 \times 120 = 0.43 \times 10^8$ kcal	$3.9 \times 10^5 \times 120 = 0.47 \times 10^8$ kcal
daily metabolic energy expenditure at high activity (active rate, see Table 17) at puberty	3.0×10^6 kcal/day	3.3×10^6 kcal/day
15 days (or time equivalent) of continuous migration at this metabolic rate	$3.0 \times 10^6 \times 15 = 0.45 \times 10^8$ kcal	$3.3 \times 10^6 \times 15 = 0.49 \times 10^8$ kcal
Total minimum expected metabolic energy expenditure	$(0.83 + 0.43 + 0.45) \times 10^8 =$ 1.71×10^8 kcal	$(0.87 + 0.47 + 0.49) \times 10^8 =$ 1.83×10^8 kcal
(3) temporary energy storage in year (see Figure 27)	10 000 kg	10 750 kg
calorific content of fat storage at 7 560 kcal/kg (maximum fat content of tissue assumed 80 %)	$10\,000 \times 7\,560 = 0.76 \times 10^8$ kcal	$10\,750 \times 7\,560 = 0.82 \times 10^8$ kcal
Total energy storage in year	0.76×10^8 kcal	0.82×10^8 kcal
Total energy utilization (neglecting (3))	$(0.09 + 1.71) \times 10^8 = 1.80 \times 10^8$ kcal	$(0.10 + 1.83) \times 10^8 = 1.93 \times 10^8$ kcal
Energy not utilized:		
input − output = faeces	$(2.17 - 1.80) \times 10^8 = 0.37 \times 10^8$ kcal	$(2.45 - 1.93) \times 10^8 = 0.52 \times 10^8$ kcal
Gross growth efficiency	4.0 %	4.0 %
Assimilation efficiency	83 %	79 %
Net growth efficiency	5.0 %	5.2 %

tissue. This temporary energy depot becomes dissipated by metabolic expenditure throughout the year, and so cannot be considered in estimating true growth efficiency.

The amount of permanent developmental growth during the pre-pubertal year can be determined from Figs. 11 and 12 for blue and fin whales respectively. At puberty the female blue whale is 78.25 ft and weighs about 87 000 kg; the male blue whale is 74 ft and weighs about 72 000 kg. The female fin whale is 65.5 ft and weighs 47 500 kg at puberty; the male fin whale is 63.5 ft and weighs 43 500 kg at puberty. The growth increments during the year prior to attaining these sizes are shown in Tables 22 and 23. Table 5 shows the distribution of this growth in weight, so that the calorific values of the annual growth increments can be calculated as in Tables 22 and 23 for blue and fin whales. In calculating the calorific value of the increase in blubber tissue, a straight conversion of weight using the total calorific value of fat has been used in order to allow for increases in fat/oil elsewhere in the body, for example, viscera and bone. The total calorific value of the growth increment is therefore a close but inexact estimation, yet does not alter the basic calculations in the energy budget.

Metabolic Energy Expenditure

During the year, the whale probably expends most energy at an economical resting rate, though there is probably a high rate of energy use during migration. From the observations on mark returns mentioned above, it is clear that baleen whales such as blue, fin and humpback are able to migrate seasonally between extreme latitudinal limits of 20°S and 65°S, representing a distance of about 6 000 miles there and back. Brown (1974) gave evidence that a sei whale marked in the Antarctic, was recovered 10 days later, 2 200 miles distant, indicating an average sustained migratory speed of just over 9 mi/h (14.5 km/h). In Table 15 it is clear that this speed falls within the anticipated range of cruising speeds. If during migration the large rorquals blue and fin are able to swim at a faster sustained speed of about 15 mi/h (24 km/h – the speed assumed in calculating active metabolic rate in Table 16), then a period of at least 15 days would be necessary to cover the journey of 6 000 miles. In Tables 22 and 23 the energy expenditure during migration and during the rest of the year is calculated, referring to Table 17 for values of resting and active metabolic rates for the relevant sizes of whales.

Calories of Food Consumed

Referring back to the section "Food and feeding patterns", it can be assumed that blue and fin whales feed at an average rate of 35 g/kg body weight/day for 120 days in the Antarctic summer, and at an average reduced rate of 10 % of summer rates of the order of 3.5 g/kg body weight/day for the remaining 245 days (but whales may not feed at all during migration). Using a conversion rate of 1 kg wet krill to 1 000 kcal, the calorific value of food consumed can be estimated as in Tables 22 and 23.

Energy Budget and Growth Efficiencies

Tables 22 and 23 show the input and output of energy in pubertal blue and fin whales. Over a year, assimilation efficiencies appear to attain levels of about 80 %. This value is high, yet in harmony with the findings of Birge and Juday (1922) who reckoned that at least 10 % of food energy is unobtainable in primary consumers, and 5 % in tertiary consumers. The greatest efficiency of assimilation is reached in animals in the secondary consumer category upwards. Primary consumers, for example, the Indian elephant (Benedict, 1936), may only achieve 32-37 % assimilation efficiency, whereas secondary consumers such as salmon (Parson and Lebrasseur, 1970) and plaice (Birkett, 1970) may achieve 80 % and 92 % efficiency respectively. The unobtainable energy in food is trapped in indigestible tissues. In krill and Crustacea generally, most of the unobtainable energy is trapped in the chitinous integuments. According to the information on energy content and composition of the krill, at least 5 % of wet weight or 20 % of dry weight of krill is in an unutilizable form, so that about 80 % efficiency of assimilation is unlikely to be greatly exceeded.

The growth efficiencies calculated in Tables 22 and 23 show that gross efficiency is about 5 % and net efficiency only slightly higher. These values are far less than those calculated for suckling calves, yet referring to Table 21, they are of the order of magnitude to be expected for animals of this age.

In Figs. 46 and 47 schematic representations of the energy budgets for pubertal blue and fin whales are shown, based on the calculations given in Tables 22 and 23. It is clear that for blue whales of both sexes, supplementary feeding would not appear to be essential, provided that the fat stored for the

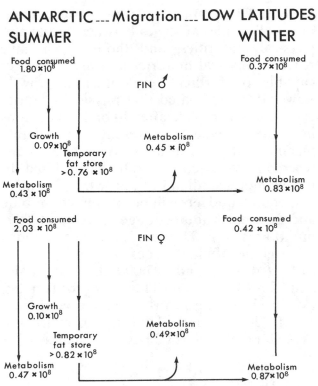

FIG. 47. – Energy pathway in juvenile fin whales.

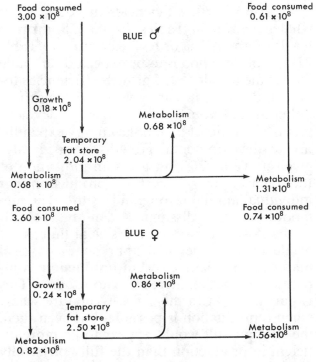

FIG. 46. – Energy pathway in juvenile blue whales.

winter season attains the levels shown. However, the fat depot in fin whales appears too small to meet the energy requirements encountered during winter, even when supplemented by occasional winter feeding. As pointed out earlier in the sub-section on seasonal variation in body weight, the time span of the fin whale weight data was insufficient to give a true indication of the extent of fattening during the summer, so that one can reasonably predict that, in fact, the 50 % increase in body weight observed in blue whales while on the Antarctic feeding grounds is also true for fin whales, because unless this was so, they would starve.

ENERGY BUDGET FOR YOUNG ADULT BLUE AND FIN WHALES

Any whale which has attained sexual maturity and can reproduce can be classified as adult. The adult population however comprises all whales at stages between sexual and physical maturities, and those which have reached physical maturity and are no longer capable of further developmental growth. Laws (1961) estimated that physical maturity is generally complete after 14 or 15 ovulations in female fin whales, representing a probably mean age at physical maturity of 25-30 years, when all the vertebral epiphyses are fused. In the examples studied here, growth efficiencies and energy budgets will be calculated for blue and fin whales adults of age 15 years. Referring to Figs. 7, 8, 11 and 12, the mean lengths and weights of blue whales at age 15 years are 81 ft (24.3 m) and 97 000 kg for males and 85 ft (25.5 m) and 114 000 kg for females. For fin whales, the mean lengths and weights are 67.5 ft (20.2 m) and 51 500 kg for males and 72 ft (21.6 m) and 62 000 kg for females.

Growth

The developmental growth increments for blue and fin whales of both sexes in the period 14 to 15 years can be ascertained from Figs. 11 and 12. They are given in Tables 24 and 25 where the calorific value of the increments is calculated in the usual way.

The temporary fat depots have been determined from Figs. 26 and 27 for the correct sizes of whales and the calorific values are calculated in Tables 24 and 25.

Metabolic Energy Expenditure

The metabolic rates can be determined from Table 17 for the same periods of time as discussed for juvenile whales. The calculations are shown in Tables 24 and 25.

Calorific Value of Food Consumption

The energy of food consumed by adult blue and fin whales has been calculated in a similar manner to that for juveniles, and is shown in Tables 24 and 25.

Energy Budget and Growth Efficiencies

Energy inputs and outputs are given in Tables 24 and 25. Assimilation efficiencies approach 80 %, and therefore do not appear greatly different from the values calculated for pubertal whales. The energy budget pathways for blue and fin whales are illustrated in Figs. 48 and 49. As observed for pubertal whales above, the energy input (feeding in winter and fat depot) required to cover the metabolic energy expenditure in low latitudes and throughout migration, is insufficient for fin whales. This finding again reinforces the argument that a body weight increase of close to 50 % in the Antarctic would be essential for the survival of fin whales; it would also provide a margin of safety in the possible event of enforced fasting during winter because of poor food availability. Weight increases may exceed 50 % body weight, as for example in humpbacks (Appendix, Table 1).

The growth efficiencies in young adult whales, shown in Tables 24 and 25, drop to levels of about 1 % or less. Referring to Table 21, this observation is to be expected. In Table 26, the mean calculated growth efficiencies for blue and fin whales are given at the suckling, pubertal and young adult stages of development. The ratio of metabolic energy expenditure to developmental growth energy is also given in Table 26, and gives an indication of the relative energy cost of production per individual animal throughout its life. The metabolic energy dissipated can be considered as lost to the system. It is clear that as the whale grows larger and approaches physical maturity, the relative cost of production, compared with calves, increases enormously. This is partly off-set in the sexually mature whale when reproduction is possible, but even then, the young adult would appear to be more efficient in production than the fully physically mature adult.

Table 24. Yearly energy budget for young adult blue whales

Item	SEX	
	male	female

Energy input:

(1) food consumed in Antarctic at 35 g/kg body weight/day
$97,000 \times 0.035 = 3\,400$ kg/day — $114\,000 \times 0.035 = 4\,000$ kg/day

calorific value of food consumed at 1 000 kcal/kg
3.4×10^6 kcal/day — 4.0×10^6 kcal/day

total calories consumed during 120 days' feeding in Antarctic
$120 \times 3.4 \times 10^6 = 4.08 \times 10^8$ kcal — $120 \times 4.0 \times 10^6 = 4.80 \times 10^8$ kcal

(2) possible reduced rate of food consumption in low latitudes at 3.5 g/kg weight/day
$97\,000 \times 0.0035 = 340$ kg/day — $114\,000 \times 0.0035 = 400$ kg/day

calorific value of food consumed at 1 000 kcal/kg
3.4×10^5 kcal/day — 4.0×10^5 kcal/day

total calories consumed during 245 days' feeding in low latitudes
$245 \times 3.4 \times 10^5 = 0.83 \times 10^8$ kcal — $245 \times 4.0 \times 10^5 = 0.98 \times 10^8$ kcal

Total annual calorie intake
$(4.08 + 0.83) \times 10^8 = 4.91 \times 10^8$ kcal — $(4.80 + 0.98) \times 10^8 = 5.78 \times 10^8$ kcal

Energy utilization and output:

(1) total growth increment between 14 and 15 years (see Figure 11)
1 000 kg — 1 500 kg

blubber weight gain as 27 % total growth increment
270 kg — 405 kg

calorific value of blubber oil and visceral fats at 9 450 kcal/kg
$270 \times 9\,450 = 2.5 \times 10^6$ kcal — $405 \times 9\,450 = 3.8 \times 10^6$ kcal

meat weight gain as 39 % total growth increment
390 kg — 585 kg

calorific value of meat at 1 500 kcal/kg
$390 \times 1\,500 = 0.6 \times 10^6$ kcal — $585 \times 1\,500 = 0.9 \times 10^6$ kcal

Total calories of growth between 14 and 15 years
$(2.5 + 0.6) \times 10^6 = 0.03 \times 10^8$ kcal — $(3.8 + 0.9) \times 10^6 = 0.05 \times 10^8$ kcal

(2) daily metabolic energy expenditure at low activity (resting rate, see Table 17)
7.8×10^5 kcal/day — 9.20×10^5 kcal/day

230 days in low latitudes at this metabolic rate
$7.8 \times 10^5 \times 230 = 1.79 \times 10^8$ kcal — $9.2 \times 10^5 \times 230 = 2.12 \times 10^8$ kcal

120 days in Antarctic at this metabolic rate
$7.8 \times 10^5 \times 120 = 0.94 \times 10^8$ kcal — $9.2 \times 10^5 \times 120 = 1.10 \times 10^8$ kcal

daily metabolic energy expenditure at high activity (active rate, see Table 17)
6.5×10^6 kcal/day — 7.7×10^6 kcal/day

15 days (or time equivalent) of continuous migration at this metabolic rate
$6.5 \times 10^6 \times 15 = 0.98 \times 10^8$ kcal — $7.7 \times 10^6 \times 15 = 1.16 \times 10^8$ kcal

Total minimum expected metabolic energy expenditure
$(1.79 + 0.94 + 0.98) \times 10^8 = 3.71 \times 10^8$ kcal — $(2.12 + 1.10 + 1.16) \times 10^8 = 4.38 \times 10^8$ kcal

(3) temporary energy storage in year (see Figure 26)
37 000 kg — 42 000 kg

calorific content of fat storage at 7 560 kcal/kg (maximum fat content of tissue assumed 80 %)
$37\,000 \times 7\,560 = 2.79 \times 10^8$ kcal — $42\,000 \times 7\,560 = 3.17 \times 10^8$ kcal

Total energy storage in year
2.79×10^8 kcal — 3.17×10^8 kcal

Total energy utilization (neglecting (3)):
$(0.03 + 3.71) \times 10^8 = 3.74 \times 10^8$ kcal — $(0.05 + 4.38) \times 10^8 = 4.43 \times 10^8$ kcal

Energy not utilized:

input – output = faeces
$(4.91 - 3.74) \times 10^8 = 1.17 \times 10^8$ kcal — $(5.78 - 4.43) \times 10^8 = 1.35 \times 10^8$ kcal

Item	male	female
Gross growth efficiency	0.6 %	0.8 %
Assimilation efficiency	76 %	77 %
Net growth efficiency	0.8 %	1.1 %

Table 25. Yearly energy budget for young adult fin whales

Item	SEX	
	male	female
Energy input:		
(1) food consumed in Antarctic at 35 g/kg body weight/day	$51\,500 \times 0.035 = 1\,800$ kg/day	$62\,000 \times 0.035 = 2\,170$ kg/day
calorific value of food consumed at 1 000 kcal/kg	1.8×10^6 kcal/day	2.2×10^6 kcal/day
total calories consumed during 120 days' feeding in Antarctic	$120 \times 1.8 \times 10^6 = 2.17 \times 10^8$ kcal	$120 \times 2.2 \times 10^6 = 2.64 \times 10^8$ kcal
(2) possible reduced rate of food consumption in low latitudes at 3.5 kg body weight/day	$51\,500 \times 0.0035 = 180$ kg/day	$62\,000 \times 0.0035 = 217$ kg/day
calorific value of food consumed at 1 000 kcal/kg	1.8×10^5 kcal/day	2.2×10^5 kcal/day
total calories consumed during 245 days' feeding in low latitudes	$245 \times 1.8 \times 10^5 = 0.44 \times 10^8$ kcal	$245 \times 2.2 \times 10^5 = 0.54 \times 10^8$ kcal
Total annual calorie intake	$(2.17 + 0.44) \times 10^8 = 2.61 \times 10^8$ kcal	$(2.64 + 0.54) \times 10^8 = 3.18 \times 10^8$ kcal
Energy utilization and output:		
(1) total growth increment between 14 and 15 years (see Figure 12)	700 kg	1 000 kg
blubber weight gain as 24 % total growth increment	168 kg	240 kg
calorific value of blubber oil and visceral fats at 9 450 kcal/kg	$168 \times 9\,450 = 1.6 \times 10^6$ kcal	$240 \times 9\,450 = 2.3 \times 10^6$ kcal
meat weight gain as 45 % total growth increment	315 kg	450 kg
calorific value of meat at 1 500 kcal/kg	$315 \times 1\,500 = 0.5 \times 10^6$ kcal	$450 \times 1\,500 = 0.7 \times 10^6$ kcal
Total calories of growth between 14 and 15 years	$(1.6 + 0.5) \times 10^6 = 0.02 \times 10^8$ kcal	$(2.3 + 0.7) \times 10^6 = 0.03 \times 10^8$ kcal
(2) daily metabolic energy expenditure at low activity (resting rate, see Table 17)	4.2×10^5 kcal/day	4.9×10^5 kcal/day
230 days in low latitudes at this metabolic rate	$4.2 \times 10^5 \times 230 = 0.97 \times 10^8$ kcal	$4.9 \times 10^5 \times 230 = 1.13 \times 10^8$ kcal
120 days in Antarctic at this metabolic rate	$4.2 \times 10^5 \times 120 = 0.51 \times 10^8$ kcal	$4.9 \times 10^5 \times 120 = 0.59 \times 10^8$ kcal
daily metabolic energy expenditure at high activity (active rate, see Table 17)	3.8×10^6 kcal/day	4.6×10^6 kcal/day
15 days (or time equivalent) of continuous migration at this metabolic rate	$3.8 \times 10^6 \times 15 = 0.57 \times 10^8$ kcal	$4.6 \times 10^6 \times 15 = 0.69 \times 10^8$ kcal
Total minimum expected metabolic energy expenditure	$(0.97 + 0.51 + 0.57) \times 10^8 =$ 2.05×10^8 kcal	$(1.13 + 0.59 + 0.69) \times 10^8 =$ 2.41×10^8 kcal
(3) temporary energy storage in year (see Figure 27)	11 500 kg	13 000 kg
calorific content of fat storage at 7 560 kcal/kg (maximum fat content of tissue assumed 80 %)	$11\,500 \times 7\,560 = 0.87 \times 10^8$ kcal	$13\,000 \times 7\,560 = 0.98 \times 10^8$ kcal
Total energy storage in year	0.87×10^8 kcal	0.98×10^8 kcal
Total energy utilization (neglecting (3))	$(0.02 + 2.05) \times 10^8 = 2.07 \times 10^8$ kcal	$(0.03 + 2.41) \times 10^8 = 2.44 \times 10^8$ kcal
Energy not utilized:		
input − output = faeces	$(2.61 - 2.07) \times 10^8 = 0.54 \times 10^8$ kcal	$(3.18 - 2.44) \times 10^8 = 0.74 \times 10^8$ kcal
Gross growth efficiency	0.7 %	0.9 %
Assimilation efficiency	79 %	77 %
Net growth efficiency	1.0 %	1.2 %

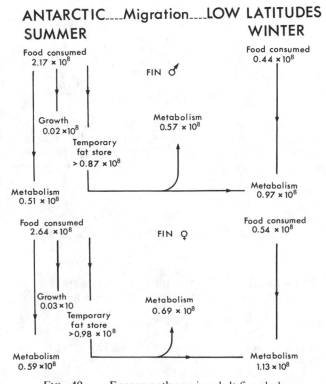

FIG. 48. – Energy pathway in adult path whales.

ANTARCTIC....Migration....LOW LATITUDES
SUMMER   WINTER

Food consumed   Food consumed
2.17 × 10⁸   0.44 × 10⁸

                    FIN ♂

Growth   Metabolism
0.02 ×10⁸   0.57 × 10⁸

        Temporary
        fat store
        > 0.87 × 10⁸

Metabolism   Metabolism
0.51 × 10⁸   0.97 × 10⁸

Food consumed   Food consumed
2.64 × 10⁸   0.54 × 10⁸

                    FIN ♀

Growth   Metabolism
0.03 × 10   0.69 × 10⁸

        Temporary
        fat store
        >0.98 × 10⁸

Metabolism   Metabolism
0.59 ×10⁸   1.13 × 10⁸

FIG. 49. – Energy pathway in adult fin whales.

Table 26. **Mean growth efficiencies and ratios of metabolic energy expenditure to growth energy in blue and fin whales**

Quantity measured	Suckling calves	Pubertal juveniles	Physically immature adults
Gross growth efficiency	30	4.6	0.8
Net growth efficiency	33	5.7	1.0
Metabolic energy			
Growth energy	2.1	16.4	98.5

ENERGY REQUIREMENTS OF PREGNANT AND LACTATING WHALES

In Table 27 the energy budgets for a 2 year breeding cycle are calculated for an 86 ft (25.8 m) female blue weighing 118 000 kg, and a 73 ft (21.9 m) female fin whale weighing 64 500 kg. To simplify calculations, the whales are assumed to be physically mature and the data are taken from Figs. 7, 8, 11 and 12.

The energy budget listed in Table 27 is mostly self-explanatory. However, there are some points which require clarification. In the predicted energy budget the female commences the pregnancy during early July (winter) in low latitudes, and the gestation, lasting 11 months, continues through one summer season in the Antarctic and part of the following season in low latitudes where the calf is born during June. The female then nurses the calf for 7 months here until the next summer season in the Antarctic when the calf is weaned. The female then possibly enters a resting phase, completing the 2 year cycle.

In Table 27 the different stages of the cycle in the first and second years have been treated separately. In the second year, allowances have been added to the female body weight to correlate with the increased weight, due to growth of the foetus. For this reason the foetal growth (Fig. 1) has been considered in 3

467

Table 27. Energy budget over 2 years for fully mature female blue and fin whales, covering entire pregnancy and lactation

Item	Blue	Fin
Energy input: — Year 1 Feeding in pre-conception phase		
(1) food consumed in Antarctic at 35 g/kg body weight/day	$118\,000 \times 0.035 = 4\,130$ kg/day	$64\,500 \times 0.035 = 2\,258$ kg/day
calorific value of food consumed at 1 000 kcal/kg	4.13×10^6 kcal/day	2.26×10^6 kcal/day
total calories consumed during 120 days feeding in Antarctic	$120 \times 4.13 \times 10^6 = 4.96 \times 10^8$ kcal	$120 \times 2.26 \times 10^6 = 2.71 \times 10^8$ kcal
Feeding in pre- and post-conception phase		
(2) possible reduced rate of food consumption in low latitudes at 3.5 g/kg body weight/day	$118\,000 \times 0.0035 = 413$ kg/day	$64\,500 \times 0.0035 = 226$ kg/day
calorific value of food consumed at 1 000 kcal/kg	4.13×10^5 kcal/day	2.26×10^5 kcal/day
total calories consumed during 245 days' feeding in low latitudes	$245 \times 4.13 \times 10^5 = 1.01 \times 10^8$ kcal	$245 \times 2.26 \times 10^5 = 0.55 \times 10^8$ kcal
Total annual calorie intake	$(4.96 + 1.01) \times 10^8 = 5.97 \times 10^8$ kcal	$(2.71 + 0.55) \times 10^8 = 3.26 \times 10^8$ kcal
Energy input: — Year 2 Feeding during pregnancy		
(1) food consumed in Antarctic at 35 g/kg body weight/day	$119\,000 \times 0.035 = 4\,170$ kg/day	$65\,300 \times 0.035 = 2\,286$ kg/day
calorific value of food consumed at 1 000 kcal/kg	4.17×10^6 kcal/day	2.28×10^6 kcal/day
total calories consumed during 120 days' feeding in Antarctic	$120 \times 4.17 \times 10^6 = 5.00 \times 10^8$ kcal	$120 \times 2.28 \times 10^6 = 2.74 \times 10^8$ kcal
Feeding during pregnancy and lactation		
(2) possible reduced rate of food consumption in low latitudes at 3.5 g/kg body weight/day	$120\,500 \times 0.0035 = 420$ kg/day	$66\,300 \times 0.0035 = 232$ kg/day
calorific value of food consumed at 1 000 kcal/kg	4.20×10^5 kcal/day	2.32×10^5 kcal/day
total calories consumed during 245 days' feeding in low latitudes	$245 \times 4.20 \times 10^5 = 1.03 \times 10^8$ kcal	$245 \times 2.32 \times 10^5 = 0.57 \times 10^8$ kcal
Total annual calorie intake	$(5.00 + 1.03) \times 10^8 = 6.03 \times 10^8$ kcal	$(2.74 + 0.57) \times 10^8 = 3.31 \times 10^8$ kcal
Calorie intake throughout 2 year breeding cycle:	$(5.97 + 6.03) \times 10^8 = 12.00 \times 10^8$ kcal	$(3.26 + 3.31) \times 10^8 = 6.57 \times 10^8$ kcal
Energy utilization and output:		
(1) growth increment, estimated from Table 26 as < 1 % gross growth efficiency	$12.00 \times 10^8 \times 10^{-2} = 0.12 \times 10^8$ kcal	$6.57 \times 10^8 \times 10^{-2} = 0.07 \times 10^8$ kcal
(2) daily metabolic energy expenditure at low activity (resting rate, see Table 17)	9.6×10^5 kcal/day	5.11×10^5 kcal/day
230 days in low latitudes at this metabolic rate	$9.6 \times 10^5 \times 230 = 2.21 \times 10^8$ kcal	$5.11 \times 10^5 \times 230 = 1.17 \times 10^8$ kcal
120 days in Antarctic at this metabolic rate	$9.6 \times 10^5 \times 120 = 1.15 \times 10^8$ kcal	$5.11 \times 10^5 \times 120 = 0.61 \times 10^8$ kcal
daily metabolic energy expenditure at high activity (active rate, see Table 17)	8.08×10^6 kcal/day	4.81×10^6 kcal/day
15 days (or time equivalent) of continuous migration at this metabolic rate	$8.08 \times 10^6 \times 15 = 1.21 \times 10^8$ kcal	$4.81 \times 10^6 \times 15 = 0.72 \times 10^8$ kcal
Total annual energy expenditure	$(2.21 + 1.15 + 1.21) \times 10^8 = 4.57 \times 10^8$ kcal	$(1.17 + 0.61 + 0.72) \times 10^8 = 2.50 \times 10^8$ kcal

Table 27. **Energy budget over 2 years for fully mature female blue and fin whales, covering entire pregnancy and lactation** *(concluded)*

Item	Blue	Fin
Minimum energy expenditure in 2 years	$4.57 \times 10^8 \times 2 = 9.14 \times 10^8$ kcal	$2.50 \times 10^8 \times 2 = 5.00 \times 10^8$ kcal
(3) temporary energy storage 2 years (see Figures 26 and 27)	(45 000 + 56 250) kg resting + pregnant	(16 000 + 20 000) kg resting + pregnant
calorific content of fat storage at 7 560 kcal/kg (maximum fat content of tissue assumed 80 %)	$101\ 250 \times 7\ 560 = 7.65 \times 10^8$ kcal	$36\ 000 \times 7\ 560 = 2.72 \times 10^8$ kcal
Total energy storage in 2 years		
(4) energy of foetal growth (from Figure 1)		
July to December	100 kg	70 kg
December to April	1 050 kg	710 kg
April to June	1 480 kg	1 020 kg
Total growth	2 630 kg	1 800 kg
calories of growth at 1 500 kcal/kg	$2\ 630 \times 1\ 500 = 0.04 \times 10^8$ kcal	$1\ 800 \times 1\ 500 = 0.03 \times 10^8$ kcal
(5) energy of lactation		
quantity/day for 210 days	90 kg/day	72.3 kg/day
at a calorific content of 4 137 kcal/kg	$90 \times 210 \times 4\ 137 = 0.78 \times 10^8$ kcal	$72.3 \times 210 \times 4\ 137 = 0.63 \times 10^8$ kcal
Total energy utilization (neglecting (3))	$(0.12 + 9.14 + 0.04 + 0.78) \times 10^8$ $= 10.08 \times 10^8$ kcal	$(0.07 + 5.00 + 0.03 + 0.63) \times 10^8$ $= 5.73 \times 10^8$ kcal
Energy not utilized:		
input − output = faeces	$(12.00 - 10.08) \times 10^8 = 1.92 \times 10^8$ kcal	$(6.57 - 5.73) \times 10^8 = 0.84 \times 10^8$ kcal
Assimilation efficiency	84 %	87 %

stages to correspond with the feeding activities and location of the female, and to permit an estimation of the approximate weight increase in her body.

The fat stored for the winter by the pregnant female is about 20-25 % above the level in resting females, judging from the blubber thickness data in Figs. 22 and 23. The amounts of fat stored are taken from Figs. 26 and 27, with 25 % added for the pregnant phase.

In the estimates of energy cost of pregnancy, the metabolic rate of the foetus has not been considered in Table 27. However, calculation of a metabolic rate for a foetus of mean weight 1 000-1 500 kg throughout pregnancy, using the formula $70.5W^{0.7325}$ shown above, gives a value of 1.34×10^4 kcal/day. Throughout a gestation time of 330-335 days this represents about 4.4×10^6 kcal. This is about equal to the actual energy cost of the foetal growth. However, because the metabolic energy expenditure of the foetus is chiefly on growth and development, the basal metabolic rate just calculated has limited meaning. The value is probably sufficiently low not to affect the basic calculations in Table 27, but should be kept in mind as another drain on the female's energy budget which is illustrated in Fig. 50.

The fat stored appears to be about balanced by the energy utilization and output in the female blue whale. This indicates that the summer increase in body weight during pregnancy probably amounts to at least 60-65 % after taking into consideration the additional 25 % over the usual 50 % observed for resting females. Lactation obviously drains the energy reserves very greatly and because the input and output are not perfectly balanced in blue

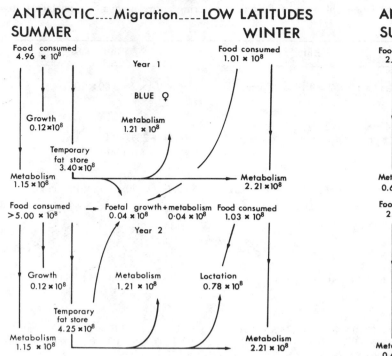

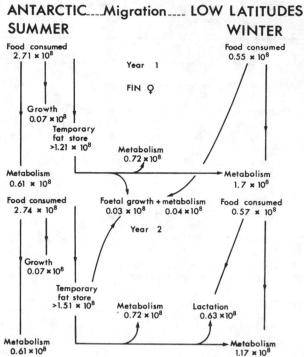

FIG. 50. — Energy pathway for blue and fin whales in pregnancy and lactation.

and fin whales, it is clear why the blubber of lactating females shown in Figs. 17 and 19 is lean and emaciated compared with that of other females. The fin females can also be considered to accumulate an extra 60-65 % body weight increase during pregnancy. The calculations in Table 27 show that the estimated weight increases from Fig. 27 are totally inadequate, as anticipated from the incompleteness of the original data in Fig. 23 over the summer period.

From Fig. 50, it is clear that food consumption in the Antarctic summer during the second year must be far higher than estimated, because the calculated calorie intake is too low to cover the fat storage, growth and metabolism. The obvious solution, for which there is some evidence given by Dawbin (1966) on apparent length of stay in the Antarctic by pregnant females, is that the pregnant whales actually remain on the feeding grounds for longer than 120 days. By so doing, they would be able to accumulate greater fat reserves. If

fat is deposited at a regular rate throughout the feeding season, then the pregnant female would probably need to feed for an extra month, making a total stay in the Antarctic of 5 months, in order to accumulate the extra 10-15 %.

Summary and Conclusion

GROWTH AND DEVELOPMENT

Foetal Growth

Published material on growth in length with time for balaenopterid foetuses is reviewed, and growth in weight with time is discussed by reference to weight at length relationships. The growth formulation of Huggett and Widdas (1951)

$$W^{1/3} = a(t - t_0),$$

(where W = body weight in grammes, a = growth velocity constant, t_0 = time constant in days since conception prior to the linear growth phase, t = time in days since conception) is found to describe foetal growth well, after the 5th month of the 11 to 12 months period of gestation. The values of a found for blue and sei whales are 0.52, 0.47 and 0.35 respectively, and agree closely with estimates given by Frazer and Huggett (1974).

Post-Natal Growth

Published data on body dimensions relative to overall body length for blue, fin, sei and humpback whales are compared. It is noted that in all species the anterior end of the body appears to increase disproportionately in size relative to the rest of the body. Conversely, the tail region is relatively smaller in adults than young whales.

Growth curves of length at age and weight at age are predicted for blue, fin and sei whales. Age has been determined from layers in ear plugs and ovarian corpora numbers, and by reference to published work on growth and age using baleen plates. It is assumed, after consideration of published evidence, that one ear plug growth layer represents one year's growth and one corpus forms approximately every 1.4 years. The ages at sexual maturity in blue, fin and sei whales are taken to be 5, 6 and 8 years respectively for the various localities within the Southern Hemisphere discussed. The mean lengths at age from the whale catches are used to predict growth curves, and it is noted that in the early years between birth and year 2, there is considerable bias owing to the minimum lengths set for the whaling operations, resulting in higher mean lengths than true. The von Bertalanffy (1938) growth equation

$$L_t = L_\infty (1 - e^{-k(t+t_0)}),$$

(where L_∞ = mean length at physical maturity in metres, t = age in years, L_t = length in metres at age t, t_0 = time constant, k = growth velocity constant), is fitted to the curves for blue, fin and sei whales as follows:

blue male: $L_t = 25.0 (1 - e^{-0.216(t+4.92)})$

blue female: $L_t = 26.2 (1 - e^{-0.240(t+4.50)})$

fin male: $L_t = 21.0 (1 - e^{-0.221(t+5.30)})$

fin female: $L_t = 22.25 (1 - e^{-0.220(t+4.80)})$

sei male: $L_t = 14.8 (1 - e^{-0.1454(t+9.36)})$

sei female: $L_t = 15.3 (1 - e^{-0.1337(t+10.00)}).$

These formulae describe the observed curves quite well except in the early years.

Data on length and apparent growth parameters for fin whales in 6 different Antarctic areas are compared, and the mean maximum sizes attained, size at sexual maturity and growth rates are found to vary with the locality.

Weight at length data for these species and also for humpback whales are reviewed, and adjustments are made to allow for an assumed blood and fluid loss of 6 % during flensing prior to piecemeal weighing.

Weight at age curves for blue, fin and sei whales are constructed from a cross-reference between length at age data and weight at length data. It is apparent that both sexes in these species have attained at least 70 % of the mature weight at sexual maturity. The modification of the von Bertalanffy formula by Laws and Parker (1968) to predict weight at age is found to describe the estimated curves of growth in weight well. The values for the formula

$$W_t = W_\infty (1 - e^{-k(t+t_0)})^3,$$

(where W_∞ = body weight in tonnes at physical maturity, W_t = body weight in tonnes at age t) are as follows:

blue male: $W_t = 102 (1 - e^{-0.216(t+4.92)})^3$

blue female: $W_t = 117 (1 - e^{-0.240(t+4.50)})^3$

471

fin male: $W_t = 55 \, (1 - e^{-0.221(t+5.30)})^3$

fin female: $W_t = 64.5 \, (1 - e^{-0.220(t+4.80)})^3$

sei male: $W_t = 18 \, (1 - e^{-0.1454(t+9.36)})^3$

sei female: $W_t = 19.5 \, (1 - e^{-0.1337(t+10.00)})^3.$

The von Bertalanffy type formula was originally used for describing growth in fish, and the reason this formulation also seems applicable to whales may be due to the simple shape of the latter.

MIGRATION

Published data on whale mark returns are reviewed. The conclusions are that balaenopterid whales of both sexes tend to migrate seasonally, moving southwards in summer and northwards in winter, and can migrate between latitudes 65°S and 22°S within only a few months or less. Whales tend to remain within the same longitudinal sector.

Analyses of whale abundance throughout the whaling seasons in summer and winter tend to show that whales reach peak densities in the Antarctic in December to February and in low latitudes around June to August.

There appears to be an ordered succession of species in the Antarctic, blue preceding fin, which are followed by sei. In recent years there has been evidence that the sei whale, perhaps for reasons of oceanographic changes in sea temperature and diminution of other balaenopterid species, is arriving earlier in the Antarctic than it was observed to do in the 1930's.

Immature whales appear to reach peak abundancies later than the mature ones, suggesting a hierarchical order of migration such as observed in humpback whales (Dawbin, 1966).

The seasonal occurrence of diatom fouling by *Cocconeis ceticola* Nelson, of the skin of whales remaining for more than 1 month in the Antarctic in summer, shows that from the steady increase in infection rate until a stable level is reached, the average stay of whales in the Antarctic is about 120 days. The pattern of infection indicates that peak arrivals in the Antarctic occur in December and January. The appearance of diatom-infected juveniles off South Africa in August indicates a definite link between the 2 localities, confirming seasonal migration.

SEASONAL VARIATION IN BLUBBER THICKNESS

Weights of blue, fin and humpback whales are analysed for the summer months in the Antarctic between November and April. The data presented for the blue whale covers the greatest time span, that for the fin whale covering only part of the feeding season. Whilst blue and fin weights are from direct weighings, those for humpbacks are estimates from cooker fillings and are for a very short time span. No data are available on body weight during winter.

All weight data are standardized in order to compare different sizes of animals at different times. For humpbacks, the body weight/body length ratio is used; for blue and fin, the body weight/bone weight ratio. Mean ratios with time show a tendency for weight increase, suggesting a doubling of body weight in humpbacks throughout the Antarctic feeding season. Trends are clarified in blue and fin whales by recognizing the fact that migration into and out of the Antarctic is continuous, so that lean and fattening whales are present together on the grounds for most of the season. The lean and fattening animals are identified by the use of arithmetical probability paper (Harding, 1949). The predicted fattening throughout the season for blue whales is about 50 % of body weight, and about 30 % for fin whales. The fin data cover a lesser period than the full feeding season, and the actual fattening in fin is probably similar to that in blue whales. These estimates of fattening are in harmony with those demonstrated for North Pacific gray whales, and appear reasonable from comparisons with the extent of fattening in hibernators, and weight losses anticipated during starvation in various mammals.

The distribution of weight increases

throughout the body are calculated in the same way as for body weight/bone weight increases. Results show that the greatest overall increases in the body occur in the musculature, then the blubber, followed by the viscera. This is chiefly because there is relatively more muscle than blubber, which in turn exceeds the weight of the visceral organs. An analysis of oil yield with time during the Antarctic whaling season shows that this exceeds expectations from blubber alone.

The possibility of oil accumulation in bone during the season is discussed, but is considered not to affect bone weight greatly because of the internal replacement of water.

FOOD AND FEEDING

Food Type

The types of food taken in different localities are reviewed, the conclusion being that feeding is generally opportunistic and that swarming planktonic Crustacea are preferred. In the true polar waters *Euphausia superba* is the chief food type for balaenopterid whales.

Incidence of Feeding

Feeding in the summer months in the Antarctic reaches an incidence level of about 85 % for animals examined in the catches. During winter in low latitudes, the incidence level of food in the stomachs falls to about 50 % or less, and only 20 % in adults.

Quantities and Rates of Feeding

A review of data on quantities of food in the stomachs of balaenopterid whales taken both in the Antarctic and in low latitudes results in the conclusion that in the former locality food intake averages 30-40 g/kg body weight/day, whereas in the latter, food consumed is only about a tenth of this quantity. If the whale feeds for 120 days at the rate of 30-40 g/kg body weight/day and at the lesser rate for the rest of the year, 3.5-5 times the body weight is consumed in a year, and an average of 12 g/kg body weight/day is eaten over the year.

Evidence for a diurnal rhythm of feeding controlled by vertical migration of the prey is discussed, together with information on digestion rates. Food passage probably takes 14-15 hours.

Method of Feeding

The balaenopterid whales, except sei, are classified as swallowers and the sei whale as a skimmer and swallower, according to the description of Nemoto (1959). The dimensions of the mouth and its component parts in these species are compared. The baleen plates seem to be shorter and narrower in the largest species when compared with the equivalent size of animal of the smaller species. The size range of krill retained by the baleen of all species seems to be 20-65 mm.

The humpback appears to have the widest jaws of all the Balaenopteridae, the sei the narrowest. The blue, minke and fin whales have very similar jaw shapes. The volume of the mouth when feeding is calculated from scale projections of photographs of the minke whale head, and extrapolated to predict mouth volumes for larger blue and fin whales. Calculations show that when feeding in waters where production is 2 kg/m³ in a band 1.22 m (4 ft) thick, the mouth volume filtered repeatedly for a few hours is sufficient to provide adequate nourishment. The bulk in krill swarms also increases the efficiency of food retention by clogging the baleen filters. However, it is clear that annual migration to the productive waters of the Antarctic is essential for survival.

Energy Content of Krill

Euphausia superba consists of about 78 % water, 5 % ash and chitin, 4 % crude fat and 13 % crude protein. From a review of published data

473

and simple calculation, the calorific value of wet whole krill is approximately 1 000 kcal/kg.

METABOLISM

Visceral Organ Weights Related to Body Weight

The heart and kidney weights of Cetacea appear to follow the usual relationship with body weight found for other mammals. The heart and kidneys are closely associated with metabolism, so that basal metabolic rate, Q, might also be expected to follow the relationship observed for other mammals,

$$Q = 70.5 W^{0.7325} \text{kcal/day}$$

where W is body weight in kilogrammes.

Lung Weight and Capacity

The average lung weight and lung capacity of whales are 1 % and 2.5-2.8 % of body weight respectively. The vital capacity of Cetacea is estimated to be a very high proportion of total capacity, resulting from the great elasticity of the lung tissue which readily collapses on deflation.

External Body Surface Area

3 methods of calculating surface area, A, are discussed. The 2 most promising methods are by Parry (1949):

$$A = 11.1 \times W^{0.67},$$

(W is body weight in grammes, A in square centimetres) and Brody (1968):

$$A = 1\,000 \times W^{0.685},$$

(W is body weight in kilogrammes, A in square centimetres), both of which give predictions close to actual observations. A third method by Guldberg (1907) assumes that the body of the whale roughly conforms to 2 cones placed end to end. The result is clearly too low and is approximately equivalent to 2 thirds of the areas calculated by the other 2 methods.

Heart Rate

Resting heart rate in whales appears to be as expected from comparisons with other mammals, but there is a possibility that bradycardia may occur during diving as observed in small marine mammals.

Respiratory Physiology

The blood weight which is almost equivalent to blood volume, in adult baleen whales is assumed to be about 10 % of body weight. The oxygen capacity of the blood appears variable from published data, perhaps mainly because of differences in the fitness of the animals examined, but generally falls in the range 20-30 %. The oxygen capacity of the muscle in Cetacea appears highest in the deep diving toothed whales.

During dives the Respiratory Quotient (R.Q.) appears to fall to 0.70-0.75, whilst on recovery, the R.Q. is about 1.00. An average R.Q. including diving and recovery periods is about 0.82. The heat equivalent of 1 litre of oxygen at this R.Q. is 4.825 kcal. The utilization of the inspired air is assumed to be about 10 % from evidence of published data.

Respiratory Rhythm and Diving

Data on blowing rates and diving times for different whale species are considered. In general, blowing rate increases with activity and smaller whales blow more frequently than larger ones. Whilst the sperm and bottlenose whales seem to dive to depths down to 1 000 m or more, rorquals do not appear to dive deeper than 300 m according to published information.

Daily Activity Pattern

Information on sleep in captive and wild Cetacea is reviewed. Patterns of sleeping and activity appear unclear, although apparent sleep is characterized by a marked fall in the respiratory rate and the adoption of certain postures in the water. Well-fed Cetacea appear to be slow and relaxed in their environment in spite of intrusion by ships, whereas unfed, hungry individuals are alert and shy of contact with ships. This suggests that sleep could possibly occur after feeding, so that like time of feeding, the sleeping period is determined be the diurnal rhythm, if any, of the prey.

Metabolic Rate

The resting metabolic rate in blue, fin, sperm and bottlenose whales has been calculated using observed estimates of lung volume and assuming that R.Q. is 0.82, utilization of inspired air is 10 % and respiratory rhythms are as observed. The predicted values thus calculated for large Cetacea are in agreement with expectations from the extrapolation of data for small Cetacea and Pinnipedia.

The active metabolic rate, representing energy expenditure at top cruising speeds, is calculated on the assumption that 0.01 hp/lb $(1.64 \times 10^8$ ergs/kg) operative musculature is developed when swimming at such speeds. The basis for such an assumption was first tested by calculating the resistance experienced when swimming at different speeds with 70 % laminar flow around the cetacean body. The efficiency of energy utilization was assumed at 20 %, so that the actual energy expended to overcome the calculated resistance is 5 times that for the resistance alone. Results suggested that 0.01 hp/lb muscle is a valid basis on which to estimate active metabolic rate using a conversion factor 1 hp = 1.54×10^4 kcal/lb/day.

The basal metabolic rate, Q, which is a theoretical standard hardly ever observed in day to day living, except perhaps during sleep and starvation, is calculated in 3 ways. One method is to use the formula

$$Q = 70.5W^{0.7325}$$

where W is body weight in kilogrammes. Another is to assume that basal rate is about 85 % of resting rate, the 15 % above basal rate being for movement essential to normal living. The third method is to calculate basal rate as 10 % of active rate. These 3 methods give results which are remarkably close in view of the differences in their derivation. The method using the formulation however generally appears to give results lower than the other 2.

ENERGY BUDGET

The heat generated within the whale body appears to be dissipated to the environment chiefly via the flukes, fin and flippers, because heat loss via the body blubber only accounts for about 10 % of total heat loss. Without an efficient means of getting rid of body heat, overheating would easily be incurred.

Energy budgets are estimated for blue and fin whales at different stages of development. The suckling calf, the pubertal juvenile, young adult and fully mature pregnant female are considered. Growth efficiencies (gross and net) are found to fall with increase in age and development, from levels of about 30 % in the unweaned calf to only 1 % in near adults. Assimilation efficiency would appear high, around 80 %, but this is estimated by indirect methods and is not necessarily accurate although assimilation is frequently as high in other carnivores.

The ratio of energy expenditure in metabolism to energy capture in terms of developmental growth is found to increase from about 2 : 1 to nearly 100 : 1 between birth and adulthood, indicating that relative production efficiency falls off dramatically with age.

In both the pubertal and adult whales, the balanced energy budget requires that an increase in body weight of at least 50 % in the form of fat tissue needs to be accumulated during summer in

475

order to survive the winter months. This supports the findings on seasonal body weight increase in blue whales and also supports the observation that the time span of the weight data for fin whales is insufficient to give the true extent of fattening. This is particularly noticeable for pregnant females which it is concluded must attain an increase of about 60-65 % body weight. In order to achieve this, they must feed for longer periods in the Antarctic relative to other classes, and observations by Dawbin (1966) on humpback whales suggest that in fact pregnant whales

do arrive in the Antarctic before and leave after the main population. This excessive fattening in pregnant whales relative to resting ones appears to be essential in order to survive the period of lactation which is a large energy drain on the female. Observations on blubber thickness already show that lactating females are exceptionally lean, and would tend to corroborate the assumptions concerning the energy expenditure of lactation. For these reasons the energy budget for pregnant females is only meaningful over a 2 year cycle.

References

ADOLPH, E.F., Quantitative relations in the physiological constitution of mammals. *Science, N.Y.*, 109:579-85.
1946

ALEYEV, Yu.G., The dolphin body as the aerofoil. *Zool. Zh.*, 44(4):626-30.
1965

—, Mobile roughness on the body-surface of nectonic organisms as a means of reducing drag. *Zool. Zh.*, 49:1173-80.
1970

ALLEN, G.M., The whalebone whales of New England. *Mem. Boston Soc. Nat. Hist.*, 8(2):107-322.
1916

ANDERSEN, H.T. Physiological adaptations in diving vertebrates. *Physiol. Rev.*, 46:212-43.
1966

ARAI, Y. and S. SAKAI, Whale meat in nutrition. *Sci. Rep. Whales Res. Inst., Tokyo*, (7):51-68.
1952

ASH, C.E. Weights of Antarctic humpback whales. *Norsk Hvalfangsttid.*, 42(7):387-91.
1953

—, Production per calculated whale. *Norsk Hvalfangsttid.*, 44(5):260-2.
1955

—, The fin whales of 1954/5: blubber thickness and factory efficiency. *Norsk Hvalfangsttid.*, 44(5):264-75.
1955a

—, Variation in blubber thickness with length in fin whales. *Norsk Hvalfangsttid.*, 45(10):550-4.
1956

—, The oil yield of fin whales. *Norsk Hvalfangsttid.*, 46(10):559-69.
1957

—, Weights and oil-yields of Antarctic humpback whales. *Norsk Hvalfangsttid.*, 46(10):569-73.
1957a

BANNISTER, J.L. and A. de C. BAKER, Observations on food and feeding of baleen whales at Durban. *Norsk Hvalfangsttid.*, 56(4):76-82.
1967

BANNISTER, J.L. and R. GAMBELL, The succession and abundance of fin, sei and other whales off Durban. *Norsk Hvalfangsttid.*, 54(3):45-60.
1965

BELKOVICH, U.M., On the question of physical thermal regulation in beluga *(Delphinapterus leucas)*. *Tr. Soveshch. Ikhtiol. Kom.*, 12:68-71.
1961

BENEDICT, F.G., The physiology of the elephant. *Publ. Carnegie Inst. Wash.*, (474):1-302.
1936

BENEDICT, F.G. and E.G. RITZMAN, The metabolism of the fasting steer. *Publ. Carnegie Inst. Wash.*, (377):1-245.
1927

BENNET, A.G., On the occurrence of diatoms on the skin of whales. *Proc. R. Soc. Lond. (B)*, 91:352.
1920

BERTALANFFY, L. von, A quantitative theory of organic growth. *Hum. Biol.*, 10(2):181-213.
1938

BEST, P.B., Distribution and feeding habits of baleen whales off the Cape Province. *Invest. Rep. Div. Fish. Union S. Afr.*, (57):1-44.
1967

—, The sperm whale *(Physeter catodon)* off the west coast of South Africa. 5. Age, growth and
1970

mortality. *Invest. Rep. Div. Fish. Union S. Afr.*, (79):1-27.

BETESHEVA, E.I., Data on the feeding of baleen whales
1954   in the Kurils region. *Tr. Inst. Okeanol.*, 11:238-45.

–, Food of whalebone whales in the Kurile Islands-re-
1955   gion. *Tr. Inst. Okeanol.*, 18:78-85.

BIBIKOV, D.I. and N.M. ZHIRNOVA, Seasonal changes in
1956   some ecologically-physiological peculiarities of *Marmota marmota baibacina. Zool. Zh.*, 35:1565-73.

BIRGE, E.A. and C. JUDAY, The inland lakes of Wiscon-
1922   sin: the plankton. Part 1. Its quantity and chemical composition. *Bull. Wis. Geol. Nat. Hist. Surv.*, 64:1-22.

BIRKETT, L., Experimental determination of food con-
1970   version and its application to ecology. *In* Marine food chains, edited by J.H. Steele. Edinburgh, Oliver and Boyd, pp. 261-4.

BOGOROV, B.G. and M.E. VINOGRADOV, Some essential
1956   features of zooplankton distribution in the northwestern Pacific. *Tr. Inst. Okeanol.*, 18:60-84.

BRODY, S., Bioenergetics and growth. New York, Hafner
1968   Publ. Co., Inc., 1023 p. (Rev. of 1945 ed.).

BROWN, S.G., Dispersal in blue and fin whales. *Dis-*
1954   *covery Rep.*, 26:355-84.

–, Whale marks recovered during the Antarctic whal-
1957   ing season 1956/57. *Norsk Hvalfangsttid.*, 46(10):555-9.

–, Whale marks recovered in the Antarctic seasons
1959   1955/56, 1958/59, and in South Africa 1958 and 1959. *Norsk Hvalfangsttid.*, 48(12):609-16.

–, Whale marks recovered in the Antarctic whaling
1960   season 1959/60. *Norsk Hvalfangsttid.*, 49(10):457-61.

–, Whale marks recovered in the Antarctic whaling
1961   season 1960/61. *Norsk Hvalfangsttid.*, 50(10):407-10.

–, A note on migration in fin whales. *Norsk Hval-*
1962   *fangsttid.*, 51(1):13-6.

–, International co-operation in Antarctic whale
1962a   marking 1957 to 1960, and a review of the

distribution of marked whales in the Antarctic. *Norsk Hvalfangsttid.*, 51(3):93-104.

–, Whale marks recovered during Antarctic seasons
1962b   1960-61, 1961/62, and in South Africa 1962. *Norsk Hvalfangsttid.*, 51(11):429-34.

–, Whale marks recovered in the Antarctic whaling
1966   season 1964/65 and some recent USSR returns. *Norsk Hvalfangsttid.*, 55(2):31-5.

–, Feeding of sei whales at South Georgia. *Norsk*
1968   *Hvalfangsttid.*, 57(6):118-25.

–, Some results of sei whale marking in the Southern
1974   Hemisphere. Paper presented to the IWC Scientific Committee, La Jolla, 1974, Sc/Sp74/Doc. 6.

CAMUS, L. and E. GLEY, Sur les variations de poids des
1901   hérissons. *C. R. Séances Soc. Biol. Paris*, 53:1019-20.

CHITTLEBOROUGH, R.G., Studies on the ovaries of the
1954   humpback whale, *Megaptera nodosa* (Bonnaterre), on the western Australian coast. *Aust. J. Mar. Freshwat. Res.*, 5(1):35-63.

–, Australian marking of humpack whales. *Norsk*
1959   *Hvalfangsttid.*, 48(2):47-55.

CLARK, A.J., Comparative physiology of the heart.
1927   Cambridge, 157 p.

CLARKE, M., Function of the spermaceti organ of the
1970   sperm whale. *Nature, Lond.*, 228(5274):873-4.

CLOWES, A.J., A note on the composition of whale milk.
1929   *Discovery Rep.*, 1:472-5.

–, A note on the oil content of blubber. *Discovery Rep.*,
1929a   1:476-8.

COLLETT, R., Norges hvirveldyr. 1. Norges pattedyr.
1911-12 Kristiania, 744 p.

COUSTEAU, J.-Y. and P. DIOLÉ, The whale – mighty
1972   monarch of the sea. London, Cassell, 304 p.

DALL, W. and D. DUNSTAN, *Euphausia superba* Dana
1957   from a humpback whale, *Megaptera nodosa* (Bonnaterre), caught off southern Queensland. *Norsk Hvalfangsttid.*, 46(1):6-9.

DAWBIN, W.H., The migrations of humpback whales
1956   which pass the New Zealand coast. *Trans. R. Soc. N. Z.*, 84:147-96.

477

—, New Zealand and South Pacific whale marking and recoveries to the end of 1958. *Norsk Hvalfangsttid.*, 48:(5):213-38.
1959

—, Movements of humpback whales marked in the South West Pacific Ocean 1952 to 1962. *Norsk Hvalfangsttid.*, 53(3):68-78.
1964

—, The seasonal migratory cycle of humpback whales. *In* Whales, dolphins and porpoises, edited by K.S. Norris. Berkeley, University of California Press, pp. 145-70.
1966

DAWBIN, W.H. and R.A. FALLA, A contribution to the study of the humpback whale based on observations at New Zealand shore stations. *Proc. Pac. Sci. Congr.*, 4:373-82.
1953

DEACON, G.E.R., The hydrology of the Southern Ocean. *Discovery Rep.*, 15:1-124.
1937

DEGERBØL, M. and P. FREUCHEN, Mammals. *Rep. Thule Exped. 1921-24*, 2(4-5).
1935

DILL, D.B., The economy of muscular exercise. *Physiol. Rev.*, 16(2):263-9.
1936

DOI, T., S. OHSUMI and T. NEMOTO, Population assessment of sei whales in the Antarctic. *Norsk Hvalfangsttid.*, 56(2):25-41.
1967

DU BOIS, D. and E.F. DU BOIS, *Arch. Int. Med.*, 17:863.
1916

ELSNER, R., D.W. KENNEY and K. BURGESS, Diving bradycardia in the trained dolphin. *Nature, Lond.*, 212:407-8.
1966

FRAZER, J.F.D. and A. St. G. HUGGETT, Species variations in the foetal growth rates of eutherian mammals. *J. Zool., Lond.*, 174:481-509.
1974

FRIANT, M., Les caractéristiques anatomiques du foie des mammifères hibernants. *C. R. Hebd. Séances Acad. Sci., Paris*, 214(1):185-7.
1942

GAMBELL, R., Seasonal cycles and reproduction in sei whales of the Southern Hemisphere. *Discovery Rep.*, 35:31-134.
1968

—, Some effects of exploitation on reproduction in whales. *J. Reprod. Fertil.*, (Suppl.) 19:531-51.
1973

GAMBELL, R., C. LOCKYER and G.J.B. ROSS, Observations on the birth of a sperm whale calf. *S. Afr. J. Sci.*, 69(5):147-8.
1973

GEPTNER, V.G., Material for a study of the geographic distribution and life history of the beluga. *Tr. Nauchno-Issled. Inst. Zool.*, 4(2):1-110.
1930

GILL, C.D. and S.E. HUGHES, A sei whale, *Balaenoptera borealis*, feeding on Pacific saury, *Cololabis saira. Calif. Fish Game*, 57(3):218-9.
1971

GILMARTIN, W.G., R.W. PIERCE and G.A. ANTONELIS, Some physiological parameters of the blood of the California gray whale. *Mar. Fish. Rev.*, 36(4):28-31.
1974

GILMORE, R.M., Whales, porpoises and the U.S. Navy. *Norsk Hvalfganfsttid.*, 50(3):1-9.
1961

GRAY, Sir J., Animal locomotion. London, Weidenfield and Nicolosn, 479 p.
1968

GRAY, W.B., Porpoise tales. New York, A.S. Barnes and Co., 111 p.
1964

GREEN, A. ARDA and A.C. REDFIELD, On the respiratory function of the blood of the porpoise. *Biol. Bull. Mar. Biol. Lab. Woods Hole, Mass.*, 64(1):44-52.
1933

GRIZZELL, R.A., Jr., A study of southern woodchuck *Marmota monax monax. Am. Midl. Nat.*, 53:257-93.
1955

GULDBERG, G., Ueber das Verfahren bei Berechnung des Rauminhaltes und Gewichtes der grossen Waltiere. *Forh. Videnskapeselskap. Krist.*, 3.
1907

GUNTHER, E.R., The habits of fin whales. *Discovery Rep.*, 25:113-42.
1949

HARDING, J.P., The use of probability paper for the graphical analysis of polymodal frequency distributions. *J. Mar. Biol. Assoc. U.K.*, 28:141-53.
1949

HARMER, Sir S.F., Southern whaling. *Proc. Linn. Soc. Land.*, 142:85-163.
1931

HART, J.S. and L. IRVING, The energetics of harbor seals in air and in water with special consideration of seasonal changes. *Can. J. Zool.*, 37:447-57.
1959

HART, J.T., On the diatoms of the skin film of whales and their possible bearing on problems of whale movements. *Discovery Rep.*, 10:247-82.
1935

HAYNES, F. and A.H. LAURIE, On the histological structure of cetacean lungs. *Discovery Rep.*, 17:1-6.
1937

HERTER, K., Gefangenschaftsbeobachtungen am euro-
1933   päischen Igel. 2. Z. Säugetierkd., 8:195-218.

HEYERDAHL, E.F., Hvalindustrien. 1. Ramaterialet.
1932   Publ. Christensens Hvalfangstmus. Sandefjord,
        (7):23-6.

HINTON, M.A.C., Report on the papers left by the late
1925   Major Barrett-Hamilton, relating to the
        whales of South Georgia. London, Crown
        Agents for the Colonies, pp. 57-209.

HIRANO, T. et al., Contents of inorganic substance and
1964   vitamin B_{12} in Euphausia. J. Tokyo Univ. Fish.,
        50:65-70.

HORVATH, S.M. et al., Respiratory and electrophoretic
1968   characteristics of hemoglobin of porpoises and
        sea lion. Comp. Biochem. Physiol., 24:1027-33.

HUGGETT, A.St.G., and W.F. WIDDAS, The relationship
1951   between mammalian foetal weight and con-
        ception age. J. Physiol., 114:306-17.

ICHIHARA, T., Criterion for determining age of fin whale
1966   with reference to ear plug and baleen plate.
        Sci. Rep. Whales Res. Inst., Tokyo, (20):17-82.

IL'ICHEV, Ye.F., The chemical composition of krill and
1967   its use for feed and food purposes. In Soviet
        fishery research on the Antarctic krill, edited
        by R.N. Burukovskiy. Kaliningrad, Atlant-
        NIRO, pp. 38-55.

INGEBRIGTSEN, A., Whales caught in the North Atlantic
1929   and other seas. Rapp. P.-V. Réun. CIEM, 2:1.

IWC, Report of the International Commission on Whal-
1952   ing. Rep. Int. Whal. Comm., (3).

–, Report of the International Commission on Whal-
1955   ing. Rep. Int. Whal. Comm., (6).

IRVING, L., The protection of whales from the danger of
1935   caisson disease. Science, N.Y., 81:560-1.

–, Respiration in diving animals. Physiol. Rev., 19:112.
1939

IRVING, L. and J.S. HART, The metabolism and insula-
1957   tion of seals as bare-skinned mammals in cold
        water. Can. J. Zool., 35:497-511.

IRVING, L., P.F. SCHOLANDER and S.W. GRUNNELL, The
1941   respiration of the porpoise, Tursiops truncatus.
        J. Cell. Comp. Physiol., 17:145-68.

IRVING, L. et al., The respiratory metabolism of the seal
1935   and its adjustment to diving. J. Cell. Comp.
        Physiol., 7:137-51.

JERDE, C.W. and R. LASKER, Moulting of Euphausid
1966   "shrimps"; shipboard observations. Limnol.
        Oceanogr., 11:120-4.

JOHNSON, G.E., Hibernation of the thirteen-lined
1928   ground squirrel, Citellus tridecimlineatus
        (Mitchell). 1. A comparison of the normal and
        hibernating states. J. Exp. Zool., 50(1):15-30.

KAKURA, Z., T. KAWAKAMI and K. IGUCHI, Biological
1953   investigation on the whales caught by the Ja-
        panese Antarctic whaling fleets in the 1951/52
        season. Sci. Rep. Whales Res. Inst., Tokyo,
        (8):147-213.

KALABUKHOV, N.I., The hibernation of animals. Char-
1956   kov, Gorki State University Press.

KANWISHER, J. and A. SENFT, Physiological measure-
1960   ments on a live whale. Science, Wash.,
        131:1379-80.

KANWISHER, J. and G. SUNDNES, Thermal regulation in
1966   cetaceans. In Whales dolphins and porpoises,
        edited by K.S. Norris. Berkeley, University of
        California Press, pp. 398-409.

KAWAMURA, A., Food of sei whale taken by Japanese
1970   whaling expeditions in the Antarctic season
        1967/68. Sci. Rep. Whales Res. Inst., Tokyo,
        (22):127-52.

–, Influence of chasing time to stomach contents of
1971   baleen and sperm whales. Sci. Rep. Whales
        Res. Inst., Tokyo, (23):27-36.

–, Food and feeding ecology in the southern sei
1974   whale. Sci. Rep. Whales Res. Inst., Tokyo,
        (26):25-144.

–, A consideration on an available source of energy and
1975   its cost for locomotion in fin whales with spe-
        cial reference to the seasonal migrations. Sci.
        Rep. Whales Res. Inst., Tokyo, (27):61-79.

KAYSER, C., La dépense d'énergie des mammifères
1952   pendant toute la durée de l'hibernation. Arch.
        Sci. Physiol., 6:193-212.

–, L'hibernation des mammifères. Ann. Biol.,
1953   29:109-50.

—, The physiology of natural hibernation. International
1961 series of monographs on pure and applied
biology. *In* Modern trends in physiological
sciences. London, Pergamon Press Ltd., vol.
8:325 p.

KELLOGG, R., What is known of the migrations of some
1929 of the whalebone whales. *Rep. Smithson. Inst.,*
1928:467.

KEMP, S. and A.G. BENNETT, On the distribution and
1932 movements of whales on the South Georgia
and South Shetlands whaling grounds. *Discovery Rep.,* 6:165-90.

KIMURA, S. and T. NEMOTO, Note on a minke whale kept
1956 alive in aquarium. *Sci. Rep. Whales Res. Inst.,
Tokyo,* (11):181-90.

KLUMOV, S.K., Plankton and the feeding of baleen
1961 whales (Mystacoceti). *Tr. Inst. Okeanol.,*
51:142-56.

—, Feeding and helminth fauna of whalebone whales
1963 (Mystacoceti) in the main whaling grounds of
the world ocean. *Tr. Inst. Okeanol.,* 71:94-194.

KOOYMAN, G.L. and W.B. CAMPBELL, Heart rates in
1972 freely diving Weddell seals, *Leptonychotes
weddelli. Comp. Biochem. Physiol. (A Comp.
Physiol.),* 43:31-6.

KOOYMAN, G.L. *et al.,* Pulmonary function in freely
1971 diving Weddell seals, *Leptonychotes weddelli.
Respir. Physiol.,* 12:271-82.

—, Pulmonary gas exchange in freely diving Weddell
1973 seals, *Leptonychotes weddelli. Respir. Physiol.,*
17:283-90.

KRAMER, M.O., Boundary layer stabilization by distrib-
1960 uted damping. *J. Am. Soc. Nav. Eng.,* February issue: 25-33.

KROGH, A., Physiology of the blue whale. *Nature, Lond.,*
1934 133:635-7.

LANG, T.G., Hydrodynamic analysis of cetacean per-
1966 formance. *In* Whales, dolphins and porpoises,
edited by K.S. Norris. Berkeley, University of
California Press, pp. 410-32.

LAURIE, A.H., Some aspects of respiration in blue and
1933 fin whales. *Discovery Rep.,* 7:363-406.

—, The age of female blue whales and the effect of

1937 whaling on the stock. *Discovery Rep.,*
15:223-84.

LAWS, R.M., The elephant seal (*Mirounga leonina*
1953 Linn.). 1. Growth and age. *Sci. Rep. Falkland
Isl. Depend. Surv.,* (8):1-62.

—, Foetal growth rates of whales with special references
1959 to the fin whale, *Balaenoptera physalus* Linn.
Discovery Rep., 29:281-308.

—, Southern fin whales. *Discovery Rep.,* 31:327-486.
1961

—, Some effects of whaling on the southern stocks of
1962 baleen whales. *In* The exploitation of natural
animal populations, edited by L.D. LeCren
and M.W. Holdgate. *Symp. Br. Ecol. Soc.,*
(2):137-58.

LAWS, R.M. and I.S.C. PARKER, Recent studies on ele-
1968 phant populations in East Africa. *Symp. Zool.
Soc. Lond.,* 21:319-59.

LOCKYER, C., The age at sexual maturity of the southern
1972 fin whale (*Balaenoptera physalus*) using annual layer counts in the ear plug. *J. Cons.
CIEM,* 34(2):276-94.

—, Investigation on the ear plug of the southern sei
1974 whale, *Balaenoptera borealis,* as a valid means
of determining age. *J. Cons. CIEM,*
36(1):71-81.

—, Body weight of some species of large whales. *J. Cons.*
1976 *CIEM,* 36(3):259-73.

—, Observations on diving behaviour of the sperm
1977 whale, *Physeter catodon. In* A voyage of discovery, edited by M. Angel. Supplement to
Deep Sea Res., Pergamon Press, pp. 591-609.

MACKINTOSH, N.A., The southern stock of whalebone
1942 whales. *Discovery Rep.,* 22:197-300.

—, The stocks of whales. London, Fishing News (Books)
1965 Ltd., 232 p.

—, Life cycle of Antarctic krill in relation to ice and
1972 water conditions. *Discovery Rep.,* 36:1-94.

—, Distribution of post-larval krill in the Antarctic.
1973 *Discovery Rep.,* 36:95-156.

MACKINTOSH, N.A. and S.G. BROWN, Preliminary esti-
1956 mates of the southern populations of the larg-

er baleen whales. *Norsk Hvalfangsttid.*, 43(9):469-81.

MACKINTOSH, N.A. and H.F.P. HERDMAN, Distribution
1940   of the pack-ice in the Southern Ocean. *Discovery Rep.*, 19:285-96.

MACKINTOSH, N.A. and J.F.G. WHEELER, Southern blue
1929   and fin whales. *Discovery Rep.*, 1:257-540.

MACQUILLAN, H., The Antarctic krill. *West Fish.*,
1962   63(4):20

MALM, E.N., Etyudy po biologii chernomorskikh del'fi-
1938   nov (Biological studies of Black Sea dolphins). *Priroda*, 1938(5).

MARR, J.W.S., *Euphausia superba* and the Antarctic
1956   surface currents. *Norsk Hvalfangsttid.*, 43(3):127-34.

–, The natural history and geography of the Antarctic
1962   krill, *Euphausia superba* Dana. *Discovery Rep.*, 32:33-464.

MARSHALL, S.M. and A.P. ORR, The biology of a marine
1955   copepod. Edinburgh, Oliver and Boyd, 388 p.

MATTHEWS, L.H., Lobster-krill. Anomuran Crustacea
1932   that are the food of whales. *Discovery Rep.*, 5:467-84.

–, The humpback whale, *Megaptera nodosa. Discovery*
1937   *Rep.*, 17:7-92.

–, The sei whale, *Balaenoptera borealis. Discovery Rep.*,
1938   17:183-290.

MAUCHLINE, J. and L.R. FISHER, The biology of eu-
1969   phausiids. *Adv. Mar. Biol.*, 7:1-454.

MAYER, W.F., Food consumption patterns in the Arctic
1954   ground squirrel, *Spermophilus undulatus. Anat. Rec.*, 120:760.

MIZUE, K. and T. MURATA, Biological investigation on
1951   the whales caught by the Japanese Antarctic whaling fleets season 1949/50. *Sci. Rep. Whales Res. Inst., Tokyo*, (6):73-131.

MORIMOTO, Y., M. TAKATA and M. SUDZUKI, Unter-
1921   suchungen über Cetacea. *Tohoku J. Exp. Med.*, 2:258.

MURDAUGH, H.V. *et al.*, Dissociation of bradycardia and
1968   arterial constriction during diving in the seal, *Phoca vitulina. Science, Wash.*, 162:364-5.

NEMOTO, T., Foods of baleen whales in the northern
1957   Pacific. *Sci. Rep. Whales Res. Inst., Tokyo*, (12);33-89.

–, *Cocconeis* diatoms infected on whales in the Ant-
1958   arctic. *Sci. Rep. Whales Res. Inst., Tokyo*, (13):185-92.

–, Food of baleen whales with reference to whale
1959   movements. *Sci. Rep. Whales Res. Inst, Tokyo*, (14):149-290.

–, Food of baleen whales collected in recent Japanese
1962   Antarctic whaling expeditions. *Sci. Rep. Whales Res. Inst., Tokyo*, (16):89-103.

Feeding of baleen whales and krill, and the value of
1968   krill as a marine resource in the Antarctic. *In* Proceedings of the Symposium on Antarctic oceanography, Santiago, Chile, 13-16th September 1966. Cambridge, England, Scott Polar Research Institute, pp. 240-53.

–, Feeding pattern of baleen whales in the ocean. *In*
1970   Marine food chains, edited by J.H. Steele. Edinburgh, Oliver and Boyd, pp. 241-52.

NEMOTO, T. and K. NASU, *Thysanoessa macrura* as food
1958   of baleen whales in the Antarctic. *Sci. Rep. Whales Res. Inst., Tokyo*, (13):193-200.

NEWMAN, M.A. and P.L. MACGEER, The capture and
1966   care of a killer whale, *Orcinus orca*, in British Columbia. *Zoologica, N.Y.*, 51(2):59-69.

NISHIMOTO, S., M. TOZAWA and T. KAWAKAMI, Food of
1952   sei whales (*Balaenoptera borealis*) caught in the Bonin Island waters. *Sci. Rep. Whales Res. Inst., Tokyo*, (5):79-85.

NISHIWAKI, M., The body weight of whales. *Sci. Rep.*
1950   *Whales Res. Inst., Tokyo*, (4):184-209.

–, Age determination of Mystacoceti, chiefly blue and
1952   fin whales. *Sci. Rep. Whales Res. Inst., Tokyo*, (7):87-120.

NISHIWAKI, M. and K. HAYASHI, Biological survey of fin
1950   and blue whales taken in the Antarctic season 1947-48 by the Japanese fleet. *Sci. Rep. Whales Res. Inst., Tokyo*, (3):168-9.

NISHIWAKI, M. and T. OYE, The biological investigations
1951   on blue and fin whales caught by Japanese Antarctic fleet. *Sci. Rep. Whales Res. Inst., Tokyo*, (5):134.

481

NOÉ, J., Variations de résistance du hérisson à l'inani-
1901   tion. *C. R. Séances Soc. Biol., Paris,*
        53:1009-10.

OHNO, M. and K. FUJINO, Biological investigations on
1952   the whales caught by the Japanese Antarctic
        whaling fleets, season 1950-51. *Sci. Rep.
        Whales Res. Inst., Tokyo,* (7):125-88.

OHSUMI, S., Y. MASAKI and A. KAWAMURA, Stock of the
1970   Antarctic minke whale. *Sci. Rep. Whales Res.
        Inst., Tokyo,* (22):75-126.

OHTA, K. *et al.,* Composition of fin whale milk. *Sci. Rep.
1955   Whales Res. Inst., Tokyo,* (1):151-67.

OLSEN, C.R., F.C. HALE and R. ELSNER, Mechanics of
1969   ventilation in the pilot whale. *Respir. Physiol.,*
        7:137-49.

OLSEN, Ø., Hvaler og hvalfangst i Sydafrika. *Bergens
1915   Mus. Arb.,* 5:1-56.

OMMANNEY, F.D., The vascular networks (Retia mira-
1932   bilia) of the fin whale *(Balaenoptera physalus).
        Discovery Rep.,* 5:327-62.

OMURA, H., On the body weight of sperm and sei whales
1950   located in the adjacent waters of Japan. *Sci.
        Rep. Whales Res. Inst., Tokyo,* (4):1-13.

—, Diatom infection on blue and fin whales in the
1950a   Antarctic whaling Area V (the Ross Sea area).
        Sci. Rep. Whales Res. Inst., Tokyo, (4):14-26.

—, North Pacific right whale. *Sci. Rep. Whales Res.
1958   Inst., Tokyo,* (13):1-52.

OMURA, H. *et al.,* Black right whales in the North Pacific.
1969   *Sci. Rep. Whales Res. Inst., Tokyo,* (21):13-26.

PARRY, D.A., The structure of whale blubber and
1949   its thermal properties. *Q.J. Microsc. Sci.,*
        90:13-26.

PARSONS, T.R. and R.J. LEBRASSEUR, The availability of
1970   food to different trophic levels in the marine
        food chain. *In* Marine food chains, edited by
        J.H. Steele. Edinburgh, Oliver and Boyd, pp.
        325-43.

PETERS, N., Über Grösse, Wachstum und Alter des
1939   Blauwales *(Balaenoptera musculus* L.). *Zool.
        Anz.,* 127(7/8).

PETRIDES, G.A. and W.O. SWANK, Estimating the pro-
1965   ductivity and energy relations of an African

elephant population. *Proc. Int. Grassland
Congr.,* 9:831-42.

PHILLIPSON, J., The food consumption of different
1960   instars of *Mitopus morio* (F.) (Phalangida)
        under natural conditions. *J. Anim. Ecol.,*
        29:299-307.

PONOMAREVA, L.A., On the nourishment of the plank-
1949   ton-eating whales of the Bering Sea. *Dokl.
        AN SSSR,* 68(2):401-3.

PORSILD, M.P., Scattered observations on narwhals. *J.
1922   Mammal.,* 3(1):8-13.

PURVES, P.E., The structure of the flukes in relation to
1969   laminar flow in cetaceans. *Z. Säugetierkd.,*
        34(1):1-8.

PURVES, P.E. and M.D. MOUNTFORD, Ear plug lamina-
1959   tions in relation to the age composition of a
        population of fin whales. *Bull. Br. Mus. Nat.
        Hist.,* 5(6):123-61.

RAYMONT, J.E.G., R.T. SRINIVASAGAN and J.K.B. RAY-
1971   MONT, Biochemical studies on marine zoo-
        plankton. 8. Further investigations of *Mega-
        nyctiphanes norvegica* (M. Sars). *Deep-Sea
        Res.,* 18(12):1167-78.

RAYNER, G.W., The Falkland species of the crustacean
1935   genus *Munida. Discovery Rep.,* 10:209-45.

—, Whale marking. Progress and results to December
1940   1939. *Discovery Rep.,* 19:245-84.

RICE, D.W. and A.A. WOLMAN, The life history and
1971   ecology of the gray whale *(Eschrichtius
        robustus). Spec. Publ. Am. Soc. Mammal.,*
        (3):142 p.

RIDGWAY, S.H. (ed.), Mammals of the sea — biology and
1972   medicine. Springfield, Illinois, C.C. Thomas,
        812 p.

RIDGWAY, S.H., B.L. SCRONCE and J. KANWISHER, Res-
1969   piration and deep diving in the bottlenose
        porpoise. *Science, Wash.,* 166:1651-4.

RISTING, S., Knolhvalen. *Norsk Fiskeritid.,* 31:437-49.
1912

—, Whales and whale foetuses: statistics of catch and
1928   measurement collected from the Norwegian
        Whalers' Association, 1922-25. *Rapp. P.-V.
        Réun. CIEM,* 50:1-122.

ROE, H.S.J., The rate of lamina formation in the ear
1967   plug of the fin whale. *Norsk Hvalfangsttid.,*
        56(2):41-3.

–, Seasonal formation of laminae in the ear plug of the
1967a   fin whale. *Discovery Rep.,* 25:1-30.

ROWE, M., They came from the sea: the story of Port
1968   Elizabeth's dolphins. Cape Town, Longmans,
        Southern Africa (Pty) Ltd.

RUBNER, M., Das Problem der Lebensdauer und seine
1908   Beziehungen zum Wachstum und Ernährung.
        Berlin.

RUUD, J., Fin hvalen. *Norsk Hvalfangsttid.,*
1937   26(3):98-112.

RUUD, J.T., Å. JONSGÅRD and P. OTTESTAD, Age-studies
1950   on blue whales. *Hvalråd. Skr.,* 33:1-72.

SAL'NIKOV, N.Ye, The feeding of the finback whale and
1953   blue whale in the Antarctic. Research on the
        whales of Antarctica. *Tr. Vses. Nauchno-Iss-
        led. Inst. Morsk. Rybn. Khoz. Okeanogr.,*
        25:54-67.

SCAMMON, C.M., On the cetaceans of the western coast
1869   of North America. *Proc. Acad. Nat. Sci. Phi-
        lad.,* 21:13-63.

SHOLANDER, P.F., Experimental investigations on the
1940   respiratory function in diving mammals and
        birds. *Hvalråd. Skr.,* 22(1):1-131.

SCHOLANDER, P.F. and W.E. SCHEVILL, Counter-current
1955   vascular heat exchange in the fins of whales. *J.
        Appl. Physiol.,* 8:279-92.

SCHOLANDER, P.F. *et al.,* Body insulation of some Arctic
1950   and tropical mammals and birds. *Biol. Bull.
        Mar. Biol. Lab. Woods Hole, Mass.,* 99:225-36.

SERGEANT, D.E., Feeding rates of Cetacea. *Fiskeridir.
1969   Skr. (Havunders.),* 15:246-58.

SHEVTSOV, V.V. Certain data on the Antarctic krill. *Tr.
1963   Vses. Nauchno-Issled. Inst. Morsk. Rybn.
        Khoz. Okeanogr.,* 12:22-30.

SLIJPER, E.J., On the thickness of the layer of blubber in
1948   Antarctic blue and fin whales. Pts 1-3. *K. Ned.
        Akad. Wet.,* 51(8):1033-45; (9):1114-24;
        (10):1310-6.

–, Whales. London, Hutchinson, 475 p.
1962

SMITH, N.T. and E.A. WAHRENBROCK, Ballistocardio-
1974   graphy as a technique for comparative physi-
        ology. *Mar. Fish. Rev.,* 36(4):9-14.

SPECTOR, W.S. (ed.), Handbook of biological data.
1956   Philadelphia, W.B. Saunders Co., 584 p.

SPENCER, M., Note. *Vancouver Public Aquar. Newsl.,*
1970   14(1):Jan-Feb issue.

STACY, R.W. *et. al.,* Essentials of biological and medical
1955   physics. New York, MacGraw-Hill Book Co.,
        Inc., 586 p.

SUND, P.N., Evidence of feeding during migration and
1975   of an early birth of the California gray whale
        (Eschrichtius robustus). J. Mammal.,
        56(1):265-6.

TOMILIN, A.G., Lactation and nutrition in cetaceans.
1946   *Dokl. An SSSR,* 52(3):277-9.

–, On the thermal regulation in cetaceans. *Priroda,*
1951   6:55-8.

–, Mammals of the USSR and adjacent countries. Vol.
1967   9. Cetacea. Jerusalem, Israel Program for
        Scientific Translations, IPST (1124): 742 p.

VALENTIN, G., Beiträge zur Kenntnis des Winterschlafes
1857   der Murmelthiere. 2. Abth. Wechsel der Or-
        gane während des Winterschlafes. *Unters.
        Naturl. Mansch. Tiere,* 2:1-55.

VINCENT, F., Etudes préliminaires de certaines émissions
1960   acoustiques de *Delphinus delphis* L. en capti-
        vité. *Bull. Inst. Océanogr. Monaco,* 57(1172).

VINOGRADOVA, Z.A., Study of the biochemical compo-
1960   sition of Antarctic krill *(Euphausia superba*
        Dana). *Dokl. AN USSR,* 133:680-2.

–, The biochemical composition of Antarctic plankton.
1967   *In* Biochemistry of marine organisms. Ukrai-
        nian Academy of Sciences, pp. 7-17.

WAHRENBROCK, E.A. *et al.,* Respiration and metabolism
1974   in two baleen whale calves. *Mar. Fish. Rev.,*
        36(4):3-9.

WILLIAMSON, G.R., The true body shape of rorqual
1972   whales. *J. Zool. Lond.,* 167:277-86.

WISLOCKI, G.B. and L. BELANGER, The lungs of the larg-
1960   er Cetacea compared to those of smaller spe-
        cies. *Biol. Bull. Mar. Biol. Lab. Woods Hole.
        Mass.,* 78:289-97.

483

ZENKOVICH, B.A., The food of far-eastern whales. *Dokl.*
1937   *AN SSSR*, 16(4):231-4.

–, Around the world after whales. Moskva, Izdatel'st-
1937a   vo'Molodaya Gvardiya'.

–, Whales and plankton in Antarctic waters. *In* The
1969   food and feeding behaviour of whales, edited
        by V.A. Arsen'ev, B.A. Zenkovich and K.K.
        Chapskii. Moskva, Izdatel'stvo Naurka, pp.
        150-2.

Appendix Table 1

Appendix Table 1. Body weight, tissue weights and oil yield for blue, fin and humpback whales taken in the Antarctic summer (by day and month of catch)

Species	Sex	Date	Body length (metres)	Body weight (tonnes)	Blubber weight (tonnes)	Meat weight (tonnes)	Bone weight (tonnes)	Viscera weight (tonnes)	Oil yield (tonnes)	Reference
Balaenoptera	f	7 Nov 1926	27.2	122.00	25.65	56.44	22.28	8.48	27.71	Laurie (1933)
musculus	m	24 Dec 1947	25.0	87.64	27.90	29.77	15.38	12.91	22.65	
	m	26 Dec 1948	24.7	76.91	19.92	27.45	16.69	9.14	20.10	
(Blue)	f[1]	29 Dec 1947	25.3	110.21	35.87	39.29	17.16	15.39	30.40	
	f[1]	30 Dec 1947	25.3	96.49	22.62	34.14	19.91	13.66	24.20	
	f	2 Jan 1948	24.1	76.28	18.44	29.50	14.45	10.42	19.34	
	f[1]	7 Jan 1948	25.3	94.23	27.55	37.01	16.98	10.30	25.55	
	f	8 Jan 1948	21.3	56.71	16.11	22.91	10.79	5.55	15.50	
	f	9 Jan 1948	23.5	70.06	16.71	27.15	13.61	8.20	14.70	
	m	11 Jan 1948	24.7	84.05	24.81	33.84	15.01	8.46	24.35	
	f	12 Jan 1948	23.8	68.93	22.69	24.83	11.98	7.94	17.45	
	m	14 Jan 1948	23.2	63.06	16.56	25.88	12.60	6.17	12.75	
	m	15 Jan 1948	22.3	56.48	15.43	22.38	10.83	6.07	12.55	
	m	16 Jan 1948	22.8	68.97	19.02	28.33	12.75	7.37	18.65	
	m	21 Jan 1948	22.6	77.42	22.00	34.93	11.10	7.74	18.27	
	f	21 Jan 1948	23.2	77.63	20.68	34.04	11.93	9.49	19.12	
	f	21 Jan 1948	25.3	93.50	22.03	31.61	20.08	12.68	omitted	
	m	22 Jan 1948	25.3	107.80	27.89	39.84	19.61	11.57	20.75	Nishiwaki (1950)
	f	27 Jan 1948	27.2	127.54	32.36	61.51	17.54	13.94	26.85	
	f	2 Feb 1948	22.6	55.33	15.58	22.67	10.21	5.64	12.90	
	f	7 Feb 1948	23.2	77.43	19.70	30.06	13.30	10.03	17.10	
	f	14 Feb 1948	26.3	108.62	31.24	46.41	16.26	12.60	26.85	
	m	15 Feb 1948	23.5	97.63	28.70	41.61	14.64	10.74	25.85	
	m	20 Feb 1948	22.8	86.27	22.66	30.73	16.73	10.85	18.72	
	f	20 Feb 1948	22.3	62.33	15.75	25.50	11.26	7.03	13.02	
	f	21 Feb 1948	23.5	82.29	20.33	35.80	13.67	8.70	16.98	
	f	23 Feb 1948	26.8	113.75	31.28	49.64	17.31	13.18	30.35	
	f	23 Feb 1948	23.2	84.28	20.27	33.96	13.83	9.95	17.65	
	m	25 Feb 1948	24.4	85.35	23.31	36.47	14.87	8.71	24.00	
	f	2 Mar 1949	25.0	99.71	27.20	34.00	16.24	13.12	26.95	
	f	3 Mar 1948	25.0	94.96	26.31	33.52	16.67	12.82	26.20	
	m	5 Mar 1948	25.3	99.33	26.07	35.89	16.64	11.82	23.75	
	f	7 Mar 1948	24.7	95.43	25.53	38.85	15.80	8.79	26.85	
	f	17 Mar 1947	28.7	140.00		no data			32.00	Voronin (1948) *in*
	f	20 Mar 1947	27.6	190.00	30.00	66.00	26.00	no data		Tomilin (1967)
Balaenoptera	f[1]	17 Dec 1948	20.7	49.39	13.34	22.12	8.01	5.12	7.01	
physalus	f	18 Dec 1948	22.3	57.13	15.86	24.29	10.01	6.17	7.72	
	m	19 Dec 1948	19.2	40.39	9.67	19.67	7.00	3.53	5.22	
(Fin)	f	27 Dec 1948	18.6	37.37	8.27	18.39	5.91	3.65	6.38	Nishiwaki (1950)
	f[1]	28 Dec 1948	21.7	53.46	11.32	24.02	9.51	5.95	9.50	
	f	30 Dec 1948	22.3	58.53	13.34	28.37	10.68	5.36	no data	
	m	30 Dec 1947	19.5	40.30	9.27	17.89	7.90	4.41	9.27	
	m	30 Dec 1947	20.4	45.24	10.34	20.75	8.94	4.40	9.75	

Appendix Table 1. **Body weight, tissue weights and oil yield for blue, fin and humpback whales taken in the Antarctic summer (by day and month of catch)** *(concluded)*

Species	Sex	Date	Body length (metres)	Body weight (tonnes)	Blubber weight (tonnes)	Meat weight (tonnes)	Bone weight (tonnes)	Viscera weight (tonnes)	Oil yield (tonnes)	Reference
Balaenoptera	f	4 Jan 1948	21.1	51.67	10.54	25.36	8.34	5.58	7.92	
physalus	f[1]	6 Jan 1949	22.6	58.22	18.25	25.42	8.91	5.23	11.68	
(cont'd)	m	7 Jan 1948	19.2	48.09	11.09	22.39	8.79	4.23	8.01	
	m	8 Jan 1948	20.2	40.21	9.57	19.50	6.27	3.96	8.38	
	m	10 Jan 1949	20.4	45.61	10.46	21.61	8.40	4.54	6.40	
	f	11 Jan 1948	22.6	64.26	15.60	29.93	10.49	6.97	11.56	
	f	11 Jan 1948	22.9	57.60	13.78	25.22	11.42	6.21	10.20	
	f[1]	13 Jan 1948	22.7	69.54	18.56	31.17	10.20	7.98	13.17	
	f[1]	17 Jan 1948	21.9	60.72	14.19	29.09	9.87	6.40	12.55	
	m	18 Jan 1948	19.5	41.65	10.59	19.11	7.08	3.78	8.67	
	m	21 Jan 1949	20.4	49.36	12.38	22.92	7.69	5.90	7.42	Nishiwaki (1950)
	f[1]	29 Jan 1948	21.7	57.68	14.75	28.30	7.82	5.69	12.58	
	f[1]	5 Feb 1948	21.7	58.51	15.18	27.02	8.96	6.06	11.28	
	m	6 Feb 1948	21.1	47.58	10.78	21.96	7.88	4.54	7.23	
	f	12 Feb 1948	20.7	47.37	10.03	21.18	9.07	5.16	7.48	
	f	25 Feb 1949	23.2	57.49	13.32	24.09	9.13	5.91	11.30	
	f	26 Feb 1949	21.4	60.33	12.39	26.32	9.80	7.21	10.87	
	f	26 Feb 1949	21.7	53.09	12.95	25.24	8.67	5.52	no data	
	f	2 Mar 1948	20.7	48.50	9.97	20.44	9.23	5.17	7.92	
	f	7 Mar 1948	20.7	49.70	13.70	26.65	7.45	5.89	10.05	
	m	14 Mar 1949	19.9	48.37	10.70	18.31	8.03	5.32	11.23	
Megaptera		28 Dec 1949	12.0	16.76						
novaeangliae[2]		29 Dec 1949	12.1	21.64						
		30 Dec 1949	11.9	19.30						
(Humpback)		31 Dec 1949	12.5	23.77						
		1 Jan 1950	12.6	21.44						
		2 Jan 1950	12.4	23.88						
		3 Jan 1950	12.1	24.89						
		4 Jan 1950	12.3	23.27						
		5 Jan 1950	12.9	32.00						Ash (1953)
		1 Feb 1953	12.8	29.67						
		2 Feb 1953	12.8	28.75						
		3 Feb 1953	12.7	29.26						
		4 Feb 1953	12.3	30.89						
		16 Feb 1953	11.7	34.54						
		17 Feb 1953	12.7	28.96						
		18 Feb 1953	12.6	28.14						
		19 Feb 1953	12.6	32.00						

[1] Pregnant animals.
[2] Weights and lengths for humpback whales are averages for several animals taken on same day and are estimated from cooker fillings.

ESTIMATES OF GROWTH AND ENERGY BUDGET FOR THE SPERM WHALE, *PHYSETER CATODON*

C. Lockyer

Abstract

This paper reviews the growth parameters and factors determining the energy budget of sperm whales (*Physeter catodon*). Gestation has been estimated to take between 14.5 and 16.5 months, with the length and weight of calves averaging 405 cm and 1 054 kg at birth. The calf appears to suckle for up to 2 years before being weaned, by which time it is approximately 6.70 m in length and weighs about 2 800 kg. The mean sizes, body weights and ages at sexual and physical maturity in the female are 8.70 m, 6.35 t and 9 years, and 18.90 m, 13.50 t and at least 30 years, respectively. The male undergoes a prolonged process of maturation, with mean sizes and ages during the main periods of development as follows: puberty – 9.65 m, 9.40 t and 9.5 years; sexual maturity – 12.00 m, 18.00 t and 19 years; social maturity – 13.65 m, 27.40 t and 26 years; and physical maturity – 15.85 m, 43.60 t and 45 years. The best estimate of the body weight (W) and length (L) is $W_{(t)} = 0.006648L_{(m)} 3.18$. Estimates of the metabolic rate are also given.

Sperm whales mostly eat squid. The rate of consumption averages about 3 % of body weight in the 40-50 t bulls. The daily activity pattern of the species is little-known other than observations of isolated activities. Increased potential diving ability in terms of depth and duration occurs with growth and development.

Calculation of the sperm whale's energy budget shows a decrease in growth efficiency as the animal grows older. Correspondingly, energy becomes mainly directed into maintaining "equilibrium" rather than into new growth as the whale matures. The extra energy required for pregnancy becomes significant from about the sixth month onward, demanding an increase of food intake of 10 % and 5 % for newly mature and fully adult females, respectively. Lactation, involving a calculated daily production of about 20 kg of milk (3 840 kcal/kg), requires more significant increases of 63 and 32 % respectively. The consequent increased daily feeding rate of nursing females, together with the presence of the calf, may perhaps be a contributing factor in the sexual segregation of sperm whales and the difference in diving behaviour between the sexes.

Résumé

L'auteur étudie les paramètres de croissance et les facteurs déterminant les paramètres énergétiques des cachalots, *Physeter catodon*. On a estimé que la gestation dure de 14 mois et demi à 16 mois et demi, la longueur et le poids à la naissance étant en moyenne de 405 cm et 1 054 kg. Il semble que le jeune tète jusqu'à deux ans avant d'être sevré. A ce moment, il mesure environ 6,70 m et pèse près de 2 800 kg. Chez la femelle, la taille, le poids et l'âge à la maturité sexuelle sont en moyenne de 8,70 m, 6,35 t et 9 ans; à la maturité physique, ils sont de 18,9 m, 13,5 t et au moins 30 ans. La maturation du mâle est plus lente; ses taille et âge moyens aux principaux stades du développement sont les suivants: puberté — 9,65 m, 9,4 t et 9 ans et demi; maturité sexuelle — 12 m, 18 t et 19 ans; maturité sociale — 13,65 m, 27,4 t et 26 ans; maturité physique — 15,85 m, 43,6 t et 45 ans. La meilleure estimation du rapport entre le poids corporel (W) et la longueur (L) est $W_{(t)} = 0,006648\ L_{(m)}\ 3,18$. L'auteur donne aussi des estimations du taux métabolique.

L'aliment principal du cachalot est le calmar. Le taux de consommation avoisine 3 % du poids corporel chez les animaux pesant jusqu'à 15 t; il augmente chez les sujets de plus grande taille et peut presque atteindre 3,5 % chez les mâles de 40-50 t. On est peu renseigné sur le type d'activité quotidienne de l'espèce, à part l'observation d'activités isolées. Avec la croissance et le développement, l'animal plonge plus profond et plus longtemps.

Le calcul du bilan énergétique du cachalot montre un fléchissement de l'aptitude à la croissance avec l'âge. De même, à mesure que la maturité avance, l'énergie est surtout consacrée au maintien d'un équilibre plutôt qu'à la croissance. L'énergie supplémentaire requise pour la gestation devient importante à partir du sixième mois: la ration alimentaire doit augmenter de 10 % chez la femelle qui vient d'atteindre la maturité et de 5 % chez le femelle complètement adulte. La lactation — on a calculé que la production quotidienne est d'environ 20 kg de lait (3 840 kcal/kg) — exige un accroissement encore plus important, de 63 % et 32 % respectivement. La suractivité alimentaire provoquée chez la femelle qui allaite, jointe à la présence du jeune, contribue peut-être à la ségrégation sexuelle des cachalots et pourrait expliquer la différence de comportement en plongée entre les sexes.

Extracto

Se examinan los parámetros de crecimiento y los factores que determinan el consumo de energía del cachalote, *Physeter catodon*. Se ha calculado que la gestación dura de 14,5 a 16,5 meses, y que en el momento del nacimiento las crías tienen una longitud media de 405 cm y un peso de 1 054 kg. Las crías maman hasta los dos años, y en el momento del destete tienen aproximadamente 6,70 m de longitud y pesan unos 2 800 kg. La talla media, peso corporal y edad de madurez sexual y física de las hembras son, respectivamente, 8,70 m, 6,35 t, nueve años y 18,90 m, 13,50 t y al menos unos 30 años. El macho atraviesa un período prolongado de maduración, con las siguientes tallas y edades medias en las principales fases de desarrollo: pubertad — 9,65 m, 9,40 t, 9,5 años; madurez sexual — 12,00 m, 18,00 t, 19 años; madurez social — 13,65 m, 27,40 t, 26 años; y madurez física — 15,85 m, 43,60 t, unos 45 años. La mejor estimación de la relación peso corporal (W)/talla (L) en *P. catodon* es $(W)_{(t)} = 0,006648\ L_{(m)}\ 3,18$. Se dan también estimaciones del índice metabólico.

El principal alimento de *P. catodon* son los calamares. El índice de consumo de alimentos es por término medio de 3 por ciento del peso corporal, en los animales de hasta 15 t, y aumenta en los de mayor talla, acercándose al 3,5 por ciento del peso corporal en los machos de 40-50 t. Es poco lo que se sabe sobre la actividad diaria de *P. catodon*, aparte de algunas observaciones de actividades aisladas; con el crecimiento, el animal adquiere mayor capacidad de inmersión, tanto en profundidad como en duración.

Se ha calculado el consumo de energía de *P. catodon* en diversas fases de desarrollo, y los resultados revelan una disminución de la eficiencia del crecimiento a medida que el animal aumenta en edad. En consecuencia, a medida que el animal madura la proporción de energía consumida se orienta principalmente a mantener el equilibrio más que al crecimiento. La energía adicional que es necesaria durante la gestación es importante a partir del sexto mes: en las hembras que acaban de llegar a la madurez se produce un aumento de la ingesta de alimentos del 10 por ciento, y del 5 por ciento en las hembras plenamente adultas. La lactación, que requiere una producción diaria de leche de unos 20 kg (3 840 kcal/kg), hace necesarios aumentos más importantes, del 63 y el 32 por ciento, respectivamente. El aumento del índice diario de alimentación de las hembras gestantes determinado por esa demanda de nutrición, unido a la presencia de la cría, puede tal vez contribuir a la segregación sexual de *P. catodon* y a la diferencia de comportamiento entre ambos sexos en lo relativo a inmersión.

C. Lockyer
British Antarctic Survey, Madangley Rd., Cambridge CB3 OET, England

Growth

FOETAL GROWTH

Recent estimates of foetal growth parameters and gestation times for sperm whales are shown in Table 1. All the estimates of gestation time, t_g are between 14½ and 16½ months, regardless of geographical location.

The growth parameters are those defined by Huggett and Widdas (1951) in the formula

$$W^{1/3} = a_w (t - t_0),$$

where w is body weight in g, a_w is the specific growth velocity constant in g/day, t is time in days since conception, and t_0 is the time period in days from conception to the intercept of the linear growth phase with the time axis. Because $W \propto L^3$, where L is body length, the original Huggett and Widdas formula has more often been used with the terms

$$L = a_L (t - t_0),$$

where a_L is the specific growth velocity constant in cm/day.

In Fig. 1, a growth curve of weight increase with time has been constructed from foetal data of Gambell (1972). The majority of conceptions are estimated to occur during December by Best (1968) and Gambell (1972) off South Africa, and birth generally occurs during February or March.

POST-NATAL GROWTH

The Calf

The sperm whale calf is approximately 405 cm in length, and is estimated to weigh about 1 tonne at birth. The animal is suckled by the mother for upward of a year. Best (1974), in his review of published data and from his own examination of young calves captured under special permit, and an analysis of the proportion of pregnant, resting and lactating females in the catch has evidence that lactation may continue for up to 2 years. Gambell (1972) also presents evidence for a 2-year lactation period from an analysis of the rate of regression of the diameter of the corpus albicans derived from the preceding pregnancy. The calf would thus be weaned during or toward the end of its second year of life.

The calf is about 6 m in length at 1 year, and about 6.7 m when weaned during the second year (Best, 1974; Clarke, 1956).

Age in sperm whales is estimated from the dentinal growth layers produced in the teeth (International Whaling Commission, 1969). These layers form continuously throughout life and in the mandibular teeth, the pulp cavity of which remains open to a considerable age, they constitute an almost complete record of the total age of the animal.

In newborn and foetal sperm whales, the neonatal layer is present at the crown of the tooth (Nishiwaki, Hibiya and Ohsumi, 1958), and is similar to the subsequent dentinal

491

Table 1. Recent estimates of growth parameters for sperm whale foetuses [1]

Author	t_g in days	t_0 days	L at birth cm	W at birth g	a_L cm/day	a_W g/day	Geographical Location	Comments
Clarke, Aguayo and Paliza (1964)	518	36	402		0.83		Chile & Peru	
Ohsumi (1965)	503	35	405		0.86		N. & S. Hemispheres	Data taken from the International Whaling Statistics
Bannister (1966; 1969)	479	33.5	425		0.96		W. Australia	
Best (1968)	444	40	404	*774 000	1.00		South Africa, west coast	* Large foetuses cut up for weighing
Gambell (1972)	449	40	405	1 054 000	0.99		South Africa, east coast	
Frazer and Huggett (1974)	470				0.9	0.24	Presumed South Africa	Source of data not well indicated
Lockyer (present study)	449	42	**	**	**	0.25	South Africa, mainly east coast	** See Gambell (1972). shown above

[1] Terms defined in text.

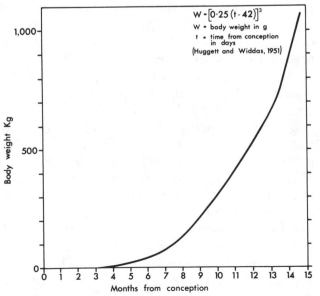

FIG. 1. – Calculated growth in weight with time for sperm whale foetuses.

growth layers. The neonatal line frequently becomes worn away with age in erupted teeth. There is still some doubt over the formation rate of the post-natal growth layers, but the general consensus of opinion (International Whaling Commission, 1969) is that 1 layer represents 1 year's growth, at least after weaning (Ohsumi, 1965; Best, 1970; Gambell, 1972), although Berzin (1972) regards the formation of 2 layers per year as possible.

At the end of the first year, the calf probably has 1 growth layer present plus the neonatal line above it. After weaning and by the end of the second year, 2 growth layers will have formed below the neonatal line. The predicted mean growth in suckling calves is included in Fig. 2.

The predicted birth weight for a 405-cm animal is 1 050 kg from examination of foetuses (see Fig. 1) by Gambell (1972), although

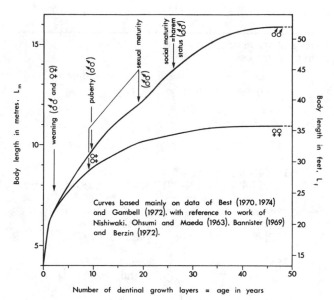

FIG. 2. – Estimated mean body length at age in the post-natal growth phase of sperm whales. These curves are most applicable to sperm whales off South Africa, west side.

Best (1968) gives a weight of about 780 kg. It is not possible to say how much the whale weighs after 1 year of growth because of the lack of published data. If weaning occurs at the end of the second year and average body length is then 670 cm (see Fig. 2), the equivalent body weight using the weight/length formula of Lockyer (1976) illustrated in Fig. 3, is about 2 800 kg. The estimated mean growth in weight of the calf with time is shown in Fig. 4.

The Post-weaning Phase

The main food items consumed when the young whale begins to take solid food are squid (see Berzin, 1972; Clarke, 1956; Tomilin, 1946). The growth of the sperm whale is not well documented in the pre-pubertal phase, chiefly because of the minimum size limits imposed by the International Whaling Commission (1950, 1952, 1974). The minimum size imposed has varied between 10.9 and 11.6 m, according to conservation requirements. The estimates of pre-pubertal growth rates are thus speculative. The sizes and ages of sperm wha-

les at sexual maturity are well documented however. Best (1974) has reviewed the available published data on age at sexual maturity in females. Generally, there is good agreement here. The information on males is not so clear, perhaps because of the prolonged process of maturation. This has been reviewed and described in 3 different phases by Best (1974):

(i) puberty, when the testis of 50 % of males have immature centres, corresponding to the onset of spermatogenic activity;

(ii) sexual maturity, when 50 % of males are immature and 50 % are maturing or mature at the periphery of the testis, corresponding to the movement of males into high latitudes for the first time;

(iii) social maturity, when 50 % of males are immature and 50 % mature at the periphery of the testis (maturing animals not included), probably corresponding to the attainment of harem bull status, and the acceleration of testis growth.

The recent published information on maturity in sperm whales is set out in Table 2. Fig. 2 shows the growth in body length with age and indicates the mean ages at the different stages of sexual maturity.

The growth in weight with age can be calculated from a cross-referencing of body length at age and weight at body length. The weight/length relationship in sperm whales in shown in Fig. 3, based on the fluid-loss corrected formula from Lockyer (1976). The predicted growth in weight with age is shown in Fig. 4. There are 2 curves shown for the male, one being that based on the weight/length relationship in North Pacific sperm whales in Fig. 3, and the other based on a weight/length relationship formulated from the actual weighings of 3 whole whales at Durban, South Africa (Lockyer, 1976). This latter is shown for information only, and to indicate that methods of weighing, geographical location and other

493

Table 2. A summary of body length, age and maturity in sperm whales

Oceanic location	Developmental stage - body length in metres and age in dentinal growth layers (ratio 1:1) in adjacent parentheses						Author
	puberty ♂	sexual maturity ♂	sexual maturity ♀	social maturity ♂	physical maturity ♂	physical maturity ♀	
Indian (east side)	8.7(11)		8.8(12-13)		16.0	10.5	Bannister (1969)
N. Pacific	9.5(10)		8.3(7)		15.9(46-50)	11.0(34)	Berzin (1972)
S. Atlantic (east side)	9.5(9-10)	12.0(19)	8.5(9)	13.7(25-27)	15.5(35-45)	10.7(28-29)	Best (1970, 1974)
S. Pacific (west side)	9.7	11.3	8.5				Clarke, Aguayo and Paliza (1964)
Indian (west side)		11.9(18)	8.8(7)	13.7(26)	15.2(40+)	11.0(45+)	Gambell (1972)
N. Pacific (west side)	9.5(9)		9.0(9)		15.9(45)	11.0(45)	Nishiwaki, Hibiya and Ohsumi (1958); Nishiwaki, Ohsumi and Maeda (1963)
N. Pacific (west side)			9.0(9.2)				Ohsumi (1965)
N. Pacific (east side)	< 10.3(7-8)		8.8(8)	14.3(22)			Pike (1966)
N. Pacific (east side)	9.2	11.0	9.2(9)				Rice and Wolman (1970)

factors may influence the final computed growth curve. The body weight at physical maturity by these two predictions is very different, and the body weight of physically mature males of length 15.85 m may reach 53 tonnes.

The important developmental stages are indicated on the curve. At physical maturity the male weighs nearly 4 times as much as the female. The point where body weights of the male and female really begin to diverge is at about age 9-10 growth layers. At this age the female is sexually mature and the growth rate begins to fall off as expected from analogy with other mammals (Brody, 1968). The male is at puberty here, and perhaps because the process of sexual maturation is very protracted, the growth rate does not slow. In fact, it would appear that the male experiences a spurt in growth rate at sexual maturity (Figs. 2 and 4), much in the same way as the elephant seal, *Mirounga leonina,* which has a harem system of social organization (Laws, 1953) similar to that of the sperm whale (Ohsumi, 1971). In the elephant seal, sexual maturity is reached at twice the age of the female. A similar relationship is also indicated in the sperm whale (Figs. 2 and 4). The age when full harem master status is attained by the elephant seal is approximately 3 times the age at sexual maturity of the female, another point of similarity with the sperm whale. The actual time scales considered for the elephant seals, however, are only a quarter of those for the sperm whale (Laws, 1953).

The female sperm whale has attained 47 % of the physically mature body weight at sexual maturity. In a similar comparison, the male has attained 22 % at puberty, 41 % at sexual maturity, and 63 % at harem master status, after which the bull continues to grow for several years.

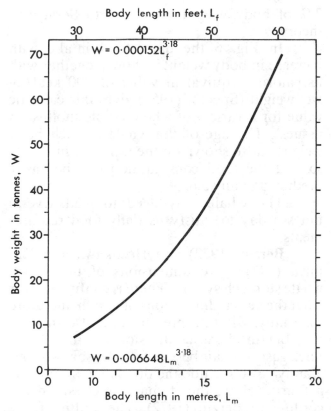

FIG. 3. — Calculated weight/length relationship in sperm whales (post-natal period), allowing for 10 % fluid loss during flensing.

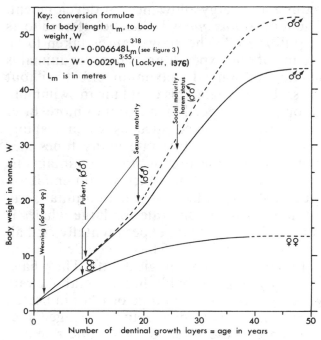

FIG. 4. — Mean body weight with age in sperm whales, cross-referencing Figs. 2 and 3.

Metabolism

The metabolic rates of sperm whales are calculated by Lockyer (1972; 1976a). These estimations of basal, resting and active metabolic rates are shown in Fig. 5. The basal rate is taken to be the minimal metabolic energy expenditure possible under conditions of starvation existence. This rate is thus rarely encountered except perhaps during sleep. The resting rate is taken to be the normal metabolic energy output when the animal is just maintaining itself in a *status quo,* that is when only usual feeding and movements necessary for life are taking place. Because of the definition of resting rate, this rate has a somewhat different meaning in young growing animals and adult fully grown ones. The true resting rate is only observed in adults, because the young animals

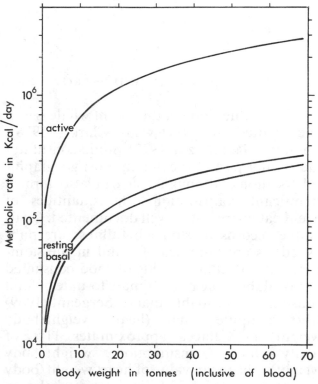

FIG. 5. — Metabolic rates estimated for sperm whales from Lockyer (1972, and MS).

require an energy allowance for development. The term *status quo* will thus be referred to as "equilibrium". The active rate is taken to be the metabolic expenditure when the animal is making prolonged maximum effort without stress. For a whale, this could mean swimming at top cruising speed for an hour or more. Brief peak effort involved such as when leaping, may exceed the active rate many times momentarily, but the animal cannot maintain this level of energy output which would entail too great a stress. Unlike some land animals, any estimate of metabolic rate for large whales is very difficult to obtain experimentally, and all the estimations in Fig. 5 have been determined indirectly from physiological data. The whale, in daily living, probably has a metabolic rate which varies over the range of rates indicated according to the level of activity engaged in at any time. These levels of metabolic rate are, therefore, only a guide as to the limits of the values which they can assume.

Food

FEEDING CAPACITY AND ENERGY

The chief food of the sperm whale is well documented and reviewed (Clarke, 1956); Tomilin, 1946; Berzin, 1972). Squid constitutes the greater part of the diet in most geographical locations, although fish can also form a significant contribution. The quantities of squid eaten are not so well documented. Qualitative records of stomach fullness are published, as are numbers of squid in the stomach, but no actual weights of food consumed are available. The best estimate to-date of food consumed is probably that of Sergeant (1969) who computes that (heart weight/body weight) in Cetacea approximates 11 % of (daily food consumption weight/body weight). The mean value of heart weight/body weight averaged 0.33 % in sperm whales so that daily food intake would constitute about

3 % of body weight according to Sergeant's theory.

In Fig. 6 the daily food intake with increase in body weight is shown together with its calorific equivalent value of 800 kcal/kg wet weight (Spector (1956) gives this calorific value for a variety of whole edible molluscan tissues). The age of the whale at each body weight is also shown on the top scale in Fig. 6, so that the food consumption can be easily predicted at any age.

The whale may need to feed several times a day to satisfy its daily food requirements.

Berzin (1972) and Hosokawa and Kamiya (1971) give dimensions of the sperm whale stomach system. There is confusion over what the actual dimensions of length and more especially, width, represent; and also the fact that Berzin classifies the stomach into 3 distinct gastric chambers, whilst Hosokawa and Kamiya also include the duodenal ampulla as a fourth. Here, the stomach is classified according to Berzin (1972). The width of the chamber has been assumed to be the diameter for the full state, and the capacity has been

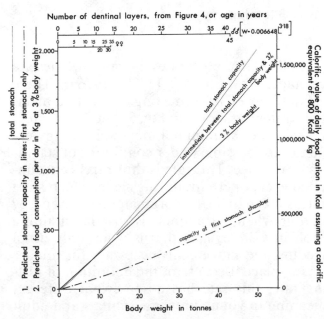

FIG. 6. — Estimates of feeding rate and stomach capacity in sperm whales.

calculated by the usual formula for a cylinder (cross-sectional area x length). However, it seems that the width given by Hosokawa and Kamiya probably represents the half circumference of the empty and flattened chamber, and necessary adjustments have been made to compare their figures with those quoted in Berzin. In Table 3 the stomach capacities are given. There is great variability but in order to gain some idea of average stomach size, a straight line relationship of $\log_{10}$ stomach capacity/$\log_{10}$ body weight was fitted to the data. The lines of best fit were calculated by the method of least squares regression analysis (Simpson, Roe and Lewontin, 1960). The resulting mean lines of regression of $\log_{10}$ stomach capacities/$\log_{10}$ body weight are shown on an arithmetic scale in Fig. 6, together with the calculated food consumption rate of 3 % body weight per day.

The total stomach capacity and calculated daily food consumption are comparable up to about a body weight of 15 tonnes, but diverge increasingly as body weight becomes heavier. It seems that whilst an estimate of the daily food consumption rate of 3 % of body weight is reasonable for females and small males, a better estimate of feeding capacity, especially for large males, might fall between the limits of total stomach capacity and 3 % body weight. In the absence of any firm evidence as to true feeding capacity the intermediate values in Fig. 6 will be taken to represent the best

available estimate of daily feeding rate in the bigger whales.

Activity

DIVING BEHAVIOUR AND GENERAL ACTIVITY

The daily activity pattern of the sperm whale is little-known. Isolated activities such as diving and incidences of ships colliding with sleeping whales are documented, but there is no continuous day – and night – long watch on the time spent on such activities for a single whale.

The newborn and suckling calves are unlikely to be very active, especially because they do not need to hunt for food. In the newborn calf, stability in swimming is a problem because of improperly unfurled tail flukes and fins (Berzin, 1972; Gambell, Lockyer and Ross, 1973). The diving ability is also less well-developed at this time. In Table 4, the potential diving ability in terms of depth, which is generally directly correlated with duration, is indicated for different sizes of whales. The size classification is broad, but serves to show that the larger animals can dive more deeply. These dives were probably all escape reactions to being chased, so that the actual frequency dis-

Table 3. Calculated capacities of sperm whale stomachs from stomach dimensions

| Sex | Body length (metres) | Calculated body weight (tonnes) | Stomach capacity in litres | | | | Reference |
| | | | Compartment | | | Total | |
			1	2	3		
♀	9.0	7.0	86.5	75.4	17.7	179.9	Berzin (1972)
♂	11.8	17.2	125.7	635.1	73.6	834.4	
♂	15.0	35.2	179.2	431.3	123.5	734.0	
♂	15.6	41.2	873.4	873.4	149.0	1 895.8	Hosokawa and Kamiya (1971)
♂	16.0	45.0	993.0	696.8	392.0	2 081.4	Berzin (1972)

497

Table 4. Percentage frequency of diving activity by groups of sperm whales of different body size at various depths

Depth (metres)	Body size (metres)		
	7.5-9.0 %	9.1-11.0 %	11.1-14.0 %
0- 99	12	15	33
100- 199	12	19	19
200- 299	13	18	11
300- 399	17	16	11
400- 499	21	10	4
500- 599	17	8	4
600- 699	8	6	4
700- 799		5	4
800- 899		3	4
900- 999			3
1 000-1 099			
1 100-1 199			3

Data collected by sonar traces from a whale catcher off Durban, South Africa, during whale marking cruises in February 1972 and 1973.

tributions of dives with depth are not relevant to this observation on potential diving ability. The larger the animal, the more able it is to dive deeply if required. This variation of diving ability with body size means that it is generally the small males and females who dive to shallow and intermediate maximum depths, and the large males who dive very deeply in excess of 1 100 m for long periods.

Berzin (1972) also discusses the increased diving ability in depth and duration of larger sperm whales. In addition to this information, he mentions that deeper and longer dives entail a subsequently greater period of respiratory recovery at the surface, a fact also personally observed for sperm whales diving off Durban in 1972 and 1973 during marking cruises.

Energy Budget

The growth, metabolic energy expenditure and food consumption are summarized for the sperm whale by sex and developmental stage in Table 5. Growth and developmental stages are identified from Figs. 3 and 4, from which the growth increment since the previous year can be estimated. The mean body weight (not shown in Table 5) during this year has been used to estimate metabolic rate from Fig. 5, and food consumption from Fig. 6.

GROWTH

The proportion of muscle and blubber in sperm whales averages 34 % and 33 % of body weight (Lockyer, 1976). The calorific value of mammal non-fatty flesh is 1 500 kcal/kg (Petrides and Swank, 1965). The calorific value of fat (in blubber) is 9 450 kcal/kg (Brody, 1968). The year's growth increment is thus converted to growth energy in Table 5. At physical maturity, developmental growth is minimal and effectively negligible.

METABOLISM

The metabolic rates for the sperm whale are estimated from Fig. 5. The total yearly metabolic energy expenditure is calculated by trial and error, using certain assumptions.

$$\text{Assimilation efficiency}$$
$$\frac{\text{Calories of food assimilated } \%}{\text{Calories of food consumed}}$$

is assumed to be 70-80 %. Assimilated energy less growth energy gives a theoretical upper limit on the metabolic energy expenditure which comprises a greater percentage at resting rate and smaller percentage at active rate. These percentages are variable, and that for active rate increases with the growth of the whale, the actual amount being determined within the consumed energy limit.

FOOD CONSUMPTION

The daily food intake is determined from Fig. 6 from the intermediate line shown.

The calorific value of the squid eaten can be determined at the same time from Fig. 6, and thence converted to an annual energy intake.

The suckling calf has been treated separately because the diet is different. The total stomach capacity of the newborn calf is about 20 litres, according to Fig. 6. This would mean that the young calf could take up to 6-7 litres of milk into the first stomach chamber, if this quantity of milk were available from the mother. Berzin (1972) quotes Sleptsov (1952), who says that the mammary glands of sperm whales contain 45 litres of milk. This seems a very great amount relative to the apparent needs of the suckling calf. Berzin says that, from observations of lactating female carcasses brought onto the flensing platform, strong pressure applied to the mammaries causes 2-3 litres of milk to gush from the nipples. This seems a more likely amount.

Berzin (1972) tabulated the chemical composition of sperm whale milk, published by various authors. Excluding two greatly different records from a total list of 10 references, the average composition was 36.21 % fat, 54.91 % water and 7.40 % dry residue. The calorific values of fat and protein are 9 450 kcal/kg and 5 650 kcal/kg respectively (Brody, 1968). The total calorific value of sperm whale milk is thus calculated to be 3 840 kcal/kg. This is comparable to the calorific values of other whale milks, 3 657-4 305 kcal/kg for balaenopterid whales and 3 700 kcal/kg for the toothed beluga (Tomilin, 1946). The quantity of milk consumed by the calf can therefore be converted to a calorific value.

GROWTH EFFICIENCIES

The gross and net growth efficiencies shown in Table 5 all show a marked fall off with age and development. The amount of energy used to maintain the animal increases greatly relative to the diminishing proportion used for growth. By adulthood, nearly all the energy consumed by the animal is directed to keeping it alive and functioning. Growth is negligible.

The net growth efficiency of about 21 % for sperm calves is similar to the high efficiencies found for other young suckling whales, for example, 38 % in blue whales and 28 % in fin whales (Lockyer, 1972, 1976a). Lockyer finds that the net growth efficiencies for pubertal blue and fin whales are about 5-7 %. In sperm whales (Table 5), the efficiency falls to about 4-6 % at puberty.

Lockyer finds that the ratio of energy of growth to energy of metabolism in blue and fin whales is about 1:2.1 for suckling calves (1:3.9 in sperm calves, Table 5) and 1:16.4 for pubertal blue and fin whales (1:15.4 for sperm males, 1:21.3 for sperm females, Table 5). Beyond puberty, comparisons between sperms and rorquals are difficult, because of their different patterns of growth.

ENERGY OF PREGNANCY AND LACTATION

The extra energy required for 14½-16½ months of gestation is unlikely to be excessive, particularly in the early stages. The effects of pregnancy are not likely to make significant energy demands upon the mother until about the sixth month, by which time the energy of growth will have cost the mother about 6.00×10^4 kcal (assuming that most energy is directed into making muscle tissue at 1 500 kgcal/kg). The extra energy required for growth by the foetus then increases exponentially. At birth, the foetus will have cost the mother at least 1.58×10^6 kcal since conception for tissue-building alone, and probably another 3.00×10^6 kcal for the metabolic expenditure of the foetus, totalling about 4.58×10^6 kcal throughout the pregnancy, nearly all of which is incurred in the last 9-10 months of gestation. In the newly mature female (Table 5) whose yearly basic expenditure is 4.43×10^7 kcal, the added cost of pregnancy brings the yearly expenditure to 4.89×10^7 kcal. In order to maintain her own equilibrium with an assimilation efficiency of about 80 % (Table 5), the whale needs to increase her daily food consumption by about 10 %, thus increasing

Table 5. Summary of energy budget in sperm whales

Sex	♂ or ♀	♀	♂	♂	♂	♀	♂
Stage of development	1st year suckling calf	puberty/ sexual maturity	puberty	sexual maturity	social maturity = harem status	physical maturity	physical maturity
Body length (m) (Fig. 3)	6.00	8.70	9.65	12.00	13.65	10.90	15.85
Body weight (t) (Fig. 4)	2.00	6.35	9.40	18.00	27.40	13.50	43.60
Growth increment since previous year (kg) (Fig. 4)	1 100 [1]	550	1 000	1 000	1 400	minimal	minimal
Calories of yearly growth (kcal) — Muscle (34% body weight) at 1 500 kcal/kg	0.56×10^6	0.28×10^6	0.51×10^6	0.51×10^6	0.71×10^6	no data	no data
Calories of yearly growth (kcal) — Blubber (33% body weight) at 9 450 kcal/kg	3.43×10^6	1.71×10^6	3.12×10^6	3.12×10^6	4.37×10^6	no data	no data
Metabolic rate (kcal/day) (Fig. 5) — Resting	2.10×10^4	5.70×10^4	7.50×10^4	1.28×10^5	1.73×10^5	1.02×10^5	2.52×10^5
Metabolic rate (kcal/day) (Fig. 5) — Active	1.70×10^5	4.50×10^5	6.00×10^5	1.04×10^6	1.37×10^6	8.20×10^5	2.02×10^6
Yearly metabolic energy expenditure (kcal) with details of the assumed % possible active rate at 70-80% assimilation efficiency [2]	1.58×10^7	4.23×10^7	5.61×10^7	11.33×10^7	17.23×10^7	8.97×10^7	31.79×10^7
	15% active	15% active	15% active	20% active	25% active	20% active	35% active
Approximate total energy utilization in a year (kcal)	1.98×10^7	4.43×10^7	5.97×10^7	11.69×10^7	17.74×10^7	$>8.97 \times 10^7$ allowing for growth	$>31.79 \times 10^7$ allowing for growth
Mean food intake/day throughout previous year (kg) (Fig. 6) [3]	20 [3]	190	270	555	870	420	1 505
Yearly food intake over 365 days (kg)	7 300	69 350	98 550	202 575	317 550	153 300	549 325
Calories of yearly food intake at 800 kcal/kg (kcal) (Fig. 6) [3]	2.80×10^7 [3]	5.55×10^7	7.88×10^7	16.21×10^7	25.40×10^7	12.26×10^7	43.95×10^7
Gross growth efficiency %, $\frac{\text{Calories of growth}}{\text{Calories of food intake}} \times 100$	14.3	3.8	4.6	2.2	2.0	probably < 1%	probably < 1%
Net growth efficiency % $\frac{\text{Calories of growth}}{\text{Calories of food assimilated}} \times 100$ (assumed assimilation efficiency ≈ 70-80 %) [2]	20.6 (71)	4.5 (80)	6.1 (76)	3.1 (72)	2.8 (70)	no data (73)	no data (72)
Ratio of energy of growth to energy of metabolism (measure of production)	1:3.9	1:21.3	1:15.4	1:30.5	1:33.9	approaching 1:∞ with age	approaching 1:∞ with age

[1] Average birth weight combining data of Gambell (1972) and Best (1968) is 900 kg.

[2] The extent of daily activity is not known accurately, so that it has been determined by trial and error, assuming that assimilation efficiencies are 70-80 % on a carnivorous diet (see text).

[3] Milk intake by calf determined by trial and error, and approximating 1 % body weight per day; calorific value of milk calculated to be 3 840 kcal/kg.

the daily intake to about 210 kg. In the fully adult female, total annual expenditure would need to reach 9.43t x 10⁷ kcal, in order to maintain equilibrium, but this represents only a 5 % increase in food consumption to cover the demands of pregnancy. The whale would thus need to eat about 440 kg daily, or an extra 20 kg, the same as the newly mature female. The energy cost of pregnancy is thus relatively small.

The energy cost of lactation at a production rate of about 20 kg milk daily, is 7.68×10^4 kcal/day. Over a year this amount is 2.80×10^7 kcal (see Table 5 for the calf). In order to maintain equilibrium, the newly mature female will need to increase her food intake by about 63 %, the fully adult female by about 32 %. In terms of extra food this represents about 120 kg and 134 kg daily for the newly and fully mature whales, respectively. Lactation is, therefore, a considerable drain on the energy resources of the female. In order to accumulate sufficient food, the first stomach would need to be filled about 4-5 times daily. This calculation indicates that nursing and pregnant females would have different feeding habits from bachelor schools. This, and the fact that the presence of calves with their mothers would cause further behavioural differences, may account, at least in part, for the observed segregation (Ohsumi, 1971) of these two groups of sperm whales.

Summary

1. Gestation of the sperm whale foetus has been estimated to take between 14½ and 16½ months. The birth length and weight average 405 cm and 1 054 kg. The growth parameters in the formula

$$W^{1/3} = a_w (t - t_0),$$

defined by Huggett and Widdas (1951), are as follows:

$a_w = 0.25$ g/day and $t_0 = 42$ days. (W = body weight in g; t = time in days.)

2. The mean sizes at sexual and physical maturities in the female are 8.70 and 10.90 m respectively, and the corresponding body weights are 6.35 and 13.50 tonnes. The average ages at sexual and physical maturities are 9 years and at least about 30 years.

4. The mean sizes and ages at puberty, sexual, social and physical maturities (as defined by Best, 1974) in the male are as follows:

(i) puberty — 9.65 m, 9.40 tonnes, 9½ years;

(ii) sexual maturity — 12.00 m, 18.00 tonnes, 19 years;

(iii) social maturity — 13.65 m, 27.40 tonnes, 26 years;

(iv) physical maturity — 15.85 m, 43.60 tonnes, ca. 45 years.

5. The body weight (W)/length (L) relationship, allowing for fluid losses in flensing, in $w_{(t)} = 0.006648 L_{(m)}^{3.18}$, for sperm whales in the North Pacific; the formula for 3 sperm whales taken off South Africa and weighed whole is $w_{(t)} = 0.0029 L_{(m)}^{3.55}$. Use of the latter formula could mean that a 15.9-m bull would weigh 53 tonnes. The former relationship is preferred, however, because of the larger sample size (Lockyer, 1976).

6. The metabolic rates for sperm whales of sizes between 10 and 50 tonnes, range as follows:

(i) basal $(6.4 \times 10^4 - 2.2 \times 10^5)$ kcal/day;

(ii) resting $(7.8 \times 10^4 - 2.7 \times 10^5)$ kcal/day;

(iii) active $(6.4 \times 10^5 - 2.2 \times 10^{10})$ kcal/day.

501

These rates have a linear logarithmic relationship with body weight.

7. The main food of sperm whales is established as a molluscan (chiefly squid) diet of calorific value 800 kcal/kg. The food consumption rate averages about 3 % body weight in animals up to 15 tonnes. In larger whales the daily food consumption rate increases and approaches 3.5 % body weight in the 40-50 tonne bulls. These conclusions are arrived at chiefly from an analysis of stomach capacities. It seems that from the size of the first stomach compartment at least 3 daily fillings must occur for satiation.

8. It is concluded from observations on diving ability in various sizes of sperm whales that increased potential activity in terms of depth and duration achieved occurs with growth and development. The large bulls are able to dive more deeply and for longer than the females and small males. From the extra respiratory recovery period required after long deep dives, it would seem that such dives are more strenuous than shallow ones.

9. The energy budget is calculated for the sperm whale at different stages of development. Briefly, the main findings are as follows:

GROWTH EFFICIENCIES

		gross[1]	net[2]
calf (♂ or ♀)		14.3 %	20.6 %
sexual maturity	♀	3.8 %	4.5 %

---

[1] $\left(\dfrac{\text{Calories of growth}}{\text{Calories of food consumed}} \right)$

[2] $\left(\dfrac{\text{Calories of growth}}{\text{Calories of food assimilated}} \right)$

puberty	♂	4.6 %	6.1 %
sexual maturity	♂	2.2 %	3.1 %
social maturity	♂	2.0 %	2.8 %

RELATIVE PRODUCTION (Calories of growth: Calories of metabolism)

calf (♂ or ♀)	1:3.9
sexual maturity	1:21.3
puberty	1:15.4
sexual maturity	1:30.5
social maturity	1:33.9

Growth efficiency decreases as the whale develops. Also the proportion of consumed energy becomes mainly directed into maintaining "equilibrium" rather than into new growth as the whale matures.

10. The extra energy required for pregnancy becomes significant from about the sixth month onward. During the year preceding birth, the newly mature female would need to increase her food intake by 10 % and the fully adult female by 5 %.

11. Lactation involves the calculated production of about 20 kg of milk daily to satisfy the calf. This demand would require the newly mature female to increase her food intake by 63 % and the fully adult female by 32 %. The calorific value of sperm whale milk is estimated to be 3 840 kcal/kg.

12. The nursing females would need to fill the first stomach 4-5 times daily in order to gain sufficient nourishment. This necessity, coupled with the presence of the calf, may perhaps be a contributing factor in the sexual segregation of sperm whales, and the difference in diving behaviour between the sexes.

References

BANNISTER, J.L., Sperm whales. *Rep. Div. Fish. Ocea-*
1966   *nogr. CSIRO, Aust.,* (1965-66): 22-3.

—, The biology and status of the sperm whale off We-
1969   stern Australia: an extended summary of re-
        sults of recent work. *Rep. IWC,* (19):70-6.

BERZIN, A.A., The sperm whale. Jerusalem, Israel Pro-
1972   gram for Scientific Translations, IPST Cat. No.
        60070 7:394 p. Translation from the Russian,
        published in Moscow by Izdatel'stvo Pishche-
        vaya Promyshlennost (1971).

BEST, P.B., The sperm whale (*Physeter catodon*) off the
1968   west coast of South Africa. 3. Reproduction in
        the female. *Invest. Rep. Div. Sea Fish., Cape
        Town,* (66):1-32.

—, The sperm whale (*Physeter catodon*) off the west
1970   coast of South Africa. 5. Age, growth and
        mortality. *Invest. Rep. Div. Sea Fish., Cape
        Town,* (79):1-27.

—, The biology of the sperm whale as it relates to stock
1974   management. *In* The whale prob-
        lem, edited by W.E. Schevill. Cam-
        bridge, Mass., Harvard University Press, pp.
        257-93.

BRODY, S., Bioenergetics and growth. New York, Hafner
1968   Publ. Co., Inc., 1023 p.

CLARKE, R., Sperm whales of the Azores. *Discovery*
1956   *Rep.,* (28):237-302.

CLARKE, R., L.A. AGUAYO and G.O. PALIZA, Progress
1964   report on sperm whale research in the south-
        east Pacific. *Norsk Hvalfangsttid.,* 53(11):
        297-302.

FRAZER, J.F.D. and A.St.G. HUGGETT, Species varia-
1974   tions in the foetal growth rates of eutherian
        mammals. *J. Zool., Lond.,* 174:481-509.

GAMBELL, R., Sperm whales off Durban. *Discovery Rep.,*
1972   (35):199-358.

GAMBELL, R., C. LOCKYER and G.J.B. ROSS, Observa-
1973   tions on the birth of a sperm whale calf. *S. Afr.
        J. Sci.,* 69(5):147-8.

HOSOKAWA, H. and T. KAMIYA, Some observations on
1971   the cetacean stomachs, with special conside-
        rations on the feeding habits of whales. *Sci.
        Rep. Whales Res. Inst., Tokyo,* (23):91-102.

HUGGETT, A.St.G. and W.F. WIDDAS, The relationship
1951   between mammalian foetal weight and con-
        ception age. *J. Physiol., Lond.,* 114: 306-17.

IWC, Report of the International Whaling Commission
1950   1950. *Rep. IWC,* (1):32 p.

—, Report of the International Whaling Commission
1952   1952. *Rep. IWC,* (3):40 p.

—, Report of the International Whaling Commission
1969   1969. *Rep. IWC,* (19):148 p.

—, Report of the International Whaling Commission
1974   1974. *Rep. IWC,* (24):236 p.

LAWS, R.M., The elephant seal (*Mirounga leonina*
1953   Linn.). 1. Growth and age. *Sci. Rep. Falkland
        Isl. Depend. Surv.,* (8):62 p.

LOCKYER, C., A review of the weights of cetaceans with
1972   estimates of the growth and energy budgets of
        the large whales. M. Phil. Thesis. University of
        London. 196 p.

LOCKYER, C., Body weights of some species of large
1976   whales. *J. Cons. CIEM,* 36(3):259-73.

—, Growth and energy budgets of large baleen whales
1976a   from the Southern Hemisphere. (MS).

NISHIWAKI, M., T HIBIYA and S. OHSUMI, Age study of
1958   sperm whale based on reading of tooth lami-
        nations. *Sci. Rep. Whales Res. Inst., Tokyo,*
        (13):135-53.

NISHIWAKI, M., S. OHSUMI and Y. MAEDA, Change of
1963   form in the sperm whale accompanied with
        growth. *Sci. Rep. Whales Res. Inst., Tokyo,*
        (17):1-14.

OHSUMI, S., Reproduction of the sperm whale in the
1965   North-West Pacific. *Sci. Rep. Whales Res.
        Inst., Tokyo,* (19):1-35.

—, Some investigations on the school structure of sperm
1971   whale. *Sci. Rep. Whales Res. Inst., Tokyo,*
        (23):1-26.

PETRIDES, G.A. and W.O. SWANK, Estimating the pro-
1965   ductivity and energy relations of an African
        elephant population. *Proc. Int. Grasslands
        Congr.,* 9:831-42.

PIKE, G., Progress report on the study of sperm whales

1966   from British Columbia. Paper presented to the Scientific Committee of the International Whaling Commission at the Sperm Whale Subcommittee Meeting, 10-18 February 1966, Honolulu, Doc. Cl:11 p. (unpubl.).

RICE, D.W. and A.A. WOLMAN, *Sperm whales in the*
1970   *eastern North Pacific: progress report on research, 1959-69.* Paper presented to the Scientific Committee of the International Whaling Commission at the Special meeting on Sperm Whale Biology and Stock Assessments, 13-24 March 1970, Honolulu, Doc. Sp. 3:18 p. (mimeo).

SERGEANT, D.E., Feeding rates of Cetacea. *Fiskeridir.*
1969   *Skr. (Havunders.),* (15):246-58.

SIMPSON, G.G., A., ROE and R.C. LEWONTIN, *Quantita-*
1960   *tive zoology.* New York, MacGraw-Hill, 440 p.

SLEPTSOV, M.M., Kitoobraznye dal'nevostochnykh
1952   morei (Whales of the Far East). *Izv. Tikhookean. Nauchno-Issled. Inst. Rybn Khoz. Okeanogr.,* 32.

SLIJPER, E.J., *Whales.* London, Hutchinson, 475 p.
1962

SPECTOR, W.S. (ed.), *Handbook of biological data.*
1956   Philadelphia, W.B. Saunders Co., 584 p.

TOMILIN, A.G., Lactation and nutrition in cetaceans.
1946   *Dokl. Akad. Nauk SSSR,* 52(3):277-9.

FAO SALES AGENTS AND BOOKSELLERS

Algeria	Société nationale d'édition et de diffusion, 92, rue Didouche Mourad, Algiers.
Argentina	Editorial Hemisferio Sur S.A., Librería Agropecuaria, Pasteur 743, 1028 Buenos Aires.
Australia	Hunter Publications, 58A Gipps Street, Collingwood, Vic. 3066; Australian Government Publishing Service, P.O. Box 84, Canberra, A.C.T. 2600; and Australian Government Service Bookshops at 12 Pirie Street, Adelaiden S.A.; 70 Alinga Street, Canberra, A.C.T.; 162 Macquarie Street, Hobart, Tas.; 347 Swanson Street, Melbourne, Vic.; 200 St. Georges Terrace, Perth, W.A.; 309 Pitt Street, Sydney, N.S.W.; 294 Adelaide Street, Brisbane, Qld.
Austria	Gerold & Co., Buchhandlung und Verlag, Graben 31, 1011 Vienna.
Bangladesh	ADAB, 79 Road 11A, P.O. Box 5045, Dhanmondi, Dacca.
Belgium	Service des publications de la FAO, M.J. de Lannoy, 202, avenue du Roi, 1060 Brussels. CCP 000-0808993-13.
Bolivia	Los Amigos del Libro, Perú 3712, Casilla 450, Cochabamba; Mercado 1315, La Paz; René Moreno 26, Santa Cruz; Junín esq. 6 de Octubre, Oruro.
Brazil	Livraria Mestre Jou, Rua Guaipá 518, São Paulo 05089; Rua Senador Dantas 19-S205/206, 20.031 Rio de Janeiro, PRODIL, Promoção e Dist. de Livros Ltda., Av. Venáncio Aires 196, Caixa Postal 4005, 90.000 Porto Alegre, A NOSSA LIVRARIA, CLS 104, Bloco C, Lojas 18/19, 70.000 Brasilia, D.F.
Brunei	SST Trading Sdn. Bhd., Bangunan Tekno No. 385, Jln 5/59, P.O. Box 227, Petaling Jaya, Selangor.
Canada	Renouf Publishing Co. Ltd. 2182 St Catherine West, Montreal, Que. H3H 1M7.
Chile	Tecnolibro S.A., Merced 753, entrepiso 15, Santiago.
China	China National Publications Import Corporation, P.O. Box 88, Beijing.
Colombia	Editorial Blume de Colombia Ltda., Calle 65 N° 16-65, Apartado Aéreo 51340, Bogotá D.E.
Costa Rica	Librería, Imprenta y Litografía Lehmann S.A., Apartado 10011, San José.
Cuba	Empresa de Comercio Exterior de Publicaciones, O'Reilly 407 Bajos entre Aguacate y Compostela, Havana.
Cyprus	MAM, P.O. Box 1722, Nicosia.
Czechoslovakia	ARTIA, Ve Smeckach 30, P.O. Box 790, 111 27 Praha 1.
Denmark	Munksgaard Export and Subscription Service, 35 Nørre Søgade, DK 1370 Copenhagen K.
Dominican Rep.	Fundación Dominicana de Desarrollo, Casa de las Gárgolas, Mercedes 4, Apartado 857, Zona Postal 1, Santo Domingo.
Ecuador	Su Librería Cía. Ltda., García Moreno 1172 y Mejía, Apartado 2556, Quito; Chimborazo 416, Apartado 3565, Guayaquil.
El Salvador	Librería Cultural Salvadoreña S.A. de C.V., Calle Arce 423, Apartado Postal 2296, San Salvador.
Finland	Akateeminen Kirjakauppa, 1 Keskuskatu, P.O. Box 128, 00101 Helsinki 10.
France	Editions A. Pedone, 13, rue Soufflot, 75005 Paris.
Germany, F.R.	Alexander Horn Internationale Buchhandlung, Spiegelgasse 9, Postfach 3340, 6200 Wiesbaden.
Ghana	Fides Enterprises, P.O. Box 14129, Accra; Ghana Publishing Corporation, P.O. Box 3632, Accra.
Greece	G.C. Eleftheroudakis S.A., International Bookstore, 4 Nikis Street, Athens (T-126); John Mihalopoulos & Son S.A., International Booksellers, 75 Hermou Street, P.O. Box 73, Thessaloniki.
Guatemala	Distribuciones Culturales y Técnicas "Artemis", 5a. Avenida 12-11, Zona 1, Apartado Postal 2923, Guatemala.
Guinea-Bissau	Conselho Nacional da Cultura, Avenida da Unidade Africana, C.P. 294, Bissau.
Guyana	Guyana National Trading Corporation Ltd. 45-47 Water Street, P.O. Box 308, Georgetown.
Haiti	Librairie "A la Caravelle", 26, rue Bonne Foi, B P. 111, Port-au-Prince.
Hong Kong	Swindon Book Co., 13-15 Lock Road, Kowloon.
Hungary	Kultura, P.O. Box 149, 1389 Budapest 62.
Iceland	Snaebjörn Jónsson and Co. h.f., Hafnarstraeti 9, P.O. Box 1131, 101 Reykjavik.
India	Oxford Book and Stationery Co., Scindia House, New Delhi 110001; 17 Park Street, Calcutta 700016.
Indonesia	P.T. Sari Agung, 94 Kebon Sirih, P.O. Box 411, Djakarta.
Iraq	National House for Publishing, Distributing and Advertising, Jamhuria Street, Baghdad.
Ireland	The Controller, Stationery Office, Dublin 4.
Italy	Distribution and Sales Section, Food and Agriculture Organization of the United Nations, Via delle Terme di Caracalla, 00100 Rome; Libreria Scientifica Dott. Lucio de Biasio "Aeiou", Via Meravigli 16, 20123 Milan; Libreria Commissionaria Sansoni S.p.A. "Licosa", Via Lamarmora 45, C.P. 552, 50121 Florence.
Japan	Maruzen Company Ltd, P.O. Box 5050, Tokyo International 100-31.
Kenya	Text Book Centre Ltd, Kijabe Street, P.O. Box 47540, Nairobi.
Kuwait	Saeed & Samir Bookstore Co. Ltd, P.O. Box 5445, Kuwait.
Luxembourg	Service des publications de la FAO, M.J. de Lannoy, 202, avenue du Roi, 1060 Brussels (Belgium).
Malaysia	SST Trading Sdn. Bhd., Bangunan Tekno No. 385, Jln 5/59, P.O. Box 227, Petaling Jaya, Selangor.
Mauritius	Nalanda Company Limited, 30 Bourbon Street, Port Louis.
Mexico	Dilitsa S.A., Puebla 182-D. Apartado 24-448, Mexico 7, D.F.
Morocco	Librairie "Aux Belles Images", 281, avenue Mohammed V, Rabat.
Netherlands	Keesing Boeken V.B., Joan Muyskenweg 22, 1096 CJ Amsterdam.
New Zealand	Government Printing Office. Government Printing Office Bookshops: Retail Bookshop, 25 Rutland Street, Mail Orders. 85 Beach Road, Private Bag C.P.O., Auckland; Retail, Ward Street, Mail Orders, P.O. Box 857. Hamilton; Retail, Mulgrave Street (Head Office), Cubacade World Trade Centre, Mail Orders, Private Bag, Wellington; Retail, 159 Hereford Street, Mail Orders, Private Bag, Christchurch; Retail, Princes Street, Mail Orders, P.O. Box 1104, Dunedin.
Nigeria	University Bookshop (Nigeria) Limited, University of Ibadan, Ibadan.
Norway	Johan Grundt Tanum Bokhandel, Karl Johansgate 41-43, P.O. Box 1177 Sentrum, Oslo 1.
Pakistan	Mirza Book Agency, 65 Shahrah-e-Quaid-e-Azam, P.O. Box 729, Lahore 3.
Panama	Distribuidora Lewis S.A., Edificio Dorasol, Calle 25 y Avenida Balboa, Apartado 1634, Panama 1.
Paraguay	Agencia de Librerías Nizza S.A., Tacuarí 144, Asunción.
Peru	Librería Distribuidora "Santa Rosa", Jirón Apurímac 375, Casilla 4937, Lima 1.
Philippines	The Modern Book Company Inc., 922 Rizal Avenue, P.O. Box 632, Manila.
Poland	Ars Polona, Krakowskie Przedmiescie 7, 00-068 Warsaw.
Portugal	Livraria Bertrand, S.A.R.L., Rua João de Deus, Venda Nova, Apartado 37. 2701 Amadora Codex; Livraria Portugal, Dias y Andrade Ltda., Rua do Carmo 70-74, Apartado 2681, 1117 Lisbon Codex; Edições ITAU, Avda. da República 46/A-r/c Esqdo., Lisbon 1.
Korea, Rep. of	Eul-Yoo Publishing Co. Ltd, 46-1 Susong-Dong, Jongro-Gu, P.O. Box Kwang-Wha-Moon 362, Seoul.
Romania	Ilexim, Calea Grivitei N° 64-66, B.P. 2001, Bucharest.
Saudi Arabia	The Modern Commercial University, P.O. Box 394, Riyadh.
Sierra Leone	Provincial Enterprises, 26 Garrison Street, P.O. Box 1228, Freetown.
Singapore	MPH Distributors (S) Pte. Ltd, 71/77 Stamford Road, Singapore 6; Select Books Pte. Ltd, 215 Tanglin Shopping Centre, 19 Tanglin Road, Singapore 1024; SST Trading Sdn. Bhd., Bangunan Tekno No. 385, Jln 5/59, P.O. Box 227, Petaling Jaya, Selangor.
Somalia	"Samater's", P.O. Box 936, Mogadishu.
Spain	Mundi Prensa Libros S.A., Castelló 37, Madrid 1; Librería Agrícola, Fernando VI 2, Madrid 4.
Sri Lanka	M.D. Gunasena & Co. Ltd, 217 Olcott Mawatha, P.O. Box 246, Colombo 11.
Sudan	University Bookshop, University of Khartoum, P.O. Box 321, Khartoum.
Suriname	VACO n.v. in Suriname, Dominee Straat 26, P.O. Box 1841, Paramaribo.
Sweden	C.E. Fritzes Kungl. Hovbokhandel, Regeringsgatan 12, P.O. Box 16356, 103 27 Stockholm.
Switzerland	Librairie Payot S.A., Lausanne et Genève; Buchhandlung und Antiquariat Heinimann & Co., Kirchgasse 17, 8001 Zurich.
Thailand	Suksapan Panit, Mansion 9, Rajadamnern Avenue, Bangkok.
Togo	Librairie du Bon Pasteur, B.P. 1164, Lomé.
Tunisia	Société tunisienne de diffusion, 5, avenue de Carthage, Tunis.
United Kingdom	Her Majesty's Stationery Office, 49 High Holborn, London WC1V 6HB (callers only); P.O. Box 569, London SE1 9NH (trade and London area mail orders); 13a Castle Street, Edinburgh EH2 3AR; 41 The Hayes, Cardiff CF1 1JW; 80 Chichester Street, Belfast BT1 4JY; Brazennose Street, Manchester M60 8AS; 258 Broad Street, Birmingham B1 2HE; Southey House, Wine Street, Bristol BS1 2BQ.
Tanzania, United Rep. of	Dar es-Salaam Bookshop, P.O. Box 9030, Dar es-Salaam; Bookshop, University of Dar es-Salaam, P.O. Box 893, Morogoro.
United States of America	UNIPUB, 345 Park Avenue South, New York, N.Y. 10010.
Uruguay	Librería Agropecuaria S.R.L., Alzaibar 1328, Casilla de Correos 1755, Montevideo.
Venezuela	Blume Distribuidora S.A., Gran Avenida de Sabana Grande, Residencias Caroni, Local 5, Apartado 50.339, 1050-A Caracas.
Yugoslavia	Jugoslovenska Knjiga, Trg. Republike 5/8, P.O. Box 36, 11001 Belgrade; Cankarjeva Zalozba, P.O. Box 201-IV, 61001 Ljubljana; Prosveta, Terazije 16, P.O. Box 555, 11001 Belgrade.
Zambia	Kingstons (Zambia) Ltd, Kingstons Building, President Avenue, P.O. Box 139, Ndola.
Other countries	Requests from countries where sales agents have not yet been appointed may be sent to: Distribution and Sales Section, Food and Agriculture Organization of the United Nations, Via delle Terme di Caracalla, 00100 Rome, Italy.

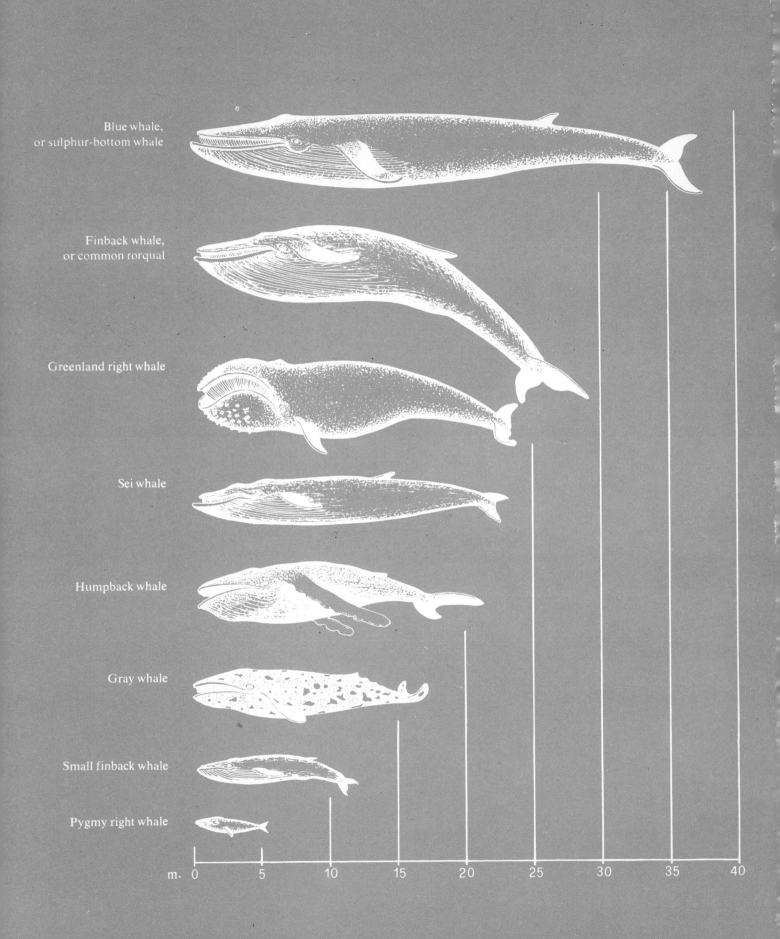

Blue whale,
or sulphur-bottom whale

Finback whale,
or common rorqual

Greenland right whale

Sei whale

Humpback whale

Gray whale

Small finback whale

Pygmy right whale

m. 0   5   10   15   20   25   30   35   40

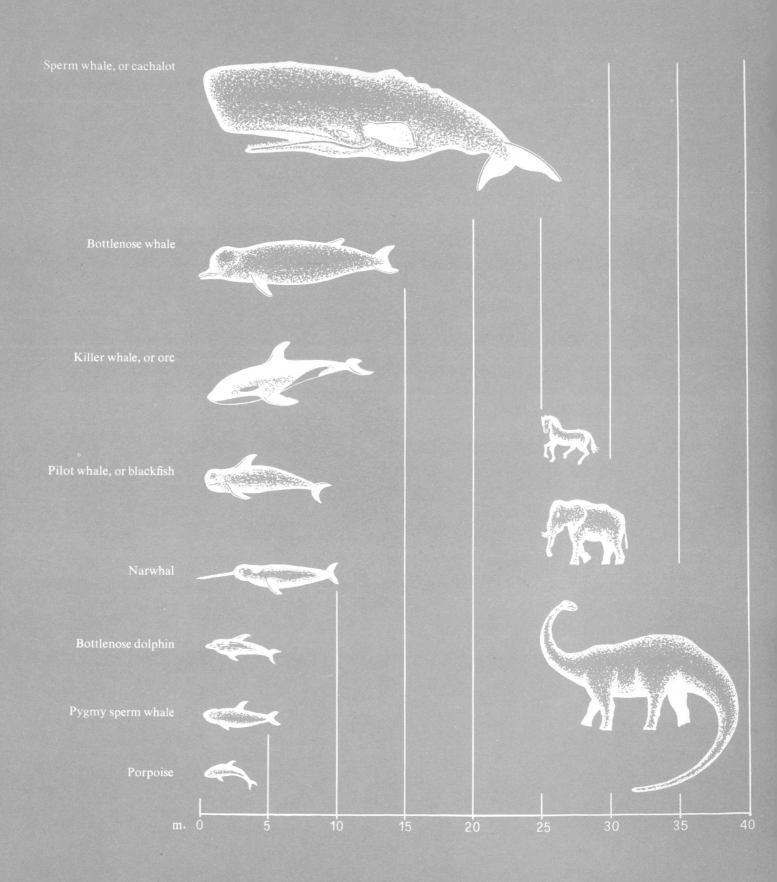

Sperm whale, or cachalot

Bottlenose whale

Killer whale, or orc

Pilot whale, or blackfish

Narwhal

Bottlenose dolphin

Pygmy sperm whale

Porpoise

m. 0   5   10   15   20   25   30   35   40